S0-AXI-978

x: All isotopes are radioactive.

( ) Indicates mass number of isotope with longest known half-life for radioactive elements.

* Number in ( ) heading each column represents the group designation recommended by the ACS Committee on Nomenclature.

			(13) III A	(14) IV A	(15) V A	(16) VI A	(17) VII A	(18) Noble Gases
								2 **He** helium 4.003
			5 **B** boron 10.81	6 **C** carbon 12.01	7 **N** nitrogen 14.01	8 **O** oxygen 16.00	9 **F** fluorine 19.00	10 **Ne** neon 20.18
(10)	(11) I B	(12) II B	13 **Al** aluminum 26.98	14 **Si** silicon 28.09	15 **P** phosphorus 30.97	16 **S** sulfur 32.06	17 **Cl** chlorine 35.45	18 **Ar** argon 39.95
28 **Ni** nickel 58.69	29 **Cu** copper 63.55	30 **Zn** zinc 65.38	31 **Ga** gallium 69.72	32 **Ge** germanium 72.59	33 **As** arsenic 74.92	34 **Se** selenium 78.96	35 **Br** bromine 79.90	36 **Kr** krypton 83.80
46 **Pd** palladium 106.4	47 **Ag** silver 107.9	48 **Cd** cadmium 112.4	49 **In** indium 114.8	50 **Sn** tin 118.7	51 **Sb** antimony 121.8	52 **Te** tellurium 127.6	53 **I** iodine 126.9	54 **Xe** xenon 131.3
78 **Pt** platinum 195.1	79 **Au** gold 197.0	80 **Hg** mercury 200.6	81 **Tl** thallium 204.4	82 **Pb** lead 207.2	83 **Bi** bismuth 209.0	84 **Po**^{x} polonium (209)	85 **At**^{x} astatine (210)	86 **Rn**^{x} radon (222)
110 **__**^{x} (271)	111 **__**^{x} (272)	112 **__**^{x} (277)						

63 **Eu** europium 152.0	64 **Gd** gadolinium 157.3	65 **Tb** terbium 158.9	66 **Dy** dysprosium 162.5	67 **Ho** holmium 164.9	68 **Er** erbium 167.3	69 **Tm** thulium 168.9	70 **Yb** ytterbium 173.0	71 **Lu** lutetium 175.0

95 **Am**^{x} americium (243)	96 **Cm**^{x} curium (248)	97 **Bk**^{x} berkelium (247)	98 **Cf**^{x} californium (251)	99 **Es**^{x} einsteinium (252)	100 **Fm**^{x} fermium (257)	101 **Md**^{x} mendelevium (260)	102 **No**^{x} nobelium (259)	103 **Lr**^{x} lawrencium (262)

Organic and Biochemistry for Today

Fourth Edition

Spencer L. Seager
Weber State University

Michael R. Slabaugh
Weber State University

Brooks/Cole
Thomson Learning™

Australia • Canada • Denmark • Japan • Mexico • New Zealand • Philippines
Puerto Rico • Singapore • Spain • United Kingdom • United States

Sponsoring Editors: *Keith Dodson and Jennifer Huber*
Marketing Team: *Steve Catalano and Dena Donnelly*
Editorial Assistant: *Dena Dowsett-Jones*
Production Editor: *Kirk Bomont*
Editorial Project Management: *GTS Publishing Services*
Manuscript Editor: *Betsy Dilernia*
Permissions Editor: *Mary Kay Hancharick*
Interior Design: *Lisa Delgado*

Cover Design: *Roy R. Neuhaus*
Cover Photo: *Art Wolfe/Tony Stone Images*
Interior Illustration: *Judy and John Waller, Scientific Illustrators*
Photo Researcher: *Laurel Anderson/Photosynthesis*
Print Buyer: *Vena Dyer*
Composition: *GTS Graphics*
Cover Printing: *Phoenix Color Corporation*
Printing and Binding: *R.R. Donnelley*

All products used herein are used for identification purposes only and may be trademarks or registered trademarks of their respective owners.

COPYRIGHT © 2000 by Brooks/Cole
A division of Thomson Learning
The Thomson Learning logo is a trademark used herein under license.

For more information about this or any other Brooks/Cole products, contact:
BROOKS/COLE
511 Forest Lodge Road
Pacific Grove, CA 93950 USA
www.brookscole.com
1-800-423-0563 (Thomson Learning Academic Resource Center)

All rights reserved. No part of this work may be reproduced, transcribed or used in any form or by any means—graphic, electronic, or mechanical, including photo-copying, recording, taping, Web distribution, or information storage and/or retrieval systems—without the prior written permission of the publisher.

For permission to use material from this work, contact us by
Web: www.thomsonrights.com
fax: 1-800-730-2215
phone: 1-800-730-2214

Printed in the United States of America

10 9 8 7 6 5 4 3

Library of Congress Cataloging-in-Publication Data
Seager, Spencer L.
    Organic and biochemistry for today / Spencer L. Seager,
  Michael R. Slabaugh—4th ed.
       p.   cm.
    Includes index.
    ISBN 0-534-37288-0
    1. Chemistry, Organic. 2. Biochemistry. I. Slabaugh,
  Michael R. II. Title.
QD253.S39   2000

To our students, who have made our teaching
and writing activities incredibly satisfying.

About the Authors

SPENCER L. SEAGER

Spencer L. Seager is a professor of chemistry at Weber State University, where he served as chemistry department chairman from 1969 until 1993. He teaches general and physical chemistry at the university and is also active in projects to help improve chemistry and other science education in local elementary schools. He received his B.S. degree in chemistry and Ph.D. degree in physical chemistry from the University of Utah. Other interests include minor home repairs, working in the garden, listening to classical music, and walking for exercise.

MICHAEL R. SLABAUGH

Michael R. Slabaugh is a professor of chemistry at Weber State University, where he teaches the year-long sequence of general, organic, and biochemistry. He received his B.S. degree in chemistry from Purdue University and his Ph.D. degree in organic chemistry from Iowa State University. His interest in plant alkaloids led to a year of postdoctoral study in biochemistry at Texas A&M University. His current professional interests are chemistry education and community involvement in science activities, particularly the State Science and Engineering Fair in Utah. He also enjoys the company of family, hiking in the mountains, and fishing the local streams.

Brief Contents

Contents

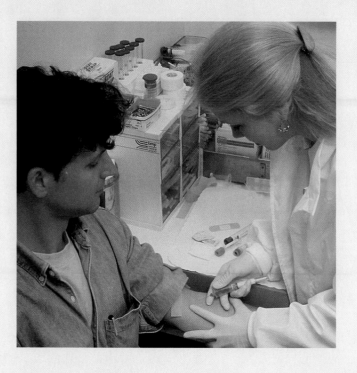

Preface

> ## THE IMAGE OF CHEMISTRY

We, as authors, are pleased that the acceptance of the previous three editions of this textbook by students and their teachers has made it possible to publish this fourth edition. In the earlier editions, we expressed our concern about the negative image of chemistry held by many of our students, and their genuine fear of working with chemicals in the laboratory. Unfortunately, little seems to have changed in the last few years. Reports in the media related to chemicals or to chemistry continue to be primarily negative, and in many cases seem to be designed to increase the fear and concern of the general public. With this edition, we continue to hope that those who use this book will gain a more positive understanding and appreciation of the important contributions that chemistry makes in their lives.

> ## THEME AND ORGANIZATION

This edition continues the theme of the positive and useful contributions made by chemistry in our world. Consistent with that theme, we continue to use the chapter opening focus on health care professionals introduced in the second edition. The photos and accompanying brief descriptions of the role of chemistry in each profession continue to emphasize positive contributions of chemistry in our lives. Web sites related to each featured profession, introduced in the third edition, are also included in this edition. These sites provide a wealth of information about each profession, including wages, work setting, required skills and education, and the current employment outlook.

This text is designed for use in a one-semester or two-quarter course of study that provides an introduction to organic chemistry and biochemistry. It is assumed that students who take the course have a background in general chemistry at least equivalent to that obtained by completing a one-term course in introductory general chemistry. Most students who take such courses are majoring in nursing, other health professions, or the life sciences, and consider biochemistry to be the most relevant part of the course of study. However, an understanding of biochemistry depends upon a sound background in organic chemistry, which in turn depends upon a good foundation in general chemistry. We have attempted to present the general and organic chemistry in sufficient depth and breadth to make the biochemistry understandable.

The decisions about what to include and what to omit from the text were based on our combined 60-plus years of teaching, input from numerous reviewers and adopters, and our philosophy that a textbook functions as a personal tutor to each student. In the role of a personal tutor, a text must be more than just a collection of facts, data, and exercises. It should also help students relate to the material they are studying, carefully guide them through more difficult material, provide them with interesting and relevant examples of chemistry in their lives, and become a reference and resource that they can use in other courses or their professions.

> ## NEW TO THIS EDITION

In this fourth edition of the text, we have retained features that received a positive reception by our own students, the students of other adopters, other teachers, and reviewers. The retained features are the *Study Skills* boxes with reaction maps, the *How Reactions Occur* boxes, and the *Chemistry Around Us* boxes. However, 50% of the *Chemistry Around Us* boxes are new to the fourth edition. We have also continued to provide periodic references to Web sites where students can further explore specific topics of interest. The *Key Chemicals* boxes of the third edition were eliminated and replaced with new *Over the Counter* boxes that contain information dealing with consumer products, mostly medications, that are available without a prescription. Another new box feature, *For Future Reference*, contains information that students will find interesting, and which they might refer to in the future. The boxes provide information about good food sources of various vitamins and minerals; deal with effects of drugs; and discuss the meaning of terms used on food labels. In addition to these feature changes, approximately 25% of the end-of-chapter problems have been changed.

In response to suggestions by reviewers, there have been some changes in content and organization of topics. One example is the increased coverage on neurotransmitters in the organic section. In addition, the art program has been enhanced to more strongly illustrate the connection between observable chemical behavior and the underlying molecular explanations. The importance of a molecular-level understanding of basic chemical principles cannot be overstated. For this reason, the fourth edition of the text also introduces *GOB 3D*, Web-based exercises that use Chime® molecular models to quiz students on their understanding of how molecular structure is related to basic chemical properties and nomenclature rules.

We have also attempted to stay up to date with topics and concepts that change rapidly with time. One example is the number of ATP molecules produced by the biochemical oxidation of one molecule of NADH in the electron transport chain. This number has been a topic of discussion and debate for some time, but the most current accepted value based on experimental measurements is 2.5, the value we use in section 13.7 of the text.

ALDEHYDES AND KETONES • 125

> **FOR FUTURE REFERENCE**   FOOD SOURCES OF VITAMIN A

Three different forms of vitamin A are active in the body: retinol (an alcohol), retinal (an aldehyde), and retinoic acid (a carboxylic acid). Collectively, these three types of vitamin A are called retinoids. Foods from animal sources contain compounds that are easily hydrolyzed to retinol during digestion. Foods derived from plants contain compounds known as carotenoids, and some of these compounds, such as beta-carotene, have vitamin A activity in the body. In the body, vitamin A plays major roles in vision, protein synthesis, cell differentiation, reproduction and growth. Because of these many functions, good health depends on an adequate supply, which is generally available from a well-balanced diet. Excessive vitamin A intake, usually from the use of supplements, poses a risk as pointed out in Chemistry Around Us 4.3.

Vitamin A from food is supplied in the form of a variety of different retinoids and carotenoids. In order to standardize the amount of the vitamin obtained in these different forms, the vitamin content is expressed in terms of retinol equivalents or RE. One RE is equivalent in vitamin A activity to 1 µg of retinol. Sometimes vitamin A content is reported in terms of the older international units or IU. The relationship between the two units is 1 IU = 0.3 RE, or 3.3 IU = 1 RE. The chart below contains a listing of foods that provide a minimum of 50 RE per serving.

Food	Serving size (kcalories)	Micrograms RE per serving
		0 100 200 300 400 500 600 700 800 900 1000
Corn flakes, fortified	1 oz (108 kcal)	
Spinach	1 c raw (12 kcal)	
Broccoli	1/2 c cooked (22 kcal)	

ORGANIC COMPOUNDS: ALKANES • 13

> **FIGURE 1.12** Propane is a common fuel for gas grills.

formulas, but they have different atom-to-atom bonding sequences. The straight-chain isomer is called a **normal alkane** and the other is a **branched alkane** (■ Figure 1.13).

The number of possible structural isomers increases dramatically with the number of carbon atoms in an alkane, as shown in ■ Table 1.3.

Normal alkane
Any alkane in which all the carbon atoms are aligned in a continuous chain.

Branched alkane
An alkane in which at least one carbon atom is not part of a continuous chain.

➤ FEATURES

Each chapter has features especially designed to help students organize, study effectively, understand, and enjoy the material in the course.

Chapter Opening Photos. Each chapter opens with a photo of one of the many health care professionals that provide us with needed services. These professionals represent some of the numerous professions that require an understanding of chemistry.

Chapter Outlines/Learning Objectives. At the beginning of each chapter, a chapter outline provides students with a convenient overview of the topics and concepts they will be studying. At the end of the chapter, they encounter a list of learning objectives that indicate skills and concepts they should have acquired or mastered from their study of the chapter. Thus, students begin each chapter with a topical preview and end with an indication of what they should have learned from the topics.

CHAPTER **6**

Amines and Amides

Individuals who require basic life support before arriving at a health care facility often receive it from an emergency medical technician (EMT), or paramedic. Paramedics function as extensions of an emergency room physician and must be familiar with a number of medications. More pharmaceuticals belong to the amine (or amide) family of compounds than to any other functional class.

http://www.brookscole.com/chemistry/seager_chemistry
At Web site 6.1, you will find further information on a career as an emergency medical technician.

CHAPTER OUTLINE
➤ Classification of Amines
➤ The Nomenclature of Amines

Key Terms. Identified within the text by the use of bold type, key terms are defined in the margin near the place where they are introduced. Students reviewing a chapter can quickly identify the important concepts on each page with this marginal glossary. A full glossary of key terms and concepts appears at the end of the text.

Chemistry Around Us. These boxed features present everyday applications of chemistry that emphasize in a real way the important role of chemistry in our lives. More than half of these are new to this edition and emphasize health-related applications of chemistry.

Over the Counter. These boxed features contain useful information about health-related products that are readily available to consumers without a prescription. The information in each box provides a connection between the chemical behavior of the product and its effect on the body.

OVER THE COUNTER • 5.1

ALPHA HYDROXY ACIDS IN COSMETICS

Alpha hydroxy acids occur in soured milk as lactic acid and in many fruits. Tartaric acid occurs in grapes, malic acid is found in grapes and apples, citric acid is a common component of citrus fruits, and glycolic acid is found in sugar cane and sugar beets.

$$CH_3CH-\overset{\overset{\displaystyle O}{\|}}{C}-OH$$
$$\underset{OH}{|}$$

lactic acid

$$HO-\overset{\overset{\displaystyle O}{\|}}{C}-\underset{\underset{OH}{|}}{CH}-\underset{\underset{OH}{|}}{CH}-\overset{\overset{\displaystyle O}{\|}}{C}-OH$$

tartaric acid

$$CH_2-\overset{\overset{\displaystyle O}{\|}}{C}-OH$$
$$\underset{OH}{|}$$

glycolic acid

$$HO-\overset{\overset{\displaystyle O}{\|}}{C}-\underset{\underset{OH}{|}}{CH}CH_2-\overset{\overset{\displaystyle O}{\|}}{C}-OH$$

malic acid

Recently, alpha hydroxy acids have been advertised as active ingredients in many cosmetics that claim to improve skin texture and tone.

perienced twice the cell damage to their skin than others who were treated similarly with a non–alpha hydroxy acid product.

Because of such test results, cosmetics companies have set guidelines suggesting that concentrations of alpha hydroxy acids in skin products should be no more than 10% for over-the-counter use and no more than 30% for professional salon use. The acid content is included on the labels of many products, and manufacturers can also provide the information.

As people age, their skin naturally becomes more wrinkled. Several factors contribute to this result of the aging process. Skin becomes drier, thinner, and less elastic with age, and people with fair skin are more predisposed to these effects than are those with darker skin. Also, smoking and long-term overexposure to sunlight dramatically accelerate the wrinkling process. Thus, one of the best ways to slow down skin wrinkling is to protect the skin from the sun by wearing long-sleeved clothing, a hat, and a sunscreen that has at least an SPF 15 rating (see Chemistry Around Us 2.3). Because many antiwrinkling cosmetics contain alpha hydroxy acids, a sunscreen should always be worn when these products are used. Adequate protection from the sun is still the best wrinkle-prevention method available.

Examples. To reinforce students in their problem-solving skill development, carefully worked out solutions in numerous examples are included in each chapter. The number of these examples that are based on health-related topics has been increased in this edition.

Learning Checks. Short self-check exercises follow examples and discussions of key or difficult concepts. A complete set of solutions is included in Appendix C. These allow students to immediately measure their understanding and progress.

Study Skills. Most chapters contain a *Study Skills* feature in which a challenging topic, skill, or concept of the chapter is addressed. Study suggestions, analogies, and approaches are provided to help students master these ideas.

How Reactions Occur. The mechanisms of representative organic reactions are presented in four boxed inserts to help students dispel the mystery of how these reactions take place.

For Future Reference. Each chapter concludes with a box containing interesting information that should be especially useful to students as they continue their studies, practice their professions, or just live their lives. For example, the boxes provide information about good food sources of various vitamins and minerals, deal with the effects of drugs, and discuss the meaning of terms used on food labels.

STUDY SKILLS 4.1

ALDEHYDES AND KETONES • 119

A REACTION MAP FOR ALDEHYDES AND KETONES

This reaction map is designed to help you master organic reactions. Whenever you are trying to complete an organic reaction, use these two basic steps: (1) Identify the functional group that is to react, and (2) identify the reagent that is to react with the functional group. If the reacting functional group is an aldehyde or a ketone, find the reagent in the summary diagram, and use the diagram to predict the correct products.

Concept Summary. Located at the end of each chapter, this feature provides a concise review to reinforce the major ideas.

Key Terms and Concepts. These are listed at the end of the chapter for easy review.

Key Equations. This boxed feature provides a useful summary of general equations and reactions from the chapter. This feature is particularly helpful to students in the organic chemistry chapters.

Web Sites. This feature allows students to further explore specific topics on the Internet, where a wealth of information on chemical topics and health-related subjects can be found. Each chapter location of this feature is identified with a red Web address and a description of the information to be found at that site. Specific Internet addresses can be obtained at the home page, where an updated list of these addresses will be maintained: http://www.brookscole.com/chemistry/seager_chemistry.

Exercises. Nearly 1700 end-of-chapter exercises are arranged by section. Color coded for easy identification, approximately half of the exercises are answered in the back of the text. Completely worked out solutions to these answered exercises are included in the Student Study Guide. Solutions and answers to the remaining exercises are provided in the Instructor's Manual. We have included a significant number of clinical and other familiar applications of chemistry in the exercises.

Points to Ponder and Pursue. Included at the end of each chapter are special questions designed to encourage students to expand their reasoning skills. Some of these exercises are based on photographs found in the chapter, others emphasize clinical or other useful applications of chemistry, and a number are based on information found at various Internet sites given in each chapter.

➤ POSSIBLE COURSE OUTLINES

This text may be used effectively in either a one-semester or two-quarter course of study:

One semester: Chapters 1–15 (organic chemistry and biochemistry)

First quarter: Chapters 1–8 (organic chemistry)
Second quarter: Chapters 9–15 (biochemistry)

➤ ANCILLARIES

The following ancillaries have been prepared for use with this text.

Print Resources

Safety-Scale Laboratory Experiments for General, Organic, and Biochemistry, 4th Edition. Prepared by Spencer L. Seager and Michael R. Slabaugh, this well-tested collection of experiments has been developed during more than 25 years of laboratory instruction with students at Weber State University. This manual provides a blend of training in laboratory skills and experiences illustrating concepts from the authors' textbook. The experiments are designed to use small quantities of chemicals, and emphasize safety and proper disposal of used materials. ISBN 0-534-37235-X

Instructor's Guide for Safety-Scale Laboratory Experiments. Prepared by the authors of the laboratory manual, this useful resource gives complete directions for preparing the reagents and other materials used in each experiment. It also contains useful comments concerning the experiments, answers to questions included in the experiments, and suggestions for the proper disposal of used materials. ISBN 0-534-37237-6

Study Guide and Solutions Manual. Prepared by Peter Krieger of Palm Beach Community College, each chapter contains a chapter outline, learning objectives, a programmed review of important topics and concepts, detailed solutions to the even-numbered exercises answered in the text, and self-test questions. ISBN 0-534-37238-4

Instructor's Manual and Testbank. Prepared by Peter Krieger of Palm Beach Community College, each chapter contains a summary chapter outline, learning objectives, instructor resource materials, solutions to *Points to Ponder and Pursue* questions, answers and solutions to odd-numbered exercises not answered in the text, and more than 1,300 exam questions. ISBN 0-534-37226-8

Transparency Acetates. The publisher provides 200 full-color transparencies illustrating key figures from the text for use in class. ISBN 0-534-37234-1

Media Resources

GOB 3D. http://www.brookscole.com/gob_3d. Tailored specifically for the general, organic, and biochemistry course, this set of Web-based exercises uses Chime® molecular models to quiz students on their understanding of how molecular structure is related to basic chemical properties and nomenclature rules. Among the resources available is a tutorial instructing students on how to use Chime, a molecule library of over 400 Chime molecules including commonly studied compounds, an insulin tutorial that examines the structure of this protein, and chapter-by-chapter exercises with associated worksheets.

InfoTrac College Edition is an online library available FREE with each new copy of *Chemistry for Today*. (Due to license restrictions, *InfoTrac College Edition* is only available free with the purchase of a new book to college students in North America.) It gives students full-length articles—not simply abstracts—from more than 700 popular and scholarly periodicals, updated daily and dating back as far as four years. Student subscribers receive a personalized account ID that gives them four months of unlimited access—at any hour of the day.

CNN Chemistry Videos, produced by Turner Learning, can stimulate and engage classes by launching a lecture, sparking a discussion, or demonstrating an application. Each chemistry-related segment from recent CNN broadcasts clearly demonstrates the relevancy of chemistry to everyday life. ISBN 0-534-36924-3

ChemLink is a cross-platform CD-ROM containing all art from the text as well as that from other Brooks/Cole chemistry titles. Using multi-tiered indexing, search capabilities, and a comprehensive resource bank that includes glossary, graphs, tables, illustrations, photographs, and animations, instructors can conduct a quick search to incorporate these materials into presentations and tests. And, any *ChemLink* file can be posted to the Web for easy student downloading. ISBN 0-534-37190-6

Thomson Learning Testing Tools™ is a fully integrated suite of test creation, delivery, and classroom management tools on a dual-platform CD-ROM. It consists of three software programs—Test, Test Online, and Manager—which enable instructors to quickly and easily assess student performance. For more information— http://www.itped.com. ISBN 0-534-37298-8

Thomson World Class Learning Course™. This tool allows professors to easily post course information, office hours, lesson information, assignments, sample tests, and links to rich Web content, including review and enrichment material from Brooks/Cole. Updates are quick and easy, and customer support is available 24 hours a day, seven days a week. For more information—http://www.itped.com.

Brooks/Cole Chemistry Resource Center. http://www.brookscole.com/ chemistry. At Brooks/Cole's chemistry Web site, instructors and students can access a homepage for *Chemistry for Today,* Fourth Edition. All information is arranged according to the *Chemistry for Today* table of contents. Students can access flash cards for all glossary terms, practice quizzes for every chapter, and hyperlinks that relate to each chapter's contents. In addition, students can research the accomplishments of past and present contributors to the field of chemistry in timeline format.

Discover Chemistry 2.0. Prepared by Jeffrey R. Appling and David L. Frank, this dynamic new multimedia CD-ROM provides students with an exciting and interesting way to obtain individual tutoring in general chemistry topics. ISBN 0-534-36134-X

➤ ACKNOWLEDGMENTS

We express our sincere appreciation to the following reviewers, who read and commented on the third edition and offered helpful advice and suggestions for improving this edition:

Gabriele Backes
Portland Community College

Deb Breiter
Rockford College

Sharon Cruse
Northern Louisiana University

Harold Fisher
University of Rhode Island

John W. Francis
Columbus State Community

Galen George
Santa Rosa Junior College

Linda Thomas-Glover
Guilford Technical Community College

James Hardy
University of Akron

Claudia Hein
Diablo Valley College

Laura Kibler-Herzog
Georgia State University

Jim Johnson
Sinclair Community College

Fredrick Jury
Collin County Community College

James F. Kirby
Quinnipiac College

Peter J. Krieger
Palm Beach Community College

Evan McHugh
Pikes Peak Community College

Melvin Merken
Worcester State College

James A. Petrich
San Antonio College

J. Donald Smith
University of Massachusetts-Dartmouth

Cary Willard
Grossmont College

Don Williams
Hope College

Les Wynston
California State University-Long Beach

We also express appreciation to the following reviewers, who helped us revise the first three editions:

Hugh Akers
Lamar University-Beaumont

Johanne I. Artman
Del Mar College

Bruce Banks
University of North Carolina-Greensboro

David Boykin
Georgia State University

Lorraine C. Brewer
University of Arkansas

Martin Brock
Eastern Kentucky University

Christine Brzezowski
University of Utah

Sybil K. Burgess
University of North Carolina-Wilmington

Sharmaine S. Cady
East Stroudsburg University

Linda J. Chandler
Salt Lake Community College

Thomas D. Crute
Augusta College

Jack L. Dalton
Boise State University

Lorraine Deck
University of New Mexico

Kathleen A. Donnelly
Russell Sage College

Jan Fausset
Front Range Community College

Patricia Fish
The College of St. Catherine

Wes Fritz
College of DuPage

Jane D. Grant
Florida Community College

James K. Hardy
University of Akron

Leland Harris
University of Arizona

Robert H. Harris
University of Nebraska-Lincoln

John Henderson
Jackson Community College

Arthur R. Hubscher
Ricks College

Kenneth J. Hughes
University of Wisconsin-Oshkosh

Jeffrey A. Hurlbut
Metropolitan State College of Denver

Richard F. Jones
Sinclair Community College

Lidija Kampa
Kean College of New Jersey

Terrie L. Lampe
De Kalb College-Central Campus

Leslie J. Lovett
Fairmont State College

Armin Mayr
El Paso Community College

Trudy McKee
Thomas Jefferson University

W. Robert Midden
Bowling Green State University

Pamela S. Mork
Concordia College

Phillip E. Morris, Jr.
University of Alabama-Birmingham

Robert N. Nelson
Georgia Southern University

Elva Mae Nicholson
Eastern Michigan University

H. Clyde Odom
Charleston Southern University

Howard K. Ono
California State University-Fresno

Thomas G. Richmond
University of Utah

James Schreck
University of Northern Colorado

William M. Scovall
Bowling Green State University

Jean M. Shankweiler
El Camino Community College

Francis X. Smith
King's College

James A. Thomson
University of Waterloo

Malcolm P. Stevens
University of Hartford

Mary Lee Trawick
Baylor University

Eric R. Taylor
University of Southwestern Louisiana

Katherin Vafeades
University of Texas-San Antonio

We also give special thanks to Keith Dodson, developmental editor for Brooks/Cole, who guided and encouraged us in the preparation of this fourth edition. His suggestions on special features, content changes, and organization of topics were invaluable. Jennifer Huber, chemistry editor, Steve Catalano, marketing manager, Kirk Bomont, production editor, and Melissa Henderson, assistant editor, were also essential to the team and contributed greatly to the success of the project. We are very grateful for the superb work of GTS Graphics, especially to Katy Spining, editorial project manager, for outstanding work in coordinating the production. We appreciate the significant help of six Weber State University students: Douglas Richards, Tony Hembrock, Abel Keogh, Krista Keogh, Maren Slabaugh, and Michael C. Slabaugh, who did excellent work in researching special topics, working problems, proofreading, and shooting photos.

Finally, we extend our love and heartfelt thanks to our families for their patience, support, encouragement, and understanding during a project that occupied much of our time and energy.

Spencer L. Seager

Michael R. Slabaugh

Organic Compounds: Alkanes

Pharmacists are responsible for the appropriate dispensing of drugs and other medications. Pharmacists must have a broad knowledge of the many drugs and medications available, as well as the effects they have on the human body when administered individually and in combinations. The great majority of drugs are organic compounds. This and the next several chapters provide the basis for understanding the structure, properties, and physiological effects of organic compounds.

http://www.brookscole.com/chemistry/
seager_chemistry
At Web site 1.1 you will find further information on a career as a pharmacist.

CHAPTER OUTLINE

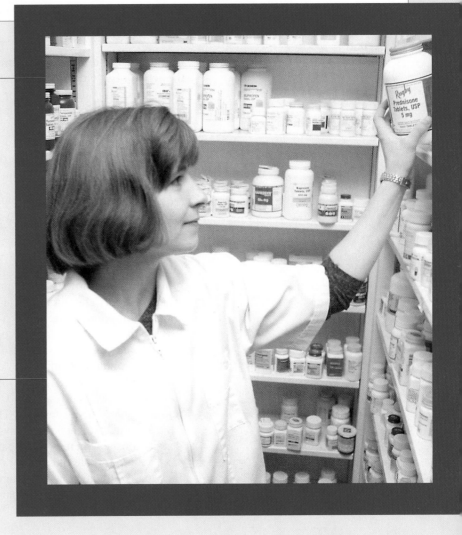

The word *organic* is used in several different contexts. Scientists of the eighteenth and nineteenth centuries studied compounds extracted from plants and animals and labeled them "organic" because they had been obtained from organized (living) systems. Organic fertilizer is organic in the original sense that it comes from a living organism. There is no universal definition of organic foods, but the term is generally taken to mean foods grown without the application of pesticides or synthetic fertilizers. When referring to organic chemistry, however, we mean the chemistry of carbon-containing compounds.

➤ 1.1 Carbon: The Element of Organic Compounds

Early chemists thought organic compounds could be produced only through the action of a "vital force," a special force active only in living organisms. This idea was central to the study of organic chemistry until 1828, because up to that time, no one had been able to synthesize an organic compound from its elements or from naturally occurring minerals. In that year, Friedrich Wöhler, a German chemist, heated an inorganic salt called ammonium cyanate and produced urea. This compound, normally found in blood and urine, was unquestionably organic, and it had come from an inorganic source. The reaction is

$$\text{NH}_4^+\text{NCO}^- \xrightarrow{\text{Heat}} \begin{array}{c} \quad\ \text{H} \quad\ \text{O} \quad\ \text{H} \\ \quad\ | \quad\ || \quad\ | \\ \text{H}-\text{N}-\text{C}-\text{N}-\text{H} \end{array} \qquad (1.1)$$

ammonium
cyanate   urea

Wöhler's urea synthesis discredited the "vital force" theory, and his success prompted other chemists to attempt to synthesize organic compounds. Today, organic compounds are being synthesized in thousands of laboratories, and most of the synthetics have never been isolated from natural sources.

Organic compounds share one unique feature: They all contain carbon. Therefore, **organic chemistry** is defined as the study of carbon-containing compounds. There are a few exceptions to this definition; a small number of carbon compounds—such as CO, CO_2, carbonates, and cyanides—were studied before Wöhler's urea synthesis. These were classified as inorganic because they were obtained from nonliving systems, and even though they contain carbon, we still consider them to be a part of **inorganic chemistry**.

The importance of carbon compounds to life on earth cannot be overemphasized. If all carbon compounds were removed from the earth, its surface would be somewhat like the barren surface of the moon (see ■ Figure 1.1).

Organic compound
A compound that contains the element carbon.

Organic chemistry
The study of carbon-containing compounds.

Inorganic chemistry
The study of the elements and all noncarbon compounds.

➤ **FIGURE 1.1** Organic chemistry makes a tremendous difference between the earth and the moon.

There would be no animals, plants, or any other form of life. If carbon-containing compounds were removed from the human body, all that would remain would be water, a very brittle skeleton, and a small residue of minerals. Many of the essential constituents of living matter—such as carbohydrates, fats, proteins, nucleic acids, enzymes, and hormones—are organic chemicals.

The essential needs of daily human life are food, fuel, shelter, and clothing. The principal components of food (with the exception of water) are organic. The fuels we use (e.g., wood, coal, and natural gas) are mixtures of organic compounds. Our homes typically involve wood construction, and our clothing, whether made of natural or synthetic fibers, is organic.

Besides the major essentials, many of the smaller everyday things often taken for granted are also derived from carbon and its compounds. Consider an ordinary pencil. The "lead" (actually graphite), the wood, the rubber eraser, and the paint on the surface are all either carbon or carbon compounds. The paper in this book, the ink on its pages, and the glue holding it all together are also made of carbon compounds.

➤ 1.2 Organic and Inorganic Compounds Compared

It is interesting that the subdivision of chemistry into its organic and inorganic branches results in one branch that deals with compounds composed mainly of one element and another branch that deals with compounds formed by the more than 100 remaining elements. However, this classification seems more reasonable when we recognize that known organic compounds are much more numerous than inorganic compounds. An estimated 250,000 inorganic compounds have been identified, but more than 6 million organic compounds are known, and thousands of new ones are synthesized or isolated each year.

One of the reasons for the large number of organic compounds is the unique ability of carbon atoms to form stable covalent bonds with other carbon atoms and with atoms of other elements. The resulting covalently bonded molecules may contain as few as one or more than a million carbon atoms.

In contrast, inorganic compounds are often characterized by the presence of ionic bonding. Covalent bonding also may be present, but it is less common. These differences generally cause organic and inorganic compounds to differ physically (see ■ Figure 1.2) and chemically, as shown in ■ Table 1.1.

TABLE 1.1 Properties of typical organic and inorganic compounds

Property	Organic compounds	Inorganic compounds
Bonding within molecules	Usually covalent	Often ionic
Forces between molecules	Generally weak	Quite strong
Normal physical state	Gases, liquids, or low melting point solids	Usually high melting point solids
Flammability	Often flammable	Usually nonflammable
Solubility in water	Often low	Often high
Conductivity of water solutions	Nonconductor	Conductor
Rate of chemical reactions	Usually low	Usually high

➤ **FIGURE 1.2** Many organic compounds, such as ski wax, have relatively low melting points. What does this fact reveal about the forces between organic molecules?

Learning Check 1.1

Classify each of the following compounds as organic or inorganic:

a. NaCl *org* **b.** CH_4 *org*
c. C_6H_6 *inorg* **d.** Na *inorg*
e. CH_3OH *org* **f.** $Mg(NO_3)_2$ *inorg*

Learning Check 1.2

Decide whether each of the following characteristics most likely describes an organic or inorganic compound:

a. Flammable **b.** Low boiling point **c.** Soluble in water
  org   *org*   *inorg*

➤ **STUDY SKILLS 1.1**

CHANGING GEARS FOR ORGANIC CHEMISTRY

Y ou will find that organic chemistry is very different from general or inorganic chemistry. By quickly picking up on the changes, you will help yourself prepare for quizzes and exams.

There is almost no math in these next six chapters or in the biochemistry section. Very few mathematical formulas need to be memorized. The problems you will encounter fall mainly into four categories: naming compounds and drawing structures, describing physical properties of substances, writing reactions, and identifying typical uses of compounds. This pattern holds true for all six of the organic chemistry chapters.

The naming of compounds is introduced in this chapter, and the rules developed here will serve as a starting point in the next five chapters. Therefore, it is important to master naming in this chapter. A well-developed skill in naming will help you do well on exams covering the coming chapters.

Only a few reactions are introduced in this chapter, but many more will be in future chapters. Writing organic reactions is just as important (and challenging) as naming, and Study Skills 2.1 will help you. Identifying the uses of compounds can best be handled by making a list as you read the chapter or by highlighting compounds and their uses so that they are easy to review. All four categories of problems are covered by numerous end-of-chapter exercises to give you practice.

➤ **FIGURE 1.3** The mixing of $2s$ and three $2p$ orbitals of a carbon atom to generate four sp^3 carbon orbitals, each with energy intermediate between $2s$ and $2p$ orbitals.

➤ 1.3 Bonding Characteristics and Isomerism

There are two major reasons for the astonishing number of organic compounds: the bonding characteristics of carbon atoms, and the isomerism of carbon-containing molecules. As a group IVA(14) element, a carbon atom has four valence electrons. Two of these outermost shell electrons are in an s orbital, and two are in p orbitals:

With only two unpaired electrons, we might predict that carbon would form just two covalent bonds with other atoms. Yet, we know from the formula of methane (CH_4) that carbon forms four bonds.

Linus Pauling (1901–1994), winner of the Nobel Prize in chemistry (1954) and Nobel Peace Prize (1963), developed a useful model to explain the bonding characteristics of carbon. Pauling found that a mathematical mixing of the $2s$ and three $2p$ orbitals could produce four new, equivalent orbitals (see ■ Figure 1.3). Each of these **hybrid orbitals** has the same energy and is designated sp^3. An sp^3 orbital has a two-lobed shape, similar to the shape of a p orbital but with different-sized lobes (see ■ Figure 1.4). Each of the four sp^3 hybrid orbitals contains a single unpaired electron available for covalent bond formation. Thus, carbon forms four bonds.

Each carbon–hydrogen bond in methane arises from an overlap of a C (sp^3) and an H ($1s$) orbital. The sharing of two electrons in this overlap region creates a sigma (σ) bond. The four equivalent sp^3 orbitals point toward the corners of a regular tetrahedron (see ■ Figure 1.5).

Hybrid orbital
An orbital produced from the combination of two or more nonequivalent orbitals of an atom.

A 2p orbital

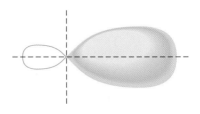

An sp^3 hybrid orbital

➤ **FIGURE 1.4** A comparison of unhybridized p and sp^3 hybridized orbital shapes. The atomic nucleus is at the junction of the lobes in each case.

➤ **FIGURE 1.5** Directional characteristics of sp^3 hybrid orbitals of carbon and the formation of C—H bonds in methane (CH_4). The hybrid orbitals point toward the corners of a regular tetrahedron. Hydrogen $1s$ orbitals are illustrated in position to form bonds by overlap with the major lobes of the hybrid orbitals.

Carbon atoms also have the ability to bond covalently to other carbon atoms to form chains and networks. This means that two carbon atoms can join by sharing two electrons to form a single covalent bond:

$$\cdot \dot{\underset{\cdot}{C}} \cdot \; + \; \cdot \dot{\underset{\cdot}{C}} \cdot \; \longrightarrow \; \cdot \dot{\underset{\cdot}{C}} \! : \! \dot{\underset{\cdot}{C}} \cdot \qquad \text{or} \qquad \cdot \dot{\underset{\cdot}{C}} - \dot{\underset{\cdot}{C}} \cdot \qquad (1.2)$$

A third carbon atom can join the end of this chain:

$$\cdot \dot{\underset{\cdot}{C}} - \dot{\underset{\cdot}{C}} \cdot \; + \; \cdot \dot{\underset{\cdot}{C}} \cdot \; \longrightarrow \; \cdot \dot{\underset{\cdot}{C}} - \dot{\underset{\cdot}{C}} - \dot{\underset{\cdot}{C}} \cdot \qquad (1.3)$$

This process can continue and form carbon chains of almost any length, such as

$$\cdot \dot{\underset{\cdot}{C}} - \dot{\underset{\cdot}{C}} - \dot{\underset{\cdot}{C}} - \dot{\underset{\cdot}{C}} - \dot{\underset{\cdot}{C}} - \dot{\underset{\cdot}{C}} - \dot{\underset{\cdot}{C}} - \dot{\underset{\cdot}{C}} - \dot{\underset{\cdot}{C}} \cdot$$

The electrons not involved in forming the chain can be shared with electrons of other carbon atoms (to form chain branches) or with electrons of other elements such as hydrogen, oxygen, or nitrogen. Carbon atoms may also share more than one pair of electrons to form multiple bonds:

$$\cdot \dot{\underset{\cdot}{C}} - \dot{C} = \dot{C} - \dot{\underset{\cdot}{C}} \cdot \qquad\qquad \cdot \dot{\underset{\cdot}{C}} - C \equiv C - \dot{\underset{\cdot}{C}} \cdot$$

Chain with   Chain with
double bond   triple bond

In principle, there is no limit to the number of carbon atoms that can bond covalently. Thus, organic molecules range from the simple molecules such as methane (CH_4) to very complicated molecules containing thousands of carbon atoms.

The variety of possible carbon atom arrangements is even more important than the size range of the resulting molecules. The carbon atoms in all but the very simplest organic molecules can bond in more than one arrangement, giving rise to different compounds with different structures and properties. This property, called **isomerism,** is characterized by compounds that have identical molecular formulas but different arrangements of atoms. One type of isomerism is characterized by compounds called **structural isomers.** Other types of isomerism are covered in Chapters 2 and 7.

Isomerism
A property in which two or more compounds have the same molecular formula but different arrangements of atoms.

Structural isomers
Compounds that have the same molecular formula but in which the atoms bond in different patterns.

➤ Example 1.1

....

Use the usual rules for covalent bonding to show that a compound with the molecular formula C_2H_6O demonstrates the property of isomerism. Draw formulas for the isomers, showing all covalent bonds.

Solution
Carbon forms four covalent bonds by sharing its four valence-shell electrons. Similarly, oxygen should form two covalent bonds, and hydrogen a single bond. On the basis of these bonding relationships, two structural isomers are possible:

```
    H   H                       H        H
    |   |                       |        |
H — C — C — O — H        H — C — O — C — H
    |   |                       |        |
    H   H                       H        H

  ethyl alcohol              dimethyl ether
```

Which one of the structures below represents a structural isomer of

$$
\begin{array}{c}
\text{H} \quad \text{O} \quad \text{H} \\
| \quad\ \| \quad\ | \\
\text{H}-\text{C}-\text{C}-\text{C}-\text{H} \\
| \qquad\ | \\
\text{H} \qquad\ \text{H}
\end{array}
$$

a.
$$
\begin{array}{c}
\text{H} \quad \text{H} \quad \text{O}-\text{H} \\
| \quad\ | \quad\ | \\
\text{H}-\text{C}=\text{C}-\text{C}-\text{H} \\
| \\
\text{H}
\end{array}
$$

c.
$$
\begin{array}{c}
\text{H} \quad \text{H} \quad \text{O} \\
| \quad\ | \quad\ \| \\
\text{H}-\text{C}-\text{C}-\text{C}-\text{O}-\text{H} \\
| \quad\ | \\
\text{H} \quad\ \text{H}
\end{array}
$$

b.
$$
\begin{array}{c}
\text{H} \quad \text{H} \quad \text{O} \\
| \quad\ | \quad\ \| \\
\text{H}-\text{C}=\text{C}-\text{C}-\text{H}
\end{array}
$$

The two isomers of Example 1.1 are quite different. Ethyl alcohol (grain alcohol) is a liquid at room temperature, whereas dimethyl ether is a gas. Thus, the structural differences exert a significant influence on properties. From this example, we can see that molecular formulas such as C_2H_6O provide much less information about a compound than do structural formulas. ■ Figure 1.6 shows ball-and-stick models of these two molecules.

As the number of carbon atoms in the molecular formula increases, the number of possible isomers increases dramatically. For example, 366,319 different isomers are possible for a molecular formula of $C_{20}H_{42}$. No one has prepared all these isomers or even drawn their structural formulas, but the number helps us understand why so many organic compounds have been either isolated from natural sources or synthesized.

➤ 1.4 Functional Groups: The Organization of Organic Chemistry

Because of the enormous number of possible compounds, the study of organic chemistry might appear to be hopelessly difficult. However, the arrangement of organic compounds into a relatively small number of classes can simplify the study a great deal. This organization is done on the basis of characteristic structural features called **functional groups.** For example, compounds with a carbon–carbon double bond

Functional group
A unique reactive combination of atoms that differentiates molecules of organic compounds of one class from those in another.

ethyl alcohol   dimethyl ether

➤ **FIGURE 1.6** Ball-and-stick models of the isomers of C_2H_6O. Ethyl alcohol is a liquid at room temperature and completely soluble in water, whereas dimethyl ether is a gas at room temperature and only partially soluble in water.

TABLE 1.2 Classes and functional groups of organic compounds

Class	Functional group	Example of expanded structural formula	Example of condensed structural formula	Common name
Alkane	None	$H-\underset{\underset{H}{\vert}}{\overset{\overset{H}{\vert}}{C}}-\underset{\underset{H}{\vert}}{\overset{\overset{H}{\vert}}{C}}-H$	CH_3CH_3	ethane
Alkene	$\text{C}=\text{C}$	$\underset{H}{\overset{H}{C}}=\underset{H}{\overset{H}{C}}$	$H_2C=CH_2$	ethylene
Alkyne	$-C\equiv C-$	$H-C\equiv C-H$	$HC\equiv CH$	acetylene
Aromatic		$H-CC-H$ (benzene ring)	(benzene ring symbol)	benzene
Alcohol	$-\overset{\vert}{\underset{\vert}{C}}-O-H$	$H-\underset{\underset{H}{\vert}}{\overset{\overset{H}{\vert}}{C}}-\underset{\underset{H}{\vert}}{\overset{\overset{H}{\vert}}{C}}-O-H$	CH_3CH_2-OH	ethyl alcohol
Ether	$-\overset{\vert}{\underset{\vert}{C}}-O-\overset{\vert}{\underset{\vert}{C}}-$	$H-\underset{\underset{H}{\vert}}{\overset{\overset{H}{\vert}}{C}}-O-\underset{\underset{H}{\vert}}{\overset{\overset{H}{\vert}}{C}}-H$	CH_3-O-CH_3	dimethyl ether
Amine	$-N-H$ ($\overset{H}{\vert}$)	$H-\underset{\underset{H}{\vert}}{\overset{\overset{H}{\vert}}{C}}-\underset{H}{\overset{H}{N}}-H$	CH_3-NH_2	methylamine
Aldehyde	$-\overset{\overset{O}{\|}}{C}-H$	$H-\underset{\underset{H}{\vert}}{\overset{\overset{H}{\vert}}{C}}-\overset{\overset{O}{\|}}{C}-H$	$CH_3-\overset{\overset{O}{\|}}{C}-H$	acetaldehyde
Ketone	$-\overset{\vert}{\underset{\vert}{C}}-\overset{\overset{O}{\|}}{C}-\overset{\vert}{\underset{\vert}{C}}-$	$H-\underset{\underset{H}{\vert}}{\overset{\overset{H}{\vert}}{C}}-\overset{\overset{O}{\|}}{C}-\underset{\underset{H}{\vert}}{\overset{\overset{H}{\vert}}{C}}-H$	$CH_3-\overset{\overset{O}{\|}}{C}-CH_3$	acetone
Carboxylic acid	$-\overset{\overset{O}{\|}}{C}-O-H$	$H-\underset{\underset{H}{\vert}}{\overset{\overset{H}{\vert}}{C}}-\overset{\overset{O}{\|}}{C}-O-H$	$CH_3-\overset{\overset{O}{\|}}{C}-OH$	acetic acid
Ester	$-\overset{\overset{O}{\|}}{C}-O-\overset{\vert}{\underset{\vert}{C}}-$	$H-\underset{\underset{H}{\vert}}{\overset{\overset{H}{\vert}}{C}}-\overset{\overset{O}{\|}}{C}-O-\underset{\underset{H}{\vert}}{\overset{\overset{H}{\vert}}{C}}-H$	$CH_3-\overset{\overset{O}{\|}}{C}-O-CH_3$	methyl acetate
Amide	$-\overset{\overset{O}{\|}}{C}-\underset{}{\overset{H}{N}}-H$	$H-\underset{\underset{H}{\vert}}{\overset{\overset{H}{\vert}}{C}}-\overset{\overset{O}{\|}}{C}-\overset{H}{N}-H$	$CH_3-\overset{\overset{O}{\|}}{C}-NH_2$	acetamide

are classified as alkenes. The major classes and functional groups are given in ■ Table 1.2. Notice that each functional group in Table 1.2 (except for alkanes) contains a multiple bond or at least one oxygen or nitrogen atom.

In Table 1.2, we have used both expanded and condensed structural formulas for the compounds. **Expanded structural formulas** show all covalent bonds, whereas **condensed structural formulas** show only specific bonds. You should become familiar with both types, but especially with condensed formulas because they will be used often.

Expanded structural formula
A structural molecular formula showing all the covalent bonds.

Condensed structural formula
A structural molecular formula showing the general arrangement of atoms but without showing all the covalent bonds.

➤ Example 1.2

Write a condensed structural formula for each of the following compounds:

a.
$$H-\underset{\underset{H}{|}}{\overset{\overset{H}{|}}{C}}-\underset{\underset{H}{|}}{\overset{\overset{H}{|}}{C}}-\underset{\underset{H}{|}}{\overset{\overset{H}{|}}{C}}-\underset{\underset{H}{|}}{\overset{\overset{H}{|}}{C}}-H$$

CH₃CH₂CH₂CH₃ (handwritten)

b.
$$H-\underset{\underset{H}{|}}{\overset{\overset{H}{|}}{C}}-\underset{\underset{H}{|}}{\overset{\overset{H-C-H}{|}}{C}}-\underset{\underset{H}{|}}{\overset{\overset{H}{|}}{C}}-\underset{\underset{H}{|}}{\overset{\overset{H}{|}}{C}}-\underset{\underset{H}{|}}{\overset{\overset{H}{|}}{C}}-H$$

CH₃ / CH₃-CH-CH₂CH₂CH₃ (handwritten)

Solution

a. Usually the hydrogens belonging to a carbon are grouped to the right. Thus, the group

$$H-\underset{\underset{H}{|}}{\overset{\overset{H}{|}}{C}}-$$

condenses to $CH_3—$, and

$$-\underset{\underset{H}{|}}{\overset{\overset{H}{|}}{C}}-$$

condenses to $—CH_2—$. Thus, the formula condenses to

$$CH_3—CH_2—CH_2—CH_3$$

Other acceptable condensations are

$$CH_3CH_2CH_2CH_3 \quad \text{and} \quad CH_3(CH_2)_2CH_3$$

Parentheses are used here to denote a series of two $—CH_2—$ groups.

b. The group

$$-\underset{\underset{H}{|}}{\overset{\overset{|}{}}{C}}-$$

condenses to $-\overset{|}{C}H-$. The condensed formula is therefore

$$\underset{\underset{CH_3-CH-CH_2-CH_2-CH_3}{}}{\overset{CH_3}{|}} \quad \text{or} \quad \underset{\underset{CH_3CHCH_2CH_2CH_3}{}}{\overset{CH_3}{|}}$$

or

$$(CH_3)_2CH(CH_2)_2CH_3$$

In the last form, the first set of parentheses indicates two identical CH_3 groups attached to the same carbon atom, and the second set of parentheses denotes two CH_2 groups.

Learning Check 1.4

Write a condensed structural formula for each of the following compounds. Retain the bonds to and within the functional groups.

a.

$$\begin{array}{ccccccccc} & H & OH & & H & H & H \\ & | & | & & | & | & | \\ H-&C-&C&-&-&C-&C-&C&-H \\ & | & | & & | & | & | \\ & H & H & & H & H & H \end{array}$$

b.

$$\begin{array}{ccccccccc} & H & & H & & H & O \\ & | & & | & & | & \| \\ H-&C&-&-&C&-&-&C-&C-OH \\ & | & & | & & | \\ & H & & H-C-H & & H \\ & & & | \\ & & & H \end{array}$$

(handwritten answers)

a. $CH_3CH \cdot (CH_2)_2 - CH_3$ with OH

b. $CH_3 - CH - CH_2 - COH$ (with O above final C) and CH_3 below

---

CHEMISTRY AROUND US • 1.1

COMBINATORIAL CHEMISTRY

Until recently, a goal of organic chemists involved in synthesizing new organic compounds was to produce specific compounds one at a time in high-percentage yields. Any compounds produced along with the one that was desired were considered to be undesirable side products. Recently, especially in pharmaceutical research, this attitude has changed. The goal of *combinatorial chemistry* is to synthesize a large variety of organic compounds, and then rapidly test them for desirable properties. This approach has been made possible by the development of a variety of sophisticated instruments. Some instruments make possible the rapid separation of the individual compounds that are found in complex mixtures of compounds. Other instruments make possible rapid assays to determine whether the individual compounds have desirable properties.

A reason that combinatorial chemistry is attractive to pharmaceutical companies is the potential to speed up the discovery and development of new drugs at a net reduced cost. For example, 10 years ago, a medicinal chemist would be able to synthesize 50–100 new compounds in a year. Today that same chemist could make thousands to tens of thousands of new compounds a year using combinatorial chemistry. However, although the net cost for developing new drugs is reduced, combinatorial chemistry laboratories are very expensive to set up and operate. Instruments for separating and assaying the compounds are costly, and the screening of thousands of compounds a year is still labor-intensive and therefore expensive. Because of these economic factors, combinatorial chemistry research takes place primarily in commercial rather than academic laboratories.

Most combinatorial laboratories maintain a large "library" of compounds that serve as a starting point for the synthesis of other compounds. A compound that is known to have desirable properties might be selected from the library and reacted with other compounds having different desirable properties. The many resulting compounds would be separated and assayed to find any with a combination of desirable properties that might make a useful new drug.

The large libraries of compounds used in combinatorial chemistry research can be created in a number of ways. One method that is useful in pharmaceutical research is called *combinatorial biocatalysis*. This technique makes use of enzymatic and microbial reactions to produce large numbers of compounds similar to compounds that can be isolated from natural sources such as plants. The resulting mixture is separated into individual compounds, which are tested for properties that would make them useful in the development of new drugs. Any compounds that have such properties are added to the library.

Another process for synthesizing large numbers of compounds makes use of miniaturization to reduce costs. Orchid Biocomputer has developed a synthesis system based on a microfluidic chip. A 5-layer, 2-inch-square chip contains 144 small wells in which reactions can be done using volumes of reactants of less than a microliter. According to the manufacturer, the reduction in the amount of reagents used results in a significant reduction in the cost of producing a library of compounds. For example, a 100,000-compound library costing more than $200,000 to produce using a popular solid-phase organic synthesis system would cost only $100 to produce with the microchip.

http://www.brookscole.com/chemistry/seager_chemistry
Web site 1.2 provides further information on combinatorial chemistry.

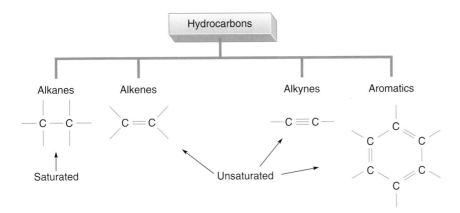

➤ 1.5 Alkane Structures

Hydrocarbons, the simplest of all organic compounds, contain only two elements, carbon and hydrogen. **Saturated hydrocarbons** or **alkanes** are organic compounds in which carbon is bonded to four other atoms by single bonds; there are no double or triple bonds in the molecule. Unsaturated hydrocarbons, studied later, are called alkenes, alkynes, and aromatics and contain double bonds, triple bonds, or six-carbon rings, as shown in ■ Figure 1.7.

Most life processes are based on the reactions of functional groups. Since alkanes have no functional group, they are not abundant in the human body. However, most compounds in human cells contain parts consisting solely of carbon and hydrogen that behave very much like hydrocarbons. Thus, to understand the chemical properties of the more complex biomolecules, it is useful to have some understanding of the structure, physical properties, and chemical behavior of hydrocarbons.

Another important reason for becoming familiar with the characteristics of hydrocarbons is the crucial role they play in modern industrial society. We use naturally occurring hydrocarbons as primary sources of energy and as important sources of raw materials for the manufacture of plastics, synthetic fibers, drugs, and hundreds of other compounds used daily (see ■ Figure 1.8).

Alkanes can be represented by the general formula C_nH_{2n+2}, where n is the number of carbon atoms in the molecule. The simplest alkane, methane, contains one carbon atom and therefore has the molecular formula CH_4. The carbon atom is at the center, and the four bonds of the carbon atom are directed toward the hydrogen atoms at the corners of a regular tetrahedron; each hydrogen atom is geometrically equivalent to the other three in the molecule (see ■ Figure 1.9). A tetrahedral orientation of bonds with bond angles

Hydrocarbon
An organic compound that contains only carbon and hydrogen.

Saturated hydrocarbon
Another name for an alkane.

Alkane
A hydrocarbon that contains only single bonds.

➤ **FIGURE 1.8** This racquet has the strength and light weight of aluminum yet is made from graphite (carbon) fibers reinforced with plastic. What other sports equipment is made from graphite fibers?

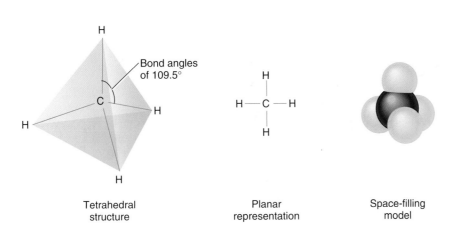

Tetrahedral structure

Planar representation

Space-filling model

➤ **FIGURE 1.9** Structural representations of methane, CH_4.

Ball-and-stick model   Space-filling model

of 109.5° is typical for carbon atoms that form four single bonds. Methane is the primary compound in natural gas. Tremendous quantities of natural gas are consumed worldwide because methane is an efficient, clean-burning fuel. It is used to heat homes, cook food, and power factories.

The next alkane is ethane, which has the molecular formula C_2H_6 and the structural formula CH₃—CH₃. This molecule may be thought of as a methane molecule with one hydrogen removed and a —CH₃ put in its place. Again, the carbon bonds have a tetrahedral geometry as shown in ■ Figure 1.10. Ethane is a minor component of natural gas.

Propane, the third alkane, has the molecular formula of C_3H_8 and the structural formula CH₃—CH₂—CH₃. Again, we can produce this molecule by removing a hydrogen atom from the preceding compound (ethane) and substituting a —CH₃ in its place (see ■ Figure 1.11). Since all six hydrogen atoms of ethane are equivalent, it makes no difference which one is replaced. Propane is used extensively as an industrial fuel, as well as for home heating and cooking (see ■ Figure 1.12).

The fourth member of the series, butane, with molecular formula C_4H_{10}, can also be produced by removing a hydrogen atom (this time from propane) and adding a —CH₃. However, all the hydrogen atoms of propane are not geometrically equivalent, and more than one position is available for substitution. Replacing a hydrogen atom on one of the end carbons of propane with —CH₃ produces a butane molecule that has the structural formula CH₃—CH₂—CH₂—CH₃. If, however, substitution is made on the central carbon atom of propane, the butane produced is

$$CH_3-CH-CH_3$$
$$|$$
$$CH_3$$

Notice that both butanes have the same molecular formula, C_4H_{10}. These two possible butanes are structural isomers because they have the same molecular

Ball-and-stick model   Space-filling model

➤ **FIGURE 1.12** Propane is a common fuel for gas grills.

formulas, but they have different atom-to-atom bonding sequences. The straight-chain isomer is called a **normal alkane** and the other is a **branched alkane** (■ Figure 1.13).

The number of possible structural isomers increases dramatically with the number of carbon atoms in an alkane, as shown in ■ Table 1.3.

a. Determine the molecular formula of the alkane containing eight carbon atoms.
b. Draw the condensed structural formula of the normal isomer of the compound in part a.

➤ 1.6 Conformations of Alkanes

Remember that planar representations such as CH_3—CH_2—CH_3 or

$$H-\overset{\overset{\displaystyle H}{|}}{\underset{\underset{\displaystyle H}{|}}{C}}-\overset{\overset{\displaystyle H}{|}}{\underset{\underset{\displaystyle H}{|}}{C}}-\overset{\overset{\displaystyle H}{|}}{\underset{\underset{\displaystyle H}{|}}{C}}-H$$

are given with no attempt to accurately portray correct bond angles or molecular geometries. Structural formulas are usually written horizontally simply because it is convenient. It is also important to know that actual organic

Normal alkane
Any alkane in which all the carbon atoms are aligned in a continuous chain.

Branched alkane
An alkane in which at least one carbon atom is not part of a continuous chain.

Learning Check 1.5

TABLE 1.3 Molecular formulas and possible structural isomers of alkanes

Molecular formula	Number of possible structural isomers
C_4H_{10}	2
C_5H_{12}	3
C_6H_{14}	5
$C_{10}H_{22}$	75
$C_{20}H_{42}$	366,319
$C_{30}H_{62}$	4,111,846,763

CH_3—CH_2—CH_2—CH_3

n-butane

CH_3—CH—CH_3
$\quad\quad\ |$
$\quad\quad CH_3$

isobutane

➤ **FIGURE 1.13** Space-filling models of the isomeric butanes.

Rotation about single bonds (*n*-butane molecule).

molecules are in constant motion—twisting, turning, vibrating, and bending. Groups connected by a single bond are capable of rotating about that bond much like a wheel rotates around an axle (see ■ Figure 1.14). As a result of such rotation about single bonds, a molecule can exist in many different orientations, called **conformations.** In a sample of butane containing billions of identical molecules, there are countless conformations present at any instant, and each conformation is rapidly changing into another. Two of the possible conformations of butane are shown in ■ Figure 1.15. We must be sure to recognize that these different conformations do not represent different structural isomers. In each case, the four carbon atoms are bonded in a continuous (unbranched) chain. Since the order of bonding is not changed, the conformations correspond to the same molecule. Two structures would be structural isomers only if bonds had to be broken and remade to convert one into the other.

Conformations
The different arrangements of atoms in space achieved by rotation about single bonds.

➤ **Example 1.3**

Which of the following pairs are structural isomers, and which are simply different representations of the same molecule?

same
a. $CH_3-CH_2-CH_2-CH_3$

$$CH_2-CH_2$$
with CH_3 above and CH_3 below

same
b. $CH_3-CH-CH_2-CH_3$ (with CH_3 on the CH) $CH_3-CH_2-CH-CH_3$ (with CH_3 on the CH)

c. $CH_3-CH-CH_2-CH_2-CH_3$ (with CH_3 on the CH) $CH_3-CH_2-CH-CH_2-CH_3$ (with CH_3 on the CH)

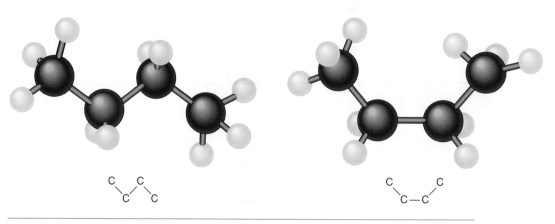

➤ **FIGURE 1.15** Perspective models and carbon skeletons of two conformations of *n*-butane.

Solution

a. Same molecule: In both molecules, the four carbons are bonded in a continuous chain.

$$\overset{1}{CH_3}-\overset{2}{CH_2}-\overset{3}{CH_2}-\overset{4}{CH_3}$$

$$\overset{4}{CH_3} \\ | \\ \overset{2}{CH_2}-\overset{}{CH_2} \\ | \quad {3} \\ \overset{1}{CH_3}$$

b. Same molecule: In both molecules, there is a continuous chain of four carbons with a branch at position 2. The molecule has simply been turned around.

$$\underset{1}{CH_3}-\underset{2}{\overset{\overset{\displaystyle CH_3}{|}}{CH}}-\underset{3}{CH_2}-\underset{4}{CH_3}$$

$$\underset{4}{CH_3}-\underset{3}{CH_2}-\underset{2}{\overset{\overset{\displaystyle CH_3}{|}}{CH}}-\underset{1}{CH_3}$$

c. Structural isomers: Both molecules have a continuous chain of five carbons, but the branch is located at different positions.

$$\underset{1}{CH_3}-\underset{2}{\overset{\overset{\displaystyle CH_3}{|}}{CH}}-\underset{3}{CH_2}-\underset{4}{CH_2}-\underset{5}{CH_3}$$

$$\underset{1}{CH_3}-\underset{2}{CH_2}-\underset{3}{\overset{\overset{\displaystyle CH_3}{|}}{CH}}-\underset{4}{CH_2}-\underset{5}{CH_3}$$

Learning Check 1.6

Which of the following pairs represent structural isomers, and which are simply the same compound?

a. $CH_3-CH_2-CH_2$
     $\quad\quad\quad\quad\quad |$
     $\quad\quad\quad\quad CH_2-CH_3$

   CH_3-CH_2
   $\quad\quad\quad |$
   $\quad\quad CH_2-CH_2$
   $\quad\quad\quad\quad\quad |$
   $\quad\quad\quad\quad CH_3$

b. $CH_3-\overset{\overset{\displaystyle CH_3}{|}}{CH}-CH_2-CH_3$

   $\quad\quad\quad\quad\quad\quad\overset{\displaystyle CH_2-CH_3}{|}$
   $CH_3-\overset{}{CH}-CH_3$

c. $CH_3-CH_2-\overset{\overset{\displaystyle CH_3}{|}}{CH}-CH_2-CH_3$

   $CH_3-CH_2-CH_2-\overset{\overset{\displaystyle CH_3}{|}}{CH}-CH_3$

➤ 1.7 Alkane Nomenclature

When only a relatively few organic compounds were known, chemists gave them what are today called trivial or common names, such as methane, ethane, propane, and butane. The names for the larger alkanes were derived from the Greek prefixes that indicate the number of carbon atoms in the molecule. Thus, *pent*ane contains five carbons, *hex*ane has six, *hept*ane has seven, and so forth, as shown in ■ Table 1.4.

As more compounds and isomers were discovered, however, it became increasingly difficult to devise unique names and much more difficult to commit them to memory. Obviously, a systematic method was needed. Such a method is now in use, but so are several common methods.

The names for the two isomeric butanes (*n*-butane and isobutane) illustrate the important features of the common nomenclature system used for alkanes. The stem *but*- indicates that four carbons are present in the molecule.

TABLE 1.4	Names of alkanes		
Number of carbon atoms	Name	Molecular formula	Structure of normal isomer
1	methane	CH_4	CH_4
2	ethane	C_2H_6	CH_3CH_3
3	propane	C_3H_8	$CH_3CH_2CH_3$
4	butane	C_4H_{10}	$CH_3CH_2CH_2CH_3$
5	pentane	C_5H_{12}	$CH_3CH_2CH_2CH_2CH_3$
6	hexane	C_6H_{14}	$CH_3CH_2CH_2CH_2CH_2CH_3$
7	heptane	C_7H_{16}	$CH_3CH_2CH_2CH_2CH_2CH_2CH_3$
8	octane	C_8H_{18}	$CH_3CH_2CH_2CH_2CH_2CH_2CH_2CH_3$
9	nonane	C_9H_{20}	$CH_3CH_2CH_2CH_2CH_2CH_2CH_2CH_2CH_3$
10	decane	$C_{10}H_{22}$	$CH_3CH_2CH_2CH_2CH_2CH_2CH_2CH_2CH_2CH_3$

The *-ane* ending signifies the alkane family. The prefix *n-* indicates that all carbons form an unbranched chain. The prefix *iso-* refers to compounds in which all carbons except one are in a continuous chain and in which that one carbon is branched from a next-to-the-end carbon, as shown:

This common naming system has limitations. Pentane has three isomers, and hexane has five. The more complicated the compound, the greater the number of isomers, and the greater the number of special prefixes needed to name all the isomers. It would be extremely difficult and time-consuming to try to identify each of the 75 isomeric alkanes containing 10 carbon atoms by a unique prefix or name.

To devise a system of nomenclature that could be used for even the most complicated compounds, committees of chemists have met periodically since 1892. The system resulting from these meetings is called the IUPAC (International Union of Pure and Applied Chemistry) system. This system is much the same for all classes of organic compounds. The IUPAC name for an organic compound consists of three component parts:

The *root* of the IUPAC name specifies the longest continuous chain of carbon atoms in the compound. The roots for the first 10 normal hydrocarbons are based on the names given in Table 1.4: C_1 *meth-*, C_2 *eth-*, C_3 *prop-*, C_4 *but-*, C_5 *pent-*, C_6 *hex-*, C_7 *hept-*, C_8 *oct-*, C_9 *non-*, C_{10} *dec-*.

The *ending* of an IUPAC name specifies the functional class or the major functional group of the compound. The ending *-ane* specifies an alkane. Each of the other functional classes has a characteristic ending; for example, *-ene* is the ending for alkenes, and the *-ol* ending designates alcohols.

Prefixes are used to specify the identity, number, and location of atoms or groups of atoms that are attached to the longest carbon chain. ■ Table 1.5 lists several common carbon-containing groups referred to as **alkyl groups.** Each alkyl group is a collection of atoms that can be thought of as an alkane minus one hydrogen atom. Alkyl groups are named simply by dropping *-ane* from the name of the corresponding alkane and replacing it with *-yl*. For example, CH_3— is called a methyl group and CH_3—CH_2— an ethyl group:

CH_4	CH_3—CH_3	CH_3—	CH_3—CH_2—
methane	ethane	methyl group	ethyl group

Two different alkyl groups can be derived from propane, depending on which hydrogen is removed. Removal of a hydrogen from an end carbon results in a propyl group:

CH_3—CH_2—CH_3	CH_3—CH_2—CH_2—
propane	propyl group

Removal of a hydrogen from the center carbon results in an isopropyl group:

CH_3—CH_2—CH_3	CH_3—$\overset{\mid}{C}H$—CH_3
propane	isopropyl group

http://www.brookscole.com/chemistry/ seager_chemistry
Further information about IUPAC is available at Web site 1.3.

Alkyl group
A group differing by one hydrogen from an alkane.

TABLE 1.5 Common alkyl groups

Parent alkane	Structure of parent alkane	Structure of alkyl group	Name of alkyl group
methane	CH_4	CH_3—	methyl
ethane	CH_3CH_3	CH_3CH_2—	ethyl
propane	$CH_3CH_2CH_3$	$CH_3CH_2CH_2$—	propyl
		$CH_3\overset{\mid}{C}HCH_3$	isopropyl
n-butane	$CH_3CH_2CH_2CH_3$	$CH_3CH_2CH_2CH_2$—	butyl
		$CH_3CH_2\overset{\mid}{C}HCH_3$	*sec*-butyl (secondary-butyl)[a]
isobutane	$CH_3\overset{\overset{\textstyle CH_3}{\mid}}{C}HCH_3$	$CH_3\overset{\overset{\textstyle CH_3}{\mid}}{C}HCH_2$—	isobutyl
		$CH_3\overset{\overset{\textstyle CH_3}{\mid}}{\underset{\mid}{C}}CH_3$	*t*-butyl (tertiary-butyl)[a]

[a]For an explanation of *secondary* and *tertiary,* see Section 3.2.

TABLE 1.6	Common nonalkyl groups
Group	**Name**
—F	fluoro
—Cl	chloro
—Br	bromo
—I	iodo
—NO$_2$	nitro
—NH$_2$	amino

An isopropyl group also can be represented by (CH$_3$)$_2$CH—. As shown in Table 1.5, four butyl groups can be derived from butane, two from the straight-chain, or normal, butane, and two from the branched-chain isobutane. A number of nonalkyl groups are also commonly used in naming organic compounds (see ■ Table 1.6).

The following steps are useful when the IUPAC name of an alkane is written on the basis of its structural formula:

Step 1. Name the longest chain. The longest continuous carbon-atom chain is chosen as the basis for the name. The names are those given in Table 1.4.

Example	Longest Chain	Comments
CH$_3$ — CH$_2$ — CH — CH$_3$ CH$_3$	C — C — C — C C	This compound is a butane.
CH$_2$ — CH$_3$ CH$_3$ — CH — CH$_2$ CH$_3$	C — C C — C — C C	This compound is a pentane.
CH$_2$ — CH$_3$ CH$_3$ — CH$_2$ — C — CH$_3$ CH$_2$ CH$_3$ — CH$_2$	C — C C — C — C — C C C — C C — C C — C — C — C C C — C	Often there is more than one way to designate the longest chain. Either way, in this case, the compound is a hexane.

Learning Check 1.7

Identify the longest carbon chain in the following:

a. CH$_3$
 CH$_2$ — CH$_2$ — CH$_3$
 butane

b. CH$_3$ CH$_3$
 CH$_2$ — CH — CH$_2$
       CH$_3$ *pentane*

c. CH$_3$ — CH — CH$_3$
 CH$_3$ — CH — CH$_3$
 propane

Step 2. Number the longest chain. The carbon atoms in the longest chain are numbered consecutively from the end that will give the lower number to any carbon to which a group is attached.

$\overset{4}{C}H_3 - \overset{3}{C}H_2 - \overset{2}{C}H - \overset{1}{C}H_3$ and not $\overset{1}{C}H_3 - \overset{2}{C}H_2 - \overset{3}{C}H - \overset{4}{C}H_3$
          CH$_3$ CH$_3$

$\overset{2}{C}H_2 - \overset{1}{C}H_3$ $\overset{4}{C}H_2 - \overset{5}{C}H_3$
$\overset{3|}{C}H_3 - \overset{}{C}H - \overset{4}{C}H_2$ the chain may also   CH$_3$ — $\overset{}{C}H - \overset{2}{C}H_2$
        $\overset{5|}{C}H_3$ be numbered $\overset{3}{}$ $\overset{1|}{C}H_3$

If two or more alkyl groups are attached to the longest chain and more than one numbering sequence is possible, the chain is numbered to get the lowest series of numbers. An easy way to follow this rule is to number from the end of the chain nearest a branch:

$$\overset{1}{C}H_3-\overset{2}{C}H-\overset{3}{C}H_2-\overset{4}{C}H-\overset{5}{C}H_2\overset{6}{C}H_3 \qquad \text{and not} \qquad \overset{6}{C}H_3-\overset{5}{C}H-\overset{4}{C}H_2-\overset{3}{C}H-\overset{2}{C}H_2-\overset{1}{C}H_3$$

with CH_3 and CH_2-CH_3 branches

Groups are located at positions 2 and 4 · · · · · · · · · · · · · Groups are located at positions 3 and 5

$$\overset{5}{C}H_3-\overset{4}{C}H-\overset{3}{C}H_2-\overset{2}{C}-\overset{1}{C}H_3 \qquad \text{and not} \qquad \overset{1}{C}H_3-\overset{2}{C}H-\overset{3}{C}H_2-\overset{4}{C}-\overset{5}{C}H_3$$

with CH_3 branches

Groups are located at positions 2,2,4 · · · · · · · Groups are located at positions 2,4,4

Here, a difference occurs with the second number in the sequence, so positions 2,2,4 is the lowest series and is used rather than positions 2,4,4.

Decide how to correctly number the longest chain in the following according to IUPAC rules:

Learning Check 1.8

a. $\overset{1}{C}H_3-\overset{2}{C}H-\overset{3}{C}H_2-\overset{4}{C}H_2-\overset{5}{C}H_3$
with CH_2 and CH_3

b. $\overset{8}{C}H_3-\overset{7}{C}H_2-\overset{6}{C}H_2-\overset{5}{C}H_2-\overset{4}{C}-\overset{3}{C}H_2-\overset{2}{C}H-\overset{1}{C}H_3$
with CH_3, CH_3, and CH_3

Step 3. Locate and name the attached alkyl groups. Each group is located by the number of the carbon atom to which it is attached on the chain.

$$\overset{4}{C}H_3-\overset{3}{C}H_2-\overset{2}{C}H-\overset{1}{C}H_3$$
with CH_3

The attached group is located on carbon 2 of the chain, and it is a methyl group.

$$\overset{1}{C}H_3-\overset{2}{C}H-\overset{3}{C}H_2-\overset{4}{C}H-\overset{5}{C}H_2-\overset{6}{C}H_3$$
with CH_3 and CH_2-CH_3

The one-carbon group at position 2 is a methyl group. The two-carbon group at position 4 is an ethyl group.

Identify the alkyl groups attached to the dashed line, which symbolizes a long carbon chain. Refer to Table 1.5 if necessary.

Learning Check 1.9

Step 4. Combine the longest chain and the branches into the name. The position and the name of the attached alkyl group are added to the name of the longest chain and written as one word:

$$\overset{4}{C}H_3 - \overset{3}{C}H_2 - \overset{2}{C}H - \overset{1}{C}H_3$$
$$|$$
$$CH_3$$

2-methylbutane

Additional steps are needed when more than one alkyl group is attached to the longest chain.

Step 5. Indicate the number and position of attached alkyl groups. If two or more of the same alkyl group occur as branches, the number of them is indicated by the prefixes di-, tri-, tetra-, penta-, etc., and the location of each is again indicated by a number. These position numbers, separated by commas, are put just before the name of the group, with hyphens before and after the numbers when necessary:

$$\overset{1}{C}H_3 - \overset{2}{C}H - \overset{3}{C}H_2 - \overset{4}{C}H - \overset{5}{C}H_3$$
$$| \qquad\qquad\quad |$$
$$CH_3 \qquad\quad CH_3$$

2, 4-dimethylpentane

$$\overset{1}{C}H_3 - \overset{2}{C}H_2 - \overset{3}{C} - \overset{4}{C}H_2 - \overset{5}{C}H_3$$
$$|$$
$$CH_3$$

3, 3-dimethylpentane

If two or more *different* alkyl groups are present, their names are alphabetized and added to the name of the basic alkane, again as one word. For purposes of alphabetizing, the prefixes *di-, tri-,* and so on are ignored, as are the italicized prefixes secondary *(sec)* and tertiary *(t)*. The prefix *iso-* is an exception and is used for alphabetizing:

$$\overset{1}{C}H_3 - \overset{2}{C}H - \overset{3}{C}H - \overset{4}{C}H_2 - \overset{5}{C}H - \overset{6}{C}H_2 - \overset{7}{C}H_2 - \overset{8}{C}H_3$$
$$| \qquad | \qquad\qquad\quad |$$
$$CH_3 \quad CH_3 \qquad\quad CH - CH_3$$
$$|$$
$$CH_3$$

5-isopropyl-2,3-dimethyloctane

Learning Check 1.10

Give the correct IUPAC name to each of the following:

a.
$$CH_3 - CH_2 - CH_2 - CH_2 - \overset{2}{C}H - \overset{1}{C}H_3$$
$$\overset{|}{C}H_3$$

2-methylhexane

b.
$$CH_2 - CH_3$$
$$\overset{7}{C}H_3 - \overset{6}{C}H_2 - \overset{5}{C}H_2 - \overset{4}{C}H - \overset{3}{C}H_2 - \overset{2}{C}H - \overset{1}{C}H_3$$
$$\overset{|}{C}H_2 - CH_3$$

2, methyl heptane

c.
$$CH_3$$
$$\overset{7}{C}H_3 - \overset{6}{C}H_2 - \overset{5}{C}H_2 - \overset{4}{C}H - \overset{3}{C}H - \overset{2}{C}H - \overset{1}{C}H_3$$
$$\overset{|}{C}H - CH_3 \quad \overset{|}{C}H_3$$
$$\overset{|}{C}H_3$$

4-isopropyl 3,2dimethyl heptane

Naming compounds is a very important skill, as is the reverse process of using IUPAC nomenclature to specify a structural formula. The two processes are very similar. To obtain a formula from a name, determine the longest chain, number the chain, and add any attached groups.

➤ **Example 1.4**

Draw a condensed structural formula for 3-ethyl-2-methylhexane.

Solution

Use the last part of the name to determine the longest chain. Draw a chain of six carbons. Then, number the carbon atoms.

$$C - C - C - C - C - C$$
1 2 3 4 5 6

Attach a methyl group at position 2 and an ethyl group at position 3.

$$CH_3$$
|
$$C - C - C - C - C - C$$
1 2 |3 4 5 6
$$CH_2CH_3$$

Complete the structure by adding enough hydrogen atoms so that each carbon has four bonds.

$$CH_3$$
|
$$CH_3 - CH - CH - CH_2 - CH_2 - CH_3$$
1   2 |3   4   5   6
$$CH_2CH_3$$

Draw a condensed structural formula for each of the following compounds:

a. 2,2,4-trimethylpentane
b. 3-isopropylhexane
c. 3-ethyl-2,4-dimethylheptane

➤ **1.8 Cycloalkanes**

From what we have said so far, the formula C_3H_6 cannot represent an alkane. Not enough hydrogens are present to allow each carbon to form four bonds, unless there are multiple bonds. For example, the structural formula $CH_3 - CH = CH_2$ fits the molecular formula but cannot represent an alkane because of the double bond. The C_3H_6 formula does become acceptable for an alkane if the carbon atoms form a ring, or cyclic, structure rather than the open-chain structure shown:

$$C - C - C$$
Open chain

$$\overset{C}{\overset{/\,\backslash}{C - C}}$$
Cyclic

The resulting saturated cyclic compound, called cyclopropane, has the structural formula

$$\overset{CH_2}{\overset{/\,\,\backslash}{CH_2 - CH_2}}$$

Alkanes containing rings of carbon atoms are called **cycloalkanes.** Like the other alkanes, cycloalkanes are not found in human cells. However, several important molecules in human cells do contain rings of five or six atoms, and the study of cycloalkanes will help you better understand the chemical behavior of these complex molecules.

Learning Check 1.11

Cycloalkane
An alkane in which carbon atoms form a ring.

TABLE 1.7 Structural formulas and symbols for common cycloalkanes

Name	Structural formula	Condensed formula
cyclopropane	CH_2 / \ $H_2C — CH_2$	△
cyclobutane	$H_2C — CH_2$ | | $H_2C — CH_2$	▢
cyclopentane	H_2 C H_2C　CH_2 \ / $H_2C — CH_2$	⬠
cyclohexane	H_2 C H_2C　CH_2 | | H_2C　CH_2 C H_2	⬡

According to IUPAC rules, cycloalkanes are named by placing the prefix *cyclo-* before the name of the corresponding alkane with the same number of carbon atoms. Chemists often abbreviate the structural formulas for cycloalkanes and draw them as geometric figures (triangles, squares, etc.) in which each corner represents a carbon atom. The hydrogens are omitted (see ■ Table 1.7). It is important to remember that each carbon atom still possesses four bonds, and that hydrogen is assumed to be bonded to the carbon atoms unless something else is indicated. When substituted cycloalkanes are named, the position of a single attached group does not need to be specified in the name because all positions in the ring are equivalent. However, when two or more groups are attached, their positions of attachment are indicated by numbers, just as they were for alkanes. The ring numbering begins with the carbon attached to the first group alphabetically and proceeds around the ring in the direction that will give the lowest numbers for the location of the other attached groups.

methylcyclopentane

1,2-dimethylcyclopentane
not 1,5-dimethylcyclopentane

1-chloro-3-methylcyclopentane
not 1-chloro-4-methylcyclopentane
not 3-chloro-1-methylcyclopentane

➤ **Example 1.5**

Represent each of the following cycloalkanes by a geometric figure, and name each compound:

a.
H_2
C
H_2C　CH_2
$H_2C — CH$
|
CH_3

b.
H_2
C
H_2C　$CH — CH_3$
H_2C　CH_2
CH
|
CH_3

Solution

a. A pentagon represents a five-membered ring, which is called cyclopentane. This compound has a methyl group attached, so the name is methylcyclopentane. The position of a single alkyl group is not indicated by a number because the positions of all carbons in the ring are equivalent.

b. A hexagon represents a six-carbon ring, which is called cyclohexane. Two methyl groups are attached; thus we have a dimethylcyclohexane.

   However, the positions of the two alkyl groups must be indicated. The ring is numbered beginning with a carbon to which a methyl group is attached, counting in the direction giving the lowest numbers. The correct name is 1,3-dimethylcyclohexane. Notice that a reverse numbering beginning at the same carbon would have given 1,5-dimethylcyclohexane. The number 3 in the correct name is lower than the 5 in the incorrect name.

Give each of the following compounds the correct IUPAC name:

a. CH₃ ... CH₃

b. CH₂—CH₃

c. CH₃ ... CH₂—CH₃

Learning Check 1.12

A. 1,4-dimethylcyclohexane

B. ethylcyclopropane

C. 1-methyl-2-ethylcyclopentane

▶ 1.9 The Shape of Cycloalkanes

Recall from Section 1.5 that a tetrahedral orientation of bonds with bond angles of 109.5° is characteristic of carbon atoms that form four single bonds. A tetrahedral arrangement is the most stable because it results in the least crowding of the atoms. In certain cycloalkanes, however, a tetrahedral arrangement for all carbon-to-carbon bonds is not possible. For example, the bond angles between adjacent carbon–carbon bonds in planar cyclopropane molecules must be 60° (see ■ Figure 1.16). In cyclobutane, they are close to 90°. As a result, cyclopropane and cyclobutane rings are much less stable than compounds with bond angles of about 109°. Both cyclobutane and cyclopentane bend slightly from a planar structure to reduce the crowding of hydrogen atoms (Figure 1.16). In larger cycloalkanes, the bonds to carbon atoms can be tetrahedrally arranged only when the carbon atoms do not lie in the same plane. For example, cyclohexane can assume several nonplanar shapes. The chair and boat forms, where all the bond angles are 109.5°, are shown in Figure 1.16.

cyclopropane

cyclobutane

cyclopentane

cyclohexane
(chair form)

cyclohexane
(boat form)

➤ **FIGURE 1.16** Ball-and-stick models for common cycloalkanes.

Free rotation

Free rotation
not possible

➤ **FIGURE 1.17** Rotation about C—C single bonds occurs in open-chain compounds but not within rings.

Stereoisomers
Compounds with the same structural formula but different spatial arrangements of atoms.

Geometric isomers
Molecules with restricted rotation around C—C bonds that differ in the three-dimensional arrangements of their atoms in space and not in the order of linkage of atoms.

The free rotation that can take place around C—C single bonds in alkanes (Section 1.6) is not possible for the C—C bonds of cycloalkanes. The ring structure allows bending or puckering but prevents free rotation (see ■ Figure 1.17). Any rotation of one carbon atom 180° relative to an adjacent carbon atom in a cycloalkane would require a single carbon–carbon bond to be broken somewhere in the ring. The breaking of such bonds would require a large amount of energy.

The lack of free rotation around C—C bonds in disubstituted cycloalkanes leads to an extremely important kind of isomerism called *stereoisomerism.* Two different compounds that have the same molecular formula and the same structural formula but different spatial arrangements of atoms are called **stereoisomers.** For example, consider a molecule of 1,2-dimethylcyclopentane. The cyclopentane ring is drawn in ■ Figure 1.18 as a planar pentagon with the heavy lines indicating that two of the carbons are in front as one views the structure. The groups attached to the ring project above or below the plane of the ring. Two stereoisomers are possible: either both groups may project in the same direction from the plane, or they may project in opposite directions from the plane of the ring. Since the methyl groups cannot rotate from one side of the ring to the other, molecules of the two compounds represented in Figure 1.18 are distinct.

These two compounds have physical and chemical properties that are quite different and therefore can be separated from each other. Stereoisomers of this type, in which the spatial arrangement or geometry of their groups is maintained by rings, are called **geometric isomers** or *cis-trans* isomers. The

➤ **FIGURE 1.18** Two geometric isomers of 1,2-dimethylcyclopentane.

CH₃   CH₃

cis-1,2-dimethylcyclopentane

CH₃

CH₃

trans-1,2-dimethylcyclopentane

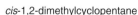

prefix *cis-* denotes the isomer in which both groups are on the same side of the ring, and *trans-* denotes the isomer in which they are on opposite sides. To exist as geometric isomers, a disubstituted cycloalkane must be bound to groups at two different carbons of the ring. For example, there are no geometric isomers of 1,1-dimethylcyclohexane:

cis-
On the same side (as applied to geometric isomers).

trans-
On opposite sides (as applied to geometric isomers).

➤ Example 1.6

Name and draw structural formulas for all the isomers of dimethylcyclobutane. Indicate which ones are geometric isomers.

Solution

There are three possible locations for the two methyl groups: positions 1,1, positions 1,2, and positions 1,3.

Geometric isomerism is not possible in this case with the two groups bound to the same carbon of the ring

1,1-dimethylcyclobutane

Two groups on opposite sides of the planar ring

trans-1,2-dimethylcyclobutane

Two groups on the same side of the ring

cis-1,2-dimethylcyclobutane

trans-1,3-dimethylcyclobutane

cis-1,3-dimethylcyclobutane

a. Identify each of the following cycloalkanes as a *cis-* or *trans-* compound:

Learning Check 1.13

(1)

(2)

(3)

b. Draw the structural formula for *cis*-1,2-dichlorocyclobutane.

➤ 1.10 Physical Properties of Alkanes

Since alkanes are composed of nonpolar carbon–carbon and carbon–hydrogen bonds, alkanes are nonpolar molecules. Alkanes have lower melting and boiling points than other organic compounds of comparable molecular weight (see

TABLE 1.8   Physical properties of some normal alkanes

Number of carbon atoms	IUPAC name	Condensed structural formula	Melting point (°C)	Boiling point (°C)	Density (g/mL)
1	methane	CH_4	−182.5	−164.0	0.55
2	ethane	CH_3CH_3	−183.2	−88.6	0.57
3	propane	$CH_3CH_2CH_3$	−189.7	−42.1	0.58
4	butane	$CH_3CH_2CH_2CH_3$	−133.4	−0.5	0.60
5	pentane	$CH_3CH_2CH_2CH_2CH_3$	−129.7	36.1	0.63
6	hexane	$CH_3CH_2CH_2CH_2CH_2CH_3$	−95.3	68.9	0.66
7	heptane	$CH_3CH_2CH_2CH_2CH_2CH_2CH_3$	−90.6	98.4	0.68
8	octane	$CH_3CH_2CH_2CH_2CH_2CH_2CH_2CH_3$	−56.8	125.7	0.70
9	nonane	$CH_3CH_2CH_2CH_2CH_2CH_2CH_2CH_2CH_3$	−53.5	150.8	0.72
10	decane	$CH_3CH_2CH_2CH_2CH_2CH_2CH_2CH_2CH_2CH_3$	−29.7	174.1	0.73

Homologous series

Compounds of the same functional class that differ by a —CH_2— group.

■ Table 1.8). This is because their nonpolar molecules exert very weak attractions for each other. Alkanes are odorless compounds.

The normal, or straight-chain, alkanes make up what is called a **homologous series**. This term describes any series of compounds in which each member differs from a previous member only by having an additional —CH_2— unit. The physical and chemical properties of compounds making up a homologous series are usually closely related and vary in a systematic and predictable way. For example, the melting and boiling points of normal alkanes increase smoothly as the length of the carbon chain increases (see ■ Figure 1.19). This pattern results from increasing dispersion forces as molecular weight increases. At ordinary temperatures and pressures, normal alkanes with 1 to 4 carbon atoms are gases, those with 5 to 20 carbon atoms are liquids, and those with more than 20 carbon atoms are waxy solids.

Because they are nonpolar, alkanes and other hydrocarbons are insoluble in water, which is a highly polar solvent. They are also less dense than water and thus float on it. These two properties of hydrocarbons are partly responsible for the well-known serious effects of oil spills from ships (see ■ Figure 1.20).

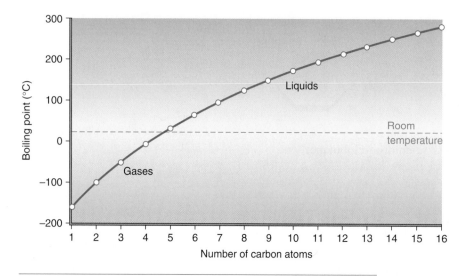

➤ **FIGURE 1.19** Normal alkane boiling points depend on chain length.

➤ **FIGURE 1.20** Oil spills can have serious and long-lasting effects on the environment because of the insolubility of hydrocarbons in water.

CHEMISTRY AROUND US • 1.2

PETROLEUM

Petroleum, the most important of the fossil fuels used today, is sometimes called "black gold" in recognition of its importance in the twentieth century. At times, the need for petroleum to keep society fueled has seemed second only to our need for food, air, water, and shelter.

It is generally believed that this complex mixture of hydrocarbons was formed over eons through the gradual decay of ocean-dwelling microscopic animals. The resulting crude oil, a viscous black liquid, collects in vast underground pockets in sedimentary rock. It is brought to the surface via drilling and pumping.

Useful products are obtained from crude oil by heating it to high temperatures to produce various fractions according to boiling point (see table). Most petroleum products are eventually burned as a fuel, but about 2% are used to synthesize organic compounds. This seemingly small amount is quite large in actual tonnage because of the huge volume of petroleum that is refined annually. In fact, more than half of all industrial synthetic organic compounds are made from this source. These industrial chemicals are eventually converted into dyes, drugs, plastics, artificial fibers, detergents, insecticides, and other materials deemed indispensable by many in industrialized nations.

Fraction	Boiling point range (°C)	Molecular size range	Typical uses
Gas	−164–30	C_1–C_4	Heating, cooking
Gasoline	30–200	C_5–C_{12}	Motor fuel
Kerosene	175–275	C_{12}–C_{16}	Fuel for stoves and diesel and jet engines
Heating oil	Up to 375	C_{15}–C_{18}	Furnace oil
Lubricating oils	350 and up	C_{16}–C_{20}	Lubrication, mineral oil
Greases	Semisolid	C_{18}–up	Lubrication, petroleum jelly
Paraffin (wax)	Melts at 52–57	C_{20}–up	Candles, toiletries
Pitch and tar	Residue in boiler	High	Roofing, asphalt paving

Asphalt for paving roads is a petroleum product.

Liquid alkanes of higher molecular weight behave as emollients (skin softeners) when applied to the skin. An alkane mixture known as mineral oil is sometimes used to replace natural skin oils washed away by frequent bathing or swimming. Petroleum jelly (Vaseline is a well-known brand name) is a semisolid mixture of alkanes that is used as both an emollient and a protective film. Water and water solutions such as urine don't dissolve or penetrate the film, and the underlying skin is protected. Many cases of diaper rash have been prevented or treated this way.

The word **hydrophobic** (literally "water fearing") is often used to refer to molecules or parts of molecules that are insoluble in water. Many biomolecules, the large organic molecules associated with living organisms, contain nonpolar (hydrophobic) parts. Thus, such molecules are not water-soluble. Palmitic acid, for example, contains a large nonpolar hydrophobic portion and is insoluble in water.

Hydrophobic
Molecules or parts of molecules that repel (are insoluble in) water.

Nonpolar portion

$$CH_3-CH_2-CH_2-CH_2-CH_2-CH_2-CH_2-CH_2-CH_2-CH_2-CH_2-CH_2-CH_2-CH_2-CH_2-\overset{\displaystyle O}{\overset{\displaystyle \|}{C}}-OH$$

palmitic acid

OVER THE COUNTER • 1.1

HYDRATING THE SKIN

Healthy, normal human skin is kept moist by natural body oils contained in sebum, a secretion of the skin's sebaceous glands. The oily sebum helps the epidermis, or outer layer of the skin, retain the 10–30% of water it normally contains. However, some people suffer from skin that is naturally dry, or dry because of aging or contact with materials like paint thinner that dissolve and remove the sebum. Such individuals may find some relief by using OTC products called *moisturizers*.

Two types of skin moisturizers are commonly available. One type, which behaves like natural sebum, contains nonpolar oily substances that form a barrier and prevent water from passing through and evaporating from the skin. These barrier-forming products often contain combinations of materials from various sources, including mineral oil and petroleum jelly from petroleum, vegetable oils such as apricot oil, sesame seed oil, palm kernel oil, olive oil, safflower oil, and lanolin, an animal fat from sheep oil glands and wool. While they effectively keep water from leaving the skin, these products are somewhat messy to use and leave the skin feeling greasy.

A second, more popular type of moisturizer works by attracting water from the air and skin. These products form a water-rich layer that adheres to the skin without giving it a greasy feel. The substances that attract water are called humectants, and are compounds capable of forming hydrogen bonds with water. Some examples of humectants used in products of this type are glycerol, urea, lactic acid, and propylene glycol.

If you are shopping for a moisturizer, remember that the main characteristic you should look for in a product is the ability to form a barrier to prevent water evaporation, or the ability to act as a humectant. Some expensive products advertise that in addition to moisturizing, they also beautify the skin and even reverse aging because they contain proteins such as collagen and elastin, vitamins, hormones, or even DNA. It is unlikely that such substances can pass through the epidermis of the skin in sufficient amounts to provide the advertised benefits.

http://www.brookscole.com/chemistry/seager_chemistry
You can go to Web site 1.4 for further information about skin care.

Alkane products marketed to soften and protect the skin.

➤ 1.11 Alkane Reactions

The alkanes are the least reactive of all organic compounds. In general, they do not react with strong acids (such as sulfuric acid), strong bases (such as sodium hydroxide), most oxidizing agents (such as potassium dichromate), and most reducing agents (such as sodium metal). This unreactive nature is reflected in the name *paraffins*, sometimes used to identify alkanes (from Latin words that mean "little affinity"). Paraffin wax, sometimes used to seal jars of homemade preserves, is a mixture of solid alkanes. Paraffin wax is also used in the preparation of milk cartons and wax paper. The inertness of the compounds in the wax makes it ideal for these uses. Alkanes do undergo reactions with halogens such as chlorine and bromine, but these are not important for our purposes.

Perhaps the most significant reaction of alkanes is the rapid oxidation called combustion. In the presence of ample oxygen, alkanes burn to form carbon dioxide and water, liberating large quantities of heat:

$$CH_4 + 2O_2 \rightarrow CO_2 + 2H_2O + 212.8 \text{ kcal/mol} \qquad (1.4)$$

It is this type of reaction that accounts for the wide use of hydrocarbons as fuels.

Natural gas contains methane (80–95%), some ethane, and small amounts of other hydrocarbons. Propane and butane are extracted from natural gas and sold in pressurized metal containers (bottled gas). In this form, they are used for heating and cooking in campers, trailers, boats, and rural homes.

Gasoline is a mixture of hydrocarbons (primarily alkanes) that contain 5 to 12 carbon atoms per molecule. Diesel fuel is a similar mixture, except the molecules contain 12 to 16 carbon atoms. The hot CO_2 and water vapor gen-

CHEMISTRY AROUND US • 1.3

CARBON MONOXIDE: SILENT AND DEADLY

The incomplete combustion of alkanes and other carbon-containing fuels may produce carbon monoxide, CO. It is a common component of furnace, stove, and automobile exhaust. Normally, these fumes are emitted into the air and dispersed. However, whenever CO gas is released into closed spaces, lethal concentrations may develop. All of the following are known to have caused CO poisoning deaths: poor ventilation of a furnace or stove exhaust, running the engine of an automobile for a period of time in a closed garage, idling the engine of a snowbound car to keep the heater working, and using a barbecue grill as a source of heat in a house or camper (burning charcoal generates a large amount of carbon monoxide). Recently, the exhaust systems of snow-blocked vehicles have been identified as another cause of carbon monoxide poisoning. The problem is created when the end of a vehicle's exhaust pipe is obstructed or plugged by snow. CO-containing exhaust can leak through cracks in the exhaust system, penetrate the floorboard, and enter the passenger compartment of the vehicle. An estimated 1000 Americans die each year from unintentional CO poisoning, and as many as 10,000 require medical treatment.

Carbon monoxide is dangerous because of its ability to bind strongly with hemoglobin in red blood cells. When this occurs, the ability of the blood to transport oxygen (O_2) is reduced because the CO molecules occupy sites on hemoglobin molecules that are normally occupied by O_2 molecules.

Because it is colorless and odorless, carbon monoxide gas is a silent and stealthy potential killer. The following symptoms of CO poisoning are given in order of increasing severity and seriousness:

1. Slight headache, dizziness, drowsiness
2. Headache, throbbing temples
3. Weakness, mental confusion, nausea
4. Rapid pulse and respiration, fainting
5. Possibly fatal coma

All except the most severe cases of CO poisoning are reversible. The most important first aid treatment is to get the victim fresh air. Any person who feels ill and suspects CO poisoning might be the cause should immediately evacuate the building, get fresh air, and summon medical assistance. Victims are often treated with 100% oxygen delivered from a mask.

Consumer safety experts stress that combustion appliances should be inspected regularly for leaks or other malfunctions. Also, it is recommended that drivers should inspect exhaust pipes and clear any obstructions before starting vehicles that have been parked in snow. The Consumer Product Safety Commission recommends that CO detectors be installed near bedrooms in residences. These detectors sound an alarm before the gas reaches potentially lethal concentrations in the air.

Fireplaces must be properly constructed to carry away carbon monoxide fumes.

http://www.brookscole.com/chemistry/seager_chemistry
Web site 1.5 provides further information about carbon monoxide poisoning.

---

erated during combustion in an internal combustion engine have a much greater volume than the air and fuel mixture. It is this sudden increase in gaseous volume and pressure that pushes the pistons and delivers power to the crankshaft.

If there is not enough oxygen available, incomplete combustion of hydrocarbons occurs, and some carbon monoxide (CO) or even carbon (see ■ Figure 1.21) may be produced (Reactions 1.5 and 1.6):

$$2CH_4 + 3O_2 \rightarrow 2CO + 4H_2O \qquad (1.5)$$

$$CH_4 + O_2 \rightarrow C + 2H_2O \qquad (1.6)$$

These reactions are usually undesirable because CO is toxic, and carbon deposits hinder engine performance. Occasionally, however, incomplete combustion is deliberately caused; specific carbon blacks (particularly lampblack, a pigment for ink) are produced by the incomplete combustion of natural gas.

➤ **FIGURE 1.21** A luminous yellow flame from a laboratory burner produces a deposit of carbon when insufficient air (O_2) is mixed with the gaseous fuel.

> ## FOR FUTURE REFERENCE   MEDICINES REMOVED FROM THE MARKET

eople generally recognize the Food and Drug Administration (FDA) as the federal agency with the authority to require that drugs be proved safe and effective before they are approved for sale and marketed. However, it is not always understood that the FDA also has the authority to withdraw approved drugs from the marketplace if the use of the drugs shows them to be ineffective or hazardous. This authority, while not used very often, provides an added protection for consumers.

The table below lists drugs that were removed from the market during the 1998. In each case the general category to which the drug belongs is given along with its generic name, at least one example of a trade name, and the reason for its removal.

Category	Generic name	Trade name	Reason for removal
Analgesic	bromfenac	Duract	Caused serious liver damage if taken for more than 10 days.
Psoriasis treatment	etretinate	Tegison	Newer drugs became available.
Cholesterol-reduction	probucol	Lorelco	More effective drugs became available.
High blood pressure treatment	mibefradil	Posicor	Serious drug interactions with multiple drugs.
Minimally sedating antihistamine	terfenadine	Seldane	Voluntarily withdrawn because of serious drug interactions with multiple drugs.
Weight-loss agents	dexfenfluramine fenfluramine	Redux Pondimin	Caused heart valve damage. Caused heart valve damage.

SOURCE: Adapted from Rybacki, J.J.; Long, J.W. *The Essential Guide to Prescription Drugs.* New York: Harper Perennial, 1999.

CONCEPT SUMMARY

Carbon: The Element of Organic Compounds. Organic compounds contain carbon, and organic chemistry is the study of those compounds. Inorganic chemistry is the study of the elements and all noncarbon compounds. Carbon compounds are of tremendous everyday importance to life on earth and are the basis of all life processes.

Organic and Inorganic Compounds Compared. The properties of organic and inorganic compounds often differ, largely as a result of bonding differences. Organic compounds contain primarily covalent bonds, whereas ionic bonding is more prevalent in inorganic compounds.

Bonding Characteristics and Isomerism. Large numbers of organic compounds are possible because carbon atoms link to form chains and networks. An additional reason for the existence of so many organic compounds is the phenomenon of isomerism. Isomers are compounds that have the same molecular formula but different arrangements of atoms.

Functional Groups: The Organization of Organic Chemistry. All organic compounds are grouped into classes based on characteristic features called functional groups. Compounds with their functional groups are represented by two types of structural formulas. Expanded structural formulas show all covalent bonds, whereas condensed structural formulas show no covalent bonds or only selected bonds.

Alkane Structures. Alkanes are hydrocarbons that contain only single covalent bonds and can be represented by the formula C_nH_{2n+2}. Alkanes possess a three-dimensional geometry in which each carbon is surrounded by four bonds directed to the corners of a tetrahedron. Methane, the simplest alkane, is an important fuel (natural gas) and a chemical feedstock for the preparation of other organic compounds. The number of structural isomers possible for an alkane increases dramatically with the number of carbon atoms present in the molecule.

Conformations of Alkanes. Rotation about the single bonds between carbon atoms allows alkanes to exist in many different conformations. When an alkane is drawn using only two dimensions, the structure can be represented in a variety of ways as long as the order of bonding is not changed.

Alkane Nomenclature. Some simple alkanes are known by common names. More complex compounds are usually named using the systematic IUPAC system. The characteristic IUPAC ending for alkanes is *-ane*.

Cycloalkanes. These are alkanes in which the carbon atoms form a ring. The prefix *cyclo-* is used in the names of these compounds to indicate their cyclic nature.

The Shape of Cycloalkanes. The carbon atom rings of cycloalkanes are usually shown as planar, although only cyclopropane is planar. Because rotation about the single bonds in the ring is restricted, certain disubstituted cycloalkanes can exist as geometric (*cis-trans*) isomers.

Physical Properties of Alkanes. The physical properties of alkanes are typical of all hydrocarbons: nonpolar, insoluble in water, less dense than water, and increasing melting and boiling points with increasing molecular weight.

Alkane Reactions. Alkanes are relatively unreactive and remain unchanged by most reagents. The reaction that is most significant is combustion.

LEARNING OBJECTIVES

When you have completed your study of this chapter, you should be able to:

1. Show that you understand the general importance of organic chemical compounds.
2. Be able to recognize the molecular formulas of organic and inorganic compounds.
3. Explain some general differences between inorganic and organic compounds.
4. Be able to use structural formulas to identify compounds that are isomers of each other.
5. Write condensed or expanded structural formulas for compounds.
6. Classify alkanes as normal, branched, or cycloalkanes.
7. Use structural formulas to determine whether compounds are structural isomers.
8. Assign IUPAC names and draw structural formulas for alkanes and cycloalkanes.
9. Name and draw structural formulas for geometric isomers of cycloalkanes.
10. Describe the key physical properties of alkanes.
11. Write alkane combustion reactions.

KEY TERMS AND CONCEPTS

Alkane (1.5)
Alkyl group (1.7)
Branched alkane (1.5)
cis- (1.9)
Condensed structural formula (1.4)
Conformations (1.6)
Cycloalkane (1.8)
Expanded structural formula (1.4)

Functional group (1.4)
Geometric isomers (1.9)
Homologous series (1.10)
Hybrid orbital (1.3)
Hydrocarbon (1.5)
Hydrophobic (1.10)
Inorganic chemistry (1.1)
Isomerism (1.3)

Normal alkane (1.5)
Organic chemistry (1.1)
Organic compound (1.1)
Saturated hydrocarbon (1.5)
Stereoisomers (1.9)
Structural isomers (1.3)
trans- (1.9)

KEY REACTIONS

1. Complete combustion of alkanes (Section 1.11):

$$\text{Alkane} + O_2 \rightarrow CO_2 + H_2O$$

2. Incomplete combustion of alkanes (Section 1.11):

$$\text{Alkane} + O_2 \rightarrow CO \text{ (or C)} + H_2O$$

CARBON: THE ELEMENT OF ORGANIC COMPOUNDS (SECTION 1.1)

1.1 Name at least six items you recognize to be composed of organic compounds.

1.2 Why were the compounds of carbon originally called organic compounds?

1.3 Describe what Wöhler did that made the vital force theory highly questionable.

1.4 What is the unique structural feature shared by all organic compounds?

1.5 Classify each of the following compounds as organic or inorganic:
- **a.** KBr
- **b.** H_2O
- **c.** $H-C\equiv C-H$
- **d.** Li
- **e.** CH_3-NH_2

ORGANIC AND INORGANIC COMPOUNDS COMPARED (SECTION 1.2)

1.6 What kind of bond between atoms is most prevalent among organic compounds?

1.7 Are the majority of all compounds that are insoluble in water organic or inorganic? Why?

1.8 Indicate for each of the following characteristics whether it more likely describes an inorganic or an organic compound. Give one reason for your answer.
- **a.** This compound is a liquid that readily burns.
- **b.** A white solid upon heating is found to melt at 735°C.
- **c.** A liquid added to water floats on the surface and does not dissolve.
- **d.** This compound exists as a gas at room temperature and ignites easily.
- **e.** A solid substance melts at 65°C.

1.9 Devise a test, based on the general properties in Table 1.1, that you could use to quickly distinguish between the substances in each of the following pairs:
- **a.** Gasoline (liquid, organic) and water (liquid, inorganic)
- **b.** Naphthalene (solid, organic) and sodium chloride (solid, inorganic)
- **c.** Methane (gaseous, organic) and hydrogen chloride (gaseous, inorganic)

BONDING CHARACTERISTICS AND ISOMERISM (SECTION 1.3)

1.10 Give two reasons for the existence of the tremendous number of organic compounds.

1.11 How many of carbon's electrons are unpaired and available for bonding according to an sp^3 hybridization model?

1.12 Describe what atomic orbitals overlap to produce a carbon–hydrogen bond in CH_4.

1.13 What molecular geometry exists when a central carbon atom bonds to four other atoms?

1.14 Compare the shapes of unhybridized p and hybridized sp^3 orbitals.

1.15 Use Example 1.1 and Tables 1.6 and 1.2 to determine the number of covalent bonds formed by atoms of the following elements: carbon, hydrogen, oxygen, nitrogen, and bromine.

1.16 Complete the following structures by adding hydrogen atoms where needed:
- **a.** $C-C=C$
- **b.** $C-\overset{\displaystyle O}{\overset{\displaystyle \|}{C}}-C$
- **c.** $C-\overset{\displaystyle O}{\overset{\displaystyle \|}{C}}-O$
- **d.** $C-N-C$

1.17 Complete the following structures by adding hydrogen atoms where needed:
- **a.** $\begin{array}{c} C \\ | \\ C-C-C \end{array}$
- **b.** $C-C\equiv C$
- **c.** $C-\overset{\displaystyle O}{\overset{\displaystyle \|}{C}}-N$
- **d.** $O-\overset{\displaystyle O}{\overset{\displaystyle \|}{C}}-C-C$

1.18 Which of the following pairs of compounds are structural isomers?
- **a.** $CH_3-CH=CH-CH_3$ and $CH_3-CH_2-CH_2-CH_3$
- **b.** $CH_3-CH_2-CH_2-CH_2-CH_3$ and $CH_3-\overset{\displaystyle CH_3}{\underset{\displaystyle CH_3}{\overset{\displaystyle |}{\underset{\displaystyle |}{C}}}}-CH_3$
- **c.** $CH_3-CH_2-\overset{\displaystyle CH_3}{\overset{\displaystyle |}{C}}H-OH$ and $CH_3-\overset{\displaystyle O}{\overset{\displaystyle \|}{C}}-CH_2-CH_3$
- **d.** $CH_3-CH_2-\overset{\displaystyle O}{\overset{\displaystyle \|}{C}}-H$ and $CH_3-\overset{\displaystyle O}{\overset{\displaystyle \|}{C}}-CH_3$
- **e.** $CH_3-CH_2-CH_2-NH_2$ and $CH_3-CH_2-NH-CH_3$

1.19 Group all the following compounds together that represent structural isomers of each other:
- **a.** $CH_3-CH_2-\overset{\displaystyle O}{\overset{\displaystyle \|}{C}}-OH$
- **b.** $CH_3-\overset{\displaystyle O}{\overset{\displaystyle \|}{C}}-CH_2-OH$
- **c.** $CH_3-CH_2-CH_2-OH$
- **d.** $CH_3-\overset{\displaystyle O}{\overset{\displaystyle \|}{C}}-CH_2-CH_3$

e.
$$CH_3 - \overset{\overset{\displaystyle OH}{|}}{CH} - \overset{\overset{\displaystyle O}{\|}}{C} - H$$

f.
$$CH_3 - \overset{\overset{\displaystyle O}{\|}}{C} - O - CH_3$$

g. $HO - CH_2 - CH_2 - CH_2 - OH$

1.20 On the basis of the number of covalent bonds possible for each atom, determine which of the following structural formulas are correct. Explain what is wrong with the incorrect structures.

a.
$$H - \overset{\overset{\displaystyle H}{|}}{C} - \overset{|}{C} = \overset{|}{C} - \overset{|}{N} - H$$
with H below each

b.
$$H - \overset{\overset{\displaystyle H}{|}}{C} - \overset{|}{C} - H$$

c.
$$H - \overset{\overset{\displaystyle H}{|}}{C} - \overset{|}{C} = \overset{|}{C} - \overset{|}{C} - H$$

d.
$$CH_3 - \overset{\overset{\displaystyle O}{\|}}{C} - \overset{\overset{\displaystyle O}{\|}}{C} - CH_3$$

e.
$$CH_3 - \overset{}{CH} - \overset{\overset{\displaystyle CH_3}{|}}{CH} = CH - C = CH_2$$
with CH₃ below first CH

FUNCTIONAL GROUPS: THE ORGANIZATION OF ORGANIC CHEMISTRY (SECTION 1.4)

1.21 Identify each of the following as a condensed structural formula, expanded structural formula, or molecular formula:

a. $CH_3CH_2CH_2 - NH_2$

b.
$$H - \overset{\overset{\displaystyle H}{|}}{C} - \overset{\overset{\displaystyle H}{|}}{C} - \overset{\overset{\displaystyle H}{|}}{C} - N - H$$

c. $CH_3 - CH_2 - CH_2 - NH_2$

d. C_3H_9N

1.22 Write a condensed structural formula for the following compounds:

a.
$$H - \overset{\overset{\displaystyle H}{|}}{C} - \overset{\overset{\displaystyle H}{|}}{C} - \overset{\overset{\displaystyle H}{|}}{C} - \overset{\overset{\displaystyle O}{\|}}{C} = C - \overset{\overset{\displaystyle H}{|}}{C} - \overset{\overset{\displaystyle H}{|}}{C} - H$$

b.
$$H - \overset{\overset{\displaystyle H}{|}}{C} - \overset{\overset{\displaystyle H}{|}}{C} - \overset{\overset{\displaystyle O}{\|}}{C} - N - H$$

1.23 Write a condensed structural formula for the following compounds:

a.
$$H - \overset{\overset{\displaystyle H}{|}}{C} - \overset{\overset{\displaystyle H}{|}}{C} - \overset{\overset{\displaystyle H}{|}}{C} - \overset{\overset{\displaystyle O}{\|}}{C} - \overset{\overset{\displaystyle H}{|}}{N} - H$$

b.
$$H - \overset{}{C} - \overset{}{C} - \overset{}{C} - C - OH$$ (with CH₃ branches)

1.24 Write an expanded structural formula for the following:

a. $CH_3 - CH_2 - CH_2 - NH_2$

b.
$$CH_3 - CH_2 - \overset{\overset{\displaystyle O}{\|}}{C} - CH_3$$

ALKANE STRUCTURES (SECTION 1.5)

1.25 Classify each of the following compounds as a normal alkane or a branched alkane:

a. $CH_3 - CH_2 - CH_2 - CH_2 - CH_3$

b.
$$CH_3 - CH_2$$
$$CH_2 - CH_2 - CH_3$$

c.
$$CH_3 - CH - CH_3$$
$$CH_2 - CH_3$$

d.
$$CH_3 - \overset{\overset{\displaystyle CH_3}{|}}{C} - CH_3$$
with CH₃ below

e.
$$CH_2 - CH_2 - CH_2 - CH_2 - CH_2$$
with CH₃ on each end below

f.
$$CH_3 - CH_2 - CH - CH_3$$
$$CH_2 - CH_2 - CH_3$$

1.26 The name of the normal alkane containing 9 carbon atoms is nonane. What are the molecular and condensed structural formulas for nonane?

CONFORMATIONS OF ALKANES (SECTION 1.6)

1.27 Which of the following pairs represent structural isomers, and which are simply the same compound?

a. $CH_3 - \overset{\underset{\displaystyle |}{CH_3}}{CH} - \overset{\underset{\displaystyle |}{CH_3}}{CH_3}$ and $CH_3 - \overset{\underset{\displaystyle |}{CH_3}}{CH} - CH_3$

b. $CH_3 - CH_2 - \overset{\underset{\displaystyle |}{CH_3}}{CH} - CH_3$ and $CH_3 - \overset{\overset{\displaystyle CH_3}{|}}{\underset{\underset{\displaystyle CH_3}{|}}{C}} - CH_3$

c. $CH_3 - \overset{\underset{\displaystyle |}{CH_2}}{\underset{\underset{\displaystyle |}{CH_3}}{CH}} - CH_3$ and $CH_3 - \overset{\overset{\displaystyle CH_3}{|}}{CH} - CH_2 - CH_3$

d. $CH_3 - CH_2 - \overset{\underset{\displaystyle |}{CH_3}}{CH} - CH_2 - CH_3$ and

$CH_3 - \overset{\underset{\displaystyle |}{CH_3}}{CH} - CH_2 - CH_2 - CH_3$

ALKANE NOMENCLATURE (SECTION 1.7)

1.28 For each of the following carbon skeletons, give the number of carbon atoms in the longest continuous chain:

a. $\begin{array}{c} C \quad C \\ | \quad | \\ C \quad C \\ | \quad | \\ C - C - C - C \\ | \\ C - C \end{array}$
b. $\begin{array}{c} C - C \\ | \\ C - C - C - C \end{array}$

c. $\begin{array}{c} C - C \qquad C - C \\ | \qquad\qquad | \\ C - C - C - C - C \end{array}$

1.29 For each of the following carbon skeletons, give the number of carbon atoms in the longest continuous chain:

a. $\begin{array}{c} C - C \\ | \\ C - C - C - C - C \end{array}$
b. $\begin{array}{c} C \\ | \\ C - C - C - C \\ \qquad | \\ \qquad C - C - C \end{array}$

c. $\begin{array}{c} C - C \\ | \quad | \\ C \quad C - C \\ | \quad | \\ C - C \quad C - C \end{array}$

1.30 Identify the following alkyl groups:

a. $CH_3 - CH_2 - CH_2 -$

b. $CH_3 - \overset{\underset{\displaystyle |}{CH_3}}{\underset{\underset{\displaystyle CH_3}{|}}{C}} - CH_3$

c. $- CH_2 - CH_3$

d. $CH_3 - \overset{\underset{\displaystyle |}{CH_3}}{CH} - CH_2 - CH_3$

1.31 Identify the following alkyl groups:

a. $CH_3 - \overset{\underset{\displaystyle |}{CH_3}}{CH} -$

b. $CH_3 - CH_2 - CH_2 - CH_2 -$

c. $CH_3 - \overset{\underset{\displaystyle |}{CH_3}}{CH} - CH_2 -$

d. $CH_3 -$

1.32 Give the correct IUPAC name for each of the following alkanes:

a. $CH_3 - CH_2 - \overset{\underset{\displaystyle |}{CH_3}}{CH} - CH_3$

b. $CH_3 - \overset{\overset{\displaystyle CH_3}{|}}{\underset{\underset{\displaystyle CH_3}{|}}{C}} - CH_3$

c. $CH_3 - \overset{\overset{\displaystyle CH_2 - CH_3}{|}}{\underset{\underset{\displaystyle CH_3}{|}}{CH}} - CH_3$

d. $CH_2 - \overset{\overset{\displaystyle CH_3}{|}}{\underset{\underset{\displaystyle CH_3}{|}}{CH}} - \overset{\overset{\displaystyle CH_2}{|}}{\underset{\underset{\displaystyle CH_3}{|}}{C}} - CH_2 - \overset{\underset{\underset{\displaystyle CH - CH_3}{|}}{\underset{\displaystyle |}{CH_2}}}{CH} - CH_2 - CH_3$

e. $CH_3 - CH_2 - \overset{\underset{\displaystyle |}{CH_3}}{CH} - \overset{\underset{\underset{\displaystyle CH_2 - CH_2 - CH_2 - CH_3}{}}{|}}{CH} - \overset{\underset{\displaystyle |}{CH_3}}{CH} - CH_2 - \overset{\underset{\displaystyle |}{CH_3}}{CH} - CH_3$

1.33 Give the correct IUPAC name for each of the following alkanes:

a. $CH_3 - CH_2 - \overset{\underset{\underset{\displaystyle CH_2 - CH_3}{}}{|}}{CH} - CH_2 - CH_2 - CH_3$

b. $CH_3 - CH_2 - \overset{\underset{\displaystyle |}{CH_3}}{CH} - \overset{\underset{\underset{\displaystyle CH_2 - CH_3}{}}{|}}{CH} - CH_2 - CH_3$

c.
$$CH_3 - CH_2 - CH_2 - CH - CH - CH_2 - CH_2 - CH_3$$
$$CH_3$$
$$|$$
$$CH - CH_3$$
$$|$$
and below the main chain at the second CH:
$$CH_3 - C - CH_3$$
$$|$$
$$CH_3$$

d.
$$CH_3 - CH - CH_2 - CH_2 - CH - CH - CH_3$$
$$|\qquad\qquad\qquad\qquad |$$
$$CH_3\qquad\qquad\qquad CH_2$$
with on the right:
$$CH_3$$
at top, and below:
$$CH_2 - CH_3$$

e.
$$CH_3 - CH_2 - CH_2 - C - CH_2 - CH_2 - CH_3$$
with above:
$$CH_2 - CH_2 - CH_3$$
and below:
$$CH - CH_3$$
$$|$$
$$CH_3$$

1.34 Draw condensed structural formulas for each of the following compounds:
   a. 2,3-dimethylbutane
   b. 3-isopropylheptane
   c. 5-t-butyl-2,3-dimethyloctane
   d. 4-ethyl-4-methyloctane

1.35 Draw a condensed structural formula for each of the following compounds:
   a. 2,2,4-trimethylpentane
   b. 4-isopropyloctane
   c. 3,3-diethylhexane
   d. 5-*t*-butyl-2-methylnonane

1.36 Draw the condensed structural formula for each of the three structural isomers of C_5H_{12}, and give the correct IUPAC names.

1.37 Isooctane is 2,2,4-trimethylpentane. Draw structural formulas for and name a branched heptane, hexane, pentane, and butane that are structural isomers of isooctane.

1.38 Draw structural formulas for the compounds and give correct IUPAC names for the five structural isomers of C_6H_{14}.

1.39 The following names are incorrect, according to IUPAC rules. Draw the structural formulas and tell why each name is incorrect. Write the correct name for each compound.
   a. 2,2-methylbutane
   b. 4-ethyl-5-methylheptane
   c. 2-ethyl-1,5-dimethylhexane

1.40 The following names are incorrect, according to IUPAC rules. Draw the structural formulas and tell why each name is incorrect. Write the correct name for each compound.
   a. 1,2-dimethylpropane
   b. 3,4-dimethylpentane
   c. 2-ethyl-4-methylpentane
   d. 2-bromo-3-ethylbutane

CYCLOALKANES (SECTION 1.8)

1.41 The general formula for alkanes is C_nH_{2n+2}. Write a general formula for cycloalkanes.

1.42 Write the correct IUPAC name for each of the following:

   a. (triangle)   c. (cyclohexane with CH_3 and CH_2CH_3)

   b. (cyclopentane with CH_3 and CH_3)   d. (cyclobutane with CH_3 and $CHCH_3$ bearing CH_3)

1.43 Write the correct IUPAC name for each of the following:

   a. (cyclopentane with CH_3 and CH_2CH_3)

   b. (cyclohexane with $C - CH_3$ bearing CH_3 above and CH_3 below)

   c. (cyclopentane with CH_3, Cl, and CH_3)

   d. (triangle with $CH_2CH_2CH_3$ and CH_2CH_3)

1.44 Draw the structural formulas corresponding to each of the following IUPAC names:
   a. isopropylcyclopentane
   b. 1,1-dimethylcyclobutane
   c. 1-isobutyl-3-isopropylcyclohexane

1.45 Draw the structural formulas corresponding to each of the following IUPAC names:
   a. 1,2-diethylcyclopentane
   b. 1,2,4-trimethylcyclohexane
   c. propylcyclobutane

1.46 Which of the following pairs of cycloalkanes represent structural isomers?

   a. (cyclobutane with CH_3 and CH_3)   (cyclobutane with CH_3 and CH_3)

   b.

c. CH$_2$CH$_3$

d.

1.47 Draw structural formulas for the five structural isomers of C$_5$H$_{10}$ that are cycloalkanes.

THE SHAPE OF CYCLOALKANES (SECTION 1.9)

1.48 Why does cyclohexane assume a chain form rather than a planar hexagon?

1.49 Explain the difference between geometric and structural isomers.

1.50 Which of the following cycloalkanes could show geometric isomerism? For each that could, draw structural formulas, and name both the *cis-* and the *trans-* isomers.

a.

c.

b.

d.

1.51 Draw structural formulas for *cis-* and *trans-* 1,3-dibromo-cyclobutane.

1.52 Using the prefix *cis-* or *trans-*, name each of the following:

a.

c.

b.

d.

1.53 For each of the following molecular formulas, give a structural formula requested. In most cases, there are several possible structures.
a. C$_6$H$_{14}$, a normal alkane
b. C$_6$H$_{14}$, a branched alkane
c. C$_5$H$_{12}$, a pair of conformations
d. C$_5$H$_{12}$, a pair of structural isomers
e. C$_5$H$_{10}$, a cyclic hydrocarbon
f. C$_6$H$_{12}$, two cycloalkane geometric isomers
g. C$_6$H$_{12}$, a cycloalkane that has no geometric isomers

PHYSICAL PROPERTIES OF ALKANES (SECTION 1.10)

1.54 The compound decane is a straight-chain alkane. Predict the following:
a. Is decane a solid, liquid, or gas at room temperature?
b. Is it soluble in water?
c. Is it soluble in hexane?
d. Is it more or less dense than water?

1.55 Explain why alkanes of low molecular weight have lower melting and boiling points than water.

1.56 Suppose you have a sample of 2-methylhexane and a sample of 2-methylheptane. Which sample would you expect to have the higher melting point? Boiling point?

1.57 Identify (circle) the alkanelike portions of the following molecules:

a.
isoleucine (an amino acid)

b.
sodium palmitate (a soap)

c.
lipoic acid (a component of a coenzyme)

d.
menthol (a flavoring)

ALKANE REACTIONS (SECTION 1.11)

1.58 Write a balanced equation to represent the complete combustion of each of the following:

a. propane

b. CH$_3$CHCH$_3$
       |
       CH$_3$

c.

1.59 Write a balanced equation to represent the complete combustion of each of the following:

    a. ethane

    b. $CH_3 - CH - CH_2 - CH_3$
                  |
                CH_3

    c.

1.60 Write a balanced equation for the *incomplete* combustion of hexane, assuming the formation of carbon monoxide and water as the products.

1.61 Why is it dangerous to relight a furnace when a foul odor is present?

CHEMISTRY AROUND US AND OVER THE COUNTER

1.62 What is the goal of combinatorial chemistry?

1.63 What general name is given to substances within skin moisturizers that attract water from the air?

1.64 What petroleum products are sometimes found in skin moisturizers?

1.65 Give a major use for each of the following petroleum fractions:

    a. $C_1 - C_4$
    b. $C_5 - C_{12}$
    c. $C_{12} - C_{16}$
    d. $C_{16} - C_{20}$
    e. $C_{20} -$ up

1.66 Give another reason why petroleum is so valuable to our society besides serving as a source of fuel.

1.67 In what way does carbon monoxide interact with the body that makes the gas dangerous?

1.68 What is the antidote for CO poisoning?

▸ POINTS TO PONDER AND PURSUE

1.1 Would you expect a molecule of urea produced in the body to have any different physical or chemical properties from a molecule of urea prepared in a laboratory?

1.2 Why might the study of organic compounds be important to someone interested in the health or life sciences?

1.3 Why do very few aqueous solutions of organic compounds conduct electricity?

1.4 The ski wax being applied in Figure 1.2 has a relatively low melting point. What does that fact reveal about the forces between molecules?

1.5 Charcoal briquettes sometimes burn with incomplete combustion when the air supply is limited. Why would it be hazardous to place a charcoal grill inside a home or a camper in an attempt to keep warm?

1.6 If carbon did not form hybridized orbitals, what would you expect to be the formula of the simplest compound of carbon and hydrogen?

1.7 What types of sports equipment are made from graphite fibers besides that shown in Figure 1.8?

1.8 According to Web site 1.2, what are the advantages of using combinatorial chemistry in new drug synthesis?

1.9 According to Web site 1.3, when was IUPAC organized?

1.10 Using Web site 1.4, name the two layers of tissue that make up skin.

1.11 According to the information at Web site 1.5, what is the most common source of carbon monoxide in the workplace?

http://www.brookscole.com/chemistry/seager_chemistry

Unsaturated Hydrocarbons

Dental technicians make dental prostheses such as false teeth, crowns, and bridges, according to orders placed by dentists. In some large dental laboratories, the demand is great enough for certain types of devices that technicians become specialists in working with specific materials such as plastics, gold, or porcelain. In this chapter, you will be introduced to the molecular nature of one type of plastic and the versatility of these useful products of human ingenuity.

http://www.brookscole.com/chemistry/
seager_chemistry
You can refer to Web site 2.1 for further information on a career as a dental laboratory technician.

CHAPTER OUTLINE

- ➤ The Nomenclature of Alkenes
- ➤ The Geometry of Alkenes
- ➤ Properties of Alkenes
- ➤ Addition Polymers
- ➤ Alkynes
- ➤ Aromatic Compounds and the Benzene Structure
- ➤ The Nomenclature of Benzene Derivatives
- ➤ Properties and Uses of Aromatic Compounds

Unsaturated hydrocarbon
A hydrocarbon containing a C=C or C≡C group.

Alkene
A hydrocarbon containing one or more double bonds.

Alkyne
A hydrocarbon containing one or more triple bonds.

Aromatic hydrocarbons
Any organic compound that contains the characteristic benzene ring or similar feature.

Unsaturated **hydrocarbons** contain one or more double or triple bonds between carbon atoms and belong to one of three classes: alkenes, alkynes, or aromatic hydrocarbons. **Alkenes** contain one or more double bonds, **alkynes** contain one or more triple bonds, and **aromatic hydrocarbons** contain three double bonds alternating with three single bonds in a six-carbon ring. Ethylene (the simplest alkene), acetylene (the simplest alkyne), and benzene (the simplest aromatic) are represented by the following structural formulas:

ethylene
(an alkene)

acetylene
(an alkyne)

benzene
(an aromatic)

Alkenes and alkynes are called *unsaturated* because more hydrogen atoms can be added in somewhat the same sense that more solute can be added to an unsaturated solution. Benzene and other aromatic hydrocarbons also react to add hydrogen atoms; in general, however, they have chemical properties very different from those of alkenes and alkynes.

➤ 2.1 The Nomenclature of Alkenes

The general formula for alkenes is C_nH_{2n} (the same as that for cycloalkanes). The simplest members are well known by their common names, ethylene and propylene:

$$CH_2{=}CH_2 \qquad CH_3{-}CH{=}CH_2$$
$$\text{ethylene, } C_2H_4 \qquad \text{propylene, } C_3H_6$$

Three structural isomers have the formula C_4H_8:

$$CH_3{-}CH_2{-}CH{=}CH_2 \qquad CH_3{-}CH{=}CH{-}CH_3 \qquad CH_3{-}\underset{\underset{CH_3}{|}}{C}{=}CH_2$$

The number of structural isomers increases rapidly as the number of carbons increases because, besides variations in chain length and branching, variations occur in the position of the double bond. IUPAC nomenclature is extremely useful in differentiating among these many alkene compounds.

The IUPAC rules for naming alkenes are similar to those used for the alkanes, with a few additions to indicate the presence and location of double bonds.

Step 1. Name the longest chain that contains the double bond. The characteristic name ending is *-ene*.

Step 2. Number the longest chain of carbon atoms so that the carbon atoms joined by the double bond have numbers as low as possible.

Step 3. Locate the double bond by the lower-numbered carbon atom bound by the double bond.

Step 4. Locate and name attached groups.

Step 5. Combine the names for the attached groups and the longest chain into the name.

➤ Example 2.1

Name the following alkenes:

a. $CH_3-CH=CH-CH_3$

b. $\overset{4}{C}H_3-\overset{3}{C}H-\overset{2}{C}H=\overset{1}{C}H_2$
     $\quad\quad\quad |$
     $\quad\quad\quad CH_3$

butene

c. $\quad\quad CH_3-CH_2$
   $\quad\quad\quad\quad |$
   $\quad\quad\quad\quad C=CH_2$
   $\quad\quad\quad\quad |$
   $CH_3-CH_2-CH_2$

d.

   $\quad\quad CH_2$
   $\quad\quad\quad |$
   $\quad\quad CH_3$

Solution

a. The longest chain containing a double bond has four carbon atoms. The four-carbon alkane is butane. Thus, the compound is a butene:

$$\overset{1}{C}H_3-\overset{2}{C}H=\overset{3}{C}H-\overset{4}{C}H_3$$

The chain can be numbered from either end because the double bond will be between carbons 2 and 3 either way. The position of the double bond is indicated by the lower-numbered carbon atom that is double bonded, carbon 2 in this case. The name is 2-butene.

b. To give lower numbers to the carbons bound by the double bond, the chain is numbered from the right:

$\overset{4}{C}H_3-\overset{3}{C}H-\overset{2}{C}H=\overset{1}{C}H_2$ not $\overset{1}{C}H_3-\overset{2}{C}H-\overset{3}{C}H=\overset{4}{C}H_2$
$\quad\quad |$ $\quad\quad |$
$\quad\quad CH_3$ $\quad\quad CH_3$

Thus, the compound is a 1-butene with an attached methyl group on carbon 3. Therefore, the name is 3-methyl-1-butene.

c. Care must be taken to select the longest chain containing the double bond. This compound is named as a pentene and not as a hexene because the double bond is not contained in the six-carbon chain:

$\quad\quad CH_3-CH_2$ $\quad\quad \overset{1}{C}H_3-\overset{2}{C}H_2$
$\quad\quad\quad\quad |$ $\quad\quad\quad\quad\quad |$
$\quad\quad\quad\overset{2}{C}=\overset{1}{C}H_2$ not $\quad\quad\quad\overset{3}{C}=CH_2$
$\quad\quad\quad\quad |$ $\quad\quad\quad\quad\quad |$
$CH_3-CH_2-CH_2$ $CH_3-CH_2-CH_2$
$\quad\,{}_5\quad\quad{}_4\quad\quad{}_3$ $\quad\,{}_6\quad\quad{}_5\quad\quad{}_4$

The compound is a 1-pentene with an ethyl group at position 2. Therefore, the name is 2-ethyl-1-pentene.

d. In cyclic alkenes, the ring is numbered so as to give the lowest possible numbers to the double-bonded carbons (they become carbons 1 and 2). The numbering direction around the ring is chosen so that attached groups are located on the lowest-numbered carbon atoms possible. Thus, the name is 3-ethylcyclopentene:

$$\text{(ring structure numbered 1, 2, 3, 4, 5)}$$
$$CH_2CH_3$$

Notice that it is not called 3-ethyl-1-cyclopentene because the double bond is always between carbons 1 and 2, and therefore its position need not be indicated.

Learning Check 2.1

Give the IUPAC name for each of the following:

Handwritten: 3,bromo 1 propylene

Handwritten: 3,4 dimethyl cyclohexene

a. Br
$$\overset{3}{CH_2}-\overset{2}{CH}=\overset{1}{CH_2}$$

c. $CH_3 \quad CH_3$

b. $\overset{1}{CH_2}=\overset{2}{C}-\overset{3}{CH_2}-\overset{4}{CH_2}-\overset{5}{CH_3}$
$$|$$
$$CH_2$$
$$|$$
$$CH_3$$

Handwritten: 2 ethyl 1 pentene

Some compounds contain more than one double bond per molecule. Molecules of this type are important components of natural and synthetic rubber and other useful materials. The nomenclature of these compounds is the same as for the alkenes with one double bond, except that the endings *-diene*, *-triene*, and the like are used to denote the number of double bonds. Also, the locations of all the multiple bonds must be indicated in all molecules, including those with rings:

$$\overset{1}{CH_2}=\overset{2}{CH}-\overset{3}{CH}=\overset{4}{CH_2}$$
1,3-butadiene

1,3-cyclohexadiene

CHEMISTRY AROUND US • 2.1

BETA-CAROTENE: SUPPLEMENTS ARE NOT THE ANSWER

The yellow compound that gives carrots their color is called beta-carotene. In the body, this compound is transformed into vitamin A. The connection between "eating your carrots" and having healthy vision has been understood for a number of years; vitamin A is essential for good vision. Other good sources of beta-carotene are spinach and other dark leafy greens, broccoli, deep-orange fruits (apricots and cantaloupe), and other yellow vegetables (squash, sweet potatoes, pumpkin).

Eating fruits and vegetables that contain beta-carotene seems to diminish the incidence of some types of human cancer. However, two recent long-term studies have shown that the use of beta-carotene dietary supplements does not reduce the risk of cancer and heart disease in adults. In fact, one study of smokers, former smokers, and workers exposed to asbestos showed a greater chance of developing lung cancer or heart disease among those who took beta-carotene supplements compared to those who took a placebo. Therefore, whereas eating generous portions of fruits and vegetables is still linked to good health, the role played by the beta-carotene the food contains is unclear.

Fruits and vegetables are good sources of beta-carotene.

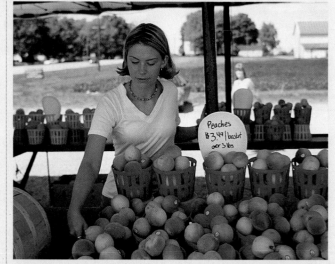

beta-carotene, the yellow pigment in carrots and yams

http://www.brookscole.com/chemistry/seager_chemistry
Web site 2.2 provides a detailed description of the health benefits of beta-carotene.

➤ Example 2.2

Name the following compounds:

a. $CH_2=\overset{\overset{\displaystyle CH_3}{|}}{\underset{\underset{\displaystyle CH_3}{|}}{C}}-C=CH-CH=CH_2$

(handwritten) 2,3methyl 1 hextriene

b. *(structure: cyclohexadiene ring with CH₃ and CH₂CH₃ substituents)*

(handwritten) 5 ethyl, 1methyl, 3 cyclohexdiene

Solution

a. This compound is a methyl-substituted hexatriene. The chain is numbered from the end nearest the branch because the direction of numbering, again, makes no difference in locating the double bonds correctly. The name is 2,3-dimethyl-1,3,5-hexatriene:

$\overset{1}{CH_2}=\overset{2}{\underset{\underset{\displaystyle CH_3}{|}}{\overset{\overset{\displaystyle CH_3}{|}}{C}}}-\overset{3}{C}=\overset{4}{CH}-\overset{5}{CH}=\overset{6}{CH_2}$ and not $\overset{6}{CH_2}=\overset{5}{\underset{\underset{\displaystyle CH_3}{|}}{\overset{\overset{\displaystyle CH_3}{|}}{C}}}-\overset{4}{C}=\overset{3}{CH}-\overset{2}{CH}=\overset{1}{CH_2}$

2,3-dimethyl-1,3,5-hexatriene   4,5-dimethyl-1,3,5-hexatriene

b. This compound is a substituted cyclohexadiene. The ring is numbered as shown. The name is 5-ethyl-1-methyl-1,3-cyclohexadiene:

(structure)   and not   *(structure)*

5-ethyl-1-methyl-1,3-cyclohexadiene   6-ethyl-4-methyl-1,3-cyclohexadiene

Give the IUPAC name for each of the following:

a. $\overset{4}{CH_2}=\overset{3}{CH}-\overset{2}{\underset{}{\overset{\overset{\displaystyle CH_3}{|}}{C}}}=\overset{1}{CH_2}$

(handwritten) 2-methyl 1,3 butadiene

b. $\overset{1}{CH_2}=\overset{2}{\underset{}{\overset{\overset{\displaystyle CH_3}{|}}{C}}}-\overset{3}{CH}=\overset{4}{CH}-\overset{5}{CH_2}-\overset{6}{CH}=\overset{7}{CH}-\overset{8}{CH_3}$

(handwritten) 2-methyl 1,3,6 octatriene

c. *(structure: cycloheptatriene ring with Br)*

(handwritten) 7-bromo 1,3,5 cyclo heptatriene

➤ 2.2 The Geometry of Alkenes

The hybridization of atomic orbitals discussed in Section 1.3 to explain the bonding characteristics of carbon atoms bonded to four other atoms can also be used to describe alkenes, compounds in which some carbon atoms are bonded to only three atoms. This hybridization involves mixing a 2s orbital and two 2p orbitals of a carbon atom to form three hybrid sp^2 orbitals (see ■ Figure 2.1 on page 45).

CHEMISTRY AROUND US • 2.2

TERPENES AND FRAGRANCES

Essential oils are volatile components of most plants that can be extracted from the flowers, stems, fruits, leaves, or roots of the plants by steam distillation. Essential oils are responsible for many of the flavors and odors associated with plants, and some have been used as perfumes for centuries. Little was known about the chemical nature of these oils before the latter half of the nineteenth century. It is now known that essential oils consist of mixtures of a specific kind of hydrocarbon called **terpenes,** and oxygenated derivatives of terpenes called *camphors.* Most terpenes are alkenes, but some contain additional functional groups as well.

All terpenes are made up of two or more isoprene units connected together to form a variety of differently shaped and sized molecules. The chemical name of isoprene is 2-methyl-1,3-butadiene.

$$CH_3$$
$$\overset{|}{\underset{\parallel}{C}} - CH = CH_2$$
$$CH_2$$

complete isoprene
formula

skeletal isoprene
formula

Because isoprene molecules contain 5 carbon atoms, all terpene molecules contain multiples of 5 carbon atoms, the most common numbers being 10, 15, 20, 30, and 40.

Some terpenes, such as geraniol and menthol, contain alcohol functional groups and are called terpene alcohols (see diagram). Geraniol contains two isoprene units and is found in many essential oils, such as citronella, lemon-grass oil, rose oil, geranium oil, and lavender oil. Menthol, a terpene containing two isoprene units, has been known and used in Japan for an estimated 2000 years. In the Western world, it was first separated from peppermint oil in 1771. Menthol has a peppermint fragrance and is often used in medicine and pharmaceuticals as a mild antiseptic or for its cooling effect on mucous membranes. It is an ingredient in some OTC products for relieving the discomfort from insect bites or stings, or exposure to poison ivy or poison sumac.

Limonene, a terpene containing two isoprene units, is one of the most widely distributed of all terpenes. It is the principal terpene in bergamot oil, lemon oil, orange oil, caraway oil, dill oil, and celery oil. One form can also be steam distilled from the needles of Douglas fir trees. It usually has a pleasant citrus type of odor.

Terpenes are responsible for many of the odors we encounter every day both in nature and in manufactured products. So the next time you get a whiff of a "lemon-like fresh" odor, remember that a terpene is probably involved.

geraniol

menthol

limonene

Dashed lines delineate isoprene units.

http://www.brookscole.com/chemistry/seager_chemistry
Find further information about terpenes at Web site 2.3.

Terpenes are responsible for part of the characteristic fragrance of flowers.

Terpene
A biological compound comprised of five-carbon isoprene units.

The three sp^2 hybrid orbitals lie in the same plane and are separated by angles of 120°. The unhybridized $2p$ orbital of the carbon atom is located perpendicular to the plane of the sp^2 hybrid orbitals (see ■ Figure 2.2).

The bonding between the carbon atoms of ethylene results partially from the overlap of one sp^2 hybrid orbital of each carbon to form a sigma (σ) bond. The second carbon–carbon bond is formed when the unhybridized $2p$ orbitals of the carbons overlap sideways to form what is called a pi (π) bond. The

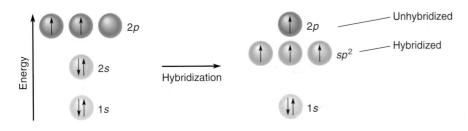

➤ **FIGURE 2.1** A representation of sp^2 hybridization of carbon. During hybridization, two of the $2p$ orbitals mix with the single $2s$ orbital to produce three sp^2 hybrid orbitals. One $2p$ orbital is not hybridized and remains unchanged.

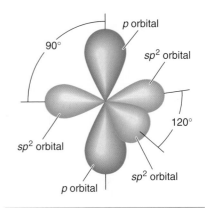

➤**FIGURE 2.2** The unhybridized p orbital is perpendicular to the plane of the three sp^2 hybridized orbitals.

remaining two sp^2 hybrid orbitals of each carbon overlap with the $1s$ orbitals of the hydrogen atoms to form sigma bonds. Thus, each ethylene molecule contains five sigma bonds (one carbon–carbon and four carbon–hydrogen) and one pi bond (carbon–carbon), as shown in ■ Figure 2.3.

Experimental data support this hybridization model. Ethylene has been found to be a planar molecule with bond angles close to 120° between the atoms (see ■ Figure 2.4).

In addition to geometry, alkenes also differ from open-chain alkanes in that the double bonds prevent the relatively free rotation that is characteristic of carbon atoms bonded by single bonds. As a result, alkenes can exhibit geometric isomerism, the same type of stereoisomerism seen earlier for the cycloalkanes (Section 1.9). There are two geometric isomers of 2-butene:

$$
\begin{array}{cc}
\underset{CH_3}{\overset{H}{\diagdown}}C=C\underset{CH_3}{\overset{H}{\diagup}} & \underset{H}{\overset{CH_3}{\diagdown}}C=C\underset{CH_3}{\overset{H}{\diagup}} \\
\textit{cis}\text{-2-butene} & \textit{trans}\text{-2-butene}
\end{array}
$$

Once again, the prefix *cis*- is used for the isomer in which the two similar or identical groups are on the same side of the double bond and *trans*- for the one in which they are on opposite sides. The two isomers *cis*- and *trans*-2-butene represent distinctly different compounds with different physical properties (see ■ Table 2.1).

Not all double-bonded compounds show cis-trans stereoisomerism. Cis-trans stereoisomerism is found only in alkenes that have two different groups

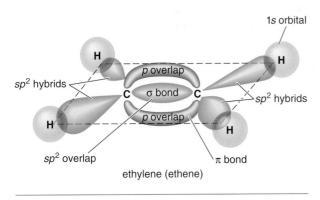

➤ **FIGURE 2.3** The bonding in ethylene is explained by combining two sp^2 hybridized carbon atoms. The C—H sigma bonds all lie in the same plane. The unhybridized p orbitals of the carbon atoms overlap to form the pi bond.

➤ **FIGURE 2.4** All six atoms of ethylene lie in the same plane.

TABLE 2.1 Physical properties of a pair of geometric isomers

Isomer	Melting point (°C)	Boiling point (°C)	Density (g/mL)
cis-2-butene	−139.9	3.7	0.62
trans-2-butene	−105.6	0.9	0.60

attached to each double-bonded carbon atom. In 2-butene, the two different groups are a methyl and a hydrogen for each double-bonded carbon:

two groups are different

two groups are different

two groups are different

cis-2-butene   *trans*-2-butene

If either double-bonded carbon is attached to identical groups, no cis-trans isomers are possible. Thus, there are no geometric isomers of ethene or propene:

two groups are the same

two groups are the same

ethene   propene

To see why this is so, let's try to draw geometric isomers of propene:

is the same as

(a)   (b)

Notice that structure (b) can be converted into (a) by just flipping it over. Thus, they are identical and not isomers.

➤ Example 2.3

Determine which of the following molecules can exhibit geometric isomerism, and draw structural formulas to illustrate your conclusions:

a. $Cl-CH=CH-Cl$
b. $CH_2=CH-Cl$
c. $Cl-CH=CH-CH_3$

Solution

a. Begin by drawing the carbon–carbon double bond with bond angles of 120° about each carbon atom:

$$\ce{C=C}$$

Next, complete the structure and analyze each carbon of the double bond to see if it is attached to two different groups:

H and Cl are *different groups* ↗↘ C=C ↖↙ H and Cl are *different groups*

(structure with H and H on top, Cl and Cl on bottom)

In this case, each carbon is attached to two different groups and geometric isomers are possible:

(cis structure: H, H top; Cl, Cl bottom) *cis*

(trans structure: Cl, H top; H, Cl bottom) *trans*

b. No geometric isomers are possible because one carbon contains two identical groups:

two identical groups ↗↘ C=C (H, H top; H, Cl bottom)

c. Geometric isomers are possible because there are two different groups on each carbon:

H and Cl are different ↗↘ C=C ↖↙ *H and CH₃ are different*

(structure: H, H top; Cl, CH₃ bottom)

(structure: H, H top; Cl, CH₃ bottom)
cis-1-chloropropene

(structure: Cl, H top; H, CH₃ bottom)
trans-1-chloropropene

Determine which of the following can exhibit geometric isomerism, and draw structural formulas for the cis and trans isomers of those that can:

Learning Check 2.3

a. $CH_2{=}C{-}CH_2{-}CH_3$
   $\quad\;\;|$
   $\quad\;\;CH_3$

b. $CH_3{-}CH{-}CH{=}CH{-}CH_3$
   $\qquad\quad|$
   $\qquad\quad CH_3$

c. $CH_3{-}C{=}CH{-}CH_3$
   $\qquad\;|$
   $\qquad\;CH_3$

➤ 2.3 Properties of Alkenes

Physical Properties

The alkenes are similar to the alkanes in physical properties—they are nonpolar compounds that are insoluble in water, soluble in nonpolar solvents, and less dense than water (see ■ Table 2.2). Alkenes containing 4 carbon atoms or fewer are gases under ordinary conditions. Those containing 5 to 17 carbon atoms are liquids, and those with 18 or more carbon atoms are solids. Low molecular weight alkenes have somewhat unpleasant, gasoline-like odors.

TABLE 2.2 Physical properties of some alkenes

IUPAC name	Structural formula	Boiling point (°C)	Melting point (°C)	Density (g/mL)
ethene	$CH_2{=}CH_2$	−104	−169	0.38
propene	$CH_2{=}CHCH_3$	−47	−185	0.52
1-butene	$CH_2{=}CHCH_2CH_3$	−6	−185	0.60
1-pentene	$CH_2{=}CHCH_2CH_2CH_3$	30	−138	0.64
1-hexene	$CH_2{=}CHCH_2CH_2CH_2CH_3$	63	−140	0.67
cyclohexene		83	−104	0.81

Chemical Properties

In contrast to alkanes, which are inert to almost all chemical reagents (Section 1.7), alkenes are quite reactive chemically. Since the only difference between an alkane and an alkene is the double bond, it is not surprising to learn that most of the reactions of alkenes take place at the double bond. These reactions follow the pattern

$$\underset{/}{\overset{\backslash}{C}}{=}\underset{\backslash}{\overset{/}{C}} \;+\; A-B \;\longrightarrow\; -\overset{\overset{\displaystyle A}{|}}{C}-\overset{\overset{\displaystyle B}{|}}{C}- \tag{2.1}$$

Addition reaction
A reaction in which a compound adds to a multiple bond.

Haloalkane or alkyl halide
A derivative of an alkane in which one or more hydrogens are replaced by halogens.

They are called **addition reactions** because a substance is added to the double bond. Addition reactions are characterized by two reactants that combine to form one product.

Halogenation is one of the most common addition reactions. For example, when bromine, a halogen, is added to an alkene, the double bond reacts, and only a carbon–carbon single bond remains in the product. Without a double bond present, the product is referred to as a **haloalkane** or **alkyl halide**. Reaction 2.2 gives the general reaction, and Reaction 2.3 is a specific example, the bromination of 1-butene.

General reaction:
$$\underset{/}{\overset{\backslash}{C}}{=}\underset{\backslash}{\overset{/}{C}} \;+\; Br-Br \;\longrightarrow\; -\underset{\underset{\displaystyle Br}{|}}{C}-\underset{\underset{\displaystyle Br}{|}}{C}- \tag{2.2}$$

Specific example:
$$CH_3-CH_2-CH{=}CH_2 \;+\; Br-Br \;\longrightarrow\; CH_3-CH_2-\underset{\underset{\displaystyle Br}{|}}{CH}-\underset{\underset{\displaystyle Br}{|}}{CH_2} \tag{2.3}$$

1-butene   1,2-dibromobutane

The addition of Br_2 to double bonds provides a simple laboratory test for unsaturation (see ■ Figure 2.5). As the addition takes place, the characteristic red-brown color of the added bromine fades as it is used up, and the colorless dibromoalkane product forms.

The addition of halogens is also used to quantitatively determine the degree of unsaturation in vegetable oils, margarines, and shortenings (Section 8.4). Chlorine reacts with alkenes to give dichloro products in an addition reaction

Dilute bromine solution added to 1-hexene loses its red-brown color immediately.

The remainder of the bromine solution is added. The last drops react as quickly as the first.

> ➤ **FIGURE 2.5** The reaction of bromine with an unsaturated hydrocarbon.

similar to that of bromine. However, it is not used as a test for unsaturation because it is difficult to see the pale green color of chlorine in solution.

Write the structural formula for the product of each of the following reactions:

Learning Check 2.4

a. $CH_3-C=CH_2 + Br_2 \longrightarrow$
     |
     CH_3

b. ⬠ + $Cl_2 \longrightarrow$

In the presence of an appropriate catalyst (such as platinum, palladium, or nickel), hydrogen adds to alkenes and converts them into the corresponding alkanes. This reaction, which is called **hydrogenation,** is illustrated in Reactions 2.4 and 2.5.

Hydrogenation
A reaction in which the addition of hydrogen takes place.

General reaction:
$$\underset{/}{\overset{\backslash}{}}C=C\underset{\backslash}{\overset{/}{}} \;+\; H-H \;\xrightarrow{Pt}\; \begin{array}{c} | \;\; | \\ -C-C- \\ | \;\; | \\ H \;\; H \end{array}$$
An alkane

(2.4)

Specific example:
$$CH_3CH=CHCH_3 + H-H \xrightarrow{Pt} \begin{array}{c} CH_3CH-CHCH_3 \\ | \quad\quad | \\ H \quad\quad H \end{array}$$
2-butene   butane

(2.5)

The hydrogenation of vegetable oils is a very important commercial process. Vegetable oils, such as soybean and cottonseed oil, are composed of long-chain organic molecules that contain many alkene bonds. The high degree of unsaturation characteristic of these oils gave rise to the term **polyunsaturated.** Upon hydrogenation, the melting point of the oils is raised, and the oils become low-melting-point solids. These products are used in the form of margarine and shortening.

Polyunsaturated
A term usually applied to molecules with several double bonds.

Write the structural formula for the product of each of the following reactions:

Learning Check 2.5

a.
$CH_3-C=CH-CH_3 + H_2 \xrightarrow{Pt}$
     |
     CH_3

b. CH_3
     ⬠ + $H_2 \xrightarrow{Pt}$

KEEPING A REACTION CARD FILE

Remembering organic reactions for exams is challenging for most students. Because the number of reactions being studied increases rapidly, it is a good idea to develop a systematic way to organize them for easy and effective review.

One way to do this is to focus on the functional group concept. When an exam question asks you to complete a reaction by identifying the product, your first step should be to identify the functional group of the reactant. Usually, only the functional group portion of a molecule undergoes reaction; in addition, a particular functional group usually undergoes the same characteristic reactions regardless of the other features of the organic molecule to which it is bound. Thus, by remembering the behavior of a functional group under specific conditions, you can predict the reactions of many compounds, no matter how complex the structures look, as long as they contain the same functional group. For example, any structure containing a C=C will undergo reactions typical of alkenes. Other functional groups will be introduced in later chapters.

Keeping a reaction card file based on the functional group concept is a good way to organize reactions for review. Write the structures and names of the reactants on one side of an index card with an arrow showing any catalyst or special conditions. Write the product structure and name on the back of the card. We recommend that you do this for the general reaction (like those in the Key Reactions section at the end of most chapters) and for a specific example. Review your cards every day (this can even be done while waiting for a bus, etc.), and add to them as new reactions are studied. As an exam approaches, put aside the reactions you know well, and concentrate on the others in what should be a dwindling deck. This is an effective way to focus on learning what you don't know.

A number of acidic compounds, such as the hydrogen halides—HF, HCl, HBr, and HI—also add to alkenes to give the corresponding alkyl halide. The reaction with HCl is illustrated as follows:

General reaction:

$$\ce{\underset{}{>}C=C\underset{}{<} + H-Cl \longrightarrow -\overset{|}{\underset{|}{C}}-\overset{|}{\underset{|}{C}}-} \tag{2.6}$$

with H and Cl on adjacent carbons.

Specific example:

$$CH_3CH=CHCH_3 + H-Cl \longrightarrow CH_3\underset{\overset{|}{H}}{C}H\underset{\overset{|}{Cl}}{C}HCH_3 \tag{2.7}$$

The addition reactions involving H_2, Cl_2, and Br_2 yield only one product because the same group (H and H or Br and Br) adds to each double-bonded carbon. However, with H—X, a different group adds to each carbon, and for certain alkenes, there are two possible products. For example, in the reaction of HBr with propene, two products might be expected: 1-bromopropane or 2-bromopropane,

$$CH_2=CH-CH_3 + H-Br \longrightarrow \underset{\text{1-bromopropane}}{CH_2-CH-CH_3} \text{ or } \underset{\text{2-bromopropane}}{CH_2-CH-CH_3} \tag{2.8}$$

with Br and H on carbons for 1-bromopropane, and H and Br for 2-bromopropane.

Markovnikov's rule
In the addition of H—X to an alkene, the hydrogen becomes attached to the carbon atom that is already bonded to more hydrogens.

It turns out that only one product, 2-bromopropane, is formed in significant amounts. This fact, first reported in 1869 by Russian chemist Vladimir Markovnikov, gave rise to a rule for predicting which product will be exclusively or predominantly formed. According to **Markovnikov's rule,** when a molecule of the form H—X adds to an alkene, the hydrogen becomes attached to the carbon atom that is already bonded to more hydrogens. A phrase to help you remember this rule is "the rich get richer." Applying this rule to propene, we find

one hydrogen attached

$$CH_2 = CH - CH_3$$

*two hydrogens
attached*

*three hydrogens but they
are not attached to the
double-bonded carbons*

Therefore, H attaches to the end carbon of the double bond and Br attaches to the second carbon, and 2-bromopropane is the major product.

➤ Example 2.4

Use Markovnikov's rule to predict the major product in the following reactions:

a. $CH_3 - \underset{\underset{CH_3}{|}}{C} = CH_2 + H - Cl$

b. $\underset{CH_3}{}$ + H — Cl ⟶

Solution

a. Analyze the C=C to see which carbon atom has more hydrogens attached:

$$CH_3 - \underset{\underset{CH_3}{|}}{C} = CH_2 \quad \textit{two hydrogens attached}$$

no hydrogens attached

The H of H—Cl will attach to the position that has more hydrogens. Thus, 2-chloro-2-methylpropane is the major product:

$$CH_3 - \underset{\underset{CH_3}{|}}{C} = CH_2 \quad \textit{H attaches here to give}$$

Cl attaches here

$$CH_3 - \underset{\underset{\substack{| \ | \\ Cl \ H}}{|}}{\overset{\overset{CH_3}{|}}{C}} - CH_2$$

2-chloro-2-methylpropane

b. The challenge with a cyclic alkene is to remember that the hydrogens are not shown. Thus,

CH_3 is the same as CH_3 H

As before, the H of H—Cl will attach to the double-bonded carbon that has more hydrogens:

CH_3 *Cl attaches here* ⟶ $\underset{\underset{H}{Cl}}{CH_3}$ or $\underset{Cl}{CH_3}$
H attaches here
H

Learning Check 2.6

Use Markovnikov's rule to predict the major product in the following reactions:

a. $CH_3 - CH = \overset{\overset{\displaystyle CH_3}{|}}{C} - CH_2 - CH_3 + HBr$

b.

+ HBr $\longrightarrow$

Hydration

The addition of water to a multiple bond.

In the absence of a catalyst, water does not react with alkenes. But, if an acid catalyst such as sulfuric acid is added, water adds to carbon–carbon double bonds to give alcohols. In this reaction, which is called **hydration**, a water molecule is split in such a way that —H attaches to one carbon of the double bond, and —OH attaches to the other carbon. In Reactions 2.9–2.11, H_2O is written H—OH to emphasize the portions that add to the double bond. Notice that the addition follows Markovnikov's rule:

$$\overset{\diagdown}{\underset{\diagup}{C}} = \overset{\diagup}{\underset{\diagdown}{C}} \quad + \quad H-OH \xrightarrow{\;H_2SO_4\;} \quad -\overset{|}{\underset{|}{C}} - \overset{|}{\underset{|}{C}} - \atop{\;\;H\quad OH} \tag{2.9}$$

An alcohol

$$CH_3CH = CHCH_3 \quad + \quad H-OH \xrightarrow{\;H_2SO_4\;} \quad CH_3\overset{|}{\underset{|}{C}}H - \overset{|}{\underset{|}{C}}HCH_3 \atop{\quad\quad H\quad\quad OH} \tag{2.10}$$

2-butene   2-butanol

$$CH_3CH = CH_2 \quad + \quad H-OH \xrightarrow{\;H_2SO_4\;} \quad CH_3\overset{|}{\underset{|}{C}}H - \overset{|}{\underset{|}{C}}H_2 \atop{\quad\quad OH\quad\quad H} \tag{2.11}$$

propene   2-propanol

The hydration of alkenes provides a convenient method for preparing alcohols on a large scale. The reaction is also important in living organisms, but the catalyst is an enzyme rather than sulfuric acid. For example, one of the steps in the body's utilization of carbohydrates for energy involves the hydration of fumaric acid, which is catalyzed by the enzyme fumarase:

$$\overset{\overset{\displaystyle O}{\|}}{HO - C} - CH = CH - \overset{\overset{\displaystyle O}{\|}}{C} - OH \; + \; H_2O \xrightarrow{\;fumarase\;} \overset{\overset{\displaystyle O}{\|}}{HO - C} - \overset{\overset{\displaystyle H}{|}}{CH} - \overset{\overset{\displaystyle OH}{|}}{CH} - \overset{\overset{\displaystyle O}{\|}}{C} - OH \tag{2.12}$$

fumaric acid   malic acid

Learning Check 2.7

Draw structural formulas for the major organic product of each of the following reactions:

a. $CH_3CH_2CH_2CH = CH_2 + H_2O \xrightarrow{\;H_2SO_4\;}$

b.

+ $H_2O \xrightarrow{\;H_2SO_4\;}$

➤ 2.4 Addition Polymers

Certain alkenes undergo a very important reaction in the presence of specific catalysts. In this reaction, alkene molecules undergo an addition reaction with one another. The double bonds of the reacting alkenes are converted to single bonds as hundreds or thousands of molecules bond and form long chains. For example, several ethylene molecules react as follows:

$$CH_2\!=\!CH_2 + CH_2\!=\!CH_2 + CH_2\!=\!CH_2 + CH_2\!=\!CH_2 \xrightarrow[\text{catalysts}]{\text{Heat, pressure,}}$$

ethylene molecules

(2.13)

$$-CH_2-CH_2-CH_2-CH_2-CH_2-CH_2-CH_2-CH_2-$$

polyethylene

The product is commonly called polyethylene even though there are no longer any double bonds present. The newly formed bonds in this long chain are shown in color. This type of reaction is called a **polymerization,** and the long-chain product made up of repeating units is a **polymer** (*poly* = many, *mer* = parts). The trade names of many polymers such as Orlon®, Plexiglas®, Lucite®, and Teflon® are familiar (see ■ Figure 2.6). These products are referred to as **addition polymers** because of the addition reaction between double-bonded compounds that is used to produce them. The starting materials that make up the repeating units of polymers are called **monomers** (*mono* = one, *mer* = part). Quite often, common names are used for both the polymer and the monomer.

It is not possible to give an exact formula for a polymer produced by a polymerization reaction because the individual polymer molecules vary in size. We could represent polymerization reactions as in Reaction 2.13. However, since this type of reaction is inconvenient, we adopt a commonly used approach: The polymer is represented by a simple repeating unit based on the monomer. For polyethylene, the unit is $+CH_2-CH_2+$. The large number of units making up the polymer is denoted by *n*, a whole number that varies from several hundred to several thousand. The polymerization reaction of ethylene is then written as

$$n\,CH_2\!=\!CH_2 \xrightarrow[\text{catalysts}]{\text{Heat, pressure,}} +CH_2-CH_2+_n$$

ethylene   polyethylene

(2.14)

➤ **FIGURE 2.6** Gore-Tex® is a thin, membranous material made by stretching Teflon fibers. Fabrics layered with Gore-Tex repel wind and rain but allow body perspiration to escape, making it an excellent fabric for sportswear.

Polymerization
A reaction that produces a polymer.

Polymer
A very large molecule made up of repeating units.

Addition polymer
A polymer formed by the linking together of many alkene molecules through addition reactions.

Monomer
The starting material that becomes the repeating units of polymers.

➤ STUDY SKILLS 2.2

A REACTION MAP FOR ALKENES

A diagram may help you visualize and remember the four common addition reactions in this section. In each case, the alkene double bond reacts, and an alkane or alkane derivative is produced. The specific reagent determines the outcome of the reaction.

▶ HOW REACTIONS OCCUR

THE HYDRATION OF ALKENES: AN ADDITION REACTION

The mechanism for the hydration of alkenes is believed to begin when H^+ from the acid catalyst is attracted to the electrons of the carbon–carbon double bond. The H^+ becomes bonded to one of the carbons by a sharing of electrons:

Step 1.

$$-C=C- + H^+ \longrightarrow -\overset{\overset{\displaystyle H}{|}}{C}-\overset{+}{C}-$$

A carbocation

This process leaves the second carbon with only three bonds about it and thus a positive charge. Such ions, referred to as **carbocations**, are extremely reactive.

As soon as it forms, the positive carbocation attracts any species that has readily available nonbonding electrons, whether it is an anion or a neutral molecule. In the case of water, the oxygen atom has two unshared pairs of electrons:

$$\overset{\cdot\cdot}{\underset{H \qquad H}{O}}$$

One pair of oxygen electrons forms a covalent bond with the carbocation:

Step 2.

$$-\overset{\overset{\displaystyle H}{|}}{C}-\overset{+}{\underset{|}{C}}- \;+\; H \overset{\cdot\cdot}{O} H \longrightarrow -\overset{\overset{\displaystyle H}{|}}{C}-\overset{\overset{\displaystyle H\;\;O^+\;\;H}{|}}{\underset{|}{C}}-$$

Carbocation
An ion of the form $-\overset{+}{\underset{|}{C}}-$.

In the third step, H^+ is lost to produce the alcohol

Step 3.

$$-\overset{\overset{\displaystyle H\;\;O^+\;\;H}{|}}{C}-\overset{|}{\underset{|}{C}}- \longrightarrow -\overset{\overset{\displaystyle H}{|}}{C}-\overset{\overset{\displaystyle OH}{|}}{\underset{|}{C}}- + H^+$$

An alcohol

Notice that the acid catalyst, H^+, which initiated the reaction, is recovered unchanged in the final step of the mechanism.

By applying Step 2 of the mechanism for hydration, we can understand how HCl and HBr react with alkenes:

$$-\overset{\overset{\displaystyle H}{|}}{C}-\overset{+}{\underset{|}{C}}- + Br^- \longrightarrow -\overset{\overset{\displaystyle H}{|}}{C}-\overset{\overset{\displaystyle Br}{|}}{\underset{|}{C}}-$$

$$-\overset{\overset{\displaystyle H}{|}}{C}-\overset{+}{\underset{|}{C}}- + Cl^- \longrightarrow -\overset{\overset{\displaystyle H}{|}}{C}-\overset{\overset{\displaystyle Cl}{|}}{\underset{|}{C}}-$$

The lowercase n in Reaction 2.14 represents a large, unspecified number. From this reaction, we see that polyethylene is essentially a very long chain alkane. As a result, it has the chemical inertness of alkanes, a characteristic that makes polyethylene suitable for food storage containers, garbage bags, eating utensils, laboratory apparatus, and hospital equipment (see ■ Figure 2.7). Polymer characteristics are modified by using alkenes with different groups attached to either or both of the double-bonded carbons. For example, the polymerization of vinyl chloride gives the polymer poly (vinyl chloride), PVC:

$$n CH_2 = \overset{\overset{\displaystyle Cl}{|}}{CH} \; \xrightarrow[\text{catalysts}]{\text{Heat,}} \; +CH_2 - CH +_n \qquad (2.15)$$

vinyl chloride   poly (vinyl chloride)

➤ **FIGURE 2.7** A large variety of items are made from polyethylene, including plastic containers, garbage bags, syringes, and laboratory equipment.

The commercial product Saran Wrap is an example of a **copolymer,** which is a polymer made up of two different monomers (Reaction 2.16):

$$n\text{CH}_2 = \text{CH} + n\text{CH}_2 = \overset{\overset{\displaystyle \text{Cl}}{|}}{\underset{\underset{\displaystyle \text{Cl}}{|}}{\text{C}}} \xrightarrow{\text{Catalyst}} \left(\text{CH}_2 - \underset{\underset{\displaystyle \text{Cl}}{|}}{\text{CH}} - \text{CH}_2 - \overset{\overset{\displaystyle \text{Cl}}{|}}{\underset{\underset{\displaystyle \text{Cl}}{|}}{\text{C}}} \right)_n \quad (2.16)$$

$$\underset{\displaystyle |}{\text{Cl}}$$

vinyl   vinylidene   Saran Wrap
chloride   chloride

Copolymer
An addition polymer formed by the reaction of two different monomers.

A number of the more important addition polymers are shown in ■ Table 2.3. As you can tell by looking at some of the typical uses, addition polymers have become nearly indispensable in modern life (see ■ Figure 2.8).

Draw the structural formula of a portion of polypropylene containing four repeating units of the monomer propylene,

$$\overset{\overset{\displaystyle \text{CH}_3}{|}}{\text{CH}_2 = \text{CH}}$$

Learning Check 2.8

Automobile safety glass contains a sheet of poly (vinyl acetate) layered between two sheets of glass to prevent the formation of sharp fragments.

Chewing gum comes in many varieties.

➤ **FIGURE 2.8** Two uses of poly (vinyl acetate). What properties of poly (vinyl acetate) are exhibited in both of these products?

TABLE 2.3	Common addition polymers		
Chemical name and trade name(s)	**Monomer**	**Polymer**	**Typical uses**
polyethylene	$CH_2 = CH_2$	$\style{}{+}\ CH_2 - CH_2\ \style{}{\overset{}{+}_n}$	Bottles, plastic bags, film
polypropylene	$CH_2 = CH$ | CH_3	$\style{}{+}\ CH_2 - CH\ \style{}{+_n}$ | CH_3	Carpet fiber, pipes, bottles, artificial turf
poly (vinyl chloride) (PVC)	$CH_2 = CH$ | Cl	$\style{}{+}\ CH_2 - CH\ \style{}{+_n}$ | Cl	Synthetic leather, floor tiles, garden hoses, water pipe
polytetraflouroethylene (Teflon®)	$CF_2 = CF_2$	$\style{}{+}\ CF_2 - CF_2\ \style{}{+_n}$	Pan coatings, plumbers' tape, heart valves, fabrics
poly (methyl methacrylate) (Lucite®, Plexiglas®)	CH_3 | $CH_2 = C$ | $C - O - CH_3$ ‖ O	CH_3 | $\style{}{+}\ CH_2 - C\ \style{}{+_n}$ | $C - O - CH_3$ ‖ O	Airplane windows, paint, contact lenses, fiber optics
poly (vinyl acetate)	$CH_2 = CH$ | $O - C - CH_3$ ‖ O	$\style{}{+}\ CH_2 - CH\ \style{}{+_n}$ | $O - C - CH_3$ ‖ O	Adhesives, latex paint, chewing gum
polyacrylonitrile (Orlon®, Acrilan®)	$CH_2 = CH$ | CN	$\style{}{+}\ CH_2 - CH\ \style{}{+_n}$ | CN	Carpets, fabrics
polystyrene (Styrofoam®)	$CH_2 = CH$ | (phenyl ring)	$\style{}{+}\ CH_2 - CH\ \style{}{+_n}$ | (phenyl ring)	Food coolers, drinking cups, insulation

➤ 2.5 Alkynes

The characteristic feature of alkynes is the presence of a triple bond between carbon atoms. Thus, alkynes are also unsaturated hydrocarbons. Only a few compounds containing the carbon–carbon triple bond are found in nature. The simplest and most important compound of this series is ethyne, more commonly called acetylene (C_2H_2):

$$H - C \equiv C - H$$
acetylene

Acetylene is used in torches for welding steel and in making plastics and synthetic fibers.

Once again, orbital hybridization provides an explanation for the bonding of the carbon atoms. Structurally, the hydrogen and carbon atoms of acetylene molecules lie in a straight line. This same linearity of the triple bond and the two atoms attached to the triple-bonded carbons is found in all alkynes. These characteristics are explained by mixing a $2s$ and a single $2p$ orbital of each carbon to form a pair of sp hybrid orbitals. Two of the $2p$ orbitals of each carbon are unhybridized (see ■ Figures 2.9 and 2.10).

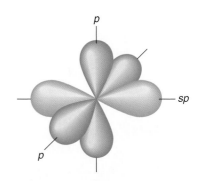

> **FIGURE 2.9** *sp* hybridization occurs when one of the 2*p* orbitals of carbon mixes with the 2*s* orbital. Two 2*p* orbitals remain unhybridized.

> **FIGURE 2.10** The unhybridized *p* orbitals are perpendicular to the two *sp* hybridized orbitals.

A carbon–carbon sigma bond in acetylene forms by the overlap of one *sp* hybrid orbital of each carbon. The other *sp* hybrid orbital of each carbon overlaps with a 1*s* orbital of a hydrogen to form a carbon–hydrogen sigma bond. The remaining pair of unhybridized *p* orbitals of each carbon overlap sideways to form a pair of pi bonds between the carbon atoms. Thus, each acetylene molecule contains three sigma bonds (two carbon–hydrogen and one carbon–carbon) and two pi bonds (both are carbon–carbon). This is shown in ■ Figure 2.11.

Alkynes are named in exactly the same ways as alkenes, except the ending -*yne* is used:

$$\overset{4}{C}H_3\overset{3}{C}H_2-\overset{2}{C}\equiv\overset{1}{C}-H \qquad H-\overset{1}{C}\equiv\overset{2}{C}-\overset{3}{C}H-\overset{4}{C}H_3$$
$$\underset{CH_3}{|}$$

1-butyne   3-methyl-1-butyne

Give the IUPAC name for each of the following:

Learning Check 2.9

a. $CH_3-CH_2-C\equiv C-CH_3$ **b.** $CH_3-\overset{\overset{\textstyle CH_3}{\textstyle |}}{CH}-CH_2-C\equiv C-CH_3$

The physical properties of the alkynes are nearly the same as those of the corresponding alkenes and alkanes: They are insoluble in water, less dense than water, and have relatively low melting and boiling points. Alkynes also resemble alkenes in their addition reactions. The same substances (Br_2, H_2, HCl, etc.) that add to double bonds also add to triple bonds. The one significant difference is that alkynes consume twice as many moles of addition reagent as alkenes in addition reactions that go on to completion.

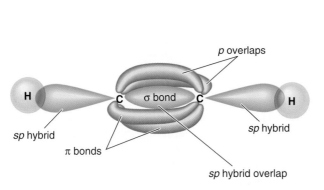

acetylene (ethyne)

> **FIGURE 2.11** The shape of acetylene is explained by sigma bonding between *sp* hybrid orbitals and pi bonding between unhybridized *p* orbitals.

➤ **FIGURE 2.12** Friedrich August Kekulé (1829–1896).

Aliphatic compound

Any organic compound that is not aromatic.

➤ 2.6 Aromatic Compounds and the Benzene Structure

Some of the early researchers in organic chemistry became intrigued by fragrant oils that could be extracted from certain plants. Oil of wintergreen and the flavor component of the vanilla bean are examples. The compounds responsible for the aromas had similar chemical properties. As a result, they were grouped together and called aromatic compounds.

As more and more aromatic compounds were isolated and studied, chemists gradually realized that aromatics contained at least six carbon atoms, had low hydrogen-to-carbon ratios (relative to other organic hydrocarbons), and were related to benzene (C_6H_6). For example, toluene, an aromatic compound from the bark of the South American tolu tree, has the formula C_7H_8.

Chemists also learned that the term *aromatic* was not always accurate. Many compounds that belong to the class because of chemical properties and structures are not at all fragrant. Conversely, there are many fragrant compounds that do not have aromatic compound properties or structures. Today, the old class name is used but with a different meaning. Aromatic compounds are those that contain the characteristic benzene ring or its structural relatives. Compounds that do not contain this structure (nonaromatic compounds) are referred to as **aliphatic compounds.** Alkanes, alkenes, and alkynes are, therefore, aliphatic compounds.

The molecular structure of benzene presented chemists with an intriguing puzzle after the compound was discovered in 1825 by Michael Faraday. The formula C_6H_6 indicated that the molecule was highly unsaturated. However, the compound did not show the typical reactivity of unsaturated hydrocarbons. Benzene underwent relatively few reactions, and these proceeded slowly and often required heat and catalysts. This was in marked contrast to alkenes, which reacted rapidly with many reagents, in some cases almost instantaneously. This apparent discrepancy between structure and reactivity plagued chemists until 1865, when Freidrich August Kekulé von Stradonitz (see ■ Figure 2.12), a German chemist, suggested that the benzene molecule might be represented by a ring arrangement of carbon atoms with alternating single and double bonds between the carbon atoms:

He later suggested that the double bonds alternate in their positions between carbon atoms to give two equivalent structures:

Kekulé structures

A modern interpretation of the benzene structure based on hybridization enables chemists to better understand and explain the chemical properties of benzene and other aromatic compounds. Each carbon atom in a benzene ring has three sp^2 hybrid orbitals and one unhybridized p orbital.

A single sigma bond between carbons of the benzene ring is formed by the overlap of two sp^2 orbitals, one from each of the double-bonded carbons. Because each carbon forms two single bonds, two of the sp^2 hybrid orbitals

Each carbon atom forms three single bonds and has one electron in a nonhybridized *p* orbital.

Two lobes represent the delocalized pi bonding of the six *p* electrons.

of each carbon are involved. The third sp^2 hybrid orbital of each carbon forms a single sigma bond with a hydrogen by overlapping with a $1s$ orbital of hydrogen. The unhybridized p orbitals of each carbon overlap sideways above and below the plane of the carbon ring to form two delocalized pi lobes that run completely around the ring (see ■ Figure 2.13).

This interpretation leads to the conclusion that six bonding electrons move freely around the ring in the overlapping delocalized pi lobes. Because of this, the benzene structure is often represented by the symbol

with the circle representing the evenly distributed electrons in the pi lobes. All six carbon and six hydrogen atoms in benzene molecules lie in the same plane (see Figure 2.13). Therefore, substituted aromatic compounds do not exhibit *cis-trans* isomerism.

As you draw the structure of aromatic compounds, remember that only one hydrogen atom or group can be attached to a particular position on the benzene ring. For example, compounds (a) and (b) below exist, but (c) does not. Examination of the Kekulé structure of compound (c) shows that the carbon has five bonds attached to it in violation of the octet rule:

(a)
exists

(b)
exists

(c)
does not
exist

Kekulé
structure

*five bonds
attached
(not possible)*

➤ 2.7 The Nomenclature
of Benzene Derivatives

The chemistry of the aromatic compounds developed somewhat haphazardly for many years before systematic nomenclature schemes were developed. Some of the common names used earlier have acquired historical respectability and are used today; some have even been incorporated into the modern systematic nomenclature system.

The following rules are all based on the IUPAC aromatic nomenclature system. They are not complete, and you will not be able to name all aromatic compounds by using them. However, you will be able to name and recognize those used in this book.

> *Rule 1.* When a single hydrogen of the benzene ring is replaced, the compound can be named as a derivative of benzene:

ethylbenzene   nitrobenzene   bromobenzene   chlorobenzene

> *Rule 2.* A number of benzene derivatives are known by common names that are also IUPAC-accepted and are used preferentially over other possibilities. Thus, toluene is favored over methylbenzene, and aniline is used rather than aminobenzene:

toluene   phenol   aniline   benzoic acid

> *Rule 3.* Compounds formed by replacing a hydrogen of benzene with a more complex hydrocarbon group can be named by designating the benzene ring as the group. When this is done, the benzene ring, C_6H_5—, or

Phenyl group
A benzene ring with one hydrogen absent, C_6H_5—.

is called a **phenyl group**:

$$CH_3 - CH_2 - CH_2 - CH - CH_2 - CH_2 - CH_3$$

4-phenylheptane   1,1-diphenylcyclobutane

It's easy to confuse the words *phenyl* and *phenol*. The key to keeping them straight is the ending: *-ol* means an alcohol (C_6H_5—OH), and *-yl* means a group (C_6H_5—).

> *Rule 4.* When two groups are attached to a benzene ring, three isomeric structures are possible. They can be designated by the prefixes *ortho (o)*, *meta (m)*, and *para (p)*:

o-dibromobenzene   *m*-dibromobenzene   *p*-dibromobenzene

> *Rule 5.* When two or more groups are attached to a benzene ring, their positions can be indicated by numbering the carbon atoms of the ring so as to obtain the lowest possible numbers for the attachment positions.

CHEMISTRY AROUND US • 2.3

TOO MUCH SUN—NOT A BRIGHT IDEA

Many people find deep, rich suntans attractive and desirable, and they expose as much skin as possible to natural or artificial sunlight. Unfortunately, sunburn results almost as often as suntan. There is also increasing evidence that exposure to the sun's rays over a long period of time may cause skin cancer. In fact, it has been found that excessive sun exposure during childhood clearly increases the risk of developing skin cancer and promotes the premature aging of the skin (photoaging). The main culprit is repeated exposure to ultraviolet rays.

Although the stratospheric ozone layer absorbs most of the short-wave ultraviolet (UV) radiation that is extremely harmful, plenty of UV rays of longer wavelengths (greater than 285 nm) reach the earth's surface. Long-wavelength ultraviolet radiation in sunlight causes both suntan and sunburn. Most of this radiation that strikes the skin is absorbed by melanin, a dark-colored skin pigment. Carefully limited and repeated exposure to sunlight results in the production of more melanin, which causes the skin to gradually darken.

The sun's rays that cause tanning or burning can be blocked by clothing, an umbrella, or opaque materials on the skin such as zinc oxide (white) in a cream formulation. Several popular lotions contain *p*-aminobenzoic acid (PABA), which, like other compounds derived from benzene, absorbs UV radiation and therefore acts as a sunscreen:

$$H_2N \text{---} \langle \bigcirc \rangle \text{---} COOH$$

p-aminobenzoic acid

However, PABA absorbs more of the UV rays that tend to cause burning than the slightly longer wavelength rays that tend to cause tanning. Thus, the skin tends to tan without burning. In some lotions, the concentration of PABA is indicated by skin-protection-factor (SPF) ratings. What do these ratings mean?

Most people can be exposed to the sun's ultraviolet radiation for a limited time without getting burned. The amount of time that can be spent in the sun without burning depends on such factors as skin tone (fair skin burns more easily than darker skin) and the intensity of the sun during the time of exposure. Therefore, the time can vary from 15 minutes to several hours. SPF rat-

ings provide a rough indication of how much longer the sunscreen's use will allow a person to be in the sun without getting a sunburn. For example, if someone usually begins to burn after about 20 minutes' exposure, the proper use of an SPF 2 lotion extends the time to about 40 minutes (2×20 min = 40 min), and an SPF 15 lotion provides protection for about 5 hours (15×20 min = 300 min = 5 h).

However, SPF ratings don't take into account factors such as altitude, wind, the reflection from surfaces like water and snow, and the angle of the sun—all of which can reduce the effectiveness of sunscreen. For this reason, most doctors recommend using a product with an SPF rating of at least 15 when exposure to the sun is likely.

Like the sun, sunlamps also give off UV radiation and can be harmful. Exposure to UV rays from sunlamps adds to the total amount of UV radiation accumulated by the skin during a lifetime, and increases the risk of skin cancer. According to the Food and Drug Administration (FDA), the use of artificial tanning devices is not recommended for anyone.

SPF ratings indicate the level of protection of the sunscreen.

http://www.brookscole.com/chemistry/seager_chemistry
Web site 2.4 provides more information about the detrimental effects of suntanning.

Groups are arranged in alphabetical order. If there is a choice of identical sets of numbers, the group that comes first in alphabetical order is given the lower number. IUPAC-acceptable common names may be used:

m-bromochlorobenzene
or 1-bromo-3-chlorobenzene

1,2,4-trichlorobenzene

3,5-dichlorobenzoic acid

➤ **Example 2.5**

Name each of the following aromatic compounds:

a. (structure: CH₃ and CH₂CH₃ substituted benzene)   b. (structure: CH₂CH₂CH₃ substituted benzene)   c. (structure: benzene with C(=O)—OH, CH₃CH₂—, and Cl substituents)

Solution

a. This compound is named as a substituted toluene. Both of the following are correct: 3-ethyltoluene and *m*-ethyltoluene. Note that in 3-ethyltoluene the methyl group, which is a part of the toluene structure, must be assigned to position number one.
b. This compound may be named as a substituted benzene or a substituted propane: propylbenzene, 1-phenylpropane.
c. When three groups are involved, the ring-numbering approach must be used: 3-chloro-5-ethylbenzoic acid.

Learning Check 2.10

Name the following aromatic compounds:

a. CH₂CH₃ (benzene ring with CH₂CH₃)   b. Cl (cyclopentane with Cl and phenyl)

c. Br (benzene with three Br)   d. CH₃ and C(=O)—OH substituted benzene

➤ 2.8 Properties and Uses of Aromatic Compounds

The physical properties of benzene and other aromatic hydrocarbons are similar to those of alkanes and alkenes. They are nonpolar and thus insoluble in water. This hydrophobic characteristic plays an important role in the chemistry of some proteins (Chapter 9).

Aromatic rings are relatively stable chemically. Because of this, benzene often reacts in such a way that the aromatic ring remains intact. Thus, benzene does not undergo the addition reactions that are characteristic of alkenes and alkynes. The predominant type of reaction of aromatic molecules is substitution, in which one of the ring hydrogens is replaced by some other group. Such aromatic reactions are of lesser importance for our purposes and are not shown here.

All aromatic compounds discussed to this point contain a single benzene ring. There are also substances called **polycyclic aromatic compounds,** which contain two or more benzene rings sharing a common side, or "fused"

Polycyclic aromatic compound
A derivative of benzene in which carbon atoms are shared between two or more benzene rings.

OVER THE COUNTER • 2.1

SMOKING: IT'S QUITTING TIME

Smoking is a difficult habit to break, especially if the attempt is made by stopping abruptly—going "cold turkey." Mark Twain described the more reasonable gradual approach when he said, "Habit is habit, and not to be flung out of the window by any man, but coaxed downstairs a step at a time." Smokers do not have to go "cold turkey" because there are several OTC aids available, as well as some new prescription products to help them gradually overcome the strong urge to smoke—and eventually quit.

The transdermal (absorbed through the skin) nicotine patch is available over the counter in doses of 7–22 mg. When used as directed, this method delivers a steady supply of nicotine to the bloodstream and helps minimize withdrawal symptoms. Nicotine gum helps reduce withdrawal symptoms when used correctly. The gum should be chewed briefly and then held next to the cheek, allowing the lining of the mouth to absorb the nicotine.

With a prescription, smokers can also obtain a nasal spray that provides a small dose of nicotine each time it is used. This product, called Nicotrol NS®, was approved by the FDA even though inhaling the nicotine poses a small risk that smokers will become as dependent on the mist as they are on cigarettes. The nicotine from the nasal spray gets into the bloodstream faster than nicotine from the gum or patch, providing immediate relief from cigarette craving. A squirt into each nostril gives a smoker 1 mg of nicotine, but it is not supposed to be used more than five times per hour.

The Nicotrol nicotine inhalation system has also received FDA approval. This inhaler, available only by prescription, consists of a plastic cylinder about the size of a cigarette that en-

closes a cartridge containing nicotine. When a smoker "puffs" on the device, nicotine vapors are absorbed through the lining of the mouth and throat. It takes about 80 puffs to deliver the amount of nicotine obtained from a single cigarette. An advantage of using the system is that a smoker still mimics the hand-to-mouth behavior of smoking, a part of the smoking habit that will be easier to break once nicotine withdrawal symptoms subside.

One of the newest prescription products designed to help break the smoking habit does not contain any nicotine. It is an antidepressant called bupropion that has been shown to be effective in the treatment of nicotine addiction. It is believed that bupropion mimics some of the action of nicotine by releasing the brain chemicals norepinephrine and dopamine, but it is not completely understood how it works. During treatment, a bupropion tablet is taken once a day for 3 days and then twice daily during the week before smoking is stopped. Usually, the treatment is continued for the next 6 to 12 weeks to help curb the craving for cigarettes.

If a smoker truly wants to quit, these aids alone will not do it. A smoker must have some kind of support from a formal program, or at least informal support from family and friends. A smoker should also get rid of all tobacco products and avoid smoking triggers, such as other smokers, stress, and alcohol. Exercise can also be a distraction from smoking and can minimize the weight gain that sometimes accompanies giving up smoking.

http://www.brookscole.com/chemistry/seager_chemistry
Information about quitting smoking is provided at Web site 2.5.

together. The simplest of these compounds, naphthalene, is the original active ingredient in mothballs:

naphthalene

A number of more complex polycyclic aromatic compounds are known to be carcinogens—chemicals that cause cancer. Two of these compounds are

benzo[*a*]pyrene

dibenz[*a,h*]anthracene

➤ **FIGURE 2.14** Cigarette smoke contains carcinogenic polycyclic aromatic compounds.

These cancer-producing compounds are often formed as a result of heating organic materials to high temperatures. They are present in tobacco smoke (see ■ Figure 2.14), automobile exhaust, and sometimes in burned or heavily browned food. Such compounds are believed to be at least partially responsible for the high incidence of lung and lip cancer among cigarette smokers.

| TABLE 2.4 | Some important aromatic compounds |

Name	Structural formula	Use
benzene	(benzene ring structure)	Industrial solvent and raw material
toluene	(benzene ring with CH_3)	Industrial solvent and raw material
phenol	(benzene ring with OH)	Manufacture of Bakelite® and Formica®
aniline	(benzene ring with NH_2)	Manufacture of drugs and dyes
styrene	(benzene ring with $CH=CH_2$)	Preparation of polystyrene products
phenylalanine	(benzene ring with $CH_2CH-\overset{\overset{O}{\|\|}}{C}-OH$ and NH_2)	An essential amino acid
riboflavin	(riboflavin structure with $CH_2-CH-CH-CH-CH_2$ bearing OH groups, fused ring system with CH_3 groups, N, NH, O)	A vitamin

Those who smoke heavily face an increased risk of getting cancer. Chemists have identified more than 4000 compounds in cigarette smoke, including 43 known carcinogens. The Environmental Protection Agency (EPA) considers tobacco smoke a Class A carcinogen.

The major sources of aromatic compounds are petroleum and coal tar, a sticky, dark-colored material derived from coal. As with many classes of organic compounds, the simplest structures are the most important commercial materials.

Benzene and toluene are excellent laboratory and industrial solvents. In addition, they are the starting materials for the synthesis of hundreds of other valuable aromatic compounds that are intermediates in the manufacture of a wide variety of commercial products, including the important polymers Bakelite® and polystyrene (see ■ Table 2.4).

A number of aromatic compounds are important in another respect: They must be present in our diet for proper nutrition. Unlike plants, which have the ability to synthesize the benzene ring from simpler materials, humans must obtain any necessary aromatic rings from their diet. This helps explain why certain amino acids, the building blocks of proteins, and some **vitamins** are dietary essentials (Table 2.4).

Vitamin
An organic nutrient that the body cannot produce in the small amounts needed for good health.

> **FOR FUTURE REFERENCE**   DRUGS THAT MAY CAUSE PHOTOSENSITIVITY FROM SUN EXPOSURE

Normal skin will tan or burn when exposed to ultraviolet light from the sun (see Chemistry Around Us 2.3). The use of some drugs can increase the sensitivity of the skin to the effects of UV light. The skin of sensitized individuals reacts by developing a rash or burning excessively. The following list contains the names of drugs that may cause such photosensitivity. Anyone taking one or more of these drugs should consult with a physician about the possibility of developing photosensitivity, and the precautions to take to prevent it.

acetazolamide
acetohexamide
alprazolam
amantadine
amiloride
aminobenzoic acid
amiodarone
amitriptyline
amoxapine
barbiturates
bendroflumethiazide
benzocaine
benzophenones
benzoyl peroxide
benzthiazide
captopril
carbamazepine
chlordiazepoxide
chloroquine
chlorothiazide
chlorpromazine
chlorpropamide
chlortetracycline
chlorthalidone
ciprofloxacin
clindamycin
clofazimine
clofibrate
clomipramine
cyproheptadine
dacarbazine
dapsone
demeclocycline
desipramine
desoximetasone
diethylstilbestrol
diflunisal
diltiazem
diphenhydramine
disopyramide

doxepin
doxycycline
enoxacin
estrogen
eretinate
flucytosine
fluorescein
fluorouracil
fluphenazine
flutamide
furosemide
glipizide
glyburide
gold preparations
griseofulvin
haloperidol
hexachlorophene
hydrochlorothiazide
hydroflumethiazide
ibuprofen
imipramine
indomethacin
isotretinoin
ketoprofen
lincomycin
lomefloxacin
maprotiline
mesoridazine
methacycline
methotrexate
methyclothiazide
methyldopa
metolazone
minocycline
minoxidil
nabumetone
nalidixic acid
naproxen
nifedipine
norfloxacin

nortriptyline
ofloxacin
oral contraceptives
oxyphenbutazone
oxytetracycline
para-aminobenzoic acid
perphenazine
phenelzine
phenobarbital
phenylbutazone
phenytoin
piroxicam
polythiazide
prochlorperazine
promazine
promethazine
protriptyline
pyrazinamide
quinidine
quinine
sulfonamides
sulindac
tetracycline
thiabendazole
thioridazine
thiothixene
tolazamide
tolbutamide
tranylcypromine
trazodone
tretinoin
triamterene
trichlormethiazide
trifluoperazine
triflupromazine
trimeprazine
trimethoprim
trimipramine
triprolidine
vinblastine

SOURCE: Adopted from Rybacki, J.J.; Long, J.W. *The Essential Guide to Prescription Drugs.* New York: Harper Perennial, 1999.

CONCEPT SUMMARY

The Nomenclature of Alkenes. Compounds containing double or triple bonds between carbon atoms are said to be unsaturated. The alkenes contain double bonds, alkynes triple bonds. In the IUPAC nomenclature system, alkene names end in *-ene*, alkynes end in *-yne*.

The Geometry of Alkenes. In alkenes, the double-bonded carbons and the four groups attached to these carbons lie in the same plane. Because rotation about the double bond is restricted, alkenes may exist as geometric, or *cis-trans*, isomers. This type of stereoisomerism is possible when each double-bonded carbon is attached to two different groups.

Properties of Alkenes. The physical properties of alkenes are very similar to those of the alkanes. They are nonpolar, insoluble in water, less dense than water, and soluble in nonpolar solvents.

Alkene Reactions. Alkenes are quite reactive, and their characteristic reaction is addition to the double bond. Three important addition reactions are bromination (an example of halogenation) to give a dibrominated alkane, hydration to produce an alcohol, and the reaction with H—X to give an alkyl halide. The addition of H_2O and H—X are governed by Markovnikov's rule.

Addition Polymers. Addition polymers are formed from alkene monomers that undergo repeated addition reactions with each other. Many familiar and widely used materials, such as fibers and plastics, are addition polymers.

Alkynes. The alkynes contain triple bonds and possess a linear geometry of the two carbons and the two attached groups. Alkyne names end in *-yne*. The physical and chemical properties of alkynes are very similar to those of the alkenes.

Aromatic Compounds and the Benzene Structure. Benzene, the simplest aromatic compound, and other members of the aromatic class contain a six-membered ring with three double bonds. This aromatic ring is often drawn as a hexagon containing a circle, which represents the six electrons of the double bonds that move freely around the ring. All organic compounds that do not contain an aromatic ring are called aliphatic compounds.

The Nomenclature of Benzene Derivatives. Several acceptable IUPAC names are possible for many benzene compounds. Some IUPAC names are based on widely used common names such as toluene and aniline. Other compounds are named as derivatives of benzene or by designating the benzene ring as a phenyl group.

Properties and Uses of Aromatic Compounds. Aromatic hydrocarbons are nonpolar and have physical properties similar to those of the alkanes and alkenes. Benzene resists addition reactions typical of alkenes. Benzene and toluene are key industrial chemicals. Other important aromatics include phenol, aniline, and styrene.

LEARNING OBJECTIVES

When you have completed your study of this chapter, you should be able to:

1. Classify unsaturated hydrocarbons as alkenes, alkynes, or aromatics.
2. Write the IUPAC names of alkenes and alkynes from their molecular structures.
3. Predict the existence of geometric *(cis-trans)* isomers from formulas of compounds.
4. Write the names and structural formulas for geometric isomers.
5. Write equations for addition reactions of alkenes, and use Markovnikov's rule to predict the major products of certain reactions.
6. Write equations for addition polymerization, and list uses for addition polymers.
7. Name and draw structural formulas for aromatic compounds.

KEY TERMS AND CONCEPTS

Addition polymer (2.4)
Addition reaction (2.3)
Aliphatic compound (2.6)
Alkene (Introduction)
Alkyne (Introduction)
Aromatic hydrocarbon (Introduction)
Carbocation (2.4)

Copolymer (2.4)
Haloalkane or alkyl halide (2.3)
Hydration (2.3)
Hydrogenation (2.3)
Markovnikov's rule (2.3)
Monomer (2.4)
Phenyl group (2.7)

Polycyclic aromatic compound (2.8)
Polymer (2.4)
Polymerization (2.4)
Polyunsaturated (2.3)
Terpene (2.1)
Unsaturated hydrocarbon (Introduction)
Vitamin (2.8)

KEY REACTIONS

1. Halogenation of an alkene (Section 2.3):

$$\text{C=C} \quad + \quad \text{Br}-\text{Br} \quad \longrightarrow \quad -\underset{\underset{\text{Br}}{|}}{\overset{|}{\text{C}}}-\underset{\underset{\text{Br}}{|}}{\overset{|}{\text{C}}}-$$

Reaction 2.2

2. Hydrogenation of an alkene (Section 2.3):

$$\text{C=C} \quad + \quad \text{H}-\text{H} \quad \xrightarrow{\text{Pt}} \quad -\underset{\underset{\text{H}}{|}}{\overset{|}{\text{C}}}-\underset{\underset{\text{H}}{|}}{\overset{|}{\text{C}}}-$$

Reaction 2.4

3. Addition of H—X to an alkene (Section 2.3):

$$\text{C=C} \quad + \quad \text{H}-\text{Cl} \quad \longrightarrow \quad -\underset{\underset{\text{H}}{|}}{\overset{|}{\text{C}}}-\underset{\underset{\text{Cl}}{|}}{\overset{|}{\text{C}}}-$$

Reaction 2.6

4. Hydration of an alkene (Section 2.3):

$$\text{C=C} \quad + \quad \text{H}-\text{OH} \quad \xrightarrow{\text{H}_2\text{SO}_4} \quad -\underset{\underset{\text{H}}{|}}{\overset{|}{\text{C}}}-\underset{\underset{\text{OH}}{|}}{\overset{|}{\text{C}}}-$$

Reaction 2.9

5. Addition polymerization of an alkene (Section 2.4):

$$n\text{CH}_2\text{=CH}_2 \quad \xrightarrow[\text{catalysts}]{\text{Heat, pressure}} \quad \text{(CH}_2-\text{CH}_2\text{)}_n$$

Reaction 2.14

EXERCISES

THE NOMENCLATURE OF ALKENES (SECTION 2.1) AND ALKYNES (SECTION 2.5)

2.1 Select those compounds that can be correctly called *unsaturated* and classify each one as an *alkene* or an *alkyne*:

a. $\text{CH}_3-\text{CH}_2-\text{CH}_3$ b. $\text{CH}_3\text{CH=CHCH}_3$

c. $\text{H}-\text{C}\equiv\text{C}-\underset{\underset{\text{CH}_3}{|}}{\text{CH}}-\text{CH}_3$ d.

e. cyclohexane with CH_3 f.

cyclopentane with CH=CH_2

g. $\underset{\text{CH}_2-\text{CH}_2}{\overset{\text{CH=CH}}{|\qquad|}}$ h. $\text{CH}_2\text{=CHCH}_2\text{CH}_3$

i. $\text{CH}_3\underset{\underset{\text{CH}_3}{|}}{\text{CH}}\text{CH}_3$

2.2 Give the IUPAC name for the following compounds:

a. $\text{CH}_3\text{CH=CHCH}_3$

b. $\text{CH}_3\text{CH}_2-\underset{\underset{\text{CH}_2\text{CH}_3}{|}}{\text{C}}\text{=CHCH}_3$

c. $\text{CH}_3-\text{C}\equiv\text{C}-\underset{\underset{\text{CH}_3}{|}}{\overset{\overset{\text{CH}_3}{|}}{\text{C}}}-\text{CH}_2\text{CH}_3$

d. cyclopentene with CH_3

e. $\text{CH}_3\underset{\underset{\text{Br}}{|}}{\text{CH}}\text{CH}_2-\text{C}\equiv\text{C}-\underset{\underset{\text{CH}_3}{|}}{\text{CH}}-\text{CH}_3$

f. cyclopropene ring with CH_3, CH_3, CH_2CH_3

g. $\text{CH}_3\underset{\underset{\text{CH}_3}{|}}{\text{CH}}-\text{CH=CHCH}_2\text{CH=CH}_2$

2.3 Give the IUPAC name for the following compounds:

a. $CH_3CHCH=CHCH_2CH_3$
   $\quad\quad |$
   $\quad\quad CH_3$

b. $CH_3CH=CHCH=CHCHCH_3$
   $\quad\quad\quad\quad\quad\quad\quad\quad |$
   $\quad\quad\quad\quad\quad\quad\quad\quad CH_3$

c.

d. $CH_3-C\equiv C-CH_2CH_3$

e. $\quad\quad CH_3$
   $\quad\quad\ |$
   CH_2CHCH_3

f. $\quad CH_2CH_2CH_3$

g. $\quad\quad\quad CH_2CH_2CH_2CH_3$
   $\quad\quad\quad\ |$
   $CH_2=C-CH-CH=CHCH_2CH_3$
   $\quad\quad\quad\quad\quad |$
   $\quad\quad\quad\quad\quad CH_2CH_3$

2.4 Draw structural formulas for the following compounds:
a. 4-methyl-2-hexene
b. 4,4-dimethyl-1-pentyne
c. 1,3-butadiene
d. 1-ethyl-3-methylcyclopentene
e. 1,6-dimethylcyclohexene

2.5 Draw structural formulas for the following compounds:
a. 4,4,5-trimethyl-2-heptyne
b. 1,3-cyclohexadiene
c. 2-*t*-butyl-4,4-dimethyl-1-pentene
d. 4-isopropyl-3,3-dimethyl-1,5-octadiene
e. 2-methyl-1,3-cyclopentadiene
f. 3-*sec*-butyl-3-*t*-butyl-1-heptyne

2.6 A compound has the molecular formula C_5H_8. Draw a structural formula for a compound with this formula that would be classified as (a) an alkyne, (b) a diene, and (c) a cyclic alkene. Give the IUPAC name for each compound.

2.7 Draw structural formulas and give IUPAC names for the 13 alkene isomers of C_6H_{12}. Ignore geometric isomers and cyclic structures.

2.8 α-farnesene is a constituent of the natural wax found on apples. Given that a 12-carbon chain is named as a dodecane, what is the IUPAC name of α-farnesene?

$\quad\quad CH_3 \quad\quad\quad\quad CH_3 \quad\quad\quad\quad CH_3$
$\quad\quad | \quad\quad\quad\quad\quad\quad | \quad\quad\quad\quad\quad\quad |$
$CH_3C=CHCH_2CH_2C=CHCH_2CH=CCH=CH_2$
$\quad\quad\quad\quad\quad\quad \alpha\text{-farnesene}$

2.9 What is wrong with each of the following names? Give the structure and correct name for each compound.
a. 3-pentene
b. 3-methyl-2-butene
c. 2-ethyl-3-pentyne

2.10 What is wrong with each of the following names? Give the structure and correct name for each compound.
a. 2-methyl-4-hexene
b. 3,5-heptadiene
c. 4-methylcyclobutene

THE GEOMETRY OF ALKENES (SECTION 2.2)

2.11 What type of hybridized orbitals is present on carbon atoms bonded by a double bond? How many of these hybrid orbitals are on each carbon atom?

2.12 What type of orbitals overlap to form a pi bond in an alkene? What symbol is used to represent a pi bond? How many electrons are in a pi bond?

2.13 Describe the geometry of the carbon–carbon double bond and the two atoms attached to each of the double-bonded carbon atoms.

2.14 Explain the difference between geometric and structural isomers of alkenes.

2.15 Draw structural formulas and give IUPAC names for all the isomeric pentenes (C_5H_{10}) that are
a. Alkenes that do not show geometric isomerism. There are four compounds.
b. Alkenes that do show geometric isomerism. There is one *cis* and one *trans* compound.

2.16 Which of the following alkenes can exist as *cis-trans* isomers? Draw structural formulas and name the *cis* and *trans* isomers.
a. $CH_3CH_2CH_2CH=CHCH_2CH_2CH_3$

b. $CH_3CH=C-CH_2CH_3$
   $\quad\quad\quad\quad\ |$
   $\quad\quad\quad\quad\ CH_2CH_3$

c. $BrCH_2CH=CHCH_2Br$

2.17 Which of the following alkens can exist as *cis-trans* isomers? Draw structural formulas and name the *cis* and *trans* isomers.
a. $H_2C=CH-CH_3$

b. $\quad\quad Br$
   $\quad\quad\ |$
   $CH_3C=CHCH_3$

c. $\quad Cl$
   $\quad\ |$
   $HC=CHCH_3$

2.18 Draw structural formulas for the following:
a. *trans*-2-pentene
b. *cis*-2-hexene

2.19 Draw structural formulas for the following:
a. *trans*-3,4-dibromo-3-heptene
b. *cis*-1,4-dichloro-2-methyl-2-butene

PROPERTIES OF ALKENES (SECTION 2.3)

2.20 In what ways are the physical properties of alkenes similar to those of alkanes?

2.21 Which of the following reactions is an addition reaction?
a. $A_2 + C_3H_6 \rightarrow C_3H_6A_2$
b. $A_2 + C_6H_6 \rightarrow C_6H_5A + HA$
c. $HA + C_4H_8 \rightarrow C_4H_9A$
d. $3O_2 + C_2H_4 \rightarrow 2CO_2 + 2H_2O$
e. $C_7H_{16} \rightarrow C_7H_8 + 4H_2$

2.22 State Markovnikov's rule, and write a reaction that illustrates its application.

2.23 Complete the following reactions. Where more than one product is possible, show only the one expected according to Markovnikov's rule.

a.

b. $CH_3CH_2CH{=}CH_2 + Cl_2 \rightarrow$

c.

$$CH_3CH_2\overset{\overset{\displaystyle CH_3}{|}}{C}{=}CH_2 + H_2O \xrightarrow{\;H_2SO_4\;}$$

d.

$$CH_3\overset{\overset{\displaystyle CH_3}{|}}{C}{=}CHCH_2CH_3 + HBr \longrightarrow$$

2.24 Complete the following reactions. Where more than one product is possible, show only the one expected according to Markovnikov's rule.

a. $CH_2{=}CHCH_2CH_3 + H_2 \xrightarrow{\;Pt\;}$

b. $CH_2{=}CHCH_2CH_3 + Br_2 \longrightarrow$

c.

d.

2.25 Draw the structural formula for the alkenes with molecular formula C_5H_{10} that will react to give the following products. Show all correct structures if more than one starting material will react as shown.

a. $C_5H_{10} + Br_2 \longrightarrow CH_3CH_2\underset{\underset{\displaystyle Br}{|}}{C}H\underset{\underset{\displaystyle Br}{|}}{C}HCH_3$

b. $C_5H_{10} + H_2 \xrightarrow{Pt} CH_3CH_2CH_2CH_2CH_3$

c.

$$C_5H_{10} + H_2O \xrightarrow{\;H_2SO_4\;} CH_3\overset{\overset{\displaystyle CH_3}{|}}{\underset{\underset{\displaystyle OH}{|}}{C}}CH_2CH_3$$

d. $C_5H_{10} + HBr \longrightarrow CH_3\underset{\underset{\displaystyle Br}{|}}{C}HCH_2CH_2CH_3$

2.26 What reagents would you use to prepare each of the following from cyclohexene?

a.

b.

c.

d.

2.27 What is an important commercial application of hydrogenation?

2.28 Cyclohexane and 2-hexene both have the molecular formula C_6H_{12}. Describe a simple chemical test that would distinguish one from the other.

2.29 Terpin hydrate is used medicinally as an expectorant for coughs. It is prepared by the following addition reaction. What is the structure of terpin hydrate?

$$+\ 2H_2O \xrightarrow{\;H_2SO_4\;} \text{terpin hydrate}$$

ADDITION POLYMERS (SECTION 2.4)

2.30 Explain what is meant by each of the following terms: *monomer, polymer, addition polymer,* and *copolymer.*

2.31 A section of polypropylene containing three units of monomer can be shown as

$$-CH_2-\overset{\overset{\displaystyle CH_3}{|}}{C}H-CH_2-\overset{\overset{\displaystyle CH_3}{|}}{C}H-CH_2-\overset{\overset{\displaystyle CH_3}{|}}{C}H-$$

Draw structural formulas for comparable three-unit sections of
a. Teflon
b. Orlon
c. Lucite

2.32 Identify a structural feature characteristic of all monomers listed in Table 2.3.

2.33 Rubber cement contains a polymer of 2-methylpropene (isobutylene) called polyisobutylene. Write an equation for the polymerization reaction.

2.34 Much of today's plumbing in newly built homes is made from a plastic called poly (vinyl chloride), or PVC. Using Table 2.3, write a reaction for the formation of poly (vinyl chloride).

2.35 Identify a major use for each of the following addition polymers:
a. Styrofoam
b. Acrilan
c. Plexiglas
d. PVC
e. polypropylene

ALKYNES (SECTION 2.5)

2.36 What type of hybridized orbitals are present on carbon atoms bonded by a triple bond? How many of these hybrid orbitals are on each carbon atom?

2.37 How many sigma bonds and how many pi bonds make up the triple bond of an alkyne?

2.38 Describe the geometry in an alkyne of the carbon–carbon triple bond and the two attached atoms.

2.39 Explain why geometric isomerism is not possible in alkynes.

2.40 Give the common name and major uses of the simplest alkyne.

2.41 Write the structural formulas and IUPAC names for all the isomeric alkynes with the formula C_5H_8.

2.42 Describe the physical and chemical properties of alkynes.

AROMATIC COMPOUNDS AND THE BENZENE STRUCTURE (SECTION 2.6)

2.43 What type of hybridized orbitals are present on the carbon atoms of a benzene ring? How many sigma bonds are formed by each carbon atom in a benzene ring?

2.44 What type of orbitals overlap to form the pi bonding in a benzene ring?

2.45 What does the circle within the hexagon represent in the structural formula for benzene?

2.46 Define the terms *aromatic* and *aliphatic*.

2.47 Limonene, which is present in citrus peelings, has a very pleasant lemonlike fragrance. However, it is not classified as an aromatic compound. Explain.

limonene

2.48 A disubstituted cycloalkane such as (a) exhibits *cis-trans* isomerism, whereas a disubstituted benzene (b) does not. Explain.

a.

b.

THE NOMENCLATURE OF BENZENE DERIVATIVES (SECTION 2.7)

2.49 Give an IUPAC name for each of the following hydrocarbons as a derivative of benzene:

a.

b.

2.50 Give an IUPAC name for each of the following hydrocarbons as a derivative of benzene:

a.

b.

2.51 Give an IUPAC name for the following as hydrocarbons with the benzene ring as a group:

a. CH_3CHCH_3

b. CH_3
$CH_3CH_2CHCHCH_3$

2.52 Give the IUPAC name for the following as hydrocarbons with the benzene ring as a group:

$CH_2=C-CH_2CH_3$

a.

b.

2.53 Name the following compounds, using the prefixed abbreviations for *ortho*, *meta*, and *para* and assigning IUPAC-acceptable common names:

a.

b.

2.54 Name the following compounds, using prefixed abbreviations for *ortho*, *meta*, and *para* and assigning IUPAC-acceptable common names:

a.

b.

2.55 Name the following by numbering the benzene ring. IUPAC-acceptable common names may be used where appropriate:

a.

b.

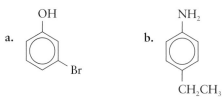

2.56 Name the following by numbering the benzene ring. IUPAC-acceptable common names may be used where appropriate:

a.

b.

2.57 Draw structural formulas for the following:
a. 2,4-diethylaniline

b. 4-ethyltoluene

c. *p*-ethyltoluene

2.58 Draw structural formulas for the following:
  a. *p*-propylphenol
  b. *o*-bromobenzoic acid
  c. 3-methyl-2-phenylbutane

2.59 There are three bromonitrobenzene derivatives. Draw their structures and give an IUPAC name for each one.

PROPERTIES AND USES OF AROMATIC COMPOUNDS (SECTION 2.8)

2.60 Describe the chief physical properties of aromatic hydrocarbons.

2.61 Why does benzene not readily undergo addition reactions characteristic of other unsaturated compounds?

2.62 Compare the chemical behavior of benzene and cyclo-hexene.

2.63 For each of the following uses, list an appropriate aromatic compound:
  a. A solvent
  b. A vitamin
  c. An essential amino acid
  d. Starting material for dyes

2.64 For each of the following uses, list an appropriate aromatic compound:
  a. Used in the production of Formica®
  b. A starting material for polystyrene
  c. Used to manufacture drugs
  d. A starting material for Bakelite®

CHEMISTRY AROUND US, OVER THE COUNTER, AND HOW REACTIONS OCCUR

2.65 What vitamin is formed within the body from beta-carotene?

2.66 Spinach, broccoli, cantaloupe, and pumpkins are dietary sources for what important nutrient?

2.67 What structure is the building block for all terpenes?

2.68 Why do terpene molecules contain a multiple of five carbon atoms?

2.69 What is the principal terpene found in orange oil?

2.70 What is a carbocation? How is it formed in the hydration of an alkene?

2.71 What disease may result from exposure to the sun's rays over a long period of time?

2.72 What is the active ingredient in patches, gum, and sprays designed to help break the smoking habit?

➤ POINTS TO PONDER AND PURSUE

2.1 Napthalene is the simplest polycyclic aromatic compound:

Draw a Kekulé structure for this compound like that shown for benzene in Section 2.6.

2.2 Why does propene not exhibit geometric isomerism?

2.3 Limonene is present in the rind of lemons and oranges. Based on its structure, would you consider it to be a solid, liquid, or gas at room temperature?

limonene

2.4 If the average molecular weight of polyethylene is 5.0×10^4 u, how many ethylene monomers ($CH_2{=}CH_2$) are contained in a molecule of the polymer?

2.5 Reactions to synthesize the benzene ring of aromatic compounds do not occur within the human body, and yet many essential body components involve the benzene structure. How does the human body get its supply of aromatic compounds?

2.6 Answer the question in the caption to Figure 2.8 pertaining to poly (vinyl acetate).

2.7 Why can't alkanes undergo addition polymerization?

2.8 According to the study at Web site 2.2, are beta-carotene supplements effective in the prevention of lung cancer?

2.9 According to Web site 2.3, what name is given to terpenes containing six isoprene units?

2.10 Using Web site 2.4, is UVA or UVB the main cause of deep wrinkling of the skin?

2.11 According to Web site 2.5, how many Americans quit smoking each year?

http://www.brookscole.com/chemistry/seager_chemistry

Alcohols, Phenols, and Ethers

Nurse anesthetists help administer anesthesia and closely monitor patients during surgery. In large hospitals, they usually work with a physician who specializes in anesthesiology (an anesthesiologist) during the procedure. However, in small hospitals, a nurse anesthetist might have the sole responsibility for anesthesia under the direction of the surgeon who is doing the surgery. General anesthetics is one of the special topics discussed in this chapter.

http://www.brookscole.com/chemistry/ seager_chemistry
Refer to Web site 3.1 for further information on a career as an anesthesiologist's assistant.

CHAPTER OUTLINE

➤ **FIGURE 3.1** A promising anticancer drug, taxol, is found in the bark of the Pacific yew tree. Though taxol is a complex compound, what does its name suggest about its structure?

Alcohol
A compound in which an —OH group is connected to an aliphatic carbon atom.

Hydroxy group
The —OH functional group.

Phenol
A compound in which an —OH group is connected to a benzene ring. The parent compound is also called phenol.

A lcohols, phenols, and ethers are important organic compounds that occur naturally and are produced synthetically in significant amounts. They are used in numerous industrial and pharmaceutical applications (see ■ Figure 3.1). Ethanol, for example, is one of the simplest and best known of all organic substances; the fermentation method for producing it was known and used by ancient civilizations. Menthol, a ten-carbon alcohol obtained from peppermint oil, is widely used as a flavoring agent. Cholesterol, with its complicated multi-ring molecular structure, has been implicated in some forms of heart disease. The chemistry of all of these substances is similar because the alcohol functional group behaves essentially the same, regardless of the complexity of the molecule in which it is found.

$$CH_3-CH_2-OH$$

ethanol   menthol   cholesterol

Structurally, an **alcohol** is obtained from a hydrocarbon by replacing a hydrogen atom with a **hydroxy group** (—OH). A formula for an alcohol can be generalized as R—OH, where R— represents CH_3—, CH_3CH_2—, or any other singly bonded carbon–hydrogen group. Thus, R—OH can stand for CH_3—OH, CH_3CH_2—OH, and so on. If the replaced hydrogen was attached to an aromatic ring, the resulting compound is known as a **phenol**:

R—OH

an alcohol   a phenol

Alcohols and phenols may also be considered to be derived from water by the replacement of one of its hydrogen atoms with an alkyl group or an aromatic ring:

$$H-OH \xrightarrow{\text{Replace H with R}} R-OH$$
water   an alcohol

$$H-OH \xrightarrow{\text{Replace H with}} \bigcirc-OH$$
water   a phenol

For most people, the word *ether* generates images of surgeons and operating rooms. Chemists define an **ether** as an organic compound in which two carbon atoms are bonded to an oxygen atom. This structure results when both hydrogen atoms of water are replaced by alkyl groups:

Ether
A compound that contains a

$$-\overset{|}{\underset{|}{C}}-O-\overset{|}{\underset{|}{C}}-$$

functional group.

$$H-O-H \xrightarrow[\text{H's with R}]{\text{Replace both}} R-O-R'$$
water   an ether

R′ indicates that the two R groups can be the same or different.

➤ 3.1 The Nomenclature of Alcohols and Phenols

The simpler alcohols are often known by common names, where the alkyl group name is followed by the word *alcohol*:

$$CH_3-OH \qquad CH_3CH_2-OH \qquad CH_3CH_2CH_2-OH \qquad CH_3CHCH_3$$

methyl alcohol   ethyl alcohol   propyl alcohol

$$\overset{\displaystyle |}{OH}$$

isopropyl alcohol

The IUPAC rules for naming alcohols that contain a single hydroxy group are as follows:

Step 1. Name the longest chain to which the hydroxy group is attached. The chain name is obtained by dropping the final *-e* from the name of the hydrocarbon that contains the same number of carbon atoms and adding the ending *-ol.*

Step 2. Number the longest chain to give the lower number to the carbon with the attached hydroxy group.

Step 3. Locate the position of the hydroxy group by the number of the carbon atom to which it is attached.

Step 4. Locate and name any other groups attached to the chain.

Step 5. Combine the name and location for other groups, the hydroxy group location, and the longest chain into the final name.

➤ Example 3.1

Name the following alcohols according to the IUPAC system:

a. CH_3-OH

b. $CH_3CH_2CH_2CHCH_2CH_3$
$$\qquad\qquad\qquad\quad |$$
$$\qquad\qquad\quad CH_2-OH$$

c.
$$\bigcirc - CH_2CHCH_3$$
$$\qquad\qquad\quad |$$
$$\qquad\qquad\quad OH$$

d.
$$HO \overset{\bigcirc}{} CH_3$$

Solution

a. The longest chain has one carbon atom; thus, the alcohol will be called methanol (methane − *e* + *ol*). Since the —OH group can be attached only to the single carbon atom, no location number for —OH is needed.

b. The longest chain containing the —OH group is numbered as follows:

$$\overset{5}{C}H_3\overset{4}{C}H_2\overset{3}{C}H_2\overset{2}{C}HCH_2CH_3$$
$$\qquad\qquad\qquad |$$
$$\qquad\qquad\quad \underset{1}{C}H_2-OH$$

Note that there is a longer chain (six carbon atoms), but the —OH group is not directly attached to it. The five-carbon chain makes the alcohol a pentanol. The —OH group is on carbon number 1; hence, the compound is a 1-pentanol. An ethyl group is attached to carbon number 2 of the chain, so the final name is 2-ethyl-1-pentanol.

c. The longest chain contains three carbon atoms, with the —OH group attached to carbon number 2. Hence, the compound is a 2-propanol. A benzene ring attached to carbon number 1 is named as a phenyl group. Therefore, the complete name is 1-phenyl-2-propanol. Note that the phenyl group is on carbon number 1 rather than number 3.

d. The —OH group is given preference, so the carbon to which it is attached is carbon number 1 of the ring. The —CH$_3$ group is located on carbon number 3 (since 3 is lower than 5, which would be obtained by counting the other way). The complete name is 3-methylcyclohexanol. Note that because the —OH group is always at the number 1 position in a ring, the position is not shown in the name.

Learning Check 3.1

Provide IUPAC names for the following alcohols:

a.
$$\begin{array}{c} CH_3 \\ | \\ CH_3CHCH_2-OH \end{array}$$

b.
$$\begin{array}{c} OH \\ | \\ CH_3CHCHCH_2CHCH_3 \\ |\qquad\quad | \\ CH_2CH_3\quad CH_3 \end{array}$$

c.

Compounds containing more than one hydroxy group are known. Alcohols containing two hydroxy groups are called *diols*. Those containing three —OH groups are called *triols*. The IUPAC nomenclature rules for these compounds are essentially the same as those for the single hydroxy alcohols, except that the ending *-diol* or *-triol* is attached to the name of the parent hydrocarbon.

➤ Example 3.2

Name the following alcohols according the IUPAC system:

a.
$$\begin{array}{c} CH_2-CH_2 \\ | \qquad | \\ OH \quad OH \end{array}$$

b.
$$\begin{array}{c} CH_3 \\ | \\ CH_2CHCH_2CH_2 \\ | \qquad\qquad | \\ OH \qquad\quad OH \end{array}$$

c.
$$\begin{array}{c} CH_2-CH-CH_2 \\ | \qquad | \qquad | \\ OH \quad OH \quad OH \end{array}$$

Solution

a. The longest chain has two carbons, making this an ethanediol. The —OH groups attached to carbons 1 and 2 give the final name 1,2-ethanediol. This substance is also called ethylene glycol.

b. Because the longest chain has four carbons, the compound is a butanediol. The —OH groups are attached to carbons 1 and 4, and a methyl group is attached to carbon 2. The name is 2-methyl-1,4-butanediol.

c. The longest chain contains three carbons, so the compound is a propanetriol. The —OH groups are attached to carbons 1, 2, and 3, making the name 1,2,3-propanetriol. Common names for this substance are glycerine and glycerol.

Learning Check 3.2

Give IUPAC names to the following diols:

OH   OH
| |
a. $CH_2CH_2CH_2CH_2$ b.

Substituted phenols are usually named as derivatives of the parent compound phenol:

phenol   4-bromophenol   2,4,6-tribromophenol

Learning Check 3.3

Name this compound as a phenol:

CH_2CH_3

OH

➤ 3.2 Classification of Alcohols

The characteristic chemistry of an alcohol sometimes depends on the groups bonded to the carbon atom that bears the hydroxy group. Alcohols are classified as primary, secondary, or tertiary on the basis of these attached groups (see ■ Table 3.1). In **primary alcohols,** the hydroxy-bearing carbon atom is attached to one other carbon atom and two hydrogen atoms. The simplest alcohol, CH_3—OH, is also considered a primary alcohol. In **secondary alcohols,** the hydroxy-bearing carbon atom is attached to two other carbon atoms and one hydrogen atom. In **tertiary alcohols,** the hydroxy-bearing carbon atom is attached to three other carbon atoms. The use of R, R′, and R″ in Table 3.1 signifies that each of those alkyl groups may be different.

Primary alcohol
An alcohol in which the —OH group is attached to CH_3— or to a carbon attached to one other carbon atom.

Secondary alcohol
An alcohol in which the carbon bearing the —OH group is attached to two other carbon atoms.

Tertiary alcohol
An alcohol in which the carbon bearing the —OH group is attached to three other carbon atoms.

| TABLE 3.1 | Primary, secondary, and tertiary alcohols |

	Primary	Secondary	Tertiary
General formula	$R-CH_2-OH$	$R-\underset{\underset{R'}{\mid}}{CH}-OH$	$R'-\underset{\underset{R''}{\mid}}{\overset{\overset{R}{\mid}}{C}}-OH$
Specific example	$CH_3CH_2CH_2-OH$ 1-propanol (propyl alcohol)	$\cdot \ CH_3\underset{\underset{CH_3}{\mid}}{CH}-OH$ 2-propanol (isopropyl alcohol)	$CH_3-\underset{\underset{CH_3}{\mid}}{\overset{\overset{CH_3}{\mid}}{C}}-OH$ 2-methyl-2-propanol (*t*-butyl alcohol)

Learning Check 3.4

Classify the following alcohols as primary, secondary, or tertiary:

a. $CH_3CH_2CH_2CH_2-OH$

b. OH

c. OH
$CH_3\underset{\underset{CH_2CH_3}{\mid}}{CH}$

➤ 3.3 Physical Properties of Alcohols

The replacement of one hydrogen of water with an organic group does not cause all the waterlike properties to disappear. Thus, the lower molecular weight alcohols—methyl, ethyl, propyl, and isopropyl alcohols—are completely miscible with water. As the size of the alkyl group in an alcohol increases, the physical properties become less waterlike and more alkanelike. Long-chain alcohols are less soluble in water (see ■ Figure 3.2) and more soluble in nonpolar solvents such as benzene, carbon tetrachloride, and ether. The solubility of alcohols in water depends on the number of carbon atoms per hydroxy group in the molecule. In general, one hydroxy group can solubilize three to four carbon atoms. The high solubility of the lower molecular weight alcohols in water can be attributed to hydrogen bonding between the alcohol and water molecules (see ■ Figure 3.3).

In long-chain alcohols, the hydrophobic alkyl group does not form hydrogen bonds with water. Thus, in 1-heptanol, water molecules surround and form bonds with the —OH end but do not bond to the alkyl portion (see ■ Figure 3.4). Because the alkyl portion is unsolvated, 1-heptanol is insoluble in water. Its solubility in water is comparable to that of heptane (seven carbon atoms), as shown in Figure 3.2.

Learning Check 3.5

Arrange the following compounds in order of increasing solubility in water (least soluble first, most soluble last):

a. $CH_3CH_2CH_2-OH$ b. $CH_3CH_2CH_3$ c. $CH_3(CH_2)_3CH_2-OH$

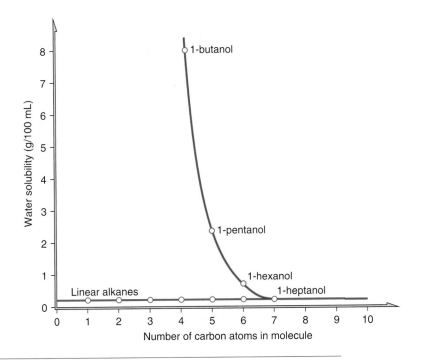

➤ **FIGURE 3.2** The solubility of alcohols and linear alkanes in water.

Hydrogen bonding also causes alcohols to have much higher boiling points than most other compounds of similar molecular weight. In this case, the hydrogen bonding is between alcohol molecules (see ■ Figure 3.5). Ethanol (CH_3CH_2OH) boils at 78°C, whereas methyl ether (CH_3—O—CH_3) (with the same molecular weight) boils at −24°C. Propane ($CH_3CH_2CH_3$) has nearly the same molecular weight but boils at −42°C (see ■ Figure 3.6).

How can you explain the observation that methanol has a boiling point of 65°C, whereas propane with a higher molecular weight has a boiling point of −42°C?

Learning Check 3.6

➤ **FIGURE 3.3** Hydrogen bonding in a water–methanol solution. A three dimensional network of molecules and hydrogen bonds is formed.

➤ **FIGURE 3.4** Water interacts only with the —OH group of 1-heptanol.

➤ **FIGURE 3.5** Hydrogen bonding in pure ethanol.

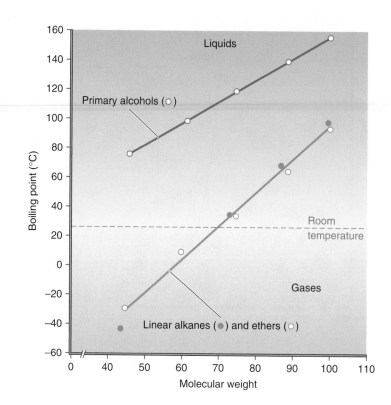

➤ **FIGURE 3.6** The boiling points of alcohols are higher than those of alkanes and ethers of similar molecular weights.

Dehydration reaction
A reaction in which water is chemically removed from a compound.

Elimination reaction
A reaction in which two or more covalent bonds are broken and a new multiple bond is formed.

➤ 3.4 Reactions of Alcohols

Alcohols undergo many reactions, but we will consider only two important ones at this time—alcohol dehydration and alcohol oxidation. In alcohol **dehydration**, water is chemically removed from an alcohol. This reaction can occur in two different ways, depending on the reaction temperature. At 140°C, the main product is an ether, whereas at higher temperatures (180°C), alkenes are the predominant products.

Dehydration to Produce an Alkene

When the elements of a water molecule are removed from a single alcohol molecule, an alkene is produced. A reaction of this type, in which two or more covalent bonds are broken and a new multiple bond is formed, is called an **elimination reaction**. In this case, a water molecule is eliminated from an alcohol. Reaction 3.1 gives a general reaction, and Reactions 3.2 and 3.3 are two specific examples:

$$\text{General reaction:} \quad -\underset{\underset{\text{H}}{|}}{\overset{|}{\text{C}}} - \underset{\underset{\text{OH}}{|}}{\overset{|}{\text{C}}} - \xrightarrow[180°\text{C}]{\text{H}_2\text{SO}_4} \quad \underset{/}{\overset{\backslash}{\text{C}}} = \underset{\backslash}{\overset{/}{\text{C}}} + \text{H}_2\text{O} \qquad (3.1)$$

$$\quad\quad\quad\quad\quad\quad \text{alcohol} \quad\quad\quad\quad\quad \text{alkene}$$

$$\text{Specific example:} \quad \underset{\underset{\text{OH}}{|}}{\text{CH}_3\text{CH}} - \underset{\underset{\text{H}}{|}}{\text{CH}_2} \xrightarrow[180°\text{C}]{\text{H}_2\text{SO}_4} \text{CH}_3\text{CH} = \text{CH}_2 + \text{H}_2\text{O} \qquad (3.2)$$

$$\quad\quad\quad\quad \text{2-propanol} \quad\quad\quad\quad\quad\quad \text{propene}$$

Specific example: $CH_3CH_2CHCH_3 \xrightarrow[180°C]{H_2SO_4} CH_3CH=CHCH_3 + CH_3CH_2CH=CH_2 + H_2O$ (3.3)

                |
               OH

2-butene   1-butene

Principal product (90%)   Minor product (10%)

(two carbon groups   (one carbon group

on double bond)   on double bond)

Notice that in Reaction 3.3 two alkenes can be produced, depending on which carbon next to the hydroxy-bearing carbon loses the hydrogen atom. The principal alkene produced in such cases will be the one in which the higher number of carbon groups is bonded to the double-bonded carbon atoms.

Alcohol dehydration is an important reaction in the human body, where enzymes (rather than sulfuric acid) function as catalysts. For example, the dehydration of citrate is part of the citric acid cycle discussed in Section 13.5. Note that the dehydration of citrate, 2-propanol (Reaction 3.2), and 2-butanol (Reaction 3.3) all involve exactly the same changes in bonding. In each reaction, an H and an —OH are removed from adjacent carbons, leaving a C=C double bond. The other molecular bonds in all three of those compounds do not change:

citrate   *cis*-aconitate   (3.4)

Predict the major products of the following reactions:

Learning Check 3.7

a.
          OH

           |

$CH_3CH_2CHCH_2CH_3 \xrightarrow[180°C]{H_2SO_4}$

b.

HOW REACTIONS OCCUR

THE DEHYDRATION OF AN ALCOHOL

A strong acid such as sulfuric acid initially reacts with an alcohol in much the same way that it reacts with water in solution. Ionization of the acid produces a proton (H$^+$) that is attracted to an oxygen atom of the alcohol:

protonated alcohol

The creation of a positive charge on the oxygen atom weakens the carbon–oxygen bond, which allows the following equilibrium reaction to take place:

a carbocation   water

(one product)

The ethyl carbocation is unstable because the carbon has only three bonds (six outer electrons) around it. To achieve an octet of electrons about this carbon, the carbocation donates an H$^+$ to a proton acceptor, such as HSO$_4^-$, and forms the double bond of the alkene product:

ethene   regenerated

catalyst

Dehydration to Produce an Ether

Alcohol dehydration can also occur when the elements of water are removed from two alcohol molecules. The resulting alcohol molecule fragments bond and form an ether. This ether-forming reaction is useful mainly with primary alcohols. Reaction 3.5 gives the general reaction, and the formation of diethyl ether (the anesthetic ether) is a specific example (Reaction 3.6):

General reaction:

$$R-O-H + H-O-R \xrightarrow[140°C]{H_2SO_4} R-O-R + H_2O \qquad (3.5)$$

an alcohol   an alcohol   an ether

Specific example:

$$CH_3CH_2-OH + HO-CH_2CH_3 \xrightarrow[140°C]{H_2SO_4} CH_3CH_2-O-CH_2CH_3 + H_2O \qquad (3.6)$$

ethanol   ethanol   diethyl ether
(ethyl alcohol)   (ethyl alcohol)

Thus, an alcohol can be dehydrated to produce either an alkene or an ether. The type of dehydration that takes place is controlled by the temperature, with sulfuric acid serving as a catalyst. Although a mixture of alkene and ether products often results, we will use these temperatures to indicate that the major dehydration product is an alkene (at 180°C) or an ether (at 140°C).

The reaction that produces ethers is an example of a dehydration synthesis in which two molecules are joined and a new functional group is generated by the removal of a water molecule. Dehydration synthesis is important in living organisms because it is part of the formation of carbohydrates, fats, proteins, and many other essential substances.

Learning Check 3.8

What catalyst and reaction temperature would you use to accomplish each of the following reactions?

a. $2CH_3CH_2CH_2-OH \rightarrow CH_3CH_2CH_2-O-CH_2CH_2CH_3 + H_2O$

b. $CH_3CH-OH \rightarrow CH_3CH=CH_2 + H_2O$
    $\quad\quad |$
    $\quad\; CH_3$

Oxidation

An oxidation reaction occurs when a molecule gains oxygen atoms or loses hydrogen atoms. Under appropriate conditions, alcohols can be oxidized in a controlled way (not combusted) by removing two hydrogen atoms per molecule. A number of oxidizing agents can be used, including potassium dichromate ($K_2Cr_2O_7$) and potassium permanganate ($KMnO_4$). The symbol (O) is used to represent an oxidizing agent. Reaction 3.7 gives the general reaction:

$$\begin{array}{ccc} | & & O \\ | & & \| \\ -C-O-H + (O) \longrightarrow & -C- & + H_2O \\ | & & \\ H & & \end{array} \qquad (3.7)$$

alcohol   aldehyde
                                     or ketone

The three classes of alcohols behave differently toward oxidizing agents.

Primary Alcohols. Reaction 3.8 gives the general reaction, and the formation of acetic acid is a specific example (Reaction 3.9):

General
reaction:

$$\underset{\substack{\text{primary}\\\text{alcohol}}}{R - \overset{\displaystyle OH}{\underset{\displaystyle H}{\overset{|}{\underset{|}{C}}}} - H} + (O) \longrightarrow \underset{\text{aldehyde}}{R - \overset{\displaystyle O}{\overset{\|}{C}} - H} + H_2O \qquad (3.8)$$

$$\xrightarrow[\;(O)\;]{\substack{\text{further}\\\text{oxidation}}} \underset{\substack{\text{carboxylic}\\\text{acid}}}{R - \overset{\displaystyle O}{\overset{\|}{C}} - OH}$$

Specific
example:

$$\underset{\text{ethanol}}{CH_3CH_2 - OH} + (O) \longrightarrow \underset{\text{acetaldehyde}}{CH_3 - \overset{\displaystyle O}{\overset{\|}{C}} - H} + H_2O \qquad (3.9)$$

$$\xrightarrow[\;(O)\;]{\substack{\text{further}\\\text{oxidation}}} \underset{\text{acetic acid}}{CH_3 - \overset{\displaystyle O}{\overset{\|}{C}} - OH}$$

The immediate product of the oxidation of a primary alcohol is an aldehyde. However, aldehydes are readily oxidized by the same oxidizing agents (Section 4.3) to give carboxylic acids. Therefore, the oxidation of a primary alcohol normally results in the corresponding carboxylic acid as the product. The oxidation of ethanol by $K_2Cr_2O_7$ is shown in ■ Figure 3.7. Because aldehydes do not hydrogen-bond (Section 4.2), they have lower boiling points than the corresponding alcohol or carboxylic acid. This makes it possible to isolate the aldehyde product before it is oxidized by maintaining the reaction temperature high enough to boil the aldehyde out of the reaction mixture before it can react.

Draw the structural formulas of the first and second products of the following reaction:

$$\text{\raisebox{-1em}{\includegraphics{benzene}}} \overset{\displaystyle CH_2 - OH}{} + (O) \longrightarrow$$

The tube on the left contains orange $K_2Cr_2O_7$ and is next to the colorless ethanol.

After mixing, the ethanol is oxidized, and chromium is reduced, forming a grayish-green precipitate.

➤ **FIGURE 3.7** Oxidation of ethanol by $K_2Cr_2O_7$. How might a color change like this one be used to measure alcohol concentration?

Learning Check 3.9

Secondary Alcohols. Reactions 3.10 and 3.11 give the general reaction and a specific example.

General
reaction:

$$\underset{\substack{\text{secondary}\\\text{alcohol}}}{R - \overset{\displaystyle OH}{\underset{\displaystyle H}{\overset{|}{\underset{|}{C}}}} - R'} + (O) \longrightarrow \underset{\text{ketone}}{R - \overset{\displaystyle O}{\overset{\|}{C}} - R'} + H_2O \qquad (3.10)$$

Specific
example:

$$\underset{\text{2-propanol}}{CH_3 - \overset{\displaystyle OH}{\underset{\displaystyle H}{\overset{|}{\underset{|}{C}}}} - CH_3} + (O) \longrightarrow \underset{\text{acetone}}{CH_3 - \overset{\displaystyle O}{\overset{\|}{C}} - CH_3} + H_2O \quad (3.11)$$

Secondary alcohols are oxidized to ketones in exactly the same way primary alcohols are oxidized to aldehydes, and by the same oxidizing agents. However, unlike aldehydes, ketones resist further oxidation, so this reaction provides an excellent way to prepare ketones.

Learning Check 3.10

Complete the following oxidation reaction:

Tertiary Alcohols. A general reaction is given in Reaction 3.12:

$$\text{General reaction:} \quad R - \underset{\underset{R''}{|}}{\overset{\overset{OH}{|}}{C}} - R' + (O) \longrightarrow \text{no reaction} \qquad (3.12)$$

Tertiary alcohols do not have any hydrogen on the —OH-bearing carbon and thus do not react with oxidizing agents.

Learning Check 3.11

Which of the following alcohols will react with an oxidizing agent?

Multistep Reactions

Alcohols are important organic compounds not only for their own commercial applications but because they can be converted by chemical reactions into a wide variety of other useful products. Most alcohols do not occur naturally in commercial quantities and therefore are prepared from alkenes.

It is very common in both laboratory and industrial processes for such preparations to involve a sequence of reactions. For example, the commercial production of vinyl plastic (PVC) begins with the alkene ethene:

Nature, too, employs multistep reactions to carry out processes essential to life. The glycolysis pathway used by the body to initiate the utilization of glucose for energy production involves a series of ten reactions (Section 13.3).

➤ **Example 3.3**

The following conversion requires more than one step. Show the reaction, reagents, and intermediate structures necessary to carry out the synthesis.

Solution

The best way to initially solve a multistep synthesis is to work backward from the final product. Identify the functional group present in the product—in this case, a ketone. Think how to prepare a ketone—by oxidizing an alcohol.

$$\text{(cyclopentanol)} + \text{(O)} \longrightarrow \text{(cyclopentanone)} + H_2O$$

Now, working backward again, think how to prepare an alcohol—from an alkene through hydration.

$$\text{(cyclopentene)} + H_2O \xrightarrow{H_2SO_4} \text{(cyclopentanol)}$$

Adding these two steps together gives

$$\text{(cyclopentene)} + H_2O \xrightarrow{H_2SO_4} \text{(cyclopentanol)} \xrightarrow{(O)} \text{(cyclopentanone)} + H_2O$$

Show the reactions necessary to carry out the following conversion.

$$CH_2{=}CH_2 \rightarrow CH_3CH_2{-}O{-}CH_2CH_3$$

Learning Check 3.12

► **STUDY SKILLS 3.1**

A REACTION MAP FOR ALCOHOLS

To solve a test question, using a stepwise approach is often helpful. First, decide what type of question is being asked. Is it a nomenclature, typical uses, physical properties, or a reaction question? Second, if you recognize it as, say, a reaction question, look for the functional group. Third, if it's an alcohol group in the starting material, identify the reagent and use the following diagram to predict the right products.

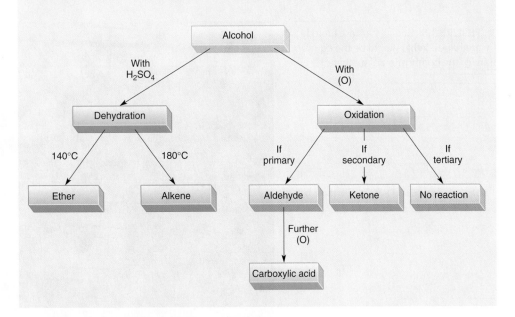

➤ 3.5 Important Alcohols

The simplest alcohol, methanol or methyl alcohol, is a very important industrial chemical; more than 1 billion gallons are produced and used annually. It is sometimes known as wood alcohol because the principal source for many years was the distillation of wood. Today, it is synthetically produced in large quantities by reacting hydrogen gas with carbon monoxide (Reaction 3.13):

$$CO + 2H_2 \xrightarrow[\text{heat, pressure}]{\text{Catalysts,}} CH_3 - OH \qquad (3.13)$$

The industrial importance of methanol is due to its oxidation product, formaldehyde, which is a major starting material for the production of plastics. Methanol is used as a fuel in Indy-style racing cars (see ■ Figure 3.8). It is also used as a fuel in products such as canned heat.

Methanol is highly toxic and causes permanent blindness if taken internally. Deaths and injuries have resulted as a consequence of mistakenly substituting methanol for ethyl alcohol in beverages.

Ethanol, or ethyl alcohol, $CH_3CH_2 - OH$, is probably the most familiar alcohol because it is the active ingredient in alcoholic beverages. It is also used in pharmaceuticals as a solvent (tinctures are ethanol solutions) and in aftershave lotions as an antiseptic and skin softener (see ■ Figure 3.9). It is an important industrial solvent as well. The gasohol now marketed at many locations around the United States is a mixture of ethanol and gasoline.

Most ethanol used industrially is produced on a large scale by the direct hydration of ethylene obtained from petroleum (Reaction 3.14):

$$H_2C = CH_2 + H - OH \xrightarrow[300°C]{70 \text{ atm}} \begin{array}{c} H_2C - CH_2 \\ | \quad\;\; | \\ H \;\;\; OH \end{array} \qquad (3.14)$$

ethylene   ethanol

Ethanol for beverages is produced by the yeast **fermentation** of carbohydrates such as sugars, starch, and cellulose. For example, a common method is the fermentation of glucose, grape sugar:

http://www.brookscole.com/chemistry/ seager_chemistry
Web site 3.2 provides more information on the use of alcohols and ethers in gasoline.

Fermentation
A reaction of sugars, starch, or cellulose to produce ethanol and carbon dioxide.

http://www.brookscole.com/chemistry/ seager_chemistry
Web site 3.3 provides more information on the effects of alcohol consumption.

➤ **FIGURE 3.8** Racing cars are fueled by methanol. What would be the equation for the combustion of methanol?

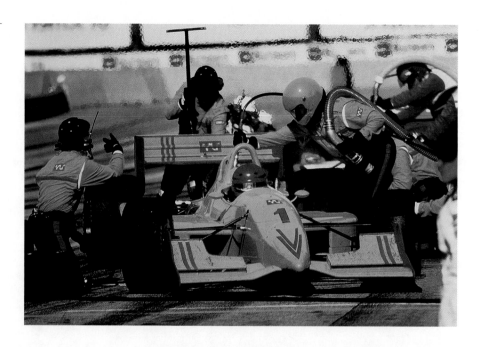

$$C_6H_{12}O_6 \xrightarrow{\text{Yeast}} 2CH_3CH_2-OH + 2CO_2$$

glucose   ethanol

2-propanol (isopropyl alcohol), the main component of rubbing alcohol, acts as an astringent on the skin. It causes tissues to contract, hardens the skin, and limits secretions. Its rapid evaporation from the skin also lowers the skin temperature. It is used as an antiseptic in 70% solutions (see ■ Figure 3.10). Isopropyl alcohol is toxic and should not be taken internally.

1,2,3-propanetriol is known more commonly as glycerol or glycerin:

$$\underset{\text{OH}}{\text{CH}_2}-\underset{\text{OH}}{\text{CH}}-\underset{\text{OH}}{\text{CH}_2}$$

1,2,3-propanetriol (glycerol)

The hydrogen bonding resulting from the three hydroxy groups causes glycerol to be a syrupy, high-boiling liquid with an affinity for water. This property and its nontoxic nature make it valuable as a moistening agent in tobacco and many food products, such as candy and shredded coconut. Florists use glycerol to retain water that maintains the freshness of cut flowers. Glycerol is also used for its soothing qualities in shaving and toilet soaps and in many cosmetics. In Chapter 8 on lipids, we will see that glycerol forms part of the structure of body fats and oils.

Due to hydrogen bonding, 1,2-ethanediol (ethylene glycol) and 1,2-propanediol (propylene glycol) are both high-boiling liquids that are completely miscible with water.

$$\underset{\text{OH}}{\text{CH}_2}-\underset{\text{OH}}{\text{CH}_2}$$

1,2-ethanediol
(ethylene glycol)

$$\underset{\text{OH}}{\text{CH}_2}-\underset{\text{OH}}{\text{CH}}-\text{CH}_3$$

1,2-propanediol
(propylene glycol)

> ▶ **FIGURE 3.9** Many aftershave lotions are 50% ethanol.

> ▶ **FIGURE 3.10** Isopropyl alcohol is used as an antiseptic.

CHEMISTRY AROUND US • 3.1

FETAL ALCOHOL SYNDROME

In adults, alcohol (ethanol) is metabolized by two biochemical steps. First, alcohol is oxidized to ethanal in the liver. Second, it is further oxidized in the liver to acetyl coenzyme A (discussed in Section 12.6). After the acetyl coenzyme A is released from the liver, it can either be used in the body to synthesize fat or eventually be oxidized to water and carbon dioxide.

$$CH_3CH_2-OH \xrightarrow{\text{oxidation}} CH_3\overset{\overset{\displaystyle O}{\|}}{C}-H \xrightarrow{\text{oxidation}} CH_3\overset{\overset{\displaystyle O}{\|}}{C}-SCoA$$

ethanol   ethanal   acetyl coenzyme A
(ethyl alcohol)   (acetaldehyde)

Unfortunately, the liver of a fetus is different than that of an adult. It does not contain the enzymes that are necessary to metabolize alcohol. When alcohol is ingested by a pregnant woman, it efficiently passes through the placental membrane and enters the fetus, where it can cause a variety of changes that appear as deformities and abnormalities in the newborn infant. Infants exhibiting these characteristics are said to suffer from *fetal alcohol syndrome*. Some common problems associated with this syndrome are mental retardation, poor growth, and/or facial deformities, such as tissue folds between the nose and lip, short eyelid tissues, a thin upper lip, and a flattened face with a short nasal bridge.

The exact amount of alcohol that causes fetal alcohol syndrome is not known, but even "social drinking" during pregnancy can have adverse effects and is discouraged. Short of full-blown fetal alcohol syndrome, some milder effects that result from fetal exposure to alcohol include substandard birth weight, slow growth and development, learning disabilities, and behavioral problems. Therefore, it is highly recommended that pregnant women completely abstain from drinking alcoholic beverages. The consequences for their infants can be severe, and there is no cure for fetal alcohol syndrome.

http://www.brookscole.com/chemistry/seager_chemistry   An overview of fetal alcohol syndrome is given at Web site 3.4.

Their main uses are as automobile antifreeze (see ■ Figure 3.11) and as a starting material for the manufacture of polyester fibers (Reaction 5.11).

Alcohols are very important biologically because the alcohol functional group occurs in a variety of compounds associated with biological systems. For example, sugars contain several hydroxy groups, and starch and cellulose contain thousands of hydroxy groups (see Chapter 7). The structures and uses (see Figure 3.11) of some common alcohols are summarized in ■ Table 3.2.

Learning Check 3.13

What is an important use for each of the following alcohols?

a. methanol   b. ethanol   c. 2-propanol
d. ethylene glycol   e. glycerol   f. menthol

➤ **FIGURE 3.11** Ethylene glycol (HO—CH_2CH_2—OH) is used as an automotive antifreeze.

➤ 3.6 Characteristics and Uses of Phenols

Phenol, C_6H_5OH, is a colorless, low-melting-point solid with a medicinal odor. The addition of a small amount of water causes the solid to liquefy at room temperature. This liquid mixture is sometimes called carbolic acid. Unlike the alcohols, phenol is a weak acid (Reactions 3.15 and 3.16), which can chemically burn the skin, and it must be handled with care.

$$\text{phenol} + H_2O \longrightarrow \text{(phenolate)} + H_3O^+ \qquad (3.15)$$

phenol
(a weak acid)

TABLE 3.2   Examples of alcohols

Name	Structural formula	Typical uses
methanol (methyl alcohol)	CH_3—OH	Solvent, making formaldehyde
ethanol (ethyl alcohol	CH_3CH_2—OH	Solvent, alcoholic beverages

2-propanol (isopropyl alcohol)	CH_3CHCH_3
	    |
	   OH
		Rubbing alcohol, solvent

1-butanol (butanol)	$CH_3CH_2CH_2CH_2$—OH	Solvent, hydraulic fluid
1,2-ethanediol (ethylene glycol)	HO—CH_2CH_2—OH	Automobile antifreeze, polyester fibers

1,2-propanediol (propylene glycol)	CH_3CH—CH_2
	   | |
	  OH OH
		Moisturizer in lotions and foods, automobile antifreeze

1,2,3-propanetriol (glycerin, glycerol)	CH_2—CH—CH_2 | | | OH OH OH	Moisturizer in foods, tobacco, and cosmetics
menthol	CH_3CHCH_3 OH	Cough drops, shaving lotion, mentholated tobacco

$$\text{(phenol)} + KOH \longrightarrow \text{(potassium phenoxide)} + H_2O \qquad (3.16)$$

In dilute solutions, phenol is used as an antiseptic and disinfectant. Joseph Lister, an English surgeon, introduced the use of phenol as a hospital antiseptic in the late 1800s. Before that time, antiseptics had not been used, and very few patients survived even minor surgery because of postoperative infections. Today, phenol has been largely replaced in this use by more effective derivatives that are less irritating to the skin. Two such compounds, 4-chloro-3,5-dimethylphenol and 4-hexylresorcinol, are shown at the top of the following page.

OVER THE COUNTER • 3.1

OUTSMARTING POISON IVY

Poison ivy is a woody, ropelike vine or bush recognizable by its three leaflets, which are green in the summer and red in the fall. The American Academy of Dermatology estimates that about 85% of those exposed to the plant will suffer some type of allergic reaction.

An allergic reaction usually takes the form of a rash, blisters, and itching. The culprit that causes the reaction is urushiol, an odorless oil with a phenolic component that is present in the sap of the plant. The urushiol-containing sap is released when the plant is damaged. The poison ivy plant is very fragile and can easily be damaged by passing animals, wind, and chewing insects. Urushiol that gets onto other objects can remain potent for years, depending on the environment. In a warm, moist environment, it can cause a reaction for up to a year. In a dry environment, it may remain active for decades.

The best way to avoid an adverse reaction is to avoid contact with the active oil by wearing long sleeves, long pants, gloves, boots, and a hat when moving through areas where exposure is possible. A new over-the-counter product called bentoquatam (Ivy Block®) also provides protection. This lotion is applied to the skin at least 15 minutes before any anticipated exposure to poison ivy. The lotion forms a visible claylike barrier against urushiol. However, it must be applied every 4 hours for continued protection.

If exposure to the plant occurs despite precautions, the following steps will help reduce the chances for an adverse reaction:

1. Cleanse the exposed skin with generous amounts of rubbing alcohol, and then wash with cool water. This should be done outdoors, and alcohol should not be used if a return to the site of the exposure the same day is likely. The alcohol removes the skin's protective oils along with the urushiol, and new contact will allow urushiol to penetrate the skin more rapidly.
2. Take a regular shower with soap and warm water.
3. With gloves on, wipe any clothes, tools, shoes, etc. that might have contacted the plant with alcohol and water.

If the skin has been exposed to urushiol and not cleansed quickly enough, redness and swelling will appear in 12–48 hours, often followed by itching and blisters. Some people also develop a rash within 7–10 days. The blisters, rash, and itch will normally disappear in 14–20 days without any treatment, but most sufferers want relief from the itch during that time. A number of OTC products have been approved by the FDA for the temporary relief of itching caused by poison ivy. The products are topical corticosteroids (commonly called hydrocortisones), with brand names such as Lanacort® and Cortaid®. For severe reactions, a doctor should be consulted. He or she may prescribe additional topical corticosteroid drugs or even oral corticosteroids.

Several OTC products can also be used to dry up the oozing blisters. These products include aluminum acetate, calamine lotion, zinc oxide, zinc carbonate, zinc acetate, kaolin, aluminum hydroxide gel, baking soda, and even an oatmeal bath.

Ivy Block® contains the compound bentoquatam which protects the skin from poison ivy.

OH

OH

OH

phenol
(Lister's original
disinfectant)

4-chloro-3,5-dimethylphenol
(nonirritating topical
antiseptic)

4-hexylresorcinol
(used in mouthwashes
and throat lozenges)

Two other derivatives of phenol, o-phenylphenol and 2-benzyl-4-chlorophenol are ingredients in Lysol®, a disinfectant for walls and furniture in hospitals and homes:

OH

OH

o-phenylphenol

2-benzyl-4-chlorophenol

Antioxidant

A substance that prevents another substance from being oxidized.

Other important phenols act as **antioxidants** by interfering with oxidizing reactions. These phenols are useful in protecting foods from spoilage and a variety of other materials from unwanted reactions. BHA and BHT are widely used as antioxidants in gasoline, lubricating oils, rubber, certain foods, and packaging materials for foods that might turn rancid (see ■ Figure 3.12). In food or containers used to package food, the amount of these antioxidants is limited to 200 ppm (parts per million) or 0.02%, based on the fat or oil content of the food. Notice that industrial nomenclature for these compounds does not follow IUPAC rules.

BHA (butylated hydroxy anisole)
2-t-butyl-4-methoxyphenol
(IUPAC name)

BHT (butylated hydroxy toluene)
2, 6-di-t-butyl-4-methylphenol
(IUPAC name)

Learning Check 3.14

What is an important use for each of the following phenols?

a. BHA   b. 4-hexylresorcinol

➤ 3.7 Ethers

Ethers were introduced at the start of this chapter as compounds whose formula is R—O—R′. Common names for ethers are obtained by first naming the two groups attached to the oxygen and then adding the word *ether*. If the

➤ **FIGURE 3.12** BHT is listed as a food ingredient that will "maintain freshness."

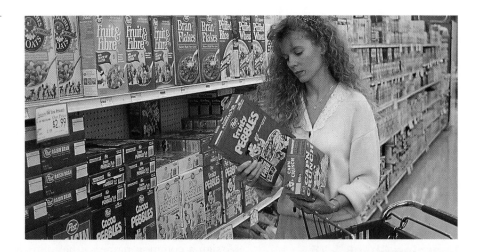

CHEMISTRY AROUND US • 3.2

VITAMIN E: AS GOOD AS IT SEEMS?

Vitamin E is a phenolic compound that is widespread in both plants and animals. Because of its antioxidant properties, it is commonly used in vegetable oils to prevent the onset of rancidity, which is thought to involve some oxidation reactions.

vitamin E

The antioxidant role of vitamin E in the human body is under extensive study. Most of the research is based on the theory that antioxidant nutrients help prevent cellular damage that leads to degenerative diseases. Recent studies suggest that vitamin E taken as a dietary supplement provides some protection against cardiovascular disease, reduces the risk for some cancers, slows the progress of Alzheimer's disease, provides some protection against Parkinson's disease, and improves the immune function of the body in elderly people.

Vitamin E is believed to be one of the body's primary defenses against oxidation, and a number of specific roles have been identified. It protects lipids and other vulnerable components of cells, including cell membranes, from oxidation damage. Red and white blood cells are important cells protected in this way. Vitamin E is especially effective in preventing the oxidation of polyunsaturated fatty acids. Protection against oxidation is important in the lungs, where the exposure of cells to oxygen is maximal. In addition to protecting against oxygen, vitamin E also protects the lungs from oxidizing air pollutants such as nitrogen dioxide and ozone.

Unlike other vitamins touted as antioxidants such as beta-carotene and vitamin C, vitamin E cannot be easily obtained from food sources. Natural vitamin E is present in fats, nuts, and oils, but in order to get a pharmacological amount (400 IU), it would be necessary to consume very large amounts of fatty foods, a practice considered to be counterproductive to good health. Despite this relative unavailability of vitamin E in the diet, many doctors still warn that proper nutrition and exercise are much more beneficial to good health than vitamin E supplements. Also, long-term studies on vitamin E have not been completed, and similar studies on the health effects of beta-carotene and vitamin C have had very mixed results. Therefore, caution should be exercised about taking vitamin E supplements until the results are in.

The research done thus far concerning vitamin E supplements has produced definite results in some areas, as mentioned earlier. It has also shown some things that vitamin E supplements do not do. The research has clearly discredited claims that vitamin E enhances sexual or physical performance, and vitamin E supplements have shown no ability to prevent skin from wrinkling or hair from turning gray with age.

Vitamin E helps protect the lungs from air pollutants, especially during periods of rapid breathing.

http://www.brookscole.com/chemistry/seager_chemistry
At Web site 3.5, you will find more information on vitamin E.

groups are the same, it is appropriate to use the prefix *di-*, as in diethyl ether, but many people omit the *di-* and simply say "ethyl ether."

$$CH_3—O—CH_2CH_3$$
ethyl methyl ether

isopropyl phenyl ether

$$CH_3—O—CH_3$$
dimethyl ether
(methyl ether)

$$CH_3CH_2—O—CH_2CH_3$$
diethyl ether
(ethyl ether)

Learning Check 3.15

Assign a common name to the following:

a. $CH_3CH_2—O$—⬡

b. $CH_3CH—O—CHCH_3$ (with CH_3 groups)

In the IUPAC naming system for ethers, the —O—R group is called an **alkoxy group**. The *-yl* ending of the smaller R group is replaced by *-oxy*. For example, —O—CH_3 is called a methoxy group, —O—CH_2CH_3 is called ethoxy, and so on:

Alkoxy group
The —O—R functional group.

$$CH_3—O—CH_3$$
methoxymethane

$$CH_3CH_2CHCH_3$$
(with $O—CH_2CH_3$)
2-ethoxybutane

p-methoxybenzoic acid

Learning Check 3.16

Give IUPAC names for the following:

a. (cyclopentane ring with O—CH_3 and O—CH_3 groups)

b. $CH_3CH_2CHCH_2CH_3$ (with $O—CH—CH_3$ and CH_3 groups)

The ether linkage is also found in cyclic structures. A ring that contains elements in addition to carbon is called a **heterocyclic ring.** Two common structures with oxygen-containing heterocyclic rings are

Heterocyclic ring
A ring in which one or more atoms are an atom other than carbon.

furan   pyran

Many carbohydrates contain the fundamental ring systems of furan and pyran (Chapter 7).

CHEMISTRY AROUND US • 3.3

GENERAL ANESTHETICS

Anesthetics are compounds that induce a loss of sensation in a specific part (local anesthetic) or all of the body (general anesthetic). A general anesthetic acts on the brain to produce unconsciousness as well as insensitivity to pain.

The word *ether* is often associated with general anesthetics because of the well-known use of ethyl ether for this purpose. Ethyl ether became generally used as an anesthetic for surgical operations in about 1850. Before that time, surgery was an agonizing procedure. The patient was strapped to a table and (if lucky) soon fainted from the pain. Divinyl ether (vinethene) is an anesthetic that acts more rapidly and is less nauseating than ethyl ether:

$$CH_2=CH-O-CH=CH_2$$
divinyl ether

Another common ether anesthetic is enflurane:

$$H-\overset{\overset{\displaystyle F}{|}}{\underset{\underset{\displaystyle H}{|}}{C}}-O-\overset{\overset{\displaystyle F}{|}}{\underset{\underset{\displaystyle H}{|}}{C}}-\overset{\overset{\displaystyle F}{|}}{\underset{\underset{\displaystyle Cl}{|}}{C}}-H$$
enflurane

Numerous inhalation anesthetics are not ethers at all. Nitrous oxide, for example, is a simple inorganic compound with the formula N_2O. Also known as "laughing gas," it is used as a general anesthetic by some dentists because its effects wear off quickly. Halothane, currently a popular general anesthetic, is a simple halogen derivative of ethane. It is nonflammable, does not cause nausea or similar upsets, and its effects wear off quickly:

$$H-\overset{\overset{\displaystyle Br}{|}}{\underset{\underset{\displaystyle Cl}{|}}{C}}-\overset{\overset{\displaystyle F}{|}}{\underset{\underset{\displaystyle F}{|}}{C}}-F$$
halothane

In modern surgical practice, the use of a single anesthetic has become quite rare. Usually, a patient is given an injection of a strong sedative that causes unconsciousness. A general anesthetic is then administered to provide insensitivity to pain and maintain the patient in an unconscious condition. A muscle relaxant may also be used to produce complete relaxation, while minimizing the need for, and hazards associated with, deep anesthesia.

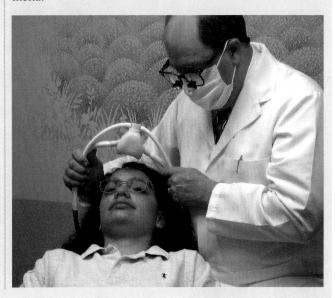

A patient receives an anesthetic prior to dental surgery.

➤ 3.8 Properties of Ethers

The oxygen atom of ethers can form hydrogen bonds with water (see ■ Figure 3.13a), so ethers are slightly more soluble in water than hydrocarbons but less soluble than alcohols of comparable molecular weight. Ethers, however, cannot form hydrogen bonds with other ether molecules in the pure state (Figure 3.13b), resulting in low boiling points close to those of hydrocarbons with similar molecular weights.

The chief chemical property of ethers is that, like the alkanes, they are inert and do not react with most reagents. It is this property that makes diethyl ether such a useful solvent. Like hydrocarbons, ethers are flammable, and diethyl ether is especially flammable because of its high volatility (see ■ Figure 3.14). Diethyl ether was the first general anesthetic used in surgery. For many years, it was the most important compound used as an anesthetic, but it has now been largely replaced by other compounds.

➤ **FIGURE 3.13** Hydrogen bonding of dimethyl ether: (a) with water and (b) no hydrogen bonding in the pure state.

(a) CH₃   H   H
        O ⋯⋯ O

(b) CH₃   CH₃

Hydrogen bond   No hydrogen bond

Learning Check 3.17

a. Explain why an ether boils at a lower temperature than an alcohol of similar molecular weight.

b. What is the most significant chemical property of ethers?

➤ 3.9 Thiols

Sulfur and oxygen belong to the same group of the periodic table, so it is not surprising that they form similar compounds. The sulfur counterparts (analogues) of alcohols contain an —SH group rather than an —OH group and are called **thiols** (an older name is mercaptans). The —SH is known as a **sulfhydryl group**.

Thiol
A compound containing an —SH group.

Sulfhydryl group
The —SH functional group.

$$R - OH \qquad\qquad R - SH$$
$$\text{an alcohol} \qquad\qquad \text{a thiol}$$

Perhaps the most distinguishing feature of thiols is their strong and disagreeable odors. Two different thiols and a disulfide are responsible for the odor associated with skunks (see ■ Figure 3.15).

trans-2-butene-1-thiol   3-methyl-1-butanethiol   methyl-1-(*trans*-2-butenyl)disulfide

Some less-offensive odors (and flavors) are also characteristic of thiols and disulfides. Freshly chopped onions emit propanethiol, and 1-propene-3-thiol and 3,3-di-(1-propenyl)disulfide are partially responsible for the odor and flavor of garlic. It is amazing that the simple substitution of a sulfur atom for an oxygen atom can cause such a dramatic change in odor.

$$CH_3CH_2CH_2 - SH \quad CH_2{=}CHCH_2 \quad CH_2{=}CHCH_2 - S - S - CH_2CH{=}CH_2$$
$$\text{SH}$$

propanethiol   1-propene-3-thiol   3,3-di-(1-propenyl)disulfide

➤ **FIGURE 3.14** The high flammability of diethyl ether makes it a useful starter fluid for extremely cold engines.

Disulfide
A compound containing an —S—S— group.

Gas companies make effective use of the odor of thiols. Natural gas (methane) has no odor, so the companies add a tiny amount of ethanethiol (CH₃CH₂—SH) to make it possible to detect leaking gas before a spark or match sets off an explosion.

Two reactions of the —SH group are important in the chemistry of proteins. When thiols are oxidized by agents such as O₂, a coupling reaction occurs (two molecules become one) to form a **disulfide**, a compound containing an —S—S— linkage:

General reaction:

$$2R-SH + (O) \rightarrow R-S-S-R + H_2O \qquad (3.17)$$

a thiol   a disulfide

Specific example:

$$2CH_3-SH + (O) \rightarrow CH_3-S-S-CH_3 + H_2O \qquad (3.18)$$

Complete the following reaction:

$$2CH_3CH-SH + (O) \rightarrow$$
$$| $$
$$CH_3$$

Learning Check 3.18

Disulfide linkages are important structural features of some proteins, especially those of hair (see Section 9.6). The coupling reaction can be reversed by treating disulfides with a reducing agent (H) such as H_2, which regenerates the thiol:

General reaction:

$$R-S-S-R + 2(H) \rightarrow 2R-SH \qquad (3.19)$$

Specific example:

$$CH_3-S-S-CH_3 + 2(H) \rightarrow 2CH_3-SH \qquad (3.20)$$

What products result from the reduction of the following disulfides?

Learning Check 3.19

a. $$CH_3CH-S-S-CHCH_3 + 2(H) \rightarrow$$
$$\quad\quad | \quad\quad\quad\quad\quad |$$
$$\quad\quad CH_3 \quad\quad\quad\quad CH_3$$

b. ⬠—S—S—⬠ + 2(H) ⟶

The reaction of thiols with heavy metal ions, such as those of lead and mercury, to form insoluble compounds has adverse biological results. The metal ions (M^{2+}) react with sulfhydryl groups in the following way:

General reation:

$$2R-SH + M^{2+} \rightarrow R-S-M-S-R + 2H^+ \qquad (3.21)$$

Specific example:

$$2CH_3-SH + Hg^{2+} \rightarrow CH_3-S-Hg-S-CH_3 + 2H^+ \qquad (3.22)$$

Many enzymes (catalysts for biological reactions—Section 10.7) contain —SH groups. Exposure to heavy metal ions ties up these groups, through Reaction 3.21, rendering the enzyme ineffective. The loss of enzymes as catalysts leads to serious biological consequences.

➤ **FIGURE 3.15** Skunks use thiol chemistry as an effective defense mechanism. Thiols are more volatile than alcohols but less soluble in water. Why are these characteristics an advantage to skunks?

Complete the following reaction:

$$2CH_3CH-SH + Pb^{2+} \rightarrow$$
$$| $$
$$CH_3$$

Learning Check 3.20

➤ 3.10 Polyfunctional Compounds

Polyfunctional compound
A compound with two or more functional groups.

So far our study of organic molecules has included several compounds with more than one functional group. Compounds of this type, such as dienes and diols, are said to be **polyfunctional compounds**. Substances with different functional groups are also referred to as polyfunctional. Compounds with multiple functional groups are very common in nature, and many play essential roles in life processes. For example, carbohydrate structures (Chapter 7) such as glucose typically involve a functional group at every carbon position.

$$\underset{\text{glucose}}{HO-CH_2-\overset{\overset{\displaystyle OH}{|}}{C}H-\overset{\overset{\displaystyle OH}{|}}{C}H-\overset{\overset{\displaystyle OH}{|}}{C}H-\overset{\overset{\displaystyle OH}{|}}{C}H-\overset{\overset{\displaystyle O}{\|}}{C}-H}$$

Even though some organic structures look quite complex, with multiple rings and long carbon chains, remember that it's the functional group or groups that determine the chemical properties of the compound. Thus, cholesterol, with two functional groups, should exhibit the reactions of both alkenes and alcohols.

cholesterol

➤ Example 3.4

Identify the functional groups in tetrahydrocannabinol, the active ingredient in marijuana.

tetrahydrocannabinol

Solution

To find functional groups, look for unique atoms or collections of atoms while ignoring ordinary rings, chains, and alkyl groups. Look for multiple bonds and atoms other than carbon. The tetrahydrocannabinol molecule has four functional groups: an ether, a phenolic —OH, the benzene ring, and a carbon–carbon double bond.

tetrahydrocannabinol

Identify the functional groups in vitamin E.

vitamin E

> **FOR FUTURE REFERENCE** GOOD FOOD SOURCES OF VITAMIN E

Vitamin E is a fat-soluble vitamin that is recognized for its antioxidant role in the human body (see Chemistry Around Us 3.2). In this role, it protects lipids and other vulnerable components of the body's cells and their membranes from destruction. Research continues to suggest additional health benefits of this antioxidant behavior. Dietary vitamin E deficiency is rare in humans, and when it is present, it is usually associated with diseases or conditions that interfere with the absorption of dietary fats during digestion. Thus, most people obtain sufficient vitamin E from their diet.

Vitamin E in foods does not occur as a single chemical compound, but as a mixture of four different compounds called alpha, beta, gamma, or delta tocopherol. These different compounds vary in their vitamin E activity, with the alpha tocopherol being the most active. Because of this, the vitamin E content of foods is represented in tocopherol equivalents (TE), where 1 TE equals the amount of vitamin E activity in 1 mg of alpha tocopherol. The chart below contains dietary sources that provide a minimum vitamin E equivalent of 0.5 mg of alpha tocopherol per serving.

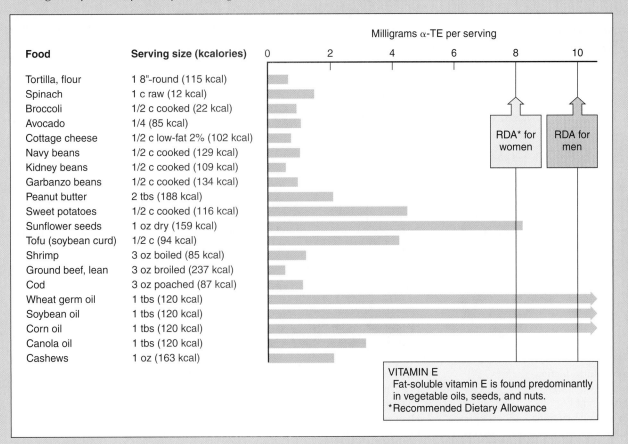

Food	Serving size (kcalories)
Tortilla, flour	1 8"-round (115 kcal)
Spinach	1 c raw (12 kcal)
Broccoli	1/2 c cooked (22 kcal)
Avocado	1/4 (85 kcal)
Cottage cheese	1/2 c low-fat 2% (102 kcal)
Navy beans	1/2 c cooked (129 kcal)
Kidney beans	1/2 c cooked (109 kcal)
Garbanzo beans	1/2 c cooked (134 kcal)
Peanut butter	2 tbs (188 kcal)
Sweet potatoes	1/2 c cooked (116 kcal)
Sunflower seeds	1 oz dry (159 kcal)
Tofu (soybean curd)	1/2 c (94 kcal)
Shrimp	3 oz boiled (85 kcal)
Ground beef, lean	3 oz broiled (237 kcal)
Cod	3 oz poached (87 kcal)
Wheat germ oil	1 tbs (120 kcal)
Soybean oil	1 tbs (120 kcal)
Corn oil	1 tbs (120 kcal)
Canola oil	1 tbs (120 kcal)
Cashews	1 oz (163 kcal)

Milligrams α-TE per serving

RDA* for women

RDA for men

VITAMIN E
Fat-soluble vitamin E is found predominantly in vegetable oils, seeds, and nuts.
*Recommended Dietary Allowance

SOURCE: Adapted from Whitney, E. N.; Rolfes, S. R. *Understanding Nutrition*, 8th ed. Belmont, CA: West/Wadsworth, 1999.

CONCEPT SUMMARY

The Nomenclature of Alcohols and Phenols. Alcohols are aliphatic compounds that contain the hydroxy functional group (—OH). Aromatic compounds with an —OH group attached to the ring are called phenols. Several alcohols are well known by common names. In the IUPAC system, the characteristic ending *-ol* is used to designate alcohols. Phenols are named as derivatives of the parent compound phenol.

Classification of Alcohols. Alcohols are classified on the basis of the number of carbon atoms bonded to the carbon that is attached to the —OH group. In primary alcohols, the OH-bonded carbon is bonded to one other carbon. In secondary alcohols, the OH-bonded carbon is bonded to two other carbons. In tertiary alcohols, the OH-bonded carbon is bonded to three other carbons.

Physical Properties of Alcohols. Hydrogen bonding can occur between alcohol molecules and between alcohol molecules and water. As a result, alcohols have higher boiling points than hydrocarbons of similar molecular weight, and low molecular weight alcohols are soluble in water.

Reactions of Alcohols. Two important reactions of alcohols are dehydration and oxidation. Alcohols may be dehydrated in two different ways depending on the reaction temperature. With H_2SO_4 at 180°C, an alkene is produced, whereas at 140°C an ether is the product. The oxidation products obtained from the alcohols depend on the class of alcohol being oxidized. A primary alcohol produces an aldehyde that is further oxidized to a carboxylic acid. A secondary alcohol produces a ketone. A tertiary alcohol does not react with oxidizing agents.

Important Alcohols. Some simple alcohols with commercial value are methanol (a solvent and fuel), ethanol (present in alco-

holic beverages and gasohol), isopropyl alcohol (rubbing alcohol), ethylene glycol (antifreeze), and glycerol (a moistening agent).

Characteristics and Uses of Phenols. Phenols are weak acids that are widely used for their antiseptic and disinfectant properties. Some phenols are used as antioxidants in foods and a variety of other materials.

Ethers. Ethers contain an oxygen attached to two carbons as the characteristic functional group. In the IUPAC system, ethers are named as alkoxy derivatives of alkanes. The —O—R group is the alkoxy group that may have names such as methoxy (—O—CH_3) or ethoxy (—O—CH_2CH_3).

Properties of Ethers. Like the alkanes, ethers are very unreactive substances. Pure ethers cannot form hydrogen bonds. As a result, their boiling points are comparable to those of hydrocarbons and lower than those of alcohols of similar molecular weight. Ethers are much less polar than alcohols but are still slightly soluble in water due to the limited formation of hydrogen bonds.

Thiols. Thiols contain an —SH group, which often imparts a strong, disagreeable odor to the compound. Two reactions of thiols that are important in protein chemistry are their oxidation to produce a disulfide and their reaction with heavy metals such as mercury. Disulfides may be converted back to thiols by a reducing agent.

Polyfunctional Compounds. It is common for organic compounds to contain more than one functional group. Many of the compounds that serve essential roles in life processes are polyfunctional.

LEARNING OBJECTIVES

When you have completed your study of this chapter, you should be able to:
1. Name and draw structural formulas for alcohols, phenols, and ethers.
2. Classify alcohols as primary, secondary, or tertiary on the basis of their structural formulas.
3. Discuss how hydrogen bonding influences the physical properties of alcohols and ethers.
4. Write equations for alcohol dehydration and oxidation reactions.
5. Recognize uses for specific alcohols, phenols, and ethers.
6. Write equations for a thiol reaction with heavy metal ions and production of disulfides.
7. Identify functional groups in polyfunctional compounds.

KEY TERMS AND CONCEPTS

Alcohol (Introduction)
Alkoxy group (3.7)
Antioxidant (3.6)
Dehydration reaction (3.4)
Disulfide (3.9)
Elimination reaction (3.4)

Ether (Introduction)
Fermentation (3.5)
Heterocyclic ring (3.7)
Hydroxy group (Introduction)
Phenol (Introduction)
Polyfunctional compound (3.10)

Primary alcohol (3.2)
Secondary alcohol (3.2)
Sulfhydryl group (3.9)
Tertiary alcohol (3.2)
Thiol (3.9)

1. Dehydration of alcohols to give alkenes (Section 3.4):

$$-\overset{\displaystyle |}{\underset{\displaystyle H}{C}}-\overset{\displaystyle |}{\underset{\displaystyle OH}{C}}- \xrightarrow[180°C]{H_2SO_4} \ \diagdown C = C \diagup + H_2O$$

Reaction 3.1

2. Dehydration of alcohols to give ethers (Section 3.4):

$$R-O-H + H-O-R \xrightarrow[140°C]{H_2SO_4} R-O-R + H_2O$$

Reaction 3.5

3. Oxidation of a primary alcohol to give an aldehyde and then a carboxylic acid (Section 3.4):

$$R-\overset{\displaystyle OH}{\underset{\displaystyle H}{\overset{\displaystyle |}{\underset{\displaystyle |}{C}}}}-H + (O) \longrightarrow R-\overset{\displaystyle O}{\overset{\displaystyle \|}{C}}-H + H_2O$$

aldehyde

primary
alcohol

$$(O) \xrightarrow{\text{further oxidation}} R-\overset{\displaystyle O}{\overset{\displaystyle \|}{C}}-OH$$

carboxylic acid

Reaction 3.8

4. Oxidation of a secondary alcohol to give a ketone (Section 3.4):

$$R-\overset{\displaystyle OH}{\underset{\displaystyle H}{\overset{\displaystyle |}{\underset{\displaystyle |}{C}}}}-R' + (O) \longrightarrow R-\overset{\displaystyle O}{\overset{\displaystyle \|}{C}}-R' + H_2O$$

secondary
alcohol

ketone

Reaction 3.10

5. Attempted oxidation of a tertiary alcohol gives no reaction (Section 3.4):

$$R-\overset{\displaystyle OH}{\underset{\displaystyle R''}{\overset{\displaystyle |}{\underset{\displaystyle |}{C}}}}-R' + (O) \longrightarrow \text{no reaction}$$

Reaction 3.12

6. Oxidation of a thiol to give a disulfide (Section 3.9):

$$2R-SH + (O) \longrightarrow R-S-S-R + H_2O$$

a thiol   a disulfide

Reaction 3.17

7. Reduction of a disulfide to give thiols (Section 3.9):

$$R-S-S-R + 2(H) \longrightarrow 2R-SH$$

Reaction 3.19

8. Reaction of thiols with heavy metals (Section 3.9):

$$2R-SH + M^{2+} \longrightarrow R-S-M-S-R + 2H^+$$

Reaction 3.21

EXERCISES

THE NOMENCLATURE OF ALCOHOLS AND PHENOLS (SECTION 3.1)

3.1 Assign IUPAC names to the following alcohols:

a. CH$_3$CHCH$_3$ (OH on middle carbon)

b. CH$_3$CH$_2$CCH$_3$ (with CH$_2$CH$_3$ above and OH below)

c. CH$_2$CH$_2$CHCH$_2$CH$_2$ — Br (with OH and CH$_2$CH$_3$ substituents)

d. CH$_3$CHCH$_3$ with cyclopropane ring bearing OH and CH$_3$

e. cyclopentane with OH and two CH$_3$ groups

f. cyclopentane with OH and OH

g. CH$_3$CH$_2$CHCH$_2$CHCH$_2$ — OH (with two OH groups)

3.2 Assign IUPAC names to the following alcohols:

a. CH$_3$CH$_2$CH$_2$—OH

b. phenyl—CH$_2$CH$_2$CH$_2$ — OH

c. CH$_3$CHCHCH$_2$ — OH (with CH$_3$ and Cl substituents)

d. CH$_2$CH$_2$CH$_2$CH$_2$ — OH (with OH)

e. cyclopentane with Cl, OH, and CH$_2$CH$_3$

f. cyclohexane with three OH groups

3.3 Several important alcohols are well known by common names. Give a common name for each of the following:
a. CH$_3$—OH
b. CH$_3$CHCH$_3$ (with OH)
c. CH$_3$CH$_2$—OH
d. CH$_2$—CH$_2$ (with OH OH)
e. CH$_2$—CH—CH$_2$ (with OH OH OH)

3.4 Give each of the structures in Exercise 3.3 an IUPAC name.

3.5 Draw structural formulas for each of the following:
a. 2-methyl-1-butanol
b. 2-bromo-3-methyl-3-pentanol
c. 1-methylcyclopentanol

3.6 Draw structural formulas for each of the following:
a. 2-methyl-2-pentanol
b. 1,3-butanediol
c. 1-ethylcyclopentanol

3.7 Name each of the following as a derivative of phenol:

a. benzene ring with OH, Cl, Cl

b. benzene ring with NO$_2$, O$_2$N, OH

3.8 Name each of the following as a derivative of phenol:

a. benzene ring with CH$_3$/CHCH$_3$ and OH

b. benzene ring with OH, CH$_3$, CH$_3$

3.9 Draw structural formulas for each of the following:
a. *p*-chlorophenol
b. 2,5-diisopropylphenol

3.10 Draw tructural formulas for each of the following:
 a. m-methylphenol
 b. 2,3-dichlorophenol

CLASSIFICATION OF ALCOHOLS (SECTION 3.2)

3.11 What is the difference between a primary, secondary, and tertiary alcohol?

3.12 Classify the following alcohols as primary, secondary, or tertiary:

 a. $CH_3CH_2\overset{\overset{\displaystyle CH_3}{|}}{\underset{\underset{\displaystyle CH_3}{|}}{C}}-OH$

 b. $CH_3CH_2CH_2CH_2-OH$

 c. $CH_3CH_2\overset{}{\underset{\underset{\displaystyle CH_3}{|}}{C}H}-OH$

3.13 Classify the following alcohols as primary, secondary, or tertiary:

 a. $CH_3\underset{\underset{\underset{\displaystyle CH_3}{|}}{\underset{\displaystyle CH_2}{|}}}{C}H-OH$

 b. $CH_3CH_2-\overset{\overset{\displaystyle CH_3}{|}}{\underset{\underset{\displaystyle OH}{|}}{C}}-CH_3$

 c. $\overset{\overset{\displaystyle CH_2CH_3}{|}}{\underset{\underset{\displaystyle CH_3}{|}}{C}}-OH$ (phenyl)

3.14 Draw structural formulas for the four aliphatic alcohols with the molecular formula $C_4H_{10}O$. Name each compound using the IUPAC system and classify it as a primary, secondary, or tertiary alcohol.

PHYSICAL PROPERTIES OF ALCOHOLS (SECTION 3.3)

3.15 Why are the boiling points of alcohols much higher than the boiling points of alkanes with similar molecular weights?

3.16 Arrange the compounds of each group in order of increasing boiling point.
 a. ethanol, 1-propanol, methanol
 b. butane, ethylene glycol, 1-propanol

3.17 Which member of each of the following pairs would you expect to be more soluble in water? Briefly explain your choices.
 a. butane or 2-butanol
 b. 2-propanol or 2-pentanol
 c. 2-butanol or 2,3-butanediol

3.18 Draw structural formulas for the following molecules and use a dotted line to show the formation of hydrogen bonds:
 a. one molecule of 1-butanol and one molecule of ethanol
 b. cyclohexanol and water

3.19 Explain why the use of glycerol (1,2,3-propanetriol) in lotions helps retain water and keep the skin moist.

REACTIONS OF ALCOHOLS (SECTION 3.4)

3.20 Draw the structures of the chief product formed when the following alcohols are dehydrated to alkenes:

 a. (cyclopentane with CH_3 and OH)

 b. $CH_3\overset{\overset{\displaystyle OH}{|}}{\underset{\underset{\displaystyle CH_2CH_3}{|}}{C}H}CH-CH_3$

3.21 Draw the structures of the chief product formed when the following alcohols are dehydrated to alkenes:

 a. (benzene ring with) CH_2CH_2-OH

 b. (cyclopentane with CH_3 and OH)

3.22 Draw the structures of the ethers that can be produced from the following alcohols:
 a. $CH_3CH_2CH_2-OH$
 b. (cyclopentane with) CH_2-OH

 c. $CH_3\overset{}{\underset{\underset{\displaystyle CH_3}{|}}{C}H}CH_2-OH$

3.23 Draw the structures of the two organic compounds that can be obtained by oxidizing $CH_3CH_2CH_2-OH$.

3.24 Give the structure of an alcohol that could be used to prepare each of the following compounds:

 a. (cyclohexanone with CH_3, O)

 b. (cyclopentane with) $C-H$, O

 c. $CH_3\overset{\overset{\displaystyle CH_3}{|}}{C}H-\overset{\overset{\displaystyle O}{||}}{C}-OH$

3.25 Give the structure of an alcohol that could be used to prepare each of the following compounds:

a. $CH_3-\overset{\displaystyle O}{\overset{\displaystyle \|}{C}}-CH_2CH_3$

b.

c.

3.26 What products would result from the following processes? Write an equation for each reaction.
   a. 2-methyl-2-butanol is subjected to controlled oxidation.
   b. 1-propanol is heated to 140°C in the presence of sulfuric acid.
   c. 3-pentanol is subjected to controlled oxidation.
   d. 3-pentanol is heated to 180°C in the presence of sulfuric acid.
   e. 1-hexanol is subjected to an *excess* of oxidizing agent.

3.27 What products would result from the following processes? Write an equation for each reaction.
   a. 1-butanol is heated to 140°C in the presence of sulfuric acid.
   b. 1-butanol is subjected to an excess of oxidizing agent.
   c. 2-pentanol is subjected to controlled oxidation.
   d. 2-pentanol is heated to 180°C in the presence of sulfuric acid.
   e. 2-methyl-2-pentanol is subjected to controlled oxidation.

3.28 Each of the following conversions requires more than one step, and some reactions studied in previous chapters may be needed. Show the reagents you would use and draw structural formulas for intermediate compounds formed in each conversion.

a. $CH_3CH_2CH=CH_2 \rightarrow CH_3CH_2\overset{\displaystyle O}{\overset{\displaystyle \|}{C}}CH_3$

b.

c. $CH_3CH_2CH_2\overset{\displaystyle OH}{\overset{\displaystyle |}{C}}H_2 \rightarrow CH_3CH_2\overset{\displaystyle O}{\overset{\displaystyle \|}{C}}CH_3$

3.29 Each of the following conversions requires more than one step, and some reactions studied in previous chapters may be needed. Show the reagents you would use and draw structural formulas for intermediate compounds formed in each conversion.

a. $CH_2=CH_2 \rightarrow CH_3CH_2-O-CH_2CH_3$

b.

c. $CH_3CH_2CH_2 \overset{\displaystyle OH}{\overset{\displaystyle |}{}} \rightarrow CH_3CHCH_3$

d. $CH_3CH_2CH_2CH_2 \overset{\displaystyle OH}{\overset{\displaystyle |}{}} \rightarrow CH_3CH_2CHCH_3$ with Cl

3.30 The three-carbon diol used in antifreeze is

$$CH_3-\overset{\displaystyle OH}{\overset{\displaystyle |}{C}}H-\overset{\displaystyle OH}{\overset{\displaystyle |}{C}}H_2$$

It is nontoxic and is used as a moisturizing agent in foods. Oxidation of this substance within the liver produces pyruvic acid, which can be used by the body to supply energy. Give the structure of pyruvic acid.

IMPORTANT ALCOHOLS (SECTION 3.65)

3.31 Methanol is fairly volatile and evaporates quickly if spilled. Methanol is also absorbed quite readily through the skin. If, in the laboratory, methanol accidentally spilled on your clothing, why would it be a serious mistake to just let it evaporate?

3.32 Why is methanol such an important industrial chemical?

3.33 Suppose you are making some chocolate cordials (chocolate-coated candies with soft fruit filling). Why might you want to add a little glycerol to the filling?

3.34 Name an alcohol used in each of the following ways:
   a. A moistening agent in many cosmetics
   b. The solvent in solutions called tinctures
   c. Automobile antifreeze
   d. Rubbing alcohol
   e. A flavoring in cough drops
   f. Present in gasohol

CHARACTERISTICS AND USES OF PHENOLS (SECTION 3.6)

3.35 Name a phenol associated with each of the following:
   a. A vitamin
   b. A disinfectant cleaner

3.36 Name a phenol used in each of the following ways:
   a. A disinfectant used for cleaning walls
   b. An antiseptic found in some mouthwashes
   c. An antioxidant used to prevent rancidity in foods

ETHERS (SECTION 3.7)

3.37 Assign a common name to each of the following ethers:
   a. $CH_3CH_2CH_2-O-CH_2CH_2CH_3$

b. $CH_3CH_2-O-\overset{\overset{\displaystyle CH_3}{|}}{\underset{\underset{\displaystyle CH_3}{|}}{C}}-CH_3$

c. $CH_3\underset{\underset{\displaystyle CH_3}{|}}{CH}-O-\langle\text{phenyl}\rangle$

3.38 Assign a common name to each of the following ethers:

   a. $CH_3-O-CH_2CH_3$

   b. $\langle\text{phenyl}\rangle O-CH_2CH_2CH_3$

   c. $CH_3CH_2\underset{\underset{\displaystyle CH_3}{|}}{CH}-O-\underset{\underset{\displaystyle CH_3}{|}}{CH}CH_2CH_3$

3.39 Assign the IUPAC name to each of the following ethers. Name the smaller alkyl group as the alkoxy group.

   a. $CH_3-O-CH_2CH_3$

   b. $\langle\text{cyclopentyl}\rangle O-CH_2CH_3$

   c. $CH_3\underset{\underset{\displaystyle CH_3}{|}}{CH}-O-\langle\text{phenyl}\rangle$

   d. $CH_3CH_2CH_2-O-\underset{\underset{\displaystyle CH_3}{|}}{CH}CH_2CH_2CH_2CH_3$

3.40 Assign the IUPAC name to each of the following ethers. Name the smaller alkyl group as the alkoxy group.

   a. $CH_3-O-CH_2CH_2CH_3$

   b. $CH_3\underset{\underset{\displaystyle CH_2CH_3}{|}}{CH}CH_2CH_2-O-CH_2CH_3$

   c. $\langle\text{cyclopropyl}\rangle O-CH_2CH_2CH_3$

   d. benzene ring with OCH_3 (top), CH_3O (bottom left), OCH_3 (bottom right)

3.41 Draw structural formulas for the following:
   a. butyl ethyl ether
   b. methyl phenyl ether
   c. diisopropyl ether
   d. 3-ethoxyhexane
   e. 1,2-dimethoxycyclobutane

3.42 Draw structural formulas for the following:
   a. methyl isopropyl ether
   b. phenyl propyl ether
   c. 2-methoxypentane
   d. 1,2-diethoxycyclopentane
   e. 2-phenoxy-2-butene

3.43 Give the IUPAC names and draw structural formulas for the six isomeric ethers of molecular formula $C_5H_{12}O$.

PROPERTIES OF ETHERS (SECTION 3.8)

3.44 What is the chief chemical property of ethers?

3.45 Why is diethyl ether hazardous to use as an anesthetic or as a solvent in the laboratory?

3.46 Arrange the following compounds in order of decreasing solubility in water. Explain the basis for your decisions.

$CH_3CH_2CH_2-OH \quad CH_3CH_2CH_2CH_3 \quad CH_3CH_2-O-CH_3$

3.47 Arrange the compounds in Exercise 3.46 in order of decreasing boiling point. Explain your answer.

3.48 Draw structural formulas and use a dotted line to show hydrogen bonding between a molecule of diethyl ether and water.

THIOLS (SECTION 3.9)

3.49 Complete the following reactions:

   a. $2CH_3\underset{\underset{\displaystyle SH}{|}}{CH_2} + Hg^{2+} \rightarrow$

   b. $\langle\text{cyclopentyl}\rangle SH + (O) \longrightarrow$

   c. $CH_3-S-S-CH_2CH_2CH_3 + 2(H) \rightarrow$

3.50 Complete the following reactions:
   a. $2CH_3CH_2CH_2-SH + Hg^{2+} \rightarrow$
   b. $2CH_3CH_2CH_2-SH + (O) \rightarrow$
   c. $\langle\text{cyclopentyl}\rangle S-S-CH_3 + 2(H) \longrightarrow$

3.51 Lipoic acid is required by many microorganisms for proper growth. As a disulfide, it functions in the living system by catalyzing certain oxidation reactions and is reduced in the process. Write the structure of the reduction product.

$\underset{\underset{\displaystyle S-S}{}}{\langle\text{ring}\rangle}CH_2CH_2CH_2CH_2-\overset{\overset{\displaystyle O}{\|}}{C}-OH + 2(H) \longrightarrow$

lipoic acid

3.52 Alcohols and thiols can both be oxidized in a controlled way. What are the differences in the products?

CHEMISTRY AROUND US, OVER THE COUNTER, AND HOW REACTIONS OCCUR

3.53 What alcohol is associated with fetal alcohol syndrome?

3.54 Write a mechanism for the dehydration reaction:

$$CH_3CHCH_3 \xrightarrow[180°C]{H_2SO_4} CH_3CH=CH_2 + H_2O$$

(with OH on the central carbon of CH_3CHCH_3)

3.55 Name an ether that is important as a modern anesthetic.

3.56 What oil is the culprit in the poison ivy plant?

3.57 How does the product bentoquatam help provide protection against poison ivy?

3.58 What is believed to be the cellular role of vitamin E?

3.59 Why is vitamin E used as an additive in vegetable oils?

POINTS TO PONDER AND PURSUE

3.1 The caption for Figure 3.9 states that many aftershaves are 50% ethanol. What do you think is the purpose of the ethanol?

3.2 A mixture of ethanol and 1-propanol is heated to 140°C in the presence of sulfuric acid. Careful analysis reveals that three ethers were formed. Write formulas for these three products, name each, and explain why three formed, rather than a single product.

3.3 The two-carbon alcohol in some antifreeze

$$CH_2-CH_2$$

(with OH on each carbon)

is toxic if ingested. Reactions within the liver oxidize the diol to oxalic acid, which can form insoluble salts and damage the kidneys. Give the structure of oxalic acid.

3.4 Although ethanol is only mildly toxic to adults, it poses greater risks for fetuses. Why do you think this is the case?

3.5 The medicinal compound taxol is mentioned in Figure 3.1. What does the name indicate about its possible structure?

3.6 When diethyl ether is spilled on the skin, the skin takes on a dry appearance. Explain whether this effect is more likely due to a removal of water or a removal of natural skin oils.

3.7 Figure 3.8 points out that methanol is used as a fuel in racing cars. Write a balanced equation for the combustion of methanol.

3.8 Figure 3.15 focuses on the use of thiol chemistry by skunks. If thiols are more volatile than alcohols and less soluble in water, why do these properties represent an advantage to skunks?

3.9 Use Web site 3.2 to explore why alcohols and ethers are sometimes added to gasoline.

3.10 According to Web site 3.3, what effect does alcohol have on the central nervous system?

3.11 Using Web site 3.5, what is another name for vitamin E?

http://www.brookscole.com/chemistry/seager_chemistry

Aldehydes and Ketones

Proper nutrition and, in some cases, a carefully monitored diet play an essential role in the recovery process for hospitalized patients. Clinical dietitians also supervise the nutritional care of people in nursing homes and other settings. In this chapter, you will become familiar with aldehydes and ketones, substances whose reactions will help you understand the chemistry of carbohydrates, a nutritionally important family of compounds.

http://www.brookscole.com/chemistry/seager_chemistry
At Web site 4.1, you will find further information on a career as a dietitian.

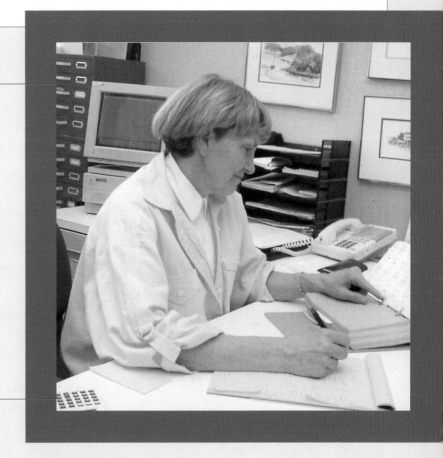

CHAPTER OUTLINE

➤ The Nomenclature of Aldehydes and Ketones

➤ Physical Properties

➤ Chemical Properties

➤ Important Aldehydes and Ketones

Carbonyl group

$$\overset{O}{\underset{\|}{}}$$

The $-\overset{O}{\underset{\|}{C}}-$ group.

he carbonyl functional group is characteristic of both aldehydes and ketones. In a **carbonyl group,** a carbon is double-bonded to an oxygen atom and single-bonded to two other atoms:

$$-\overset{O}{\underset{\|}{C}}-$$

carbonyl group

Aldehydes and ketones occur widely in nature and play important roles in living organisms. For example, the carbonyl group is found in numerous carbohydrates, including glucose and fructose. Glucose, the major source of energy in living systems, is found combined with fructose in cane sugar:

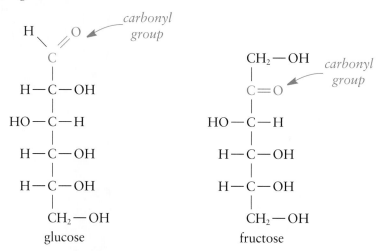

glucose   fructose

In this chapter, we will study the naming of aldehydes and ketones, as well as some of their important reactions. These concepts will be very useful when we study the chemistry of carbohydrates in Chapter 7.

➤ 4.1 The Nomenclature of Aldehydes and Ketones

When a carbonyl group is directly bonded to at least one hydrogen atom, the compound is an **aldehyde.** In **ketones,** two carbon atoms are directly bonded to the carbonyl group:

$$-\overset{O}{\underset{\|}{C}}- \qquad -\overset{O}{\underset{\|}{C}}-H \qquad -\overset{\;\;|}{\underset{\;\;|}{C}}-\overset{O}{\underset{\|}{C}}-\overset{\;\;|}{\underset{\;\;|}{C}}-$$

carbonyl group   aldehyde   ketone

Aldehyde

A compound that contains the

$$-\overset{O}{\underset{\|}{C}}-H \text{ group; the general}$$

formula is $R-\overset{O}{\underset{\|}{C}}-H$.

Ketone

A compound that contains the

$$-\overset{\;\;|}{\underset{\;\;|}{C}}-\overset{O}{\underset{\|}{C}}-\overset{\;\;|}{\underset{\;\;|}{C}}- \text{ group;}$$

the general formula is

$$R-\overset{O}{\underset{\|}{C}}-R'.$$

Structural models of the simplest aldehyde (formaldehyde) and the simplest ketone (acetone) are shown in ■ Figure 4.1.

Several aldehydes and ketones, in addition to formaldehyde and acetone, are well known by common names (see ■ Table 4.1). Notice that aldehyde and ketone common names are easily recognized by the endings *-aldehyde* and *-one,* respectively.

In the IUPAC system, aldehydes are named by selecting the longest chain that includes the carbonyl carbon. The *-e* of the alkane that has the same number of carbons is replaced by *-al.* Because the carbonyl carbon in aldehydes is always at the end of the chain, it is always designated as carbon number 1.

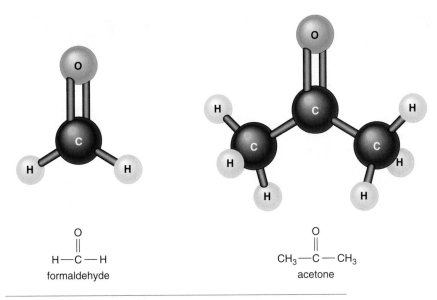

➤ **FIGURE 4.1** Ball-and-stick models of formaldehyde and acetone.

| **TABLE 4.1** | Some common aldehydes and ketones | | |

Structural formula	IUPAC name	Common name	Boiling point (°C)
Aldehydes			
$H-\overset{\overset{\displaystyle O}{\|}}{C}-H$	methanal	formaldehyde	−21
$CH_3-\overset{\overset{\displaystyle O}{\|}}{C}-H$	ethanal	acetaldehyde	21
$CH_3CH_2-\overset{\overset{\displaystyle O}{\|}}{C}-H$	propanal	propionaldehyde	49
$CH_3CH_2CH_2-\overset{\overset{\displaystyle O}{\|}}{C}-H$	butanal	butyraldehyde	76
$\bigcirc-\overset{\overset{\displaystyle O}{\|}}{C}-H$	benzaldehyde	benzaldehyde	178
Ketones			
$CH_3-\overset{\overset{\displaystyle O}{\|}}{C}-CH_3$	propanone	acetone	56
$CH_3-\overset{\overset{\displaystyle O}{\|}}{C}-CH_2CH_3$	butanone	methyl ethyl ketone (MEK)	80
$\bigcirc=O$	cyclohexanone	cyclohexanone	156

CHEMISTRY AROUND US • 4.1

FAKING A TAN

Many people believe that a suntan makes one look healthy and attractive. Studies, however, indicate that this perception is far from the truth. According to these studies, sunbathing, especially when sunburn results, ages the skin prematurely and increases the risk of skin cancer. Cosmetic companies have developed a tanning alternative for those not willing to risk using the sun but who want to be "fashionably" tan.

Tanning lotions and creams are now available that chemically darken the skin. The active ingredient in these "bronzers" is dihydroxyacetone (DHA), a colorless compound classified by the Food and Drug Administration (FDA) as a safe skin dye:

$$HO-CH_2-\overset{\overset{\displaystyle O}{\|}}{C}-CH_2-OH$$

dihydroxyacetone (DHA)

Within several hours of application, DHA produces a brown skin color by reacting with the outer layer of the skin, which consists of dead cells. Only the dead cells react with DHA, so the color gradually fades as the dead cells slough off and are replaced. This process generally leads to the fading of chemical tans within a few weeks. Another problem with chemical tans is uneven skin color. Areas of skin such as elbows and knees, which contain a thicker layer of dead cells, may absorb and react with more tanning lotion and become darker than other areas.

Perhaps the greatest problem with chemical tans is the false sense of security they might give. Some people with chemical tans think it is safe to go into the sun and get a deeper tan. This isn't true. Sunlight presents the same hazards to chemically tanned skin that it does to untanned skin.

Some DHA-containing products.

➤ **Example 4.1**
....

Name the following aldehydes according to the IUPAC system.

a. $CH_3CH_2CH_2-\overset{\overset{\displaystyle O}{\|}}{C}-H$
b. $CH_3\overset{\overset{\displaystyle Br}{|}}{C}HCH_2-\overset{\overset{\displaystyle O}{\|}}{C}-H$

c. $CH_3CH_2\overset{\overset{\displaystyle \overset{\displaystyle O}{\|}}{\underset{\displaystyle |}{C-H}}}{C}HCH_2CH_3$

Solution

a. This aldehyde contains four carbon atoms, and the name is butanal. The aldehyde group is always position number 1, so the number designating the aldehyde position is omitted from the name.

b. Carbon chain numbering begins with the carbonyl end of the molecule even though a lower number for the Br could be obtained by numbering from the left:

$$\underset{4}{C}H_3\underset{3}{\overset{\overset{\displaystyle Br}{|}}{C}}H\underset{2}{C}H_2-\underset{1}{\overset{\overset{\displaystyle O}{\|}}{C}}-H$$

The compound is 3-bromobutanal.

c. The carbonyl group must be position number 1 in the chain even though a longer carbon chain (pentane) exists:

$$\begin{array}{c} O \\ \parallel \\ {}^{1}C-H \\ | \\ \underset{4}{CH_3}\underset{3}{CH_2}\underset{2}{CH}CH_2CH_3 \end{array}$$

The name is 2-ethylbutanal.

Give IUPAC names to the following aldehydes:

Learning Check 4.1

a. $\begin{array}{c} O \\ \parallel \\ CH_3CH_2CH_2CH_2-C-H \end{array}$
b. $\begin{array}{c} CH_3 \quad\quad O \\ | \quad\quad\quad \parallel \\ CH_3CHCHCH_2-C-H \\ | \\ CH_3 \end{array}$

In naming ketones by the IUPAC system, the longest chain that includes the carbonyl group is given the name of the corresponding alkane, with the -*e* replaced by -*one*. The chain is numbered from the end closest to the carbonyl group, and the carbonyl group position is indicated by a number.

➤ Example 4.2

Give IUPAC names to the following ketones:

a. $\begin{array}{c} O \\ \parallel \\ CH_3-C-CH_3 \end{array}$
b. $\begin{array}{c} O \\ \parallel \\ CH_3-C-CHCH_2CH_3 \\ | \\ CH_3 \end{array}$
c. a cyclohexanone ring with O and CH_3

Solution

a. There are three carbons in the chain, and the correct name is propanone. The number 2 is not used in this case because the carbonyl group in a ketone cannot be on the end of a chain. Therefore, it must be in the number 2 position.
b. The chain is numbered from the left to give a lower number for the carbonyl carbon:

$$\begin{array}{c} O \\ \parallel \quad {}_{3}\;{}_{4}\;{}_{5} \\ \underset{1}{CH_3}-\underset{2}{C}-CHCH_2CH_3 \\ | \\ CH_3 \end{array}$$

The complete name is 3-methyl-2-pentanone.
c. In cyclic ketones, the carbonyl carbon will always be at position 1, and so the number is omitted from the name:

The correct name is 2-methylcyclohexanone.

Learning Check 4.2

Give IUPAC names to the following ketones:

a.   Br   Br   O
     | | ||
CH₃CHCH₂CHCH₂ — C — CH₃   **b.**

➤ 4.2 Physical Properties

The physical properties of aldehydes and ketones can be explained by an examination of their structures. First, the lack of a hydrogen on the oxygen prevents the formation of hydrogen bonds between molecules:

no H attached to the oxygen

$$
\begin{array}{cc}
O & O \\
\| & \| \\
R-C-H & R-C-R'
\end{array}
$$

Therefore, boiling points of pure aldehydes and ketones are expected to be lower than those of alcohols with similar molecular weights. Remember, alcohols can form hydrogen bonds with one another. ■ Table 4.2 shows that the boiling points of propanal and acetone are 49°C and 56°C, respectively, whereas the alcohol of comparable molecular weight, 1-propanol, has a boiling point of 97°C.

Also notice in Table 4.2 that the boiling points of the aldehyde and ketone are higher than that of the comparable alkane. This can be explained by differences in polarity. Whereas alkanes are very nonpolar, the carbonyl group of aldehydes and ketones is polar due to the more electronegative oxygen atom:

$$
\begin{array}{c}
\overset{\delta-}{O} \\
\| \\
-\underset{\delta+}{C}-
\end{array}
$$

The attractive forces between molecules possessing such dipoles are not as strong as hydrogen bonds, but they do cause aldehydes and ketones to boil at

---

TABLE 4.2   A comparison of physical properties

Class	Example	Formula	Molecular weight	Boiling point (°C)	Water solubility
Alkane	butane	CH₃CH₂CH₂CH₃	58	0	Insoluble
Aldehyde	propanal	CH₃CH₂ — C — H (with O double bonded to C)	58	49	Soluble
Ketone	propanone (acetone)	CH₃ — C — CH₃ (with O double bonded to C)	58	56	Soluble
Alcohol	1-propanol	CH₃CH₂CH₂ — OH	60	97	Soluble

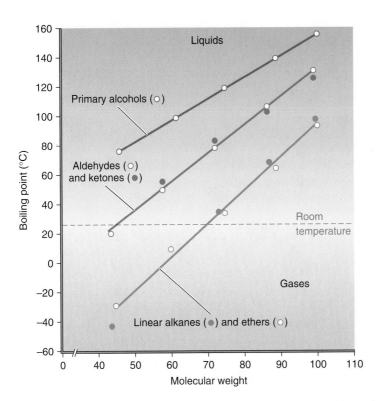

➤ **FIGURE 4.2** The boiling points of aldehydes and ketones are higher than those of alkanes and ethers but lower than those of alcohols of comparable molecular weights.

higher temperatures than the nonpolar alkanes. The graph in ■ Figure 4.2 summarizes the boiling point comparison for several families of compounds.

Water is a polar solvent and would be expected to be effective in dissolving compounds containing polar carbonyl groups. Besides exhibiting dipolar attractions, the carbonyl oxygen also forms hydrogen bonds with water:

As a result, low-molecular-weight aldehydes and ketones are water-soluble (Table 4.2).

➤ 4.3 Chemical Properties

Oxidation

We learned in Section 3.4 that when aldehydes are prepared by oxidizing primary alcohols with $KMnO_4$ or $K_2Cr_2O_7$, the reaction may continue and produce carboxylic acids. Ketones are prepared similarly, but they are much less susceptible to further oxidation, especially by mild oxidizing agents:

$$\text{General reaction: } R-CH_2-OH \xrightarrow{(O)} R-\overset{\overset{\displaystyle O}{\|}}{C}-H \xrightarrow{(O)} R-\overset{\overset{\displaystyle O}{\|}}{C}-OH \quad (4.1)$$

primary alcohol   aldehyde   carboxylic acid

➤ **FIGURE 4.3** Attempted oxidation of acetone and oxidation of benzaldehyde. As the reaction proceeds, chromium is reduced, forming a grayish green precipitate.

From left to right, three test tubes containing potassium dichromate ($K_2Cr_2O_7$), acetone, and benzaldehyde.

After the addition of equal amounts of $K_2Cr_2O_7$, the acetone remains unreacted, whereas the benzaldehyde is oxidized.

General reaction:

$$R - \underset{\underset{\text{secondary}}{\text{alcohol}}}{\overset{\overset{\displaystyle OH}{|}}{CH}} - R' \xrightarrow{\text{(O)}} R - \underset{\text{ketone}}{\overset{\overset{\displaystyle O}{\|}}{C}} - R' \xrightarrow{\text{(O)}} \text{no reaction} \qquad (4.2)$$

The difference in reactivity toward oxidation is the chief reason why aldehydes and ketones are classified in separate families (see ■ Figure 4.3).

Specific example:

benzaldehyde $+$ (O) $\longrightarrow$ benzoic acid   (4.3)

Specific example:

$$CH_3 - \overset{\overset{\displaystyle O}{\|}}{C} - CH_3 + \text{(O)} \longrightarrow \text{no reaction} \qquad (4.4)$$
acetone

The oxidation of aldehydes occurs so readily that even atmospheric oxygen causes the reaction. A few drops of liquid benzaldehyde (Reaction 4.3) placed on a watch glass will begin to oxidize to solid benzoic acid within an hour. In the laboratory, it is not unusual to find that bottles of aldehydes that have been opened for a length of time contain considerable amounts of carboxylic acids.

Learning Check 4.3

Draw the structural formula for each product. Write "no reaction" if none occurs.

a.

$$CH_3CH\underset{\overset{\displaystyle |}{CH_3}}{}CH_2 - \overset{\overset{\displaystyle O}{\|}}{C} - H + \text{(O)} \longrightarrow$$

b.

$\bigcirc\!\!=\!\!O$ + (O) $\longrightarrow$

The effectiveness of two common chemical reagents used to test for the presence of aldehydes—Tollens' reagent and Benedict's reagent—depends on the ease with which aldehydes are oxidized. In general, ketones fail to react with these same reagents.

Tollens' reagent is a mild oxidizing solution containing complexed Ag^+ ions. The oxidizing agent is the silver ion, which is reduced to metallic silver in the process and precipitates out (Reaction 4.5):

Tollens' reagent
A mild oxidizing solution containing silver ions used to test for aldehydes.

$$\underset{\text{an aldehyde}}{R-\overset{\overset{\displaystyle O}{\|}}{C}-H} + \underset{\substack{\text{Tollens'}\\\text{reagent}}}{2Ag(NH_3)_2^+} + 2OH^- \longrightarrow \underset{\substack{\text{a carboxylate}\\\text{salt}}}{R-\overset{\overset{\displaystyle O}{\|}}{C}-O^-NH_4^+} + 3NH_3 + \underset{\text{silver}}{2Ag} + H_2O \quad (4.5)$$

When the reaction is carried out in a clean, grease-free glass container, the metallic silver deposits on the glass and forms a mirror. This reaction is used commercially to silver some mirrors (see ■ Figure 4.4).

OVER THE COUNTER • 4.1

BIRTH CONTROL: PROGESTERONE SUBSTITUTES

The ketone *norethynodrel* is not a major industrial material, but in terms of social impact, it has had a profound influence since 1960. In that year, Enovid, with norethynodrel as the major component, became the first oral contraceptive for women approved by the FDA. Norethynodrel is effective because it mimics the action of another ketone, *progesterone*, a hormone that causes changes in the wall of the uterus to prepare it to accept a fertilized egg and maintain the resulting pregnancy.

norethynodrel

progesterone

Norethynodrel establishes a state of false pregnancy; a woman ceases to ovulate and therefore cannot conceive. The pill (as this product is popularly called) has been used by millions of women and appears to be relatively safe even though some women experience undesirable side effects such as acne, hypertension, abnormal bleeding, or increased blood clotting. Women who suffer from these or other serious side effects should not take the pill. Women over the age of 40, particularly those who smoke or are overweight, are advised by the FDA to use contraceptive methods other than the pill.

Doctors sometimes prescribe higher doses of combined oral contraceptives for use as "morning after" pills, which are taken within 72 hours of unprotected intercourse to prevent any possibly fertilized egg from reaching the uterus. The FDA's Advisory Committee for Reproductive Health Drugs concluded that certain oral contraceptives are safe and effective for this type of use.

Another type of contraception is injectable progestin. Depo-Provera® is a product that is injected by a health care professional into the arm muscle or buttocks every 3 months. Depo-Provera® prevents pregnancy in three ways. It changes the cervical mucus to help prevent sperm from reaching the egg, inhibits ovulation, and changes the uterine lining to prevent a fertilized egg from implanting in the uterus.

Long-term birth control for women became available in 1990 when the FDA approved under-the-skin implants that contain a slowly-released contraceptive. In these devices, levonorgestrel (Norplant®, or the newer Norplant 2®), a synthetic compound similar to progesterone, is sealed inside small rubber rods about the size of matchsticks. The rods are implanted under the skin of the upper arm. Levonorgestrel is slowly released through the walls of the rods and prevents pregnancy for up to 5 years (for Norplant) or 3 years (for Norplant 2), with a failure rate of only 0.2%. Possible side effects from Norplant are vaginal bleeding, headaches, nausea, dizziness, and nervousness.

http://www.brookscole.com/chemistry/seager_chemistry
Web site 4.2 provides further information on Norplant.

A variety of oral contraceptives contain norethynodrel.

➤ **FIGURE 4.4** This young woman can see herself because the silver coating on the back of the mirror reflects light.

Benedict's reagent
A mild oxidizing solution containing Cu^{2+} ions used to test for the presence of aldehydes.

Benedict's reagent is a mild oxidizing solution containing Cu^{2+} ions. When an aldehyde oxidation takes place in an alkaline solution containing Cu^{2+}, a red precipitate of Cu_2O is produced:

$$
\begin{array}{c}
\overset{\displaystyle O}{\overset{\displaystyle \|}{R-C-H}} + 2Cu^{2+} + 5OH^- \longrightarrow \overset{\displaystyle O}{\overset{\displaystyle \|}{R-C-O^-}} + Cu_2O + 3H_2O \quad (4.6)
\end{array}
$$

certain aldehydes   Benedict's reagent (blue)   a carboxylate ion   red precipitate

The presence of Cu^{2+} in Benedict's reagent causes this alkaline solution to have a bright blue color. A positive Benedict's test consists of the appearance of a colored (usually red) precipitate. All aldehydes give a positive Tollens' test, but only certain aldehydes and one type of easily oxidized ketone readily give a positive Benedict's test. The necessary structural features are shown.

$$
\underset{\text{(a) an aldehyde with an adjacent carbonyl group}}{\overset{\displaystyle O \quad O}{\overset{\displaystyle \| \quad \|}{R-C-C-H}}}
\qquad
\underset{\text{(b) an aldehyde with an adjacent alcohol group}}{\overset{\displaystyle O}{\overset{\displaystyle \|}{R-\underset{\underset{\displaystyle OH}{|}}{CH}-C-H}}}
\qquad
\underset{\text{(c) a ketone with an adjacent alcohol group}}{\overset{\displaystyle O}{\overset{\displaystyle \|}{R-\underset{\underset{\displaystyle OH}{|}}{CH}-C-R'}}}
$$

The structural features (b) or (c) are found in a number of sugars, including glucose. Accordingly, Benedict's reagent reacts with glucose to produce a red precipitate of Cu_2O (see ■ Figure 4.5).

The Addition of Hydrogen

In Section 2.3, we learned that addition reactions are common for compounds containing carbon–carbon double bonds. The same is true for the carbon–oxygen double bonds of carbonyl groups. An important example is the addition of H_2 in the presence of catalysts such as nickel or platinum:

$$
\text{General reaction:} \quad \underset{\text{aldehyde}}{R - \overset{\overset{\displaystyle O}{\|}}{C} - H} + H_2 \xrightarrow{\text{Pt}} \underset{\text{primary alcohol}}{R - \underset{\overset{\displaystyle |}{H}}{\overset{\overset{\displaystyle OH}{|}}{C}} - H} \qquad (4.7)
$$

$$
\text{General reaction:} \quad \underset{\text{ketone}}{R - \overset{\overset{\displaystyle O}{\|}}{C} - R'} + H_2 \xrightarrow{\text{Pt}} \underset{\text{secondary alcohol}}{R - \underset{\overset{\displaystyle |}{H}}{\overset{\overset{\displaystyle OH}{|}}{C}} - R'} \qquad (4.8)
$$

$$
\text{Specific example:} \quad \underset{\text{propanal}}{CH_3CH_2 - \overset{\overset{\displaystyle O}{\|}}{C} - H} + H_2 \xrightarrow{\text{Pt}} \underset{\text{1-propanol}}{CH_3CH_2CH_2} \overset{\overset{\displaystyle OH}{|}}{} \qquad (4.9)
$$

$$
\text{Specific example:} \quad \underset{\text{acetone}}{CH_3 - \overset{\overset{\displaystyle O}{\|}}{C} - CH_3} + H_2 \xrightarrow{\text{Pt}} \underset{\text{2-propanol}}{CH_3 - \overset{\overset{\displaystyle OH}{|}}{CH} - CH_3} \qquad (4.10)
$$

We see that the addition of hydrogen to an aldehyde produces a primary alcohol, whereas the addition to a ketone gives a secondary alcohol. Notice that these reactions are reductions that are essentially the reverse of the alcohol oxidations used to prepare aldehydes and ketones (Reactions 4.1 and 4.2).

Draw the structural formula for each product of the following reactions:

a.

$$
CH_3CH_2 - \overset{\overset{\displaystyle O}{\|}}{C} - CH_2CH_3 + H_2 \xrightarrow{\text{Pt}}
$$

b.

$$
\underset{}{\bigcirc\!\!-\!\!\overset{\overset{\displaystyle O}{\|}}{C} - H} + H_2 \xrightarrow{\text{Pt}}
$$

From left to right, three test tubes containing Benedict's reagent, 0.5% glucose solution, and 2.0% glucose solution.

The addition of Benedict's reagent produces colors (due to the red Cu_2O) that indicate the amount of glucose present.

➤ **FIGURE 4.5** The Benedict's test for glucose.

Learning Check 4.4

The Addition of Alcohols

Aldehydes react with alcohols to form **hemiacetals**, a functional group that is very important in carbohydrate chemistry (Chapter 7). Hemiacetals are usually not very stable and are difficult to isolate. With excess alcohol present and with an acid catalyst, however, a stable product called an acetal is formed:

Hemiacetal
A compound that contains the

$$
\text{functional group} \quad -\overset{\overset{\displaystyle OH}{|}}{\underset{\overset{\displaystyle |}{OR}}{C}} - H
$$

$$
\underset{\substack{\text{General} \\ \text{reaction:}}}{} \quad \underset{\text{aldehyde}}{R - \overset{\overset{\displaystyle O}{\|}}{C} - H} + \underset{\text{alcohol}}{R' - O - H} \rightleftharpoons \underset{\substack{\text{hemiacetal} \\ \text{intermediate}}}{R - \underset{\overset{\displaystyle |}{OR'}}{\overset{\overset{\displaystyle OH}{|}}{C}} - H} \underset{\substack{H^+ \text{ and} \\ R' - O - H}}{\rightleftharpoons} \underset{\text{acetal}}{R - \underset{\overset{\displaystyle |}{OR'}}{\overset{\overset{\displaystyle OR'}{|}}{C}} - H} + H_2O \qquad (4.11)
$$

Specific example:

$$CH_3 - \overset{\overset{\displaystyle O}{\|}}{C} - H \; + \; CH_3 - O - H \; \rightleftharpoons \; CH_3 - \overset{\overset{\displaystyle OH}{|}}{\underset{\underset{\displaystyle OCH_3}{|}}{C}} - H \; \underset{CH_3 - O - H}{\overset{H^+ \text{ and}}{\rightleftharpoons}} \; CH_3 - \overset{\overset{\displaystyle OCH_3}{|}}{\underset{\underset{\displaystyle OCH_3}{|}}{C}} - H \; + \; H_2O \qquad (4.12)$$

acetaldehyde   methanol   hemiacetal intermediate   acetal

Acetal

A compound that contains the

$$-\overset{\overset{\displaystyle OR}{|}}{\underset{\underset{\displaystyle OR}{|}}{C}} - H \quad \text{arrangement of atoms.}$$

Hemiacetals contain an —OH group, hydrogen, and an —OR group on the same carbon. Thus, a hemiacetal appears to have alcohol and ether functional groups. **Acetals** contain a carbon that has a hydrogen and two —OR groups attached; they look like diethers:

$$CH_3 - \overset{\overset{\displaystyle OH}{|}}{\underset{\underset{\displaystyle OCH_3}{|}}{C}} - H \qquad\qquad CH_3 - \overset{\overset{\displaystyle OCH_3}{|}}{\underset{\underset{\displaystyle OCH_3}{|}}{C}} - H$$

hemiacetal carbon   *acetal carbon*

To better understand the changes that have occurred, let's look first at hemiacetal formation and then at acetal formation. The first molecule of alcohol adds across the double bond of the aldehyde in a manner similar to other addition reactions:

—H attaches here

$$CH_3 - \overset{\overset{\displaystyle O}{\|}}{C} - H \; + \; CH_3 - O - H \; \rightleftharpoons \; CH_3 - \overset{\overset{\displaystyle OH}{|}}{\underset{\underset{\displaystyle OCH_3}{|}}{C}} - H \qquad (4.13)$$

—OCH₃ attaches here

hemiacetal

Hemiketal

A compound that contains the

$$-\overset{}{\underset{}{C}} - \overset{\overset{\displaystyle OH}{|}}{\underset{\underset{\displaystyle OR}{|}}{C}} - \overset{}{\underset{}{C}} - \quad \text{group.}$$

Ketal

A compound that contains the

$$-\overset{}{\underset{}{C}} - \overset{\overset{\displaystyle OR}{|}}{\underset{\underset{\displaystyle OR}{|}}{C}} - \overset{}{\underset{}{C}} - \quad \text{group.}$$

The second step, in which the hemiacetal is converted to an acetal, is like earlier substitution reactions. In this case, an —OCH₃ substitutes for an —OH:

$$CH_3 - \overset{\overset{\displaystyle OH}{|}}{\underset{\underset{\displaystyle OCH_3}{|}}{C}} - H \; + \; CH_3 - O - H \; \overset{H^+}{\rightleftharpoons} \; CH_3 - \overset{\overset{\displaystyle OCH_3}{|}}{\underset{\underset{\displaystyle OCH_3}{|}}{C}} - H \; + \; H - OH \qquad (4.14)$$

hemiacetal   acetal

In a very similar way, ketones can react with alcohols to form **hemiketals** and **ketals**:

hemiketal carbon   *ketal carbon*

General reaction:

$$R - \overset{\overset{\displaystyle O}{\|}}{C} - R' \; + \; R' - O - H \; \rightleftharpoons \; R - \overset{\overset{\displaystyle OH}{|}}{\underset{\underset{\displaystyle OR'}{|}}{C}} - R' \; \underset{R' - OH}{\overset{H^+ \text{ and}}{\rightleftharpoons}} \; R - \overset{\overset{\displaystyle OR'}{|}}{\underset{\underset{\displaystyle OR'}{|}}{C}} - R' \; + \; H_2O \qquad (4.15)$$

ketone   alcohol   hemiketal intermediate   ketal

Specific example:

$$CH_3-\overset{\overset{\displaystyle O}{\|}}{C}-CH_3 \ + \ CH_3CH_2OH \ \rightleftharpoons \ CH_3-\overset{\overset{\displaystyle OH}{|}}{\underset{\underset{\displaystyle OCH_2CH_3}{|}}{C}}-CH_3 \ \underset{CH_3CH_2-OH}{\overset{H^+ \text{ and}}{\rightleftharpoons}} \ CH_3-\overset{\overset{\displaystyle OCH_2CH_3}{|}}{\underset{\underset{\displaystyle OCH_2CH_3}{|}}{C}}-CH_3 \ + \ H_2O \qquad (4.16)$$

acetone   ethanol   hemiketal   ketal
           intermediate

➤ Example 4.3

Identify each of the following as a hemiacetal, hemiketal, acetal, ketal, or none of the above:

a. $CH_3-O-CH_2CH_2-OH$

b. $CH_3-O-\overset{\overset{\displaystyle OH}{|}}{C}HCH_3$

c. $CH_3CH_2-O-\overset{\overset{\displaystyle OCH_3}{|}}{\underset{\underset{\displaystyle CH_3}{|}}{C}}-CH_3$

d.
(cyclopentane ring with OH and OCH_3 on same carbon)

Solution

a. Look first at any structure to see if there is a carbon with two oxygens attached. Hemiacetals, hemiketals, acetals, and ketals are all alike in that regard. Compound a has two oxygens, but they are on different carbons; thus it is none of the possibilities.

➤ HOW REACTIONS OCCUR

HEMIACETAL FORMATION

In the first step of the mechanism, the carbonyl oxygen is protonated:

$$R-\overset{\overset{\displaystyle :\ddot{O}}{\|}}{C}-H \ \overset{H^+}{\rightleftharpoons} \ R-\overset{\overset{\displaystyle \overset{+}{O}/H}{\|}}{C}-H$$

The alcohol oxygen then bonds to the carbonyl carbon:

$$R-\overset{\overset{\displaystyle \overset{+}{O}/H}{\|}}{\underset{\underset{\displaystyle R'-\ddot{O}-H}{\uparrow}}{C}}-H \ \rightleftharpoons \ R-\overset{\overset{\displaystyle :\ddot{O}/H}{|}}{\underset{\underset{\displaystyle \overset{R'}{}\ \ H}{\overset{+}{O}}}{C}}-H$$

In the final step, a proton is given off to form the hemiacetal and regenerate the acid catalyst.

$$R-\overset{\overset{\displaystyle :\ddot{O}/H}{|}}{\underset{\underset{\displaystyle \overset{R'}{}\ H}{\overset{+}{O}}}{C}}-H \ \rightleftharpoons \ R-\overset{\overset{\displaystyle :\ddot{O}H}{|}}{\underset{\underset{\displaystyle R'}{\ddot{O}:}}{C}}-H \ + \ H^+$$

hemiacetal

b. After finding a carbon with two oxygens attached, next look to see if one is an —OH. If it is, that makes it a hemiacetal or hemiketal. The third step is to look at the carbon containing the two oxygens. If it is attached to an —H, the compound was originally an aldehyde and not a ketone. Thus, this compound is a hemiacetal:

$$OH \longleftarrow \textit{one group is —OH}$$
$$|$$
$$CH_3 - O - \underset{\nwarrow}{CHCH_3}$$

two oxygens and
one hydrogen attached

c. There is a carbon with two oxygens attached and neither is an —OH. Thus, the compound is an acetal or ketal:

$$OCH_3$$
$$|$$
$$CH_3CH_2 - O - \underset{|}{\overset{}{C}} - CH_3$$
$$|$$
$$CH_3$$

no hydrogens attached

No hydrogens are attached to the carbon bonded to two oxygens. The compound is a ketal.

d. Two oxygens are attached to the same carbon. One group is an —OH. No hydrogens are attached to the central carbon. The compound is a hemiketal.

Learning Check 4.5

Draw the structural formula for the hemiacetal and hemiketal intermediates and for the acetal and ketal products of the following reactions. Label each structure as a hemiacetal, hemiketal, ketal, or acetal.

a.
$$\overset{O}{\overset{||}{C}} - H$$
$$+ 2CH_3CH_2 - OH \xrightarrow{H^+}$$

b.
$$\overset{O}{\overset{||}{C}} - CH_3$$
$$+ 2CH_3 - OH \xrightarrow{H^+}$$

Some molecules contain both an —OH and a C=O group on different carbon atoms. In such cases, an intramolecular (within the molecule) reaction can occur, and a cyclic hemiacetal or hemiketal can be formed:

$$
\begin{array}{c}
{}^{5}CH_2 - O - H \\
{}^{4}CH_2 \quad C = O \\
CH_2 - CH_2 \quad H \\
{}^{3} \quad {}^{2}
\end{array}
\rightleftharpoons
\begin{array}{c}
{}^{5} \quad O \quad OH \\
{}^{4} \quad {}^{1}H \\
{}^{3} \quad {}^{2}
\end{array}
\qquad (4.17)
$$

open-chain molecule   cyclic structure
containing an aldehyde   containing a
and alcohol group   hemiacetal group

Cyclic hemiacetals and hemiketals are much more stable than the open-chain compounds discussed earlier. In Chapter 7, we'll see that sugars such as glu-

A REACTION MAP FOR ALDEHYDES AND KETONES

This reaction map is designed to help you master organic reactions. Whenever you are trying to complete an organic reaction, use these two basic steps: (1) Identify the functional group that is to react, and (2) identify the reagent that is to react with the functional group. If the reacting functional group is an aldehyde or a ketone, find the reagent in the summary diagram, and use the diagram to predict the correct products.

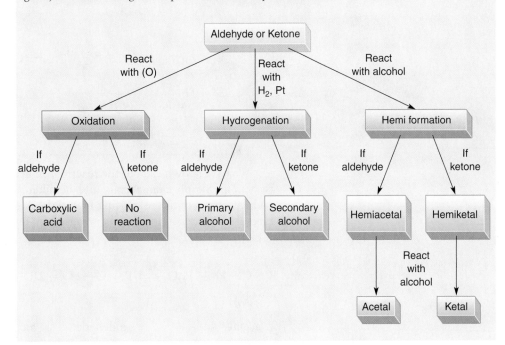

cose and fructose exist predominantly in the form of cyclic hemiacetals and hemiketals.

In another reaction important in carbohydrate chemistry (Chapter 7), cyclic acetals and ketals may form from the hemistructures. As before, all reactions are reversible:

$$\text{cyclic hemiacetal} + CH_3OH \overset{H^+}{\rightleftharpoons} \text{cyclic acetal} + H_2O \qquad (4.18)$$

➤ Example 4.4

Identify each of the following as a cyclic hemiacetal, hemiketal, acetal, or ketal:

a.

b.

c.

d.

Solution

Use the same criteria that were applied to open-chain compounds in Example 4.3. Compound a has two oxygens attached to the same carbon. One is an —OH (hemiacetal or hemiketal). The central carbon has no hydrogens. Thus, the compound is a hemiketal. These criteria give the following results for each structure:

a.

hemiketal carbon

b.

ketal carbon (—OH absent, —H absent)

c.

acetal carbon (—OH absent, —H present)

d.

hemiacetal carbon (—OH present, —H present)

We have shown each of the reactions in acetal and ketal formation as being reversible. Simple hemiacetals and hemiketals are characteristically unstable and easily revert to the starting materials. Acetals and ketals are stable structures, but the reactions may be reversed by using water and an acid catalyst:

General reaction:

$$R - \underset{\underset{OR'}{|}}{\overset{\overset{OR'}{|}}{C}} - H + H_2O \underset{}{\overset{H^+}{\rightleftharpoons}} R - \overset{\overset{O}{\|}}{C} - H + 2R' - OH \qquad (4.19)$$

acetal   aldehyde   alcohol

General reaction:

$$R - \underset{\underset{OR''}{|}}{\overset{\overset{OR''}{|}}{C}} - R' + H_2O \underset{}{\overset{H^+}{\rightleftharpoons}} R - \overset{\overset{O}{\|}}{C} - R' + 2R'' - OH \qquad (4.20)$$

ketal   ketone   alcohol

Specific example:

$$CH_3 - \underset{\underset{OCH_2CH_3}{|}}{\overset{\overset{OCH_2CH_3}{|}}{C}} - CH_3 + H_2O \underset{}{\overset{H^+}{\rightleftharpoons}} CH_3 - \overset{\overset{O}{\|}}{C} - CH_3 + 2CH_3CH_2 - OH \qquad (4.21)$$

This type of reaction, in which water causes a compound to split into component substances, is called **hydrolysis**. It will be an important reaction to remember as we discuss, in later chapters, the chemistry of biomolecules such as carbohydrates, lipids, and proteins.

Hydrolysis
Bond breakage by reaction with water.

Learning Check 4.6

Draw the structural formulas needed to complete the following hydrolysis reactions:

a. $CH_3CH_2CH_2 - \underset{\underset{OCH_2CH_3}{|}}{\overset{\overset{OCH_2CH_3}{|}}{C}} - H + H_2O \overset{H^+}{\rightleftharpoons}$

b. $\langle\!\!\!\!\!\bigcirc\!\!\!\!\!\rangle\!\!<^{OCH_3}_{OCH_3} + H_2O \overset{H^+}{\rightleftharpoons}$

➤ Example 4.5

Draw structural formulas for the products of the following reactions:

a.

$$\text{(cyclic hemiketal with } O, OH, CH_3 \text{)} + CH_3CH_2OH \xrightarrow{H^+} \rightleftharpoons$$

b.

$$\text{(ring with } O, OCH_2CH_2CH_3, H \text{)} + H_2O \xrightarrow{H^+} \rightleftharpoons$$

c.

$$\text{(ring with } O, CH_2CH_3, OCH_3 \text{)} + H_2O \xrightarrow{H^+} \rightleftharpoons$$

Solution

a. The starting material is a cyclic hemiketal. Reaction with an alcohol produces a cyclic ketal by a substitution of $-OCH_2CH_3$ for the $-OH$:

$$\text{(ring } O, OH, CH_3) + CH_3CH_2O-H \rightleftharpoons \xrightarrow{H^+} \text{(ring } O, OCH_2CH_3, CH_3) + H-OH$$

b. The starting compound is an acetal. Hydrolysis with water will form an aldehyde and two alcohol groups:

becomes an –OH   *alcohol groups*

$$\rightleftharpoons$$

becomes $-\overset{\overset{\displaystyle O}{\|}}{C}-H$

c. The starting compound is a ketal and upon hydrolysis will form a ketone and two alcohol groups:

becomes an –OH   *alcohol groups*

$$\rightleftharpoons \quad + HO-CH_3$$

becomes $-\overset{\overset{\displaystyle O}{\|}}{C}-$

Learning Check 4.7

Identify each of the following as cyclic hemiacetals, hemiketals, acetals, or ketals and show structural formulas for the hydrolysis products:

a.

b.

CHEMISTRY AROUND US • 4.2

VITAMIN A AND BIRTH DEFECTS

Three active forms of vitamin A are known. One is the aldehyde called retinal. In the other two forms, the aldehyde group is replaced by an alcohol group (retinol) or a carboxylic acid group (retinoic acid).

$$H_3C \quad CH_3$$

retinal

Vitamin A is an essential substance for healthy vision and skin, but as is sometimes true, too much of this good thing creates problems. A relationship between excessive vitamin A intake during pregnancy and birth defects has been recognized for some time. However, the minimum amount of vitamin A required to produce harmful effects is now believed to be much lower than was previously thought. According to a recent study, the consumption of vitamin A at or above 10,000 international units per day was linked to birth defects. An **international unit** (IU) is a measure of vitamin activity, determined by such biological methods as feeding a compound to vitamin-deprived animals and measuring growth. In the study, an increase in a variety of birth defects was detected in babies born to mothers who took more than 10,000 IU of vitamin A per day. The birth defects found included those of the heart, brain, limbs, kidneys, and genitals.

The active forms of vitamin A that were associated with the defects are found in fortified breakfast cereals, liver, and some vitamin supplements. Beta-carotene, a form of vitamin A that has not been linked to birth defects, is now used in many prenatal vitamin supplements. Beta-carotene is a plant-based vitamin A source that is converted into an active form of vitamin A once ingested.

A maximum dose of 5000 IU of vitamin A during pregnancy is now recommended by the American College of Obstetricians and Gynecologists. In addition, the diet should contain plenty of fruits and vegetables that are naturally rich in beta-carotene (see Chemistry Around Us 2.1).

International unit (IU)
A measure of vitamin activity, determined by biological methods.

http://www.brookscole.com/chemistry/ seager_chemistry
See Web site 4.3 for further information on the uses of formaldehyde.

➤ **FIGURE 4.6** Acetone is an excellent solvent for many organics, including nail polish. Why does acetone-based polish remover evaporate fairly quickly when used?

http://www.brookscole.com/chemistry/ seager_chemistry
Web site 4.4 provides more information on the uses of acetone.

➤ 4.4 Important Aldehydes and Ketones

Formaldehyde: $H-\overset{\overset{\displaystyle O}{\|}}{C}-H$

As seen with other functional classes, the most important compound of the class is usually the simplest member. Formaldehyde, the simplest aldehyde, is a gas at room temperature, but it is often supplied and used in the form of a 37% aqueous solution called *formalin*. This solution kills microorganisms and is effective in sterilizing surgical instruments. It is also used to embalm cadavers. Formaldehyde is a key industrial chemical in the production of plastics such as Bakelite and Formica.

Acetone: $CH_3-\overset{\overset{\displaystyle O}{\|}}{C}-CH_3$

Judging from the quantity used, acetone is by far the most important of the ketones. Over 1 billion pounds are used annually in the United States. It is particularly useful as a solvent because it dissolves most organic compounds and yet it is miscible in water. It is widely used as a solvent for coatings, which range from nail polish remover to exterior enamel paints (see ■ Figure 4.6).

A number of naturally occurring aldehydes and ketones play important roles within living systems. The carbonyl functional group is fairly common in biological compounds. For example, progesterone and testosterone, female and male sex hormones, respectively, are both ketones.

testosterone

Several naturally occurring aldehydes and ketones have very fragrant odors and are used in flavorings (see ■ Figure 4.7 and ■ Table 4.3).

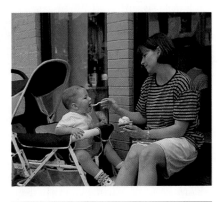

➤ **FIGURE 4.7** Vanilla ice cream owes its characteristic flavor to a special aldehyde.

TABLE 4.3	Some fragrant aldehydes and ketones	
Name	**Structural formula**	**Sources and typical uses**
Vanillin		Vanilla bean; flavoring
Cinnamaldehyde		Oil of cinnamon; flavoring
Citral		Rinds of lemons, limes, and oranges; citrus flavoring
Biacetyl	$CH_3 - \overset{\overset{O}{\|\|}}{C} - \overset{\overset{O}{\|\|}}{C} - CH_3$	Butter; flavoring for margarines
Camphor		Camphor tree; characteristic medicinal odor; an ingredient in some inhalants
Menthone		Mint plants; peppermint flavoring

CHEMISTRY AROUND US • 4.3

VANILLOIDS: HOT RELIEF FROM PAIN

The word *vanilloids* might bring to mind images of vanilla-flavored candies or ice cream. However, vanilloids have properties very different from those of vanilla, and their inclusion in foods would result in a hot and spicy flavor that might even be extremely irritating, rather than a smooth vanilla flavor.

The best known vanilloids are capsaicin and resiniferatoxin (RTX). Capsaicin is the compound in various types of red peppers that gives them their characteristic hot and spicy flavor. It is used in a variety of products ranging from personal defense pepper sprays to squirrel-proof birdseed. RTX is a powerful irritant extracted from a cactuslike plant, *Euphorbia resinifera.* The milky latex obtained from the plant is referred to in medical literature dating as far back as the first century. In these writings, it was described as a nose and skin irritant and as a treatment for chronic pain. In powdered form, RTX causes sneezing, a characteristic used by some practical jokers.

Structurally, molecules of capsaicin and RTX (not shown) are similar to those of vanillin in that they both have a homovanillyl group in their structure.

vanillin   homovanillyl group   capsaicin

When either of the compounds is applied to the skin or mouth, a burning sensation results. This sensation is caused by the compounds bonding to receptors of small sensory neurons associated with the transmission of pain. As a result of the repeated applications of capsaicin or RTX, these neurons, or C sensory nerve fibers, become desensitized to the initial irritation or pain the compounds cause. This characteristic allows the compounds to be used topically to relieve chronic pain. Capsaicin is now used as an ingredient in a few OTC pain-relieving creams such as Capzasin-P, Heet, Menthacin, and Pain Doctor.

Both capsaicin and RTX have also been tested as treatments for a condition called *bladder hyperreflexia,* which is characterized by the urge to urinate even when there is very little urine in the bladder. Capsaicin effectively relieved the problem, but the first application caused an intense burning sensation that was intolerable and unacceptable to patients. RTX fared better in the tests. It also relieved the condition, and caused only mild discomfort and an itching sensation with the first application.

Basic research has led to some important discoveries about these compounds. The gene for the capsaicin receptor has been cloned, making available the large quantities of the receptor needed to do further research and development of new drugs. For example, it has been found that the receptor responds not only to capsaicin but also to levels of heat that can damage tissue. It is suggested that possibly the reason hot peppers feel hot in the mouth is because they stimulate the same pain receptors that heat does.

Another important development is the discovery of a way to synthesize RTX in the laboratory. The availability of synthetic RTX may reduce the cost for the material and make possible further advances in the use of RTX in medicines.

http://www.brookscole.com/chemistry/seager_chemistry
Web site 4.5 gives additional information on capsaicin and RTX.

Hot peppers are a source of capsaicin.

➤ **FOR FUTURE REFERENCE** FOOD SOURCES OF VITAMIN A

Three different forms of vitamin A are active in the body: retinol (an alcohol), retinal (an aldehyde), and retinoic acid (a carboxylic acid). Collectively, these three types of vitamin A are called retinoids. Foods from animal sources contain compounds that are easily hydrolyzed to retinol during digestion. Foods derived from plants contain compounds known as carotenoids, and some of these compounds, such as beta-carotene, have vitamin A activity in the body. In the body, vitamin A plays major roles in vision, protein synthesis, cell differentiation, reproduction and growth. Because of these many functions, good health depends on an adequate supply, which is generally available from a well-balanced diet. Excessive vitamin A intake, usually from the use of supplements, poses a risk as pointed out in Chemistry Around Us 4.3.

Vitamin A from food is supplied in the form of a variety of different retinoids and carotenoids. In order to standardize the amount of the vitamin obtained in these different forms, the vitamin content is expressed in terms of retinol equivalents or RE. One RE is equivalent in vitamin A activity to 1 µg of retinol. Sometimes vitamin A content is reported in terms of the older international units or IU. The relationship between the two units is 1 IU = 0.3 RE, or 3.3 IU = 1 RE. The chart below contains a listing of foods that provide a minimum of 50 RE per serving.

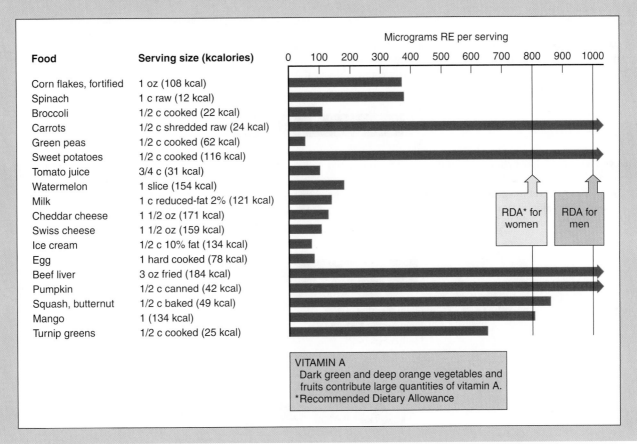

Food	Serving size (kcalories)
Corn flakes, fortified	1 oz (108 kcal)
Spinach	1 c raw (12 kcal)
Broccoli	1/2 c cooked (22 kcal)
Carrots	1/2 c shredded raw (24 kcal)
Green peas	1/2 c cooked (62 kcal)
Sweet potatoes	1/2 c cooked (116 kcal)
Tomato juice	3/4 c (31 kcal)
Watermelon	1 slice (154 kcal)
Milk	1 c reduced-fat 2% (121 kcal)
Cheddar cheese	1 1/2 oz (171 kcal)
Swiss cheese	1 1/2 oz (159 kcal)
Ice cream	1/2 c 10% fat (134 kcal)
Egg	1 hard cooked (78 kcal)
Beef liver	3 oz fried (184 kcal)
Pumpkin	1/2 c canned (42 kcal)
Squash, butternut	1/2 c baked (49 kcal)
Mango	1 (134 kcal)
Turnip greens	1/2 c cooked (25 kcal)

Micrograms RE per serving

RDA* for women
RDA for men

VITAMIN A
Dark green and deep orange vegetables and fruits contribute large quantities of vitamin A.
*Recommended Dietary Allowance

SOURCE: Adapted from Whitney, E.N.; Rolfes, S.R. *Understanding Nutrition,* 8th ed. Belmont, CA: West/Wadsworth, 1999.

The Nomenclature of Aldehydes and Ketones. The functional groups characteristic of aldehydes and ketones are very similar; they both contain a carbonyl group:

$$\begin{array}{c} O \\ \parallel \\ -C- \end{array}$$

Aldehydes have a hydrogen attached to the carbonyl carbon,

$$\begin{array}{c} O \\ \parallel \\ -C-H \end{array}$$

whereas ketones have two carbons attached to the carbonyl carbon:

$$\begin{array}{c} \quad\ O \\ |\ \ \parallel\ \ | \\ -C-C-C- \\ |\qquad\ | \end{array}$$

The IUPAC ending for aldehyde names is *-al*, whereas that for ketones is *-one*.

Physical Properties. Molecules of aldehydes and ketones cannot form hydrogen bonds with each other. As a result, they have lower boiling points than alcohols of similar molecular weight. Alde- hydes and ketones have higher boiling points than alkanes of sim- ilar molecular weight because of the presence of the polar car- bonyl group. The polarity of the carbonyl group, and the fact that aldehydes and ketones can form hydrogen bonds with water, ex- plains why the low-molecular-weight compounds of those kinds are water-soluble.

Chemical Properties. Aldehydes and ketones are prepared by the oxidation of primary and secondary alcohols, respectively. Alde- hydes can be further oxidized to carboxylic acids, but ketones re- sist oxidation. Thus, aldehydes are oxidized by Tollens' reagent (Ag^+) and Benedict's solution (Cu^{2+}), whereas ketones are not. A characteristic reaction of both aldehydes and ketones is the addi- tion of hydrogen to the carbonyl double bond to form alcohols. In a reaction that is very important in sugar chemistry, an alcohol can add across the carbonyl group of an aldehyde to produce a hemi- acetal. The substitution reaction of a second alcohol molecule with the hemiacetal produces an acetal. Ketones can undergo sim- ilar reactions to form hemiketals and ketals.

Important Aldehydes and Ketones. Formaldehyde is a key indus- trial chemical in the production of plastics such as Bakelite and Formica. Acetone is an important solvent. Some naturally occur- ring aldehydes and ketones are important in living systems. Some function as sex hormones; others are used as flavorings.

When you have completed your study of this chapter, you should be able to:
1. Recognize the carbonyl group in compounds, and classify the compounds as aldehydes or ketones.
2. Assign IUPAC names to aldehydes and ketones.
3. Compare the physical properties of aldehydes and ketones with those of compounds in other classes.
4. Write key reactions for aldehydes and ketones.
5. Give specific uses for aldehydes and ketones.

Acetal (4.3)
Aldehyde (4.1)
Benedict's reagent (4.3)
Carbonyl group (Introduction)

Hemiacetal (4.3)
Hemiketal (4.3)
Hydrolysis (4.3)
International unit (IU) (4.3)

Ketal (4.3)
Ketone (4.1)
Tollens' reagent (4.3)

1. Oxidation of an aldehyde to give a carboxylic acid (Section 4.3):

$$\begin{array}{ccc} O & & O \\ \parallel & & \parallel \\ R-C-H + (O) & \longrightarrow & R-C-OH \end{array}$$

Reaction 4.1

2. Attempted oxidation of a ketone (Section 4.3):

$$\begin{array}{c} O \\ \parallel \\ R-C-R' + (O) \longrightarrow \text{no reaction} \end{array}$$

Reaction 4.2

3. Hydrogenation of an aldehyde to give a primary alcohol (Section 4.3):

$$R-\overset{\overset{\displaystyle O}{\|}}{C}-H + H_2 \xrightarrow{\text{Pt}} R-\overset{\overset{\displaystyle OH}{|}}{\underset{\underset{\displaystyle H}{|}}{C}}-H$$

Reaction 4.7

4. Hydrogenation of a ketone to give a secondary alcohol (Section 4.3):

$$R-\overset{\overset{\displaystyle O}{\|}}{C}-R' + H_2 \xrightarrow{\text{Pt}} R-\overset{\overset{\displaystyle OH}{|}}{\underset{\underset{\displaystyle H}{|}}{C}}-R'$$

Reaction 4.8

5. Addition of an alcohol to an aldehyde to form a hemiacetal and then an acetal (Section 4.3):

$$R-\overset{\overset{\displaystyle O}{\|}}{C}-H + R'-OH \rightleftharpoons R-\overset{\overset{\displaystyle OH}{|}}{\underset{\underset{\displaystyle OR'}{|}}{C}}-H \underset{\xrightarrow{R'-OH}}{\overset{H^+ \text{ and}}{\rightleftharpoons}} R-\overset{\overset{\displaystyle OR'}{|}}{\underset{\underset{\displaystyle OR'}{|}}{C}}-H + H_2O$$

Reaction 4.11

6. Addition of an alcohol to a ketone to form a hemiketal and a ketal (Section 4.3):

$$R-\overset{\overset{\displaystyle O}{\|}}{C}-R' + R'-O-H \rightleftharpoons R-\overset{\overset{\displaystyle OH}{|}}{\underset{\underset{\displaystyle OR'}{|}}{C}}-R' \underset{\xrightarrow{R'-OH}}{\overset{H^+ \text{ and}}{\rightleftharpoons}} R-\overset{\overset{\displaystyle OR'}{|}}{\underset{\underset{\displaystyle OR'}{|}}{C}}-R' + H_2O$$

Reaction 4.15

7. Hydrolysis of an acetal to yield an aldehyde and two moles of alcohol (Section 4.3):

$$R-\overset{\overset{\displaystyle OR'}{|}}{\underset{\underset{\displaystyle OR'}{|}}{C}}-H + H_2O \overset{H^+}{\rightleftharpoons} R-\overset{\overset{\displaystyle O}{\|}}{C}-H + 2R'-OH$$

Reaction 4.19

8. Hydrolysis of a ketal to yield a ketone and two moles of alcohol (Section 4.3):

$$R-\overset{\overset{\displaystyle OR''}{|}}{\underset{\underset{\displaystyle OR''}{|}}{C}}-R' + H_2O \overset{H^+}{\rightleftharpoons} R-\overset{\overset{\displaystyle O}{\|}}{C}-R' + 2R''-OH$$

Reaction 4.20

.... **EXERCISES** ....

THE NOMENCLATURE OF ALDEHYDES AND KETONES (SECTION 4.1)

4.1 What is the structural difference between an aldehyde and a ketone?

4.2 Identify each of the following compounds as an aldehyde, ketone, or neither:

a. $CH_3CH_2-\overset{\overset{\displaystyle O}{\|}}{C}-H$ b. $CH_3-\overset{\overset{\displaystyle O}{\|}}{C}-O-CH_3$

c. (cyclopentanone structure)

d. $CH_3CH_2CH_2-\overset{\overset{\displaystyle O}{\|}}{C}-OH$

e. $CH_3CH_2-\overset{\overset{\displaystyle O}{\|}}{C}-CH_3$ f. $CH_3CH_2-\overset{\overset{\displaystyle O}{\|}}{C}-NH_2$

4.3 Assign IUPAC names to the following aldehydes and ketones:

a. $CH_3-\overset{\overset{\displaystyle O}{\|}}{C}-H$ b. $CH_3\overset{\displaystyle CHCH_2CH}{\underset{\underset{\displaystyle CH_3}{|}}{}}\overset{}{\underset{\underset{\displaystyle CH_3}{|}}{}}-\overset{\overset{\displaystyle O}{\|}}{C}-H$

c. $CH_3\overset{\displaystyle CHCH_2}{\underset{\underset{\displaystyle CH_3}{|}}{}}-\overset{\overset{\displaystyle O}{\|}}{C}-CH_3$

d.

e.

4.4 Assign IUPAC names to the following aldehydes and ketones:

a. $CH_3CH_2-\overset{\overset{\displaystyle O}{\|}}{C}-H$

b. $CH_3\overset{\overset{\displaystyle Br}{|}}{C}HCH_2-\overset{\overset{\displaystyle O}{\|}}{C}-H$

c. $\bigcirc-CH_2CH_2-\overset{\overset{\displaystyle O}{\|}}{C}-H$

d. $CH_3\overset{\overset{\displaystyle CH_3}{|}}{C}HCH_2-\overset{\overset{\displaystyle O}{\|}}{C}-CH_3$

e.

4.5 Draw structural formulas for each of the following compounds:
a. methanal
b. 3-ethyl-2-pentanone
c. 3-methylcyclohexanone
d. 2,4,6-trimethylheptanal

4.6 Draw structural formulas for each of the following compounds:
a. butanal
b. 2-methyl-3-pentanone
c. 2-ethyl-4-propylcyclohexanone
d. 2-chloro-4-phenylpentanal

4.7 Draw structural formulas and give IUPAC names for all the isomeric aldehydes and ketones that have the molecular formula $C_5H_{10}O$.

4.8 What is wrong with each of the following names? Give the structure and correct name for each compound.
a. 3-ethyl-2-methylbutanal
b. 2-methyl-4-butanone
c. 4,5-dibromocyclopentanone

PHYSICAL PROPERTIES (SECTION 4.2)

4.9 Why does hydrogen bonding *not* take place between molecules of aldehydes or ketones?

4.10 Most of the remaining water in washed laboratory glassware can be removed by rinsing the glassware with acetone (propanone). Explain how this process works (acetone is much more volatile than water).

4.11 The boiling point of propanal is 49°C, whereas the boiling point of propanol is 97°C. Explain this difference in boiling points.

4.12 Explain why propane boils at −42°C, whereas ethanal, which has the same molecular weight, boils at 20°C.

4.13 Use a dotted line to show hydrogen bonding between molecules in each of the following pairs:

a. $CH_3-\overset{\overset{\displaystyle O}{\|}}{C}-CH_3$ and $\overset{\displaystyle H}{\underset{\displaystyle H}{|}}$ O

b.

4.14 Use a dotted line to show hydrogen bonding between molecules in each of the following pairs:

a.

b.

4.15 Arrange the following compounds in order of increasing boiling point:

a. O   **b.** OH   **c.** CH₃

4.16 The compounds menthone and menthol are fragrant substances present in an oil produced by mint plants:

menthone   menthol

When pure, one of these pleasant-smelling compounds is a liquid at room temperature; the other is a solid. Identify the solid and the liquid and explain your reasoning.

CHEMICAL PROPERTIES (SECTION 4.3)

4.17 Write an equation for the formation of the following compounds from the appropriate alcohol:

a. $CH_3CH_2CH_2-\overset{\overset{\displaystyle O}{\|}}{C}-H$

b. CH$_3$

（2-methylcyclohexanone structure）

$$\underset{\text{c.}}{}\ CH_3-\overset{\overset{\displaystyle O}{\|}}{C}-CH_2CH_3$$

4.18 Write an equation for the formation of the following compounds from the appropriate alcohol:

a. （cyclopropyl ketone with CH$_3$）

b. $CH_3-\overset{\overset{\displaystyle O}{\|}}{C}-\underset{\underset{\displaystyle Br}{|}}{C}HCH_3$

c. （benzene ring with C—H double bond O and CH$_2$CH$_3$）

4.19 Identify the following structures as hemiacetals, hemiketals, or neither:

a. $CH_3-\overset{\overset{\displaystyle CH_3}{|}}{\underset{\underset{\displaystyle H}{|}}{C}}-OCH_3$

b. $CH_3CH_2\overset{\overset{\displaystyle OCH_3}{|}}{C}H-OH$

c. $CH_3-\overset{\overset{\displaystyle OH}{|}}{\underset{\underset{\displaystyle OH}{|}}{C}}-CH_2CH_3$

d. $CH_3\overset{\overset{\displaystyle OH}{|}}{\underset{\underset{\displaystyle CH_3}{|}}{C}}-OCH_2CH_3$

4.20 Identify the following structures as hemiacetals, hemiketals, or neither:

a. $CH_3-O-\overset{\overset{\displaystyle CH_3}{|}}{\underset{\underset{\displaystyle CH_3}{|}}{C}}-OH$

b. $CH_3CH_2\overset{\overset{\displaystyle OH}{|}}{C}H-OCH_2CH_3$

c. （cyclopentane with OH and OCH$_3$）

d. （benzene ring O—CH with OH and CH$_2$CH$_3$）

4.21 Label each of the following as acetals, ketals, or neither:

a. $CH_3CH_2\overset{\overset{\displaystyle OCH_3}{|}}{C}H-OCH_2CH_3$

b. $CH_3\overset{\overset{\displaystyle OCH_2CH_3}{|}}{C}H-OH$

c. （cyclopentane with CH—OCH$_3$ and OCH$_3$）

4.22 Label each of the following as acetals, ketals, or neither:

a. $CH_3-\overset{\overset{\displaystyle OCH_2CH_3}{|}}{\underset{\underset{\displaystyle OCH_2CH_3}{|}}{C}}-CH_3$

b.

c. $CH_3CH_2-\overset{\overset{\displaystyle OH}{|}}{\underset{\underset{\displaystyle OCH_3}{|}}{C}}-CH_3$

4.23 Label each of the following structures as a cyclic hemiacetal, hemiketal, acetal, ketal, or none of these:

a. （pyran ring with OH） **b.** （pyran ring with OH） **c.** （pyran ring with OCH$_3$）

4.24 Label each of the following structures as a hemiacetal, hemiketal, acetal, ketal, or none of these:

a. （pyran ring with OCH$_2$CH$_3$ and CH$_3$）

b. （cyclopentane with CH$_2$CH$_3$ and OH）

c. （cyclopentane with CH—OH and OCH$_3$）

4.25 What two functional groups react to form the following?
  a. A hemiacetal **b.** An acetal
  c. A ketal **d.** A hemiketal

4.26 Hemiacetals are sometimes referred to as potential aldehydes. Explain.

4.27 Complete the following statements:
  a. Oxidation of a secondary alcohol produces ____.
  b. Oxidation of a primary alcohol produces an aldehyde that can be further oxidized to a ____.
  c. Hydrogenation of a ketone produces ____.
  d. Hydrogenation of an aldehyde produces ____.
  e. Hydrolysis of an acetal produces ____.
  f. Hydrolysis of a ketal produces ____.

4.28 What observation characterizes a positive Tollens' test?

4.29 Which of the following compounds will react with Tollens' reagent (Ag$^+$, NH$_3$, and H$_2$O)? For those that do react, draw a structural formula for the organic product.

a. $\overset{\text{OH}}{\underset{\displaystyle |}{\text{CH}_2}} - \text{CH}_2 - \text{CH}_2 - \text{OCH}_3$

b. $\text{H} - \overset{\overset{\displaystyle O}{\parallel}}{\text{C}} - \text{CH}_2 - \text{CH}_2 - \text{CH}_3$

c. [cyclopentanone with CH₃ substituent and O]

d. [benzene ring attached to $\text{CH}_2 - \overset{\overset{\displaystyle O}{\parallel}}{\text{C}} - \text{CH}_3$]

e. [cyclohexane ring attached to $\text{CH}_2 - \overset{\overset{\displaystyle O}{\parallel}}{\text{C}} - \text{H}$]

4.30 Which of the following compounds will react with Tollens' reagent (Ag^+, NH_3, and H_2O)? For those that do react, draw a structural formula for the organic product.

a. [cyclohexanone ring with CH₂CH₃ substituent and O]

b. [benzene ring attached to $\overset{\overset{\displaystyle O}{\parallel}}{\text{C}} - \text{H}$]

c. $\text{CH}_3 - \text{O} - \overset{\overset{\text{OH}}{\displaystyle |}}{\text{CH}}\text{CH}_3$

d. $\text{CH}_3\text{CH}_2 - \overset{\overset{\displaystyle O}{\parallel}}{\text{C}} - \text{H}$

e. $\text{CH}_3 - \overset{\overset{\displaystyle O}{\parallel}}{\text{C}} - \text{CH}_3$

4.31 What is the formula of the precipitate that forms in a positive Benedict's test?

4.32 Not all aldehydes give a positive Benedict's test. Which of the following aldehydes do?

a. $\text{CH}_3\text{CH}_2 - \overset{\overset{\displaystyle O}{\parallel}}{\text{C}} - \text{H}$

b. [benzene ring attached to $\overset{\overset{\displaystyle O}{\parallel}}{\text{C}} - \text{H}$]

c. [cyclopentane ring attached to $\overset{\overset{\displaystyle O}{\parallel}}{\text{C}} - \overset{\overset{\displaystyle O}{\parallel}}{\text{C}} - \text{H}$]

d. $\text{CH}_3\overset{\overset{\text{OH}}{\displaystyle |}}{\text{CH}} - \overset{\overset{\displaystyle O}{\parallel}}{\text{C}} - \text{H}$

e. $\text{CH}_3 - \text{O} - \text{CH}_2 - \overset{\overset{\displaystyle O}{\parallel}}{\text{C}} - \text{H}$

4.33 A stockroom assistant prepares three bottles, each containing one of the following compounds:

$$\text{CH}_3\text{CH}_2 - \overset{\overset{\displaystyle O}{\parallel}}{\text{C}} - \text{CH}_3 \qquad \text{CH}_3\text{CH}_2\text{CH}_2\overset{\overset{\displaystyle O}{\parallel}}{\text{CH}} \qquad \text{CH}_3\text{CH}_2\overset{\overset{\displaystyle O}{\parallel}}{\text{CHCH}}$$
$$\underset{\displaystyle \text{OH}}{|}$$

However, the bottles are left unlabeled. Instead of discarding the samples, another assistant runs tests on the contents, using Tollens' and Benedict's reagents. Match the compound to the bottle, based on these results.

	Tollens' reagent	Benedict's reagent
Bottle A	A dark precipitate forms	A red-orange precipitate forms
Bottle B	No reaction	No reaction
Bottle C	A dark precipitate and slight mirror form	No reaction

4.34 Glucose, the sugar present within the blood, gives a positive Benedict's test. Circle the structural features that enable glucose to react.

$$\underset{\text{CH}_2 - \text{CH} - \text{CH} - \text{CH} - \text{CH} - \text{C} - \text{H}}{\overset{\text{OH} \quad\ \text{OH} \quad\ \text{OH} \quad\ \text{OH} \quad\ \text{OH} \quad\ \text{O}}{|\qquad\ |\qquad\ |\qquad\ |\qquad\ |\qquad\ \parallel}}$$

4.35 Glucose can exist as a cyclic hemiacetal because one of the alcohol groups reacts with the aldehyde:

$$\underset{\text{CH}_2 - \text{CH} - \text{CH} - \text{CH} - \text{CH} - \text{C} - \text{H}}{\overset{\text{OH} \quad\ \text{OH} \quad\ \text{OH} \quad\ \text{OH} \quad\ \text{OH} \quad\ \text{O}}{|\qquad\ |\qquad\ |\qquad\ |\qquad\ |\qquad\ \parallel}} \rightleftharpoons$$

[cyclic hemiacetal structure of glucose]

Draw an arrow to the carbon of the cyclic hemiacetal that was originally part of the aldehyde group. Circle the alcohol group in the open-chain structure that became a part of the cyclic hemiacetal.

4.36 Fructose, present with glucose in honey, reacts with Benedict's reagent. Circle the structural features that enable fructose to react.

$$\underset{\text{CH}_2 - \text{CH} - \text{CH} - \text{CH} - \text{C} - \text{CH}_2}{\overset{\text{OH} \quad\ \text{OH} \quad\ \text{OH} \quad\ \text{OH} \quad\ \text{O} \quad\ \text{OH}}{|\qquad\ |\qquad\ |\qquad\ |\qquad\ \parallel\qquad\ |}}$$

4.37 After an intramolecular reaction fructose forms a hemiketal:

$$\begin{array}{ccccccc}
OH & OH & OH & OH & O & OH \\
| & | & | & | & \| & | \\
CH_2 & - CH & - CH & - CH & - C & - CH_2
\end{array} \rightleftharpoons$$

Draw an arrow to the carbon of the cyclic hemiketal that was originally part of the ketone group. Circle the alcohol group in the open-chain structure that became part of the cyclic hemiketal.

4.38 Complete the following equations. If no reaction occurs, write "no reaction."

a.

$+ H_2 \xrightarrow{Pt}$

b.
$$\begin{array}{c}
\overset{\displaystyle OH}{|} \\
CH_3CH_2 - \underset{\underset{\displaystyle CH_3}{|}}{C} - OCH_3 + CH_3CH_2CH_2 - OH \xrightarrow{H^+}
\end{array}$$

c.
$$\begin{array}{c}
\overset{\displaystyle O}{\|} \qquad\quad \overset{\displaystyle OH}{|} \\
CH_3CH_2 - C - H + 2CH_3CHCH_3 \xrightarrow{H^+}
\end{array}$$

d.

$+ (O) \longrightarrow$

e.
$$\begin{array}{c}
\overset{\displaystyle CH_3}{|} \quad \overset{\displaystyle O}{\|} \\
CH_3CH_2CH - C - H + H_2 \xrightarrow{Pt}
\end{array}$$

4.39 Complete the following equations. If no reaction occurs, write "no reaction."

a.
$$\begin{array}{c}
\overset{\displaystyle O}{\|} \\
CH_3CH_2 - C - CH_3 + H_2 \xrightarrow{Pt}
\end{array}$$

b.
$$\begin{array}{c}
\overset{\displaystyle Br}{|} \quad \overset{\displaystyle O}{\|} \\
CH_3CHCH - C - H + (O) \longrightarrow \\
\underset{\displaystyle CH_3}{|}
\end{array}$$

c.

$+ 2CH_3 - OH \rightleftharpoons$

d.

$+ (O) \longrightarrow$

e.

$+ CH_3CH_2 - OH \xrightarrow{H^+} \rightleftharpoons$

4.40 Describe the products that result when hydrogen (H_2) is added to alkenes, alkynes, aldehydes, and ketones. You might review Chapter 2 briefly.

4.41 Draw structural formulas for the products of the following hydrolysis reactions:

a.
$$\begin{array}{c}
\overset{\displaystyle OCH_2CH_3}{|} \\
CH_3CH - OCH_3 + H_2O \xrightarrow{H^+} \rightleftharpoons
\end{array}$$

b.

$+ H_2O \xrightarrow{H^+} \rightleftharpoons$

c.

$+ H_2O \xrightarrow{H^+} \rightleftharpoons$

d.
$$\begin{array}{c}
CH_2 - OCH_2CH_3 + H_2O \xrightarrow{H^+} \rightleftharpoons \\
\overset{|}{OCH_2CH_3}
\end{array}$$

4.42 The following compounds are cyclic acetals or ketals. Write structural formulas for the hydrolysis products.

a.

$+ H_2O \xrightarrow{H^+} \rightleftharpoons$

b.

$+ H_2O \xrightarrow{H^+} \rightleftharpoons$

4.43 Write equations to show how the following conversions can be achieved. More than one reaction is required, and reactions from earlier chapters may be necessary.

a.
$$\begin{array}{c}
\qquad\qquad\qquad \overset{\displaystyle O}{\|} \\
CH_3CH=CHCH_3 \longrightarrow CH_3 - C - CH_2CH_3
\end{array}$$

b.
$$\begin{array}{c}
\qquad\qquad\qquad\qquad \overset{\displaystyle O}{\|} \\
CH_3CH_2CH_2 - OH \longrightarrow CH_3CH_2 - C - OH
\end{array}$$

4.44 Write equations to show how the following conversions can be achieved. More than one reaction is required, and reactions from earlier chapters may be necessary.

a.
$$\begin{array}{c}
\overset{\displaystyle OH}{|} \qquad\qquad\qquad \overset{\displaystyle OCH_2CH_3}{|} \\
CH_3CH_2CH_2 \longrightarrow CH_3CH_2 - CH \\
\qquad\qquad\qquad\qquad\qquad \underset{\displaystyle OCH_2CH_3}{|}
\end{array}$$

b.
$$\begin{array}{c}
\overset{\displaystyle CH_3}{|} \quad \overset{\displaystyle O}{\|} \qquad\qquad\qquad \overset{\displaystyle CH_3}{|} \\
CH_3CH - C - CH_3 \longrightarrow CH_3C=CH - CH_3
\end{array}$$

IMPORTANT ALDEHYDES AND KETONES (SECTION 4.4)

4.45 Identify the most important aldehyde and ketone (on the basis of amount used), and list at least one characteristic for each one that contributes to its usefulness.

4.46 Using Table 4.3, name an aldehyde or ketone used in the following ways:
a. Peppermint flavoring
b. Flavoring for margarines
c. Cinnamon flavoring
d. Vanilla flavoring

CHEMISTRY AROUND US, HOW REACTIONS OCCUR, AND OVER THE COUNTER

4.47 Show the mechanism for hemiacetal formation using the compounds

$$\begin{array}{c} O \\ \parallel \\ CH_3C-H \end{array} \quad \text{and} \quad CH_3-OH$$

4.48 What is the active ingredient in some self-tanning lotions?

4.49 Why might there be some cause for caution when using a self-tanning lotion?

4.50 Explain why norethynodrel is effective as a contraceptive.

4.51 Why should pregnant women use caution in taking vitamin A supplements?

4.52 What flavor is characteristic of the vanilloids?

4.53 Which vanilloid is used in some OTC pain-relieving creams?

▸ POINTS TO PONDER AND PURSUE

4.1 In the IUPAC name for the following ketone, it is not common to use a number for the position of the carbonyl group. Why not?

$$\begin{array}{c} O \\ \parallel \\ CH_3CH_2-C-CH_3 \end{array}$$

4.2 Why can formaldehyde (CH_2O) be prepared in the form of a 37% solution in water, whereas decanal cannot?

4.3 There are other addition reactions of aldehydes that also occur. Water, for example, adds to the carbonyl of trichloroacetaldehyde to form chloral hydrate, a strong hypnotic and sedative known as "knock-out drops" or (when mixed with alcohol) a "Mickey Finn." Complete the reaction by drawing the structural formula of chloral hydrate:

$$\begin{array}{c} Cl \quad O \\ \mid \quad \parallel \\ Cl-C-C-H + H-OH \longrightarrow \\ \mid \\ Cl \end{array}$$

       trichloroacetaldehyde   chloral hydrate

4.4 In the labels of some consumer products, ketone components listed with an *-one* ending can be found. However, very few aldehyde-containing (*-al* ending) products are found. How can you explain this?

4.5 In Figure 4.6, it is noted that acetone is used as a solvent in fingernail polish remover. Why do you think fingernail polish remover evaporates fairly quickly when used?

4.6 Vanilla flavoring is either extracted from a tropical orchid or synthetically produced from wood pulp by-products. What differences would you expect in these two commercial products: vanilla extract and imitation vanilla extract?

4.7 According to Web site 4.2, how long is Norplant effective?

4.8 Using Web site 4.3, identify common household sources of formaldehyde.

4.9 According to Web site 4.4, what home products may contain acetone?

Carboxylic Acids and Esters

The physician's assistant is trained to perform many tasks that traditionally were done only by doctors. These tasks include giving physical examinations, carrying out certain routine treatments of patients, suturing wounds, and prescribing medications. The two Chemistry Around Us features in this chapter deal with some commonly used medications: aspirin and nitroglycerin.

http://www.brookscole.com/chemistry/seager_chemistry
Web site 5.1 offers further information about a career as a physician's assistant.

Carboxylic acid
An organic compound that

$$\text{— C — OH}$$

with O double-bonded to C, functional group.

Carboxyl group

The

$$\text{— C — OH}$$

with O double-bonded to C, group.

> **FIGURE 5.1** Citrus fruits and juices are rich in citric acid. What flavor sensation is characteristic of acids?

Fatty acid
A long-chain carboxylic acid found in fats.

> **FIGURE 5.2** Rhubarb contains oxalic acid, a compound with two carboxyl groups. The nomenclature of dicarboxylic acids is not discussed here. What would you propose as the IUPAC ending for molecules with two carboxyl groups?

he tart flavor of foods that taste sour is generally caused by the presence of one or more carboxylic acids. Vinegar contains acetic acid, lemons and other citrus fruits contain citric acid (see ■ Figure 5.1), and the tart taste of apples is caused by malic acid.

The characteristic functional group of **carboxylic acids** is the **carboxyl group**:

$$\overset{\displaystyle O}{\underset{\displaystyle \text{— C — OH}}{\parallel}} \qquad \text{— COOH} \quad \text{or} \quad \text{— CO}_2\text{H}$$

carboxyl group   abbreviated forms

Carboxyl-containing compounds are found abundantly in nature (see ■ Figure 5.2). Examples are lactic acid, which is formed in muscle cells during vigorous exercise, and citric acid, which is an important intermediate in cellular energy production, as well as a flavoring agent in citrus fruits (see ■ Table 5.1).

➤ 5.1 The Nomenclature of Carboxylic Acids

Because of their abundance in nature, carboxylic acids were among the first organic compounds studied in detail. No systematic nomenclature system was then available, so the acids were usually named after some familiar source. Acetic acid,

$$\text{CH}_3 - \overset{\displaystyle O}{\underset{}{\overset{\parallel}{\text{C}}}} - \text{OH}$$

acetic acid

the sour constituent in vinegar (see ■ Figure 5.3) is named after *acetum*, the Latin word for vinegar. Butyric acid occurs in rancid butter (the Latin word for butter is *butyrum*). Many of these common names are still widely used today (see Table 5.1). Carboxylic acids with long hydrocarbon chains, generally 12 to 20 carbon atoms, are called **fatty acids** (Chapter 8) because they were first isolated from natural fats. These acids, such as stearic acid with 18 carbons, are known almost exclusively by their common names.

To name a carboxylic acid using the IUPAC system, the longest carbon chain including the carboxyl group is found and numbered. The numbering begins with the carboxyl carbon, so in monocarboxylic acids it is always located at the beginning of the chain, and no number is needed to locate it. The final *-e* of the parent hydrocarbon is dropped, and the ending *-oic* is added, followed by the word *acid*. Groups attached to the chain are named and

> **FIGURE 5.3** Acetic acid is the active component in vinegar and a familiar laboratory weak acid.

TABLE 5.1 Examples of carboxylic acids

Common name	IUPAC name	Structural formula	Characteristics and primary uses
Formic acid	methanoic acid	$H-\overset{\overset{O}{\|\|}}{C}-OH$	Stinging agents of certain ants and nettles; used in food preservation
Acetic acid	ethanoic acid	$CH_3-\overset{\overset{O}{\|\|}}{C}-OH$	Active ingredient in vinegar; used in food preservation
Propionic acid	propanoic acid	$CH_3CH_2-\overset{\overset{O}{\|\|}}{C}-OH$	Salts used as mold inhibitors
Butyric acid	butanoic acid	$CH_3(CH_2)_2-\overset{\overset{O}{\|\|}}{C}-OH$	Odor-causing agent in rancid butter
Caproic acid	hexanoic acid	$CH_3(CH_2)_4-\overset{\overset{O}{\|\|}}{C}-OH$	Characteristic odor of Limburger cheese
Oxalic acid	ethanedioic acid	$HO-\overset{\overset{O}{\|\|}}{C}-\overset{\overset{O}{\|\|}}{C}-OH$	Present in leaves of some plants such as rhubarb and spinach; used as a cleaning agent for rust stains on fabric and porcelain
Citric acid	2-hydroxy-1,2,3-propane-tricarboxylic acid	$HO-\overset{\overset{O}{\|\|}}{C}-CH_2-\overset{\overset{OH}{\|}}{\underset{\underset{O}{\|\|}}{\underset{C-OH}{C}}}-CH_2-\overset{\overset{O}{\|\|}}{C}-OH$	Present in citrus fruits; used as a flavoring agent in foods; present in cells
Lactic acid	2-hydroxypropanoic acid	$CH_3\overset{}{\underset{\underset{OH}{\|}}{CH}}-\overset{\overset{O}{\|\|}}{C}-OH$	Found in sour milk and sauerkraut; formed in muscles during exercise

located as before. Aromatic acids are given names derived from the parent compound, benzoic acid.

➤ Example 5.1

Use IUPAC rules to name the following carboxylic acids:

a. $CH_3CH_2-\overset{\overset{O}{\|\|}}{C}-OH$ **b.** $CH_3\overset{}{\underset{\underset{Br}{\|}}{CH}}CH_2-\overset{\overset{O}{\|\|}}{C}-OH$ **c.** $\overset{\overset{O}{\|\|}}{C}-OH$ (on benzene ring with CH_3)

Solution
a. The carbon chain contains three carbons. Thus, the name is based on the parent hydrocarbon, propane. The name is propanoic acid.

b. The chain contains four carbon atoms, and a bromine is attached to carbon number 3. The name is 3-bromobutanoic acid.

c. The aromatic acids are named as derivatives of benzoic acid. The carboxyl group is at position one, and the acid name is 3-methylbenzoic acid.

Learning Check 5.1

Give the IUPAC name to the following:

a.
$$\begin{array}{c} O \\ \parallel \\ C - OH \\ \mid \\ CH_3CH_2CHCH_3 \end{array}$$

b.
$$\begin{array}{c} O \\ \parallel \\ C - OH \end{array}$$
(benzene ring with Br at two positions, Br and Br)

➤ 5.2 Physical Properties of Carboxylic Acids

Carboxylic acids with low molecular weights are liquids at room temperature and have characteristically sharp or unpleasant odors (see ■ Table 5.2). Butyric acid, for example, is a component of perspiration and is partially responsible for the odor of locker rooms and unwashed socks. As the molecular weight of carboxylic acids increases, so does the melting point. The heavier acids (containing more than ten carbons) are waxlike solids. Stearic acid ($C_{18}H_{36}O_2$), for example, is mixed with paraffin to produce a wax used to make candles.

TABLE 5.2 Physical properties of some carboxylic acids

Common name	Structural formula	Boiling point (°C)	Melting point (°C)	Solubility (g/100 mL H_2O)
Formic acid	H—COOH	101	8	Infinite
Acetic acid	CH_3—COOH	118	17	Infinite
Propionic acid	CH_3—CH_2—COOH	141	−21	Infinite
Butryic acid	CH_3—$(CH_2)_2$—COOH	164	−5	Infinite
Valeric acid	CH_3—$(CH_2)_3$—COOH	186	−34	5
Caproic acid	CH_3—$(CH_2)_4$—COOH	205	−3	1
Caprylic acid	CH_3—$(CH_2)_6$—COOH	239	17	Insoluble
Capric acid	CH_3—$(CH_2)_8$—COOH	270	32	Insoluble
Lauric acid	CH_3—$(CH_2)_{10}$—COOH	299	44	Insoluble
Myristic acid	CH_3—$(CH_2)_{12}$—COOH	Decomposes	58	Insoluble
Palmitic acid	CH_3—$(CH_2)_{14}$—COOH	Decomposes	63	Insoluble
Stearic acid	CH_3—$(CH_2)_{16}$—COOH	Decomposes	71	Insoluble

TABLE 5.3	Boiling points of compounds with similar molecular weight		
Class	**Compound**	**Molecular weight**	**Boiling point (°C)**
Alkane	pentane	72	35
Ether	diethyl ether	74	35
Aldehyde	butanal	72	76
Alcohol	1-butanol	74	118
Carboxylic acid	propanoic acid	74	141

When boiling points of compounds with similar molecular weights are compared, the carboxylic acids have the highest boiling points of any organic compounds we have studied so far (see ■ Table 5.3 and ■ Figure 5.4).

For simple aliphatic compounds, the boiling points usually increase in this order: hydrocarbons and ethers, aldehydes and ketones, alcohols, and carboxylic acids. Carboxylic acids, like alcohols, can form intermolecular hydrogen bonds. It is this bonding that causes high boiling points. Acids have higher boiling points than alcohols with similar molecular weights because hydrogen-bonded acids are held together by two hydrogen bonds rather than the one hydrogen bond that is characteristic of alcohols. A structure where two molecules are joined together is called a **dimer:**

Dimer
Two identical molecules bonded together.

$$R-C \overset{\displaystyle O \cdots H-O}{\underset{\displaystyle O-H \cdots O}{}} C-R$$

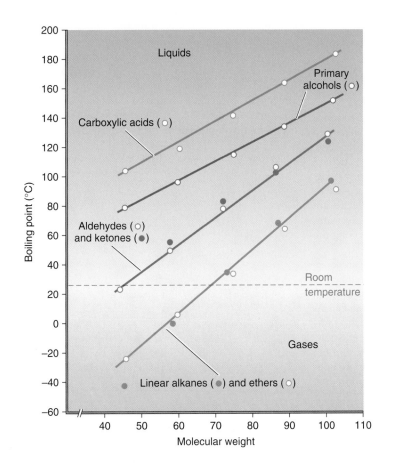

➤ **FIGURE 5.4** The boiling points of carboxylic acids are higher than those of all the other organic compounds studied thus far.

Low molecular weight acids are very soluble in water (Table 5.2) because of hydrogen bonds that form between the carboxyl group and water molecules (intermolecular hydrogen bonds). As the length of the nonpolar hydrocarbon portion (the R group) of the carboxylic acid increases, the water solubility decreases. Carboxylic acids containing eight or more carbon atoms are not appreciably soluble in water and are considered to be insoluble:

$$
\begin{array}{c}
O \cdots H - O \diagdown H \\
\parallel \qquad\qquad H \\
R - C - O - H \cdots O \\
\diagdown H
\end{array}
$$

OVER THE COUNTER • 5.1

ALPHA HYDROXY ACIDS IN COSMETICS

Alpha hydroxy acids occur in soured milk as lactic acid and in many fruits. Tartaric acid occurs in grapes, malic acid is found in grapes and apples, citric acid is a common component of citrus fruits, and glycolic acid is found in sugar cane and sugar beets.

$$
\begin{array}{cc}
\overset{\displaystyle O}{\overset{\displaystyle \parallel}{CH_3CH - C - OH}} & \overset{\displaystyle O\ \ OH\ \ OH\ \ O}{\overset{\displaystyle \parallel\ \ \mid\ \ \ \mid\ \ \parallel}{HO - C - CH - CH - C - OH}} \\
\mid & \\
OH &
\end{array}
$$

lactic acid   tartaric acid

$$
\begin{array}{cc}
\overset{\displaystyle O}{\overset{\displaystyle \parallel}{CH_2 - C - OH}} & \overset{\displaystyle O\ \ OH\qquad\ O}{\overset{\displaystyle \parallel\ \ \mid\qquad\ \parallel}{HO - C - CHCH_2 - C - OH}} \\
\mid & \\
OH &
\end{array}
$$

glycolic acid   malic acid

Recently, alpha hydroxy acids have been advertised as active ingredients in many cosmetics that claim to improve skin texture and tone, unblock or cleanse pores, improve oily skin or acne, and smooth fine lines and surface wrinkles. These cosmetics are most aggressively promoted for their antiwrinkle characteristics. The most commonly found alpha hydroxy acids used in cosmetics are lactic acid and glycolic acid. These weak acids are thought to work by loosening the cells of the epidermis (outer layer of skin) and enhancing cell turnover by accelerating the flaking off of dead skin (exfoliating).

However, even though exfoliation may reveal healthier looking skin, the acid-containing cosmetics that speed up exfoliation may irritate the exposed skin and cause burning and stinging. In addition, new evidence indicates that alpha hydroxy acids can increase the skin's sensitivity to sunlight and especially to the ultraviolet component of sunlight. In one industry-sponsored study, participants' skin was exposed to a 4% glycolic acid product twice daily for 12 weeks. Most of the participants developed minimal skin redness when exposed to 13% less UV radiation than normal, and three developed redness when exposed to 50% less UV light than normal. In another study, individuals who used an alpha hydroxy acid product in the presence of UV radiation ex-

perienced twice the cell damage to their skin than others who were treated similarly with a non–alpha hydroxy acid product.

Because of such test results, cosmetics companies have set guidelines suggesting that concentrations of alpha hydroxy acids in skin products should be no more than 10% for over-the-counter use and no more than 30% for professional salon use. The acid content is included on the labels of many products, and manufacturers can also provide the information.

As people age, their skin naturally becomes more wrinkled. Several factors contribute to this result of the aging process. Skin becomes drier, thinner, and less elastic with age, and people with fair skin are more predisposed to these effects than are those with darker skin. Also, smoking and long-term overexposure to sunlight dramatically accelerate the wrinkling process. Thus, one of the best ways to slow down skin wrinkling is to protect the skin from the sun by wearing long-sleeved clothing, a hat, and a sunscreen that has at least an SPF 15 rating (see Chemistry Around Us 2.3). Because many antiwrinkling cosmetics contain alpha hydroxy acids, a sunscreen should always be worn when these products are used. Adequate protection from the sun is still the best wrinkle-prevention method available.

http://www.brookscole.com/chemistry/seager_chemistry
See Web site 5.2 for additional information on glycolic acids.

Cosmetics products with alpha hydroxy acids.

Organic compounds may be arranged according to increasing water solubility as follows: hydrocarbons, ethers, aldehydes and ketones, alcohols, and carboxylic acids. Because forces resulting from hydrogen bonding play important roles in determining both boiling points and water solubility, it is not surprising that the same order is generally observed for these two properties.

▶ 5.3 The Acidity of Carboxylic Acids

The most important chemical property of carboxylic acids is the acidic behavior implied by their name. The hydrogen attached to the oxygen of the carboxyl group gives carboxylic acids their acidic character. In water, this proton may leave the acid, converting it into a **carboxylate ion**. Reaction 5.1 gives the general reaction, and Reaction 5.2 gives a specific example:

General reaction:

$$\underset{\text{a carboxylic acid}}{R-\overset{\displaystyle O}{\overset{\|}{C}}-O-H} + H_2O \rightleftharpoons \underset{\substack{\text{a carboxylate} \\ \text{ion}}}{R-\overset{\displaystyle O}{\overset{\|}{C}}-O^-} + \underset{\substack{\text{a hydronium} \\ \text{ion}}}{H_3O^+} \qquad (5.1)$$

Generally, carboxylic acids behave as weak acids. A 1M solution of acetic acid, for example, is only about 0.5% dissociated.

Specific example:

$$\underset{\text{acetic acid}}{CH_3-\overset{\displaystyle O}{\overset{\|}{C}}-O-H} + H_2O \rightleftharpoons \underset{\text{acetate ion}}{CH_3-\overset{\displaystyle O}{\overset{\|}{C}}-O^-} + H_3O^+ \qquad (5.2)$$

It is important to realize that Reactions 5.1 and 5.2 are reversible. According to Le Châtelier's principle, the addition of H_3O^+ (low pH) should favor formation of the carboxylic acid, and removal of H_3O^+ by adding base (high pH) should favor formation of the carboxylate ion:

$$\underset{\substack{\text{carboxylic} \\ \text{acid}}}{R-\overset{\displaystyle O}{\overset{\|}{C}}-O-H} + H_2O \underset{\substack{\text{acidic conditions} \\ \text{(low pH)}}}{\overset{\substack{\text{basic conditions} \\ \text{(high pH)}}}{\rightleftharpoons}} \underset{\substack{\text{carboxylate} \\ \text{ion}}}{R-\overset{\displaystyle O}{\overset{\|}{C}}-O^-} + H_3O^+ \qquad (5.3)$$

Thus, the pH of a solution determines the form in which a carboxylic acid exists in the solution. At pH 7.4, the normal pH of body fluids, the carboxylate form predominates. For example, citric acid in body fluids is often referred to as citrate. This is an important point for later chapters in biochemistry.

Pyruvic acid, an important intermediate in the energy conversion reactions in living organisms, is usually called pyruvate because it is commonly in the carboxylate form. Draw the structure of pyruvate.

$$\underset{\text{pyruvic acid}}{CH_3-\overset{\displaystyle O}{\overset{\|}{C}}-\overset{\displaystyle O}{\overset{\|}{C}}-OH}$$

Carboxylate ion

$$\text{The } R-\overset{\displaystyle O}{\overset{\|}{C}}-O^- \text{ ion that}$$
results from the dissociation of a carboxylic acid.

Learning Check 5.2

Although they are weak, carboxylic acids react readily with strong bases such as sodium or potassium hydroxide to form salts. Reaction 5.4 gives the general reaction, and Reactions 5.5 and 5.6 are two specific examples:

General reaction:

$$\underset{\text{a carboxylic acid}}{R - \overset{\overset{\textstyle O}{\|}}{C} - O - H} + NaOH \longrightarrow \underset{\text{a sodium carboxylate}}{R - \overset{\overset{\textstyle O}{\|}}{C} - O^- Na^+} + H_2O \qquad (5.4)$$

Specific examples:

$$\underset{\text{acetic acid}}{CH_3 - \overset{\overset{\textstyle O}{\|}}{C} - O - H} + NaOH \longrightarrow \underset{\text{sodium acetate}}{CH_3 - \overset{\overset{\textstyle O}{\|}}{C} - O^- Na^+} + H_2O \qquad (5.5)$$

benzoic acid   potassium benzoate

$$(5.6)$$

Learning Check 5.3

Write the structural formulas for the products of the following reaction:

$$CH_3CH_2CH_2 - \overset{\overset{\textstyle O}{\|}}{C} - OH + NaOH \longrightarrow$$

➤ 5.4 Salts of Carboxylic Acids

Nomenclature

Both common names and IUPAC names are assigned to carboxylic acid salts by naming the metal first and changing the *-ic* ending of the acid name to *-ate*.

➤ Example 5.2

Give both a common name and an IUPAC name to the following salts:

a. $CH_3 - \overset{\overset{\textstyle O}{\|}}{C} - O^- Na^+$ b.

Solution

a. This salt is formed from $CH_3 - \overset{\overset{\textstyle O}{\|}}{C} - OH$, which is called acetic acid (common name) or ethanoic acid (IUPAC name). Thus, the salt is sodium acetate (common name) or sodium ethanoate (IUPAC name).
b. This compound is the potassium salt of benzoic acid (both the common and IUPAC names). The name is potassium benzoate.

Give the IUPAC name for the following salts:

a.

$$\underset{\text{(ring with } CH_3 \text{ substituent)}}{\text{C}-\text{O}^-\text{Li}^+}$$

b. $CH_3CH_2CH_2-\overset{\overset{\displaystyle O}{\|}}{C}-O^-Na^+$

Carboxylic acid salts are solids at room temperature and are usually soluble in water because they are ionic. Even long-chain acids with an extensive nonpolar hydrocarbon portion can be solubilized by converting them into salts (Reaction 5.7):

$$CH_3(CH_2)_{16}-\overset{\overset{\displaystyle O}{\|}}{C}-O-H + NaOH \longrightarrow CH_3(CH_2)_{16}-\overset{\overset{\displaystyle O}{\|}}{C}-O^-Na^+ + H_2O \qquad (5.7)$$

stearic acid (insoluble)	sodium stearate (soluble)

Useful Carboxylic Acid Salts

A number of acid salts are important around the home. Sodium stearate $(CH_3(CH_2)_{16}COO^-Na^+)$ and other sodium and potassium salts of long-chain carboxylic acids are used as soaps (see ■ Figure 5.5).

Calcium and sodium propanoate are used commercially as preservatives in bread, cakes, and cheese to prevent the growth of bacteria and molds (see ■ Figure 5.6). The parent acid, CH_3CH_2COOH, occurs naturally in Swiss cheese. The labels for these bakery products often contain the common name propionate, rather than the IUPAC-acceptable name propanoate.

Sodium benzoate, another common food preservative, occurs naturally in many foods, especially cranberries and prunes. It is used in bakery products, ketchup, carbonated beverages, and a host of other foods:

$$\overset{\overset{\displaystyle O}{\|}}{\underset{\text{sodium benzoate}}{\text{C}-\text{O}^-\text{Na}^+}}$$

> **FIGURE 5.5** Soap products enable us to keep ourselves and our homes clean.

> **FIGURE 5.6** Propanoates extend the shelf life of bread, preventing the formation of mold.

➤ **FIGURE 5.7** Esters are partially responsible for the fragrance of oranges, pears, bananas, pineapples and strawberries.

http://www.brookscole.com/chemistry/
seager_chemistry
Web site 5.3 offers further information on food additives.

Carboxylic ester
A compound with the

$$\overset{\overset{\displaystyle O}{\|}}{-C} - OR$$ functional group.

Esterification
The process of forming an ester.

Ester linkage
The carbonyl carbon–oxygen single bond of the ester group.

http://www.brookscole.com/chemistry/
seager_chemistry
Web site 5.4 presents further information about the use of esters in soaps and cleaning products.

Zinc 10-undecylenate, the zinc salt of $CH_2=CH(CH_2)_8COOH$, is commonly used to treat athlete's foot. One commercial product that contains zinc 10-undecylenate is Desenex®.

A mixture of sodium citrate and citric acid

$$HO-\overset{\overset{\displaystyle O}{\|}}{C}-CH_2-\overset{\overset{\displaystyle OH}{|}}{\underset{\underset{\displaystyle O}{\|}}{\underset{\displaystyle C-OH}{C}}}-CH_2-\overset{\overset{\displaystyle O}{\|}}{C}-OH$$

citric acid

is widely used as a buffer to control pH. Products sold as foams or gels such as jelly, ice cream, candy, and whipped cream maintain their desirable characteristics only at certain pHs, which can be controlled by the citrate/citric acid buffer. This same buffer is used in medicines and in human blood used for transfusions. In blood, it also functions as an anticoagulant.

➤ 5.5 Carboxylic Esters

A very important reaction occurs when carboxylic acids are heated with alcohols or phenols in the presence of an acid catalyst. A molecule of water splits out and a **carboxylic ester** is formed:

General reaction:

$$R-\overset{\overset{\displaystyle O}{\|}}{C}-OH + H-O-R' \underset{\xrightarrow{\hspace{1cm}}}{\overset{H^+,\ heat}{\rightleftharpoons}} R-\overset{\overset{\displaystyle O}{\|}}{C}-O-R' + H_2O \qquad (5.8)$$

carboxylic   alcohol   carboxylic
acid   or phenol   ester

As before, R′ indicates that the two R groups can be the same or different. The process of ester formation is called **esterification,** and the carbonyl carbon–oxygen single bond of the ester group is called the **ester linkage:**

$$-\overset{\overset{\displaystyle O}{\|}}{C}-O-\overset{|}{\underset{|}{C}}- \qquad \textit{ester linkage}$$

ester group

The ester functional group is a key structural feature in fats, oils, and other lipids (Chapter 8). Also widely found in fruits and flowers, many esters are very fragrant and represent some of nature's most pleasant odors (see ■ Figure 5.7). Because of this characteristic, esters are commonly used as flavoring agents in foods and as scents in personal products.

Reactions 5.9 and 5.10 illustrate the formation of two widely used esters:

Specific examples:

$$CH_3CH_2CH_2-\overset{\overset{\displaystyle O}{\|}}{C}-OH + HO-CH_2CH_3 \overset{H^+}{\rightleftharpoons} CH_3CH_2CH_2-\overset{\overset{\displaystyle O}{\|}}{C}-O-CH_2CH_3 + H_2O \qquad (5.9)$$

butanoic acid   ethyl alcohol   ethyl butanoate
(strawberry flavoring)

$$
\text{salicylic acid} \quad + \text{HO} - \text{CH}_3 \underset{\text{H}^+}{\rightleftharpoons} \text{methyl salicylate (oil of wintergreen)} + H_2O \qquad (5.10)
$$

salicylic acid   methyl alcohol   methyl salicylate (oil of wintergreen)

➤ Example 5.3

Give the structure of the ester formed in the following reactions:

a. $CH_3 - \overset{O}{\underset{\|}{C}} - OH + CH_3 - OH \xrightleftharpoons{H^+, \text{ heat}}$

b. $CH_3CH_2CH_2 - \overset{O}{\underset{\|}{C}} - OH + \bigcirc - OH \xrightleftharpoons{H^+, \text{ heat}}$

Solution

a. $CH_3 - \overset{O}{\underset{\|}{C}} - OH + CH_3OH \xrightleftharpoons{H^+, \text{ heat}} CH_3 - \overset{O}{\underset{\|}{C}} - O - CH_3 + H_2O$

b. $CH_3CH_2CH_2 - \overset{O}{\underset{\|}{C}} - OH + \bigcirc - OH \xrightleftharpoons{H^+, \text{ heat}} CH_3CH_2CH_2 - \overset{O}{\underset{\|}{C}} - O - \bigcirc + H_2O$

Write equations to illustrate ester formation from the following:

a. $CH_3CH_2 - \overset{O}{\underset{\|}{C}} - OH$ and $CH_3 - OH$

b. $CH_3 - \overset{O}{\underset{\|}{C}} - OH$ and $\bigcirc - OH$

c. $\bigcirc - \overset{O}{\underset{\|}{C}} - OH$ and $CH_3CH - OH$ with CH_3

Learning Check 5.5

Polyesters, which are found in many consumer products, are just what the name implies—polymers made by an esterification reaction. The process used is an example of **condensation polymerization,** in which monomers combine and form a polymer; a small molecule such as water is formed as well. This is in contrast to addition polymerization, in which all atoms of the alkene monomer are incorporated into the polymer (Section 2.4). In many condensation polymerization reactions, each monomer has two functional groups so

Condensation polymerization
The process by which monomers combine together, with the simultaneous elimination of a small molecule.

that the chain can grow from either end. A typical example is the reaction of terephthalic acid and ethylene glycol:

$$nHO-\overset{\overset{\displaystyle O}{\|}}{C}-\hexagon-\overset{\overset{\displaystyle O}{\|}}{C}-OH \; + \; nHO-CH_2CH_2-OH \underset{\longleftarrow}{\overset{H^+, \text{ heat}}{\longrightarrow}} \tag{5.11}$$

terephthalic acid ethylene glycol

$$+O-\overset{\overset{\displaystyle O}{\|}}{C}-\hexagon-\overset{\overset{\displaystyle O}{\|}}{C}-O-CH_2CH_2 \xrightarrow{}_{n} \; + \; nH_2O$$

polyethylene terephthalate (PET)

Molecules of product continue to react with available monomers until, finally, a polyester is formed with the general structure given above, where the n means that the unit in parentheses is repeated many times.

Over 3 billion pounds of the polyester PET are produced annually. Fibers are formed by melting the polymer and forcing the liquid through tiny holes in devices called spinnerettes. The resulting fibers are spun into thread or yarn and marketed under the trade names Dacron®, Fortrel®, and Terylene®, depending on the manufacturer. Typical uses include automobile tire cord and permanent-press clothing. In medicine, Dacron thread is used for sutures, and woven fabric is used to replace damaged or diseased sections of blood vessels and the esophagus. Dacron fibers are often blended with cotton to make a fabric that is more comfortable to wear on hot, humid days than those containing 100% polyester and that retains the latter's wrinkle resistance. Besides being forced through spinnerettes, PET melts may be forced through a narrow slit to produce thin sheets or films called Mylar® (see ■ Figure 5.8). This form of polyester is used extensively in the manufacture of magnetic tapes for tape recorders.

Although many esters can be prepared by the method just outlined in Reaction 5.8, the reversible nature of the reaction does not always allow for good yields of product. Two compounds that are more reactive than carboxylic acids and that produce esters in good yield are **carboxylic acid chlorides** and **carboxylic acid anhydrides**:

Carboxylic acid chloride
An organic compond that contains the $-\overset{\overset{\displaystyle O}{\|}}{C}-Cl$ functional group.

Carboxylic acid anhydride
An organic compond that contains the $-\overset{\overset{\displaystyle O}{\|}}{C}-O-\overset{\overset{\displaystyle O}{\|}}{C}-$ functional group.

$$R-\overset{\overset{\displaystyle O}{\|}}{C}-Cl$$
carboxylic acid
chloride

$$R-\overset{\overset{\displaystyle O}{\|}}{C}-O-\overset{\overset{\displaystyle O}{\|}}{C}-R$$
carboxylic acid
anhydride

These substances react vigorously with alcohols in a nonreversible manner.

➤ **FIGURE 5.8** Mylar®, a polyester developed by E. I. duPont de Nemours Co., was used to cover the wings and pilot's compartment of the Gossamer Albatross, the first human-powered aircraft to fly across the English Channel.

General reactions:

$$
\underset{\substack{\text{acid chloride}}}{R-\overset{\displaystyle O}{\overset{\|}{C}}-Cl} + \underset{\text{alcohol}}{H-O-R'} \longrightarrow \underset{\substack{\text{carboxylic}\\\text{ester}}}{R-\overset{\displaystyle O}{\overset{\|}{C}}-O-R'} + HCl \qquad (5.12)
$$

$$
\underset{\substack{\text{carboxylic acid}\\\text{anhydride}}}{R-\overset{\displaystyle O}{\overset{\|}{C}}-O-\overset{\displaystyle O}{\overset{\|}{C}}-R} + \underset{\text{alcohol}}{H-O-R'} \longrightarrow \underset{\text{carboxylic ester}}{R-\overset{\displaystyle O}{\overset{\|}{C}}-O-R'} + \underset{\substack{\text{carboxylic}\\\text{acid}}}{R-\overset{\displaystyle O}{\overset{\|}{C}}-OH} \qquad (5.13)
$$

Reactions 5.14 and 5.15 illustrate two specific examples. In another example, the laboratory synthesis of aspirin (Reaction 5.16), using acetic anhydride, gives excellent yields of the ester.

Specific reactions:

$$
\underset{\text{acetyl chloride}}{CH_3-\overset{\displaystyle O}{\overset{\|}{C}}-Cl} + \underset{\substack{\text{ethyl}\\\text{alcohol}}}{HO-CH_2CH_3} \longrightarrow \underset{\text{ethyl acetate}}{CH_3-\overset{\displaystyle O}{\overset{\|}{C}}-O-CH_2CH_3} + HCl \qquad (5.14)
$$

$$
\underset{\text{acetic anhydride}}{CH_3-\overset{\displaystyle O}{\overset{\|}{C}}-O-\overset{\displaystyle O}{\overset{\|}{C}}-CH_3} + \underset{\text{phenol}}{\text{(phenol)}} \longrightarrow CH_3-\overset{\displaystyle O}{\overset{\|}{C}}-O \quad + \underset{\text{acetic acid}}{CH_3-\overset{\displaystyle O}{\overset{\|}{C}}-OH} \qquad (5.15)
$$

phenyl acetate

► Example 5.4

Give the structure of the ester formed in the following reactions:

a. (benzene ring)$-\overset{\displaystyle O}{\overset{\|}{C}}-Cl$ + $CH_3CH_2-OH \longrightarrow$

b. $CH_3CH_2-\overset{\displaystyle O}{\overset{\|}{C}}-O-\overset{\displaystyle O}{\overset{\|}{C}}-CH_2CH_3 + CH_3CH_2-OH \longrightarrow$

Solution

a. The products are (benzene ring)$-\overset{\displaystyle O}{\overset{\|}{C}}-O-CH_2CH_3 + HCl$

b. The products are
$$CH_3CH_2 - \overset{\overset{\displaystyle O}{\|}}{C} - O - CH_2CH_3 \; + \; CH_3CH_2 - \overset{\overset{\displaystyle O}{\|}}{C} - OH$$

The first product in each case is the ester.

Learning Check 5.6
....

Write equations to represent ester formation when the following pairs of compounds are reacted:

a.
$$\text{(phenyl)} - \overset{\overset{\displaystyle O}{\|}}{C} - O - \overset{\overset{\displaystyle O}{\|}}{C} - \text{(phenyl)} \quad \text{and} \quad CH_3 - OH$$

b.
$$CH_3CH_2 - \overset{\overset{\displaystyle O}{\|}}{C} - Cl \quad \text{and} \quad \underset{\underset{\displaystyle CH_3}{|}}{CH_3CH} - OH$$

➤ 5.6 The Nomenclature of Esters
....

Esters are named by a common system as well as by IUPAC rules; the common names are used most often. It is helpful with both IUPAC and common names to think of the esters as being formed from a carboxylic acid, the

$$R - \overset{\overset{\displaystyle O}{\|}}{C} - \text{part and an alcohol, the } - OR' \text{ part: } \underbrace{R - \overset{\overset{\displaystyle O}{\|}}{C}}_{\text{acid part}} - \overset{\overset{\textit{alcohol part}}{\diagup}}{OR'}$$

The first word of the name of an ester is the name of the alkyl or aromatic group (R′) contributed by the alcohol. The second word is the carboxylic acid name, with the *-ic acid* ending changed to *-ate*. This is similar to the method used for naming carboxylic acid salts. Thus, an ester of acetic acid becomes an acetate, one of butyric acid becomes a butyrate, one of lactic acid becomes a lactate, and so on.

➤ Example 5.5
....

Give common and IUPAC names for the following esters:

a.
$$CH_3 - \overset{\overset{\displaystyle O}{\|}}{C} - O - CH_3$$

b.
$$CH_3CH_2CH_2 - \overset{\overset{\displaystyle O}{\|}}{C} - O - \text{(phenyl)}$$

c.
$$\text{(phenyl)} - \overset{\overset{\displaystyle O}{\|}}{C} - O - CH_2CH_3$$

CHEMISTRY AROUND US • 5.1

ASPIRIN: SHOULD YOU TAKE A DAILY DOSE?

Aspirin, an ester of salicylic acid first prepared in 1893, is remarkably versatile. In the body it reduces fever but does not reduce normal body temperature. It also relieves a variety of simple pains such as headache, sprain, and toothache, and is an anti-inflammatory agent often used to treat arthritis.

Americans consume an estimated 80 billion aspirin tablets each year. The Physicians' Desk Reference lists more than 50 OTC drugs in which aspirin is the principal ingredient. Yet, despite the fact that aspirin has been in routine use for nearly a century, scientific journals continue to publish reports about new uses for this old remedy.

$$\text{salicylic acid} + \text{acetic anhydride} \longrightarrow \text{acetylsalicylic acid (aspirin)} + CH_3-C-OH \quad (5.16)$$

In recent years, researchers have discovered one particular effect of aspirin in the body that helps explain its benefits as well as some of its side effects. Aspirin interferes with the clotting action of blood. When blood vessels are injured and bleed, cells in the blood called platelets accumulate at the injury site and clump together to form a plug that seals the opening in the blood vessel. Aspirin reduces the ability of the platelets to clump together, thereby reducing the tendency of blood to clot. In this same way, aspirin may help reduce or prevent clots from forming in arteries that are narrowed by diseases such as *atherosclerosis*. A heart attack results when a coronary artery that provides blood to the heart is blocked by a blood clot. One study shows that aspirin reduces the chances of having a second heart attack by more than 30%. Other studies suggest that aspirin use helps prevent a first heart attack in healthy individuals, and can also help prevent first or recurrent strokes of the type caused by blood clots that block the blood supply to the brain.

The same process that reduces blood clotting is also responsible for some of the undesirable side effects of using aspirin. It is known that aspirin use causes increased rates of gastrointestinal bleeding and hemorrhagic stroke (stroke caused by a ruptured blood vessel in the brain). Heavy aspirin use before surgery can present a danger because it increases the risk of severe bleeding after the operation.

Recent research reported in medical journals provides evidence of another important medical use of aspirin that is not related to its ability to inhibit blood clotting: Aspirin helps prevent esophageal and colorectal cancers. This indication is very significant because colorectal cancer is the second leading cancer killer after lung cancer. Researchers theorize that the aspirin exerts this effect by blocking the production of hormonelike prostaglandins, which in turn inhibits the growth of tumors. Such research may ultimately lead to the use of aspirin therapy to prevent colorectal cancer, but no official health agency has yet made such a recommendation.

Most doctors recognize that the majority of their patients with coronary artery disease benefit from daily aspirin therapy. When such therapy is recommended, a low dose of 81 mg per day (one baby aspirin) is generally used. Because the long-term, regular use of aspirin poses some risks as well as some benefits, the decision to begin such a program should be made only after consultation with a physician.

http://www.brookscole.com/chemistry/seager_chemistry
Refer to Web site 5.5 for more details regarding aspirin and related drugs.

A variety of aspirin products are available.

Solution

a. The first word is methyl, which refers to the alcohol portion:

$$
\underline{CH_3-\overset{\displaystyle O}{\overset{\displaystyle \|}{C}}-O}-CH_3 \quad \nearrow methyl
$$

acetate (common name)
ethanoate (IUPAC name)

The second word is derived from the name of the two-carbon carboxylic acid (acetic acid or ethanoic acid). Thus, we have methyl acetate (common name) and methyl ethanoate (IUPAC name).

b.
$$
\underline{CH_3CH_2CH_2-\overset{\displaystyle O}{\overset{\displaystyle \|}{C}}-O}- \bigcirc \quad \nearrow phenyl
$$

butyrate (common name)
butanoate (IUPAC name)

Thus, the two names are phenyl butyrate (common name) and phenyl butanoate (IUPAC name).

c.
$$
\underline{\bigcirc-\overset{\displaystyle O}{\overset{\displaystyle \|}{C}}}-O-\underline{CH_2CH_3} \quad \nearrow ethyl
$$

benzoate

Thus, we have ethyl benzoate (common name and IUPAC name).

Learning Check 5.7
....

Give both a common and the IUPAC names for each of the following esters:

a. $H-\overset{\displaystyle O}{\overset{\displaystyle \|}{C}}-O-\underset{\displaystyle CH_3}{\overset{\displaystyle |}{CH}}CH_3$

b. $\bigcirc-\overset{\displaystyle O}{\overset{\displaystyle \|}{C}}-OCH_3$

➤ 5.7 Reactions of Esters
....

The most important reaction of carboxylic esters, both in commercial processes and in the body, involves the breaking of the ester linkage. The process is called either ester hydrolysis or saponification, depending on the reaction conditions.

Ester Hydrolysis

Ester hydrolysis is the reaction of an ester with water to break an ester linkage and produce an alcohol and a carboxylic acid. This process is simply the reverse of the esterification reaction (Reaction 5.8). Strong acids are frequently used as catalysts to hydrolyze esters:

General reaction:

$$\underset{\text{an ester}}{R-\overset{\displaystyle O}{\overset{\|}{C}}-OR'} + \underset{}{H-OH} \overset{H^+}{\rightleftharpoons} \underset{\text{an acid}}{R-\overset{\displaystyle O}{\overset{\|}{C}}-OH} + \underset{\substack{\text{an alcohol} \\ \text{or phenol}}}{R'-OH} \qquad (5.17)$$

Specific example:

$$\underset{\text{ethyl acetate}}{CH_3-\overset{\displaystyle O}{\overset{\|}{C}}-O-CH_2CH_3} + H-OH \overset{H^+}{\rightleftharpoons} \underset{\text{acetic acid}}{CH_3-\overset{\displaystyle O}{\overset{\|}{C}}-OH} + \underset{\text{ethyl alcohol}}{CH_3CH_2-OH} \qquad (5.18)$$

Notice that when ester linkages break during hydrolysis, the elements of water (—H and —OH) are attached to the ester fragments to form the acid and alcohol (Reaction 5.17). The breaking of a bond and the attachment of the elements of water to the fragments are characteristic of all hydrolysis reactions.

The human body and other biological organisms are constantly forming and hydrolyzing a variety of esters. The catalysts used in these cellular processes are very efficient protein catalysts called enzymes (Chapter 10). Animal fats and vegetable oils are esters. Their enzyme-catalyzed hydrolysis is an important process that takes place when they are digested.

Give the structure of the products formed in the following hydrolysis reaction:

Learning Check 5.8

$$\text{C}6\text{H}5-\overset{\displaystyle O}{\overset{\|}{C}}-O-CH_2CH_2CH_3$$
$$+ H_2O \overset{H^+}{\rightleftharpoons}$$

▶ **STUDY SKILLS 5.1**

A REACTION MAP FOR CARBOXYLIC ACIDS

We hope you have become familiar with the reaction maps of Chapters 2, 3, and 4 and have found them useful in solving problems. Another reaction map is given below. It includes the key reactions of this chapter and should be useful as you answer questions and prepare for exams.

Saponification

Saponification
The basic cleavage of an ester linkage.

Like ester hydrolysis, **saponification** results in a breaking of the ester linkage and produces an alcohol and a carboxylic acid (see Section 8.4 for more on saponification). However, unlike ester hydrolysis, saponification is done in solutions containing strong bases such as potassium or sodium hydroxide. The carboxylic acid is converted to a salt under these basic conditions.

General reaction:

$$\underset{\text{an ester}}{R-\overset{\displaystyle O}{\overset{\|}{C}}-O-R'} + NaOH \longrightarrow \underset{\substack{\text{a carboxylic} \\ \text{acid salt}}}{R-\overset{\displaystyle O}{\overset{\|}{C}}-O^-Na^+} + \underset{\text{an alcohol}}{R'-OH} \qquad (5.19)$$

Specific example:

methyl benzoate + NaOH ⟶ sodium benzoate + methyl alcohol (5.20)

We will see in Chapter 8 that saponification of fats and oils is important in the production of soaps.

Learning Check 5.9

Give structural formulas for the products of the following saponification reactions.

a. $CH_3CH_2-\overset{\displaystyle O}{\overset{\|}{C}}-O-\underset{\underset{CH_3}{|}}{CHCH_3} + NaOH \longrightarrow$

b. ⬡$-\overset{\displaystyle O}{\overset{\|}{C}}-O-CH_2CH_3 + KOH \longrightarrow$

➤ 5.8 Esters of Inorganic Acids

Ester
A compound in which the —OH of an acid is replaced by an —OR.

We have defined a carboxylic ester as the product of a reaction between a carboxylic acid and an alcohol. Alcohols can also form **esters** by reacting with inorganic acids such as sulfuric, nitric, and phosphoric acids. The most important of these in biochemistry are the esters of phosphoric acid (H_3PO_4):

General reaction:

$$\underset{\text{an alcohol}}{R-OH} + \underset{\substack{\text{phosphoric} \\ \text{acid}}}{HO-\underset{\underset{OH}{|}}{\overset{\displaystyle O}{\overset{\|}{P}}}-OH} \rightleftharpoons \underset{\substack{\text{a phosphate} \\ \text{ester}}}{R-O-\underset{\underset{OH}{|}}{\overset{\displaystyle O}{\overset{\|}{P}}}-OH} + H_2O \qquad (5.21)$$

Specific example:

$$CH_3CH_2-OH + HO-\overset{\overset{\displaystyle O}{\|}}{\underset{\underset{\displaystyle OH}{|}}{P}}-OH \rightleftharpoons CH_3CH_2-O-\overset{\overset{\displaystyle O}{\|}}{\underset{\underset{\displaystyle OH}{|}}{P}}-OH + H_2O \quad (5.22)$$

Because phosphoric acid has three —OH groups, it can form mono-, di- and triesters:

$$RO-\overset{\overset{\displaystyle O}{\|}}{\underset{\underset{\displaystyle OH}{|}}{P}}-OH \qquad RO-\overset{\overset{\displaystyle O}{\|}}{\underset{\underset{\displaystyle OH}{|}}{P}}-OR' \qquad RO-\overset{\overset{\displaystyle O}{\|}}{\underset{\underset{\displaystyle OR''}{|}}{P}}-OR'$$

| monoester, | diester, | triester, |
| one R group | two R groups | three R groups |

Mono- and diesters are essential to life and represent some of the most important biological molecules. An example of an important monoester is glucose 6-phosphate, which represents the first intermediate compound formed when glucose is oxidized to supply energy for the body (Chapter 13):

glucose 6-phosphate

HOW REACTIONS OCCUR

ESTER SAPONIFICATION

The cleavage of an ester linkage under basic conditions is initiated by the attraction of the carbonyl carbon for hydroxide:

$$CH_3-\overset{\overset{\displaystyle O}{\|}}{C}-O-CH_3 \rightleftharpoons CH_3-\overset{\overset{\displaystyle \ddot{O}:^-}{|}}{\underset{\underset{\displaystyle OH}{|}}{C}}-O-CH_3$$

$$:\overset{..}{O}H^-$$

As the —OH group bonds to the carbonyl carbon, the double bond must break (two electrons move to the carbonyl oxygen) so that carbon does not possess five bonds.

In the next step of the mechanism, the electrons reestablish the double bond. Once again, carbon will have five bonds if there is not a simultaneous bond breakage. In this case, it is the ester linkage that is cleaved:

$$CH_3-\overset{\overset{\displaystyle \ddot{O}:^-}{|}}{\underset{\underset{\displaystyle OH}{|}}{C}}-O-CH_3 \rightleftharpoons CH_3-\overset{\overset{\displaystyle O}{\|}}{C}-OH + {}^-:\overset{..}{O}-CH$$

a carboxylic acid   a strong base

A carboxylic acid is formed momentarily but it quickly donates its proton to the strongly basic anion:

$$CH_3-\overset{\overset{\displaystyle O}{\|}}{C}-OH + {}^-:\overset{..}{O}-CH_3 \longrightarrow$$

$$CH_3-\overset{\overset{\displaystyle O}{\|}}{C}-O^- + CH_3-OH$$

the carboxylate   the alcohol
product   product

If the source of hydroxide for the reaction was sodium hydroxide, the carboxylate product would be coupled with the positive sodium ion to form the salt.

CHEMISTRY AROUND US • 5.2

NITROGLYCERIN IN DYNAMITE AND IN MEDICINE

Nitroglycerin is a nitrate ester resulting from the reaction of nitric acid and glycerol:

$$
\begin{array}{l}
CH_2-OH \\
| \\
CH-OH \\
| \\
CH_2-OH
\end{array}
+\ 3HO-NO_2
\ \rightleftarrows\
\begin{array}{l}
CH_2-O-NO_2 \\
| \\
CH-O-NO_2 \\
| \\
CH_2-O-NO_2
\end{array}
+\ 3H_2O
$$

glycerol   nitric acid   nitroglycerin

First made in 1846 by the Italian chemist Sobrero, nitroglycerin was discovered to be a powerful explosive with applications in both war and peace. However, as a shock-sensitive, unstable liquid, nitroglycerin proved to be extremely dangerous to manufacture and handle. The Swedish chemist Alfred Nobel later perfected its synthesis and devised a safe method for handling it. Nobel mixed nitroglycerin with a claylike absorbent material. The resulting solid, called dynamite, is an explosive that is much less sensitive to shock than liquid nitroglycerin. Dynamite is still one of the most important explosives used for mining, digging tunnels, and blasting hills for road building.

Surprising as it may seem, nitroglycerin is also an effective medicine. It is used to treat patients with *angina pectoris*—sharp chest pains caused by an insufficient supply of oxygen to the heart muscle. Angina pectoris is usually found in patients with coronary artery diseases, such as arteriosclerosis (hardening of the arteries). During overexertion or excitement, the partially clogged coronary arteries prevent the heart from getting an ade-quate supply of oxygenated blood, and pain results. Nitroglycerin relaxes cardiac muscle and causes a dilation of the arteries, thus increasing blood flow to the heart and relieving the chest pains.

Nitroglycerin (combined with other substances to render it nonexplosive) can be administered in small tablets, which are placed under the tongue during an attack of angina. The nitroglycerin is rapidly absorbed into the bloodstream and finds its way to the heart muscle within seconds. It can also be taken orally, applied directly to the skin as a cream, or absorbed through the skin from an adhesive patch attached to the skin. This last delivery system is a recent development that allows the nitroglycerin to be continuously absorbed for 24 hours.

An adhesive patch allows nitroglycerin to be absorbed through the skin.

Phosphoric anhydride
A compound that contains the

$$
\begin{array}{cc}
O & O \\
\| & \| \\
-O-P-O-P-O- \\
| & | \\
O^- & O^-
\end{array}
$$
group.

At body pH, the two —OH groups of the phosphoric acid are ionized, and the phosphate group has a charge of -2. Examples of important diesters include the phospholipids (Chapter 8) and the nucleic acids (Chapter 11).

As we study key compounds in the storage and transfer of chemical energy in living systems (Chapter 12), we will encounter phosphate esters in which two or three phosphate groups are linked. These **phosphoric anhydrides** are referred to as diphosphates and triphosphates.

$$
\begin{array}{ll}
\begin{array}{cc}
O & O \\
\| & \| \\
R-O-P-O-P-O^- \\
| & | \\
O^- & O^-
\end{array}
&
\begin{array}{ccc}
O & O & O \\
\| & \| & \| \\
R-O-P-O-P-O-P-O^- \\
| & | & | \\
O^- & O^- & O^-
\end{array}
\end{array}
$$

a diphosphate ester   a triphosphate ester

Adenosine diphosphate (ADP) and adenosine triphosphate (ATP) are the most important of these compounds. We will study their rather complex structures in more detail in Section 12.7. For now, note the ester linkages in each structure and the presence of the di- and triphosphate groups. Again, at the pH of body fluids, the —OH groups attached to the phosphorus lose H^+ and acquire negative charges.

FOR FUTURE REFERENCE FOODS RICH IN VITAMIN C

The chemical name *ascorbic acid* reflects the fact that vitamin C molecules contain an acid functional group. In 1747 the first nutrition experiment ever performed found that British sailors afflicted with scurvy quickly recovered when they received citrus fruits as a supplemental ration to their regular diet. This discovery led to the practice of providing British sailors at sea with a daily ration of lime juice to prevent scurvy. Nearly 200 years later ascorbic acid, the substance responsible for the cure, was first isolated from lime juice.

Today, it is recognized that vitamin C plays a number of vital roles in the human body. Vitamin C is easily oxidized, and in the body protects other substances from being oxidized by becoming a sacrificial target for oxidizing agents. It also helps to form collagen, the fibrous structural protein of connective tissues. In this role it is especially important in strengthening blood vessel walls. The vitamin is also involved in the synthesis of the thyroid hormone thyroxin, and in the metabolism of amino acids.

Vitamin C supplements are inexpensive and readily available, but the RDA can be met easily by including fruits and green vegetables in the diet. As shown in the chart below, a number of these foods contain more than the RDA in a single serving. If the diet does not include such fruits and vegetables, it is almost impossible to obtain the RDA from the diet. The food sources listed in the chart below each provide a minimum of 10 mg of vitamin C per serving.

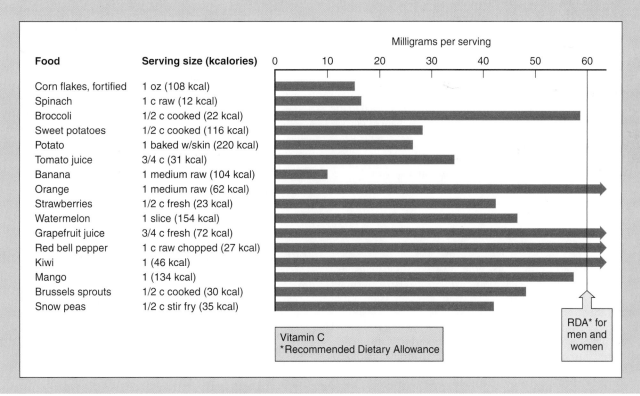

SOURCE: Adapted from Whitney, E.N.; Rolfes, S.R. *Understanding Nutrition,* 8th ed. Belmont, CA: West/Wadsworth, 1999.

ADP

ATP

CONCEPT SUMMARY

The Nomenclature of Carboxylic Acids. The characteristic functional group of carboxylic acids is the carboxyl group

$$\overset{\displaystyle O}{\underset{\displaystyle \parallel}{-C-OH}}$$

Many of the simpler carboxylic acids are well known by common names. In the IUPAC system, the ending of *-oic acid* is used in the names of these compounds. Aromatic acids are named as derivatives of benzoic acid.

Physical Properties of Carboxylic Acids. At room temperature, low molecular weight carboxylic acids are liquids with distinctively sharp or unpleasant odors. High molecular weight long-chain acids are waxlike solids. Carboxylic acids are quite effective in forming dimers, two molecules held together by hydrogen bonds. Thus, they have relatively high boiling points, and those with lower molecular weights are soluble in water.

The Acidity of Carboxylic Acids. Soluble carboxylic acids behave as weak acids; they dissociate only slightly in water to form an equilibrium mixture with the carboxylate ion. The equilibrium concentrations of the carboxylic acid and the carboxylate ion depend on pH. At low pH, the acid form predominates, and at pH 7.4 (the pH of cellular fluids) and above, the carboxylate ion predominates. Carboxylic acids react with bases to produce carboxylate salts and water.

Salts of Carboxylic Acids. The carboxylate salts are named by changing the *-ic* ending of the acid to *-ate*. The ionic nature of the salts makes them water-soluble. A number of carboxylate salts are useful as food preservatives, soaps, and medicines.

Carboxylic Esters. Carboxylic acids, acid chlorides, and acid anhydrides react with alcohols to produce esters. Polyesters result from the reaction of dicarboxylic acids and diols. Polyesters are an example of condensation polymers; these are produced when monomers react to form a polymer plus a small molecule such as water. Many esters are very fragrant and represent some of nature's most pleasant odors. Because of this characteristic, esters are widely used as flavoring agents.

The Nomenclature of Esters. Both common and IUPAC names for esters are formed by first naming the alkyl group of the alcohol portion followed by the name of the acid portion in which the *-ic acid* ending has been changed to *-ate*.

Reactions of Esters. Esters can be converted back to carboxylic acids and alcohols under either acidic or basic conditions. Hydrolysis, the reaction with water in the presence of acid, produces the carboxylic acid and alcohol. Saponification occurs in the presence of base to produce the carboxylate salt and alcohol.

Esters of Inorganic Acids. Alcohols can also form esters by reacting with inorganic acids such as phosphoric acid. Phosphate esters represent some of the most important biological compounds.

LEARNING OBJECTIVES

When you have completed your study of this chapter, you should be able to:

1. Assign IUPAC names and draw structural formulas for carboxylic acids.
2. Explain how hydrogen bonding affects the physical properties of carboxylic acids.
3. Assign common and IUPAC names to carboxylic acid salts and esters.
4. Recognize and write key reactions of carboxylic acids and esters.
5. Describe uses for carboxylic acids, carboxylate salts, and esters.
6. Write reactions for the formation of phosphate esters.

KEY TERMS AND CONCEPTS

Carboxyl group (Introduction)
Carboxylate ion (5.3)
Carboxylic acid (Introduction)
Carboxylic acid anhydride (5.5)
Carboxylic acid chloride (5.5)

Carboxylic ester (5.5)
Condensation polymerization (5.5)
Dimer (5.2)
Ester (5.8)
Esterification (5.5)

Ester linkage (5.5)
Fatty acid (5.1)
Phosphoric anhydride (5.8)
Saponification (5.7)

KEY REACTIONS

1. Dissociation of a carboxylic acid to give a carboxylate ion (Section 5.3):

$$\underset{\substack{\text{O}\\\|}}{\text{R}-\text{C}-\text{OH}} + \text{H}_2\text{O} \rightleftharpoons \underset{\substack{\text{O}\\\|}}{\text{R}-\text{C}-\text{O}^-} + \text{H}_3\text{O}^+$$

Reaction 5.1

2. Reaction of a carboxylic acid with base to produce a carboxylate salt plus water (Section 5.3):

$$\underset{\substack{\text{O}\\\|}}{\text{R}-\text{C}-\text{OH}} + \text{NaOH} \longrightarrow \underset{\substack{\text{O}\\\|}}{\text{R}-\text{C}-\text{O}^-\text{Na}^+} + \text{H}_2\text{O}$$

Reaction 5.4

3. Reaction of a carboxylic acid with an alcohol to produce an ester plus water (Section 5.5):

$$\underset{\substack{\text{O}\\\|}}{\text{R}-\text{C}-\text{OH}} + \text{HO}-\text{R}' \underset{}{\overset{\text{H}^+,\ \text{heat}}{\rightleftharpoons}} \underset{\substack{\text{O}\\\|}}{\text{R}-\text{C}-\text{O}-\text{R}'} + \text{H}_2\text{O}$$

Reaction 5.8

4. Reaction of a carboxylic acid chloride with an alcohol to produce an ester plus hydrogen chloride (Section 5.5):

$$\underset{\substack{\text{O}\\\|}}{\text{R}-\text{C}-\text{Cl}} + \text{HO}-\text{R}' \longrightarrow \underset{\substack{\text{O}\\\|}}{\text{R}-\text{C}-\text{O}-\text{R}'} + \text{HCl}$$

Reaction 5.12

5. Reaction of a carboxylic acid anhydride with an alcohol to produce an ester plus a carboxylic acid (Section 5.5):

$$\underset{\substack{\text{O}\quad\ \ \text{O}\\\|\qquad\|}}{\text{R}-\text{C}-\text{O}-\text{C}-\text{R}} + \text{HO}-\text{R}' \longrightarrow \underset{\substack{\text{O}\\\|}}{\text{R}-\text{C}-\text{O}-\text{R}'} + \underset{\substack{\text{O}\\\|}}{\text{R}-\text{C}-\text{OH}}$$

Reaction 5.13

6. Ester hydrolysis to produce a carboxylic acid and alcohol (Section 5.7):

$$\underset{\substack{\text{O}\\\|}}{\text{R}-\text{C}-\text{OR}'} + \text{H}-\text{OH} \underset{}{\overset{\text{H}^+}{\rightleftharpoons}} \underset{\substack{\text{O}\\\|}}{\text{R}-\text{C}-\text{OH}} + \text{R}'-\text{OH}$$

Reaction 5.17

7. Ester saponification to give a carboxylate salt and alcohol (Section 5.7):

$$\underset{\substack{\text{O}\\\|}}{\text{R}-\text{C}-\text{OR}'} + \text{NaOH} \longrightarrow \underset{\substack{\text{O}\\\|}}{\text{R}-\text{C}-\text{O}^-\text{Na}^+} + \text{R}'-\text{OH}$$

Reaction 5.19

8. Phosphate ester formation (Section 5.8):

$$\text{R}-\text{OH} + \underset{\substack{\text{O}\\\|\\\ \\\text{OH}}}{\text{HO}-\text{P}-\text{OH}} \rightleftharpoons \underset{\substack{\text{O}\\\|\\\ \\\text{OH}}}{\text{R}-\text{O}-\text{P}-\text{OH}} + \text{H}_2\text{O}$$

Reaction 5.21

THE NOMENCLATURE OF CARBOXYLIC ACIDS (SECTION 5.1)

5.1 What is the structure of the carboxylic acid functional group? How does it differ from the structure of an alcohol and from that of a ketone?

5.2 What structural features are characteristic of fatty acids? Why are fatty acids given that name?

5.3 What compound is responsible for the sour or tart taste of Italian salad dressing (vinegar and oil)?

5.4 What carboxylic acid is present in sour milk and sauerkraut?

5.5 Write the correct IUPAC name for each of the following:

a. $CH_3CH_2CH_2CH_2 — \overset{\overset{\displaystyle O}{\|}}{C} — OH$

b. $CH_3CH_2CHCH_2 — \overset{\overset{\displaystyle O}{\|}}{C} — OH$
　　　　|
　　　CH_3

c. $CH_3CH_2CHCH_2CH_2 — \overset{\overset{\displaystyle O}{\|}}{C} — OH$

d.

e. $CH_3CHCH_2CHCH_2 — \overset{\overset{\displaystyle O}{\|}}{C} — OH$
　　　|　　　|
　　CH_3　CH_2CH_3

5.6 Write the correct IUPAC name for each of the following:

a. $CH_3CH_2CH_2 — \overset{\overset{\displaystyle O}{\|}}{C} — OH$

b. $CH_3 — O — CH_2CH_2 — \overset{\overset{\displaystyle O}{\|}}{C} — OH$

c. $CH_3CHCHCH_2 — \overset{\overset{\displaystyle O}{\|}}{C} — OH$
　　　　|
　　　Br
　　CH_3

d.

e. $CH_3CHCH_2CH_2 — Br$

5.7 Write a structural formula for each of the following:
a. methanoic acid
b. 3, 5-dinitrobenzoic acid
c. 4-ethylheptanoic acid

5.8 Write a structural formula for each of the following:
a. hexanoic acid
b. 4-bromo-3-methylpentanoic acid
c. o-ethylbenzoic acid

PHYSICAL PROPERTIES OF CARBOXYLIC ACIDS (SECTION 5.2)

5.9 Of the classes of organic compounds studied so far, which have particularly unpleasant odors?

5.10 Which compound in each of the following pairs would you expect to have the higher boiling point? Explain your answer.
a. acetic acid or 1-propanol
b. propanoic acid or butanone
c. acetic acid or butyric acid

5.11 List the following compounds in order of increasing boiling point:
a. 2-butanone
b. propanoic acid
c. pentane
d. 1-butanol

5.12 Draw the structure of the dimer formed when two molecules of propanoic acid hydrogen bond with each other.

5.13 The two isomers propanoic acid and methyl acetate are both liquids. One boils at 141°C, the other at 57°C.

$CH_3CH_2 — \overset{\overset{\displaystyle O}{\|}}{C} — OH$　　　$CH_3 — \overset{\overset{\displaystyle O}{\|}}{C} — OCH_3$
　propanoic acid　　　　　methyl acetate

a. Predict which compound boils at 57°C and which at 141°C. Explain your reasoning.
b. Which of these two compounds would you predict to be the more soluble in water? Explain.

5.14 Caproic acid, a six-carbon acid, has a solubility in water of 1 g/100 mL of water (Table 5.2). Which part of the structure of caproic acid is responsible for its solubility in water, and which part prevents greater solubility?

5.15 Why are acetic acid, sodium acetate, and sodium caprate all soluble in water, whereas capric acid, a 10-carbon fatty acid, is not?

5.16 List the following compounds in order of increasing water solubility:
   a. 1-butanol
   b. pentane
   c. butyric acid
   d. butanal

THE ACIDITY OF CARBOXYLIC ACIDS (SECTION 5.3)

5.17 What is the most important chemical property of carboxylic acids?

5.18 As we discuss the cellular importance of lactic acid in a later chapter we will refer to this compound as lactate. Explain why.

$$
\begin{array}{cc}
OH & O \\
| & \| \\
CH_3-CH- & C-OH
\end{array}
$$

lactic acid

5.19 Write an equation to illustrate the equilibrium that is present when propanoic acid is dissolved in water. What structure predominates when OH^- is added to raise the pH to 12? What structure predominates as acid is added to lower the pH to 2?

5.20 Complete each of the following reactions:

$$
\begin{array}{c}
O \\
\|
\end{array}
$$
a. $H-C-OH + NaOH \longrightarrow$

$$
\begin{array}{c}
O \\
\|
\end{array}
$$
b. $CH_3CH_2CH_2-C-OH + KOH \longrightarrow$

5.21 Complete each of the following reactions:

$$
\begin{array}{c}
O \\
\|
\end{array}
$$
a. $CH_3(CH_2)_7-C-OH + NaOH \longrightarrow$

b.
$$
\begin{array}{c}
CH_3 \quad O \\
\quad\quad \| \\
C-OH
\end{array}
$$
$+ KOH \longrightarrow$

5.22 Give a balanced equation for the reaction of propanoic acid with each of the following:
   a. KOH
   b. H_2O
   c. NaOH

SALTS OF CARBOXYLIC ACIDS (SECTION 5.4)

5.23 Write the IUPAC name for each of the following:

$$
\begin{array}{c}
O \\
\|
\end{array}
$$
a. $CH_3CH_2-C-O^-Na^+$

$$
\begin{array}{cc}
CH_3 & O \\
| & \| \\
\end{array}
$$
b. $CH_3CH-C-O^-K^+$

c.
$$
\begin{array}{c}
NO_2 \\
\end{array}
$$
$$
\begin{array}{c}
C-O^-Na^+ \\
\|
\\
O
\end{array}
$$

5.24 Give the IUPAC name for each of the following:

$$
\begin{array}{cc}
Br & O \\
| & \| \\
\end{array}
$$
a. $CH_3CHCH_2-C-O^-Na^+$

$$
\begin{array}{c}
O \\
\|
\end{array}
$$
b. $(H-C-O^-)_2Ca^{2+}$

c.
$$
\begin{array}{c}
O \\
\| \\
O-CH_2CH_2-C-O^-K^+
\end{array}
$$

5.25 Draw structural formulas for the following:
   a. sodium formate
   b. calcium benzoate
   c. sodium 2-methylbutanoate

5.26 Draw structural formulas for the following:
   a. sodium methanoate
   b. calcium 3-methylbutanoate
   c. potassium p-propylbenzoate

5.27 Name each of the following:
   a. The sodium salt of valeric acid
   b. The magnesium salt of lactic acid
   c. The potassium salt of citric acid

5.28 Give the name of a carboxylic acid or carboxylate salt used in each of the following ways:
   a. As a soap
   b. As a general food preservative used to pickle vegetables
   c. As a preservative used in soft drinks
   d. As a treatment for athlete's foot
   e. As a mold inhibitor used in bread
   f. As a food additive noted for its pH buffering ability

CARBOXYLIC ESTERS (SECTION 5.5)

5.29 Which of the following compounds are esters?

a.
$$
CH_2-O-CH_2-OH
$$

b.
$$
\begin{array}{c}
O \\
\| \\
CH_3-C-O-CH_2CH_3
\end{array}
$$

c. $CH_3CH_2-CH-OCH_3$
$$
\quad\quad\quad\quad | \\
\quad\quad\quad OCH_3
$$

d.
$$
\begin{array}{c}
O \\
\| \\
O-C-CH_3
\end{array}
$$

e. $CH_3-O-CH_2-\overset{\displaystyle O}{\overset{\displaystyle \|}{C}}-CH_2-OH$

f. (benzene ring)$-\overset{\displaystyle O}{\overset{\displaystyle \|}{C}}-O-CH_2CH_3$

5.30 For each ester in Exercise 5.29, draw a circle around the portion that came from the acid and use an arrow to point out the *ester linkage*.

5.31 Complete the following reactions:

a. (benzene ring)$-\overset{\displaystyle O}{\overset{\displaystyle \|}{C}}-OH$ $+\ CH_3\underset{\underset{CH_3}{|}}{CH}-OH \underset{\longleftarrow}{\overset{H^+,\ heat}{\longrightarrow}}$

b. (benzene ring)$-\overset{\displaystyle O}{\overset{\displaystyle \|}{C}}-Cl$ $+\ CH_3\underset{\underset{CH_3}{|}}{CH}-OH \longrightarrow$

c. (benzene ring)$-\overset{\displaystyle O}{\overset{\displaystyle \|}{C}}-O-\overset{\displaystyle O}{\overset{\displaystyle \|}{C}}-$(benzene ring) $+\ CH_3-OH \longrightarrow$

5.32 Complete the following reactions:

a. $CH_3-\overset{\displaystyle O}{\overset{\displaystyle \|}{C}}-O-\overset{\displaystyle O}{\overset{\displaystyle \|}{C}}-CH_3 + CH_3-OH \longrightarrow$

b. $CH_3\underset{\underset{CH_3}{|}}{CH}-\overset{\displaystyle O}{\overset{\displaystyle \|}{C}}-OH +$ (benzene ring)$-CH_2-OH \underset{\longleftarrow}{\overset{H^+,\ heat}{\longrightarrow}}$

c. $CH_3\underset{\underset{CH_3}{|}}{CH}-\overset{\displaystyle O}{\overset{\displaystyle \|}{C}}-Cl +$ (benzene ring with OH) $\longrightarrow$

5.33 Using the alcohol CH_3CH_2-OH, show three different starting compounds that may be used to synthesize the ester

$$CH_3CH_2CH_2-\overset{\displaystyle O}{\overset{\displaystyle \|}{C}}-O-CH_2CH_3$$

5.34 Give the structures of the ester that forms when propanoic acid is reacted with the following:
a. methyl alcohol
b. phenol
c. 2-methyl-1-propanol

5.35 The structures of two esters used as artificial flavorings are given below. Write the structure of the acid and the alcohol from which each ester could be synthesized.

a. Pineapple flavoring,

$$CH_3CH_2CH_2-\overset{\displaystyle O}{\overset{\displaystyle \|}{C}}-O-CH_2CH_2CH_2CH_3$$

b. Apple flavoring,

$$CH_3CH_2CH_2-\overset{\displaystyle O}{\overset{\displaystyle \|}{C}}-O-CH_3$$

5.36 Heroin is formed by reacting morphine with 2 mol of $CH_3-\overset{\displaystyle O}{\overset{\displaystyle \|}{C}}-Cl$ to form the diester. Show the structure of heroin.

(structure of morphine)

morphine

THE NOMENCLATURE OF ESTERS (SECTION 5.6)

5.37 Assign common names to the following esters. Refer to Table 5.1 for the common names of the acids.

a. $CH_3-\overset{\displaystyle O}{\overset{\displaystyle \|}{C}}-O-CH_3$

b. $CH_3CH_2-\overset{\displaystyle O}{\overset{\displaystyle \|}{C}}-O-CH_2CH_2CH_3$

5.38 Assign common names to the following esters. Refer to Table 5.1 for the common names of the acids.

a. $H-\overset{\displaystyle O}{\overset{\displaystyle \|}{C}}-O-CH_2CH_3$

b. $CH_3(CH_2)_4-\overset{\displaystyle O}{\overset{\displaystyle \|}{C}}-O-\underset{\underset{CH_3}{|}}{CH}CH_3$

5.39 Give the IUPAC name to each of the following:

a. $CH_3CH_2-\overset{\displaystyle O}{\overset{\displaystyle \|}{C}}-O-CH_2CH_2CH_3$

b. (benzene ring with CH_3)$-\overset{\displaystyle O}{\overset{\displaystyle \|}{C}}-O-CH_3$

5.40 Give the IUPAC name to each of the following:

a. $CH_3CH_2CH_2 - \overset{\overset{\displaystyle O}{\|}}{C} - O - CH_2CH_3$

b. A benzene ring with a $\overset{\overset{\displaystyle O}{\|}}{C} - O - \underset{\underset{\displaystyle CH_3}{|}}{CH}CH_3$ group and a CH_3 group ortho to it.

5.41 Assign IUPAC names to the simple esters produced by a reaction between ethanoic acid and the following:
   a. ethanol
   b. 1-propanol
   c. 1-butanol

5.42 Assign common names to the simple esters produced by a reaction between methanol and the following:
   a. propionic acid
   b. butyric acid
   c. lactic acid

5.43 Draw structural formulas for the following:
   a. ethyl butyrate
   b. methyl formate
   c. lactic acid

5.44 Draw structural formulas for the following:
   a. methyl ethanoate
   b. propyl 2-bromobenzoate
   c. ethyl 3,4-dimethylpentanoate

REACTIONS OF ESTERS (SECTION 5.7)

5.45 Define and compare the terms *ester hydrolysis* and *saponification*.

5.46 Write equations for the hydrolysis and saponification of phenyl butanoate and name the component parts.

$$CH_3CH_2CH_2 - \overset{\overset{\displaystyle O}{\|}}{C} - O - \text{(benzene ring)}$$

5.47 Complete the following reactions:

a. $CH_3\underset{\underset{\displaystyle CH_3}{|}}{CH} - \overset{\overset{\displaystyle O}{\|}}{C} - O - CH_2CH_3 + H_2O \overset{H^+}{\rightleftharpoons}$

b. (benzene ring) $- \overset{\overset{\displaystyle O}{\|}}{C} - O - CH_2 -$ (benzene ring) $+ NaOH \longrightarrow$

5.48 Complete the following reactions:

a. $CH_3CH_2\underset{\underset{\displaystyle CH_3}{|}}{CH} - \overset{\overset{\displaystyle O}{\|}}{C} - O - CH_2CH_3 + NaOH \longrightarrow$

b. $CH_3(CH_2)_{10} - \overset{\overset{\displaystyle O}{\|}}{C} - O - CH_2\underset{\underset{\displaystyle \text{(benzene ring)}}{|}}{CH}CH_3 + H_2O \overset{H^+}{\rightleftharpoons}$

5.49 How do the acids and alcohols involved in *polyester* formation differ from those that commonly form simple esters?

5.50 Draw the structure of the repeating monomer unit in the polyester formed in the condensation of oxalic acid and 1,3-propanediol.

$$HO - \overset{\overset{\displaystyle O}{\|}}{C} - \overset{\overset{\displaystyle O}{\|}}{C} - OH + \underset{\underset{\displaystyle OH}{|}}{CH_2} - \underset{\underset{\displaystyle OH}{|}}{CH_2} - CH_2 \rightleftharpoons$$

ESTERS OF INORGANIC ACIDS (SECTION 5.8)

5.51 Glyceraldehyde 3-phosphate, an important compound in the cellular oxidation of carbohydrates, is an ester of phosphoric acid and glyceraldehyde.

$$\begin{array}{c} O = C - H \\ | \\ H - C - OH \qquad O \\ | \qquad\qquad \| \\ CH_2 - O - P - O^- \\ | \\ O^- \end{array}$$

   a. Give the structure of glyceraldehyde.
   b. Why is glyceraldehyde 3-phosphate shown with a -2 charge?

5.52 Dihydroxyacetone reacts with phosphoric acid to form the important monoester called dihydroxyacetone phosphate. Complete the reaction for its formation.

$$\underset{\underset{\displaystyle \text{dihydroxyacetone}}{}}{\underset{\displaystyle OH}{|}}{CH_2} - \overset{\overset{\displaystyle O}{\|}}{C} - \underset{\underset{\displaystyle OH}{|}}{CH_2} + HO - \overset{\overset{\displaystyle O}{\|}}{\underset{\underset{\displaystyle OH}{|}}{P}} - OH \longrightarrow$$

5.53 Give the structure for the following:
   a. monoethyl phosphate (with a -2 charge)
   b. monoethyl diphosphate (with a -3 charge)
   c. monoethyl triphosphate (with a -4 charge)

CHEMISTRY AROUND US, OVER THE COUNTER, AND HOW REACTIONS OCCUR

5.54 Draw the characteristic structure of an alpha hydroxy acid.

5.55 Name two of the most common alpha hydroxy acids used in cosmetics.

5.56 What serious diseases may be partially prevented or treated with aspirin?

5.57 Explain how nitroglycerin helps relieve the pain of angina pectoris.

5.58 In the saponification of esters, why is one of the products a carboxylate salt rather than a carboxylic acid?

POINTS TO PONDER AND PURSUE

5.1 Citric acid is often added to carbonated beverages as a flavoring. Why is citric acid viewed as a "safe" food additive?

5.2 Ester formation (Reaction 5.8) and ester hydrolysis (Reaction 5.17) are exactly the same reaction only written in reverse. What determines which direction the reaction proceeds and what actually forms?

5.3 Write a mechanism for the following reaction. Show the final location of the oxygen atom in the second color.

$$CH_3CH_2 - \overset{\displaystyle O}{\overset{\displaystyle \|}{C}} - O - CH_3 + NaOH \longrightarrow$$

5.4 Answer the question in Figure 5.1. What flavor sensation is characteristic of acids?

5.5 Oxalic acid, the substance mentioned in Figure 5.2, has two carboxyl groups. If you were to give this compound an IUPAC name, what ending would be appropriate?

5.6 Why is it safe for us to consume foods like vinegar that contain acetic acids?

5.7 A Dacron patch is sometimes used in heart surgery. Why do polyesters like the Dacron patch resist hydrolysis reactions that would result in deterioration of the patch?

5.8 Use Web site 5.3 to list the five main reasons why additives are used in food.

5.9 Using Web site 5.4, name a water-insoluble soap.

5.10 Use Web site 5.5 to identify the meaning of the acronym NSAID.

http://www.brookscole.com/chemistry/seager_chemistry

Amines and Amides

Individuals who require basic life support before arriving at a health care facility often receive it from an emergency medical technician (EMT), or paramedic. Paramedics function as extensions of an emergency room physician and must be familiar with a number of medications. More pharmaceuticals belong to the amine (or amide) family of compounds than to any other functional class.

http://www.brookscole.com/chemistry/
seager_chemistry
At Web site 6.1, you will find further information on a career as an emergency medical technician.

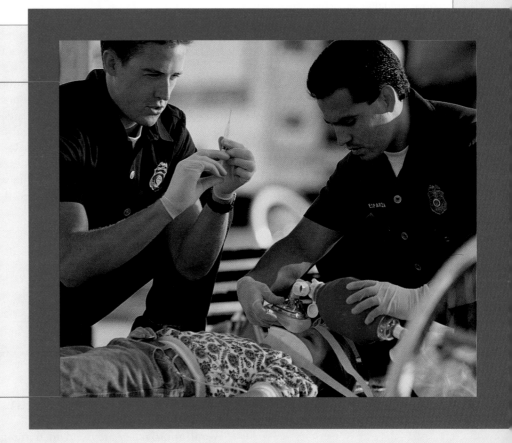

Amine

An organic compound derived by replacing one or more of the hydrogen atoms of ammonia with alkyl or aromatic groups, as in RNH_2, R_2NH, and R_3N.

Primary amine

An amine having one alkyl or aromatic group bonded to nitrogen, as in $R—NH_2$.

Secondary amine

An amine having two alkyl or aromatic groups bonded to nitrogen, as in R_2NH.

Tertiary amine

An amine having three alkyl or aromatic groups bonded to nitrogen, as in R_3N.

Learning Check 6.1

The effectiveness of a wide variety of important medicines depends either entirely or partly on the presence of a nitrogen-containing group in their molecules. Nitrogen-containing functional groups are found in more medications than any other type of functional group. In this chapter, we will study two important classes of organic nitrogen compounds, the amines and the amides.

Both amines and amides are abundant in nature, where they play important roles in the chemistry of life. Our study of these two functional classes will help prepare us for later chapters dealing with amino acids, proteins, and nucleic acids and provide a basis for understanding the structures and chemistry of a number of medicines.

➤ 6.1 Classification of Amines

Amines are organic derivatives of ammonia, NH_3, in which one or more of the hydrogens are replaced by an aromatic or alkyl group (R). Like alcohols, amines are classified as primary, secondary, or tertiary on the basis of molecular structure. However, the classification is done differently. When classifying amines, the number of groups (R) that have replaced hydrogens in the NH_3 is counted. The nitrogen atom in a **primary amine** is bonded to one R group. The nitrogen in **secondary amines** is bonded to two R groups, and that of **tertiary amines** is bonded to three R groups. These amine subclasses are summarized in ■ Table 6.1.

Classify each of the following amines as primary, secondary, or tertiary:

a. CH₃—N—CH₂CH₃ (with H on N)   b. (ring)N—H   c. (benzene ring)N—CH₃ (with CH₃ below)

➤ 6.2 The Nomenclature of Amines

Several approaches are used in naming amines. Common names are used extensively for those with low molecular weights. These are named by alphabetically linking the names of the alkyl or aromatic groups bonded to the nitro-

TABLE 6.1	Subclasses of amines	
Subclass	**General formula**	**Example**
Primary (1°)	R—N—H (with H below)	CH₃—N—H (with H below)
Secondary (2°)	R—N—R' (with H below)	CH₃—N—CH₃ (with H below)
Tertiary (3°)	R—N—R' (with R" below)	CH₃—N—CH₃ (with CH₃ below)

gen and attaching the suffix -amine. The name is written as one word, and the prefixes di- and tri- are used when identical alkyl groups are present. Some examples are

$$CH_3-NH_2 \qquad CH_3-NH-CH_3 \qquad CH_3CH_2-NH-CH_3$$
methylamine   dimethylamine   ethylmethylamine

Assign a common name to the following amines:

a. $CH_3CH_2CH_2-NH_2$ **b.** ⟨◯⟩$-NH-CH_3$ **c.** $CH_3-\underset{\underset{CH_3}{|}}{N}-CH_3$

Learning Check 6.2

An IUPAC approach is frequently used for naming primary amines. In this method, the $-NH_2$ functional group is treated as a chain substituent called the amino group, and the amine is named as a derivative of the parent compound following IUPAC rules. The following are examples:

$$CH_3CH_2CH_2\underset{\underset{NH_2}{|}}{C}HCH_3 \qquad \underset{\underset{NH_2}{|}}{C}H_2CH_2\underset{\underset{OH}{|}}{C}H_2 \qquad \underset{\underset{NH_2}{|}}{C}H_2CH_2\underset{\underset{NH_2}{|}}{C}H_2$$

2-aminopentane   3-amino-1-propanol   1,3-diaminopropane

Name the following primary amines by using the amino group as a substituent:

a. $CH_3\underset{\underset{NH_2}{|}}{C}HCH_2CH_2-NH_2$ **b.**

Learning Check 6.3

Aromatic amines are usually given names (both common and IUPAC) based on aniline, the simplest aromatic amine. They are named as substituted anilines, and an italic capital N is used to indicate that an akyl group is attached to the nitrogen and not to the ring. If two groups are attached to the nitrogen, an italic capital N is used for each one. Some examples are

aniline   N-methylaniline   2-ethyl-N-methylaniline   N-ethyl-N-methylaniline

Name the following amines as derivatives of aniline:

Learning Check 6.4

a.   **c.**

b.

OVER THE COUNTER • 6.1

THE RIGHT STUFF FOR STUFFY NOSES

It's springtime, and millions of Americans will be experiencing one or more of the following symptoms: sneezing; runny or stuffy nose; coughing; postnasal drip; itchy eyes, nose, or throat; or watery eyes. The cause of all this misery is *allergic rhinitis,* commonly called hay fever. The term *hay fever* is a misnomer, since generally fever is not a symptom. The term originated on the farm, where late-summer harvesting and haying caused widespread discomfort in farm workers.

Hay fever is caused by a hypersensitive response of the body's immune system to airborne allergens such as pollen, mold spores, yeast spores, house dust, mites, proteins on dog and cat fur, and various chemicals from air pollution, paint, carpeting, plastics, and other materials. The immune system responds to an allergen in the same way it responds to bacteria or viruses, by generating large amounts of immunoglobulin E, a type of antibody that attaches to tissue and blood cells. When the allergen is detected in the body again, the cells with the antibody attached respond. Most cells in the skin, lungs, and respiratory passages respond by releasing powerful chemicals such as histamine. *Histamine* causes blood vessels to dilate and become leaky, releasing fluid into surrounding tissues. This fluid causes the swelling, congestion, and runny nose symptoms characteristic of hay fever.

histamine

Several OTC products are available to combat the symptoms of hay fever. These products generally fall into one of three categories: decongestant, antihistamine, or allergy preventive. Decongestants relieve a stuffy nose by constricting the enlarged nasal blood vessels, thereby reducing the pressure in the ear canals and nasal passages. Unfortunately, this remedy provides only temporary relief because after 3–4 days, decongestants have a rebound effect and actually increase nasal swelling and congestion. They can also cause side effects, including nervousness, insomnia, and elevated blood pressure.

Antihistamines are the most often common hay fever remedy. Molecules of antihistamines are structurally similar to histamine and are therefore able to occupy the receptor sites normally occupied by histamine. In this way, antihistamines prevent histamines from causing their normal physiological effects in the body. Over-the-counter antihistamines found in medications such as diphenhydramine (Benadryl®) and chlorpheniramine (Teldrin®) cause drowsiness. This side effect accounts for their use in OTC sleeping aids and sedatives. In one study, motorists who were given a single dose of OTC antihistamines were found

to have impaired driving skills equivalent to someone with a blood alcohol level of 0.08. It is obvious that the warning on most antihistamines about not using them before driving or operating machinery should be followed. If possible, antihistamines should be taken at bedtime.

diphenhydramine
(Benadryl)

chlorpheniramine
(Teldrin)

Some antihistamines do not cause drowsiness, but they are available only by prescription. Fexofenadine (Allegra®) and loratadine (Claritin®) are less effective at relieving a stuffy nose, but they have no serious side effects, including drowsiness. Another prescription drug, cetirizine (Zyrtec®), is stronger than the latter two, but it does cause drowsiness in some people. Two additional prescription antihistamines called terfenadine (Seldane®) and astemizole (Hismanal®) do not cause drowsiness but have some other drawbacks that must be considered before they are used. Both of these compounds interact with the antifungal medication itraconazole and the antibiotics erythromycin, clarithromycin, and troleandomycin. The interactions cause abnormal heart rhythms and, in rare cases, fatal arrhythmia.

Nasalcrom® is an OTC nasal spray that functions as an allergy preventive. It is effective without serious side effects, but somewhat inconvenient to use. The nostrils have to be spritzed three or four times a day, and noticeable effects do not occur for a week or more. Also, the product does not ease nasal stuffiness or relieve eye irritation caused by allergies.

As a final option, allergy shots have proven to be effective as a complete cure for a significant number of allergy sufferers. This approach, called immunotherapy, involves skin tests to determine which substances cause allergic reactions, followed by a series of shots of the identified substances to desensitize the immune system.

http://www.brookscole.com/chemistry/seager_chemistry
At Web site 6.2, you will find advice from an allergist.

Antihistamine
A drug that relieves the symptoms of allergies and colds by countering the action of histamine.

TABLE 6.2	Properties of some amines		Melting point (°C)	Boiling point (°C)
Name	Structure			
methylamine	$CH_3 — NH_2$		−94	−6
ethylamine	$CH_3CH_2 — NH_2$		−81	17
dimethylamine	$CH_3 — NH — CH_3$		−93	7
diethylamine	$CH_3CH_2 — NH — CH_2CH_3$		−48	56
trimethylamine	$CH_3 — N — CH_3$ $\quad\quad\mid$ $\quad\quad CH_3$		−117	3
triethylamine	$CH_3CH_2 — N — CH_2CH_3$ $\quad\quad\quad\mid$ $\quad\quad CH_2CH_3$		−114	89

➤ 6.3 Physical Properties of Amines

The simple, low molecular weight amines are gases at room temperature (see ■ Table 6.2). Heavier, more complex compounds are liquids or solids.

Like alcohols, primary and secondary amines form hydrogen bonds among themselves:

$$CH_3 — N — H \cdots N — CH_3$$

Because nitrogen is less electronegative than oxygen, the hydrogen bonds formed by amines are weaker than those formed by alcohols. For this reason, the boiling points of primary and secondary amines are usually somewhat lower than the boiling points of alcohols with similar molecular weights. Tertiary amines cannot form hydrogen bonds among themselves (because the nitrogen has no attached hydrogen atom), and their boiling points are similar to alkanes that have about the same molecular weights.

Amines with fewer than six carbon atoms are generally soluble in water as a result of hydrogen bond formation between amine functional groups and water molecules. This is true for tertiary as well as primary and secondary amines (see ■ Figure 6.1).

1° amine   2° amine   3° amine

➤ **FIGURE 6.1** Hydrogen bonding between amines in water.

Low-molecular-weight amines have a sharp, penetrating odor similar to that of ammonia. Somewhat larger amines have an unpleasant odor reminiscent of decaying fish. In fact, some amines are partially responsible for the odor of decaying animal tissue. Two such compounds have especially descriptive common names—putrescine and cadaverine:

$$H_2N-CH_2CH_2CH_2CH_2-NH_2 \qquad H_2N-CH_2CH_2CH_2CH_2CH_2-NH_2$$

putrescine
(1,4-diaminobutane)

cadaverine
(1,5-diaminopentane)

Learning Check 6.5

Draw structural formulas to show how each of the following amines forms hydrogen bonds with water:

a. $CH_3-\overset{\displaystyle |}{\underset{\displaystyle H}{N}}-H$ **b.** $CH_3CH_2-\overset{\displaystyle |}{\underset{\displaystyle CH_3}{N}}-CH_3$

➤ 6.4 Chemical Properties of Amines

Amines undergo many chemical reactions, but we will limit our study to the two most important properties of amines: basicity and amide formation.

Basicity and Amine Salts

The single most distinguishing feature of amines is their behavior as weak bases. They are the most common organic bases. We know from inorganic chemistry that ammonia behaves as a Brønsted base by accepting protons, and in the process becomes a conjugate acid. The reaction of NH_3 with HCl gas illustrates this point (Reaction 6.1):

$$H-\overset{\displaystyle \overset{H}{|}}{\underset{\displaystyle |}{N}}: + H-Cl \longrightarrow \left[H-\overset{\displaystyle \overset{H}{|}}{\underset{\displaystyle |}{N}}-H \right]^+ + Cl^- \tag{6.1}$$

ammonia   ammonium   chloride
                 ion   ion

We also remember from inorganic chemistry that Brønsted-Lowry bases such as ammonia react with water to liberate OH^- ions (Reaction 6.2):

$$H-\overset{\displaystyle \overset{H}{|}}{\underset{\displaystyle |}{N}}: + H-OH \rightleftharpoons \left[H-\overset{\displaystyle \overset{H}{|}}{\underset{\displaystyle |}{N}}-H \right]^+ + OH^- \tag{6.2}$$

ammonia   water   ammonium   hydroxide
                        ion   ion

Because the amines are derivatives of ammonia, they react in similar ways. In water, they produce OH^- ions:

General reaction: $R-NH_2 + H_2O \rightleftharpoons R-NH_3^+ + OH^-$ (6.3)

Specific reaction: $CH_3-NH_2 + H_2O \rightleftharpoons CH_3-NH_3^+ + OH^-$ (6.4)

methylamine   methylammonium
ion

Example 6.1

Complete the following reaction:

$$CH_3-NH-CH_3 + H_2O \rightleftarrows$$

Solution

Primary, secondary, and tertiary amines are all basic and exhibit the same reaction in water.

$$CH_3-NH-CH_3 + H_2O \rightleftarrows CH_3-\overset{+}{N}H_2-CH_3 + OH^-$$

Complete the following reactions:

Learning Check 6.6

a. $CH_3-\underset{\underset{\displaystyle CH_3}{|}}{N}-CH_2CH_3 + H_2O \rightleftarrows$ b. ⬡$-NH_2 + H_2O \rightleftarrows$

All amines behave as weak bases and form salts when they react with acids such as HCl:

General reaction: $\underset{\text{amine}}{R-NH_2} + \underset{\text{acid}}{HCl} \rightarrow \underset{\text{amine salt}}{R-NH_3{}^+Cl^-}$ (6.5)

Specific example: $\underset{\text{methylamine}}{CH_3-NH_2} + HCl \rightarrow \underset{\substack{\text{methylammonium}\\\text{chloride}}}{CH_3-NH_3{}^+Cl^-}$ (6.6)

Other acids, such as sulfuric, nitric, phosphoric, and carboxylic acids (Reaction 6.7), also react with amines to form salts:

$$\underset{\text{ethylamine}}{CH_3CH_2-NH_2} + \underset{\text{acetic acid}}{CH_3COOH} \rightarrow \underset{\text{ethylammonium acetate}}{CH_3CH_2-NH_3{}^+CH_3COO^-} \quad (6.7)$$

Complete the following reactions:

Learning Check 6.7

a. ⬡$-NH-CH_3 + HCl \longrightarrow$

b. ⬠$^{NH_2} + CH_3CH_2COOH \longrightarrow$

As we have seen, amine salts may be named by changing "amine" to "ammonium" and adding the name of the negative ion derived from the acid. Some more examples are

$$CH_3-\underset{\underset{\displaystyle CH_2CH_3}{|}}{\overset{\overset{\displaystyle CH_3}{|}}{N}}-H^+Br^- \qquad\qquad (CH_3CH_2)_3NH^+CH_3COO^-$$

ethyldimethylammonium bromide   triethylammonium acetate

Amine salts have physical properties characteristic of other ionic compounds. They are white crystalline solids with high melting points. Because they are ionic, amine salts are more soluble in water than the parent amines. For this

➤ **FIGURE 6.2** Each of these OTC medications contains an amine salt as the active ingredient.

reason, amine drugs are often given in the form of salts because they will dissolve in body fluids (see ■ Figure 6.2).

$$\text{morphine (water insoluble)} + H_2SO_4 \longrightarrow \text{morphine sulfate (water soluble)} \quad (6.8)$$

Amine salts are easily converted back to amines by adding a strong base:

$$R-NH_3^+Cl^- + NaOH \rightarrow R-NH_2 + H_2O + NaCl \quad (6.9)$$

The form in which amines occur in solutions is pH dependent, just as is the case for carboxylic acids (Section 5.3). In solutions of high H^+ (low pH), the salt form of amines predominates. In solutions of high OH^- (high pH), the amine form is favored:

$$\underset{\substack{\text{amine} \\ \text{(high pH)}}}{R-NH_2} \underset{OH^-}{\overset{H^+}{\rightleftarrows}} \underset{\substack{\text{amine salt} \\ \text{(low pH)}}}{R-NH_3^+} \quad (6.10)$$

The amine cations in the salts we have discussed to this point have one, two, or three alkyl groups attached to the nitrogen:

$$\underset{\substack{\text{primary amine} \\ \text{cation}}}{R-NH_3^+} \qquad \underset{\substack{\text{secondary amine} \\ \text{cation}}}{R-\overset{+}{N}H_2-R'} \qquad \underset{\substack{\text{tertiary amine} \\ \text{cation}}}{R-\overset{+}{\underset{R''}{N}}H-R'}$$

It is also possible to have amine cations in which four alkyl groups or benzene rings are attached to the nitrogen. An example is

$$\left[CH_3CH_2-\overset{\overset{\displaystyle CH_2CH_3}{|}}{\underset{\underset{\displaystyle CH_3}{|}}{N}}-CH_2CH_3 \right]^+ Cl^-$$

triethylmethylammonium chloride

Quaternary ammonium salt
An ionic compound containing a positively charged ion in which four alkyl or aromatic groups are bonded to nitrogen, as in R_4N^+.

These salts are called **quaternary ammonium salts** and are named the same way as other amine salts. Unlike other amine salts, quaternary salts contain no hydrogen attached to the nitrogen that can be removed by adding a base. Thus, quaternary salts are present in solution in only one form, which is independent of the pH of the solution.

Choline is an important quaternary ammonium ion that is a component of certain lipids (Section 8.6); acetylcholine is a compound involved in the transmission of nerve impulses from one cell to another (Section 6.5):

$$\left[HO-CH_2CH_2 -\underset{\underset{CH_3}{|}}{\overset{\overset{CH_3}{|}}{N}}- CH_3 \right]^+ \qquad \left[CH_3 -\overset{\overset{O}{||}}{C}- O - CH_2CH_2 -\underset{\underset{CH_3}{|}}{\overset{\overset{CH_3}{|}}{N}}- CH_3 \right]^+$$

choline cation   acetylcholine cation

Some quaternary ammonium salts are important because they have disinfectant properties. For example, benzalkonium chloride (Zephiran®) is a well-known antiseptic compound that kills many pathogenic (disease-causing) bacteria and fungi on contact:

$$\left[\langle\bigcirc\rangle - CH_2 -\underset{\underset{CH_3}{|}}{\overset{\overset{CH_3}{|}}{N}}- R \right]^+ Cl^-$$

Zephiran chloride
(R represents a long alkyl chain)

Its detergent action destroys the membranes that coat and protect the microorganisms. Zephiran® chloride is recommended as a disinfectant solution for skin and hands prior to surgery and for the sterile storage of instruments. The trade names of some other anti-infectives that contain quaternary ammonium salts are Phemerol®, Bactine®, and Ceepryn®.

Amide Formation

Amines react with acid chlorides or acid anhydrides to form amides. The characteristic functional group of **amides** is a carbonyl group attached to nitrogen. The single bond linking the carbonyl carbon and nitrogen atoms in the group is called an **amide linkage**:

$$\overset{\overset{O}{||}}{-C}-\underset{\underset{|}{}}{N}- \quad \textit{— amide linkage}$$

amide functional group

Amides may be thought of as containing an ammonia or amine portion and a portion derived from a carboxylic acid:

$$\overset{\overset{O}{||}}{-C}-\underset{\underset{|}{}}{N}-$$

from a carboxylic acid
from ammonia or an amine

However, we've seen that the reaction of an amine with a carboxylic acid normally produces a salt and not an amide. For this reason, the preparation of amides requires the more reactive acid chlorides or acid anhydrides.

Both primary and secondary amines can be used to prepare amides in this manner, a method similar to the formation of esters (Section 5.5). Tertiary amines, however, do not form amides because they lack a hydrogen atom on the nitrogen.

Amide
An organic compound having the functional group

$$\overset{\overset{O}{||}}{-C}-\underset{\underset{|}{}}{N}-\cdot$$

Amide linkage
The carbonyl carbon–nitrogen single bond of the amide group.

General reactions:

$$
\underset{\substack{\text{acid chloride}}}{R-\overset{\overset{\displaystyle O}{\|}}{C}-Cl} \;+\; \underset{\substack{\text{amine}}}{R'-\overset{\displaystyle}{\underset{\displaystyle H}{N}}-H} \;\longrightarrow\; \underset{\substack{\text{amide}}}{R-\overset{\overset{\displaystyle O}{\|}}{C}-\overset{\displaystyle}{\underset{\displaystyle H}{N}}-R'} \;+\; HCl \tag{6.11}
$$

$$
\underset{\substack{\text{acid anhydride}}}{R-\overset{\overset{\displaystyle O}{\|}}{C}-O-\overset{\overset{\displaystyle O}{\|}}{C}-R} \;+\; \underset{\substack{\text{amine}}}{R'-\overset{\displaystyle}{\underset{\displaystyle H}{N}}-H} \;\longrightarrow\; \underset{\substack{\text{amide}}}{R-\overset{\overset{\displaystyle O}{\|}}{C}-\overset{\displaystyle}{\underset{\displaystyle H}{N}}-R'} \;+\; \underset{\substack{\text{carboxylic}\\\text{acid}}}{R-\overset{\overset{\displaystyle O}{\|}}{C}-OH} \tag{6.12}
$$

Specific examples:

$$
\underset{\substack{\text{acetic anhydride}}}{CH_3-\overset{\overset{\displaystyle O}{\|}}{C}-O-\overset{\overset{\displaystyle O}{\|}}{C}-CH_3} \;+\; \underset{\substack{\text{ammonia}}}{H-\overset{\displaystyle H}{\underset{\displaystyle H}{N}}-H} \;\longrightarrow\; \underset{\substack{\text{acetamide}}}{CH_3-\overset{\overset{\displaystyle O}{\|}}{C}-\overset{\displaystyle}{\underset{\displaystyle H}{N}}-H} \;+\; \underset{\substack{\text{acetic acid}}}{CH_3-\overset{\overset{\displaystyle O}{\|}}{C}-OH} \tag{6.13}
$$

$$
\underset{\substack{\text{benzoyl chloride}}}{\text{C}_6\text{H}_5-\overset{\overset{\displaystyle O}{\|}}{C}-Cl} \;+\; \underset{\substack{\text{methylamine}\\\text{(a 1° amine)}}}{H-\overset{\displaystyle H}{\underset{}{N}}-CH_3} \;\longrightarrow\; \underset{\substack{\text{N-methylbenzamide}}}{\text{C}_6\text{H}_5-\overset{\overset{\displaystyle O}{\|}}{C}-NH-CH_3} \;+\; HCl \tag{6.14}
$$

$$
\underset{\substack{\text{benzoyl chloride}}}{\text{C}_6\text{H}_5-\overset{\overset{\displaystyle O}{\|}}{C}-Cl} \;+\; \underset{\substack{\text{dimethylamine}\\\text{(a 2° amine)}}}{H-\overset{\displaystyle CH_3}{\underset{}{N}}-CH_3} \;\longrightarrow\; \underset{\substack{\text{N,N-dimethylbenzamide}}}{\text{C}_6\text{H}_5-\overset{\overset{\displaystyle O}{\|}}{C}-\overset{\displaystyle CH_3}{\underset{}{N}}-CH_3} \;+\; HCl \tag{6.15}
$$

Learning Check 6.8

Complete the following reactions:

a. $CH_3-\overset{\overset{\displaystyle O}{\|}}{C}-O-\overset{\overset{\displaystyle O}{\|}}{C}-CH_3 \;+\; CH_3CH_2CH_2-NH_2 \;\longrightarrow$

b. $CH_3CH_2-\overset{\overset{\displaystyle O}{\|}}{C}-Cl \;+\; NH_3 \;\longrightarrow$

CHEMISTRY AROUND US • 6.1

ASPIRIN SUBSTITUTES

Even though aspirin is an excellent and useful drug (see Chemistry Around Us 5.1), it occasionally produces adverse effects that cannot be tolerated by some people. These effects include gastrointestinal bleeding and such allergic reactions as skin rashes and asthmatic attacks. Children who are feverish with influenza or chickenpox should not be given aspirin because of a strong correlation with Reye's syndrome, a devastating illness characterized by liver and brain dysfunction.

The amide *acetaminophen*, marketed under trade names such as Tylenol®, Excedrin Aspirin Free®, Panadol®, and Anacin-3®, is an effective aspirin substitute available for use in such situations.

acetaminophen

Acetaminophen does not irritate the intestinal tract and yet has comparable analgesic (pain-relieving) and antipyretic (fever-reducing) effects. Acetaminophen is available in a stable liquid form suitable for administration to children and other patients who have difficulty taking tablets or capsules. Unlike aspirin, acetaminophen has virtually no anti-inflammatory effect and is therefore a poor substitute for aspirin in the treatment of inflammatory disorders such as rheumatoid arthritis. The overuse of acetaminophen has been linked to liver and kidney damage.

Ibuprofen, a carboxylic acid rather than an amide, is a drug with analgesic, antipyretic, and anti-inflammatory properties that was available for decades only as a prescription drug under the trade name Motrin®.

ibuprofen

In 1984, it became available in an OTC form under trade names such as Advil®, Ibuprin®, Mediprin®, and Motrin IB®. Ibuprofen is thought to be somewhat superior to aspirin as an anti-inflammatory drug, but it has about the same effectiveness as aspirin in relieving mild pain and reducing fever.

In 1994, the FDA approved another compound for sale as an over-the-counter pain reliever. *Naproxen*, marketed under the trade names Aleve® and Anaprox®, exerts its effects for a longer time in the body (8–12 hours per dose) than ibuprofen (4–6 hours per dose), or aspirin and acetaminophen (4 hours per dose). However, the chances for causing slight intestinal bleeding and stomach upset are greater with naproxen than with ibuprofen, and naproxen is not recommended for use by children under age 12.

naproxen

Ketoprofen, another pain reliever, became available for OTC sales in 1996. This compound, sold under the trade name Orudis KT® and Actron®, has the advantage of requiring very small doses when compared to other pain-relieving products. A dose of ketoprofen is 12.5 mg, compared to doses of 200–500 mg for the other products. The tiny, easy-to-swallow tablet is a characteristic emphasized by the manufacturers in advertising. The side effects of ketoprofen are similar to those of aspirin and ibuprofen, so ketoprofen should not be used by people known to be allergic to either of these other two compounds.

ketoprofen

http://www.brookscole.com/chemistry/seager_chemistry
At Web site 6.3, you will find additional information on naproxen.

Aspirin substitutes containing acetaminophen or ibuprofen.

▶ **FIGURE 6.3** The preparation of nylon.

The reactants occupy separate immiscible layers. Hexamethylenediamine is dissolved in water (bottom layer), and adipoyl chloride is dissolved in hexane (top layer). The two compounds react at the interface between the two layers to form a film of nylon, which is lifted by forceps.

The rotating tube winds the nylon strand, which continues to form at the interface in the beaker.

Reaction of diacid chlorides with diamines produces polyamides that, like polyesters, are condensation polymers. The repeating units in polyamides are held together by amides linkages. An example is the formation of nylon, shown in Reaction 6.16 and ■ Figure 6.3.

$$n\text{Cl} - \overset{\overset{\text{O}}{\|}}{\text{C}} - (\text{CH}_2)_4 - \overset{\overset{\text{O}}{\|}}{\text{C}} - \text{Cl} \quad + \quad n\text{H}_2\text{N} - (\text{CH}_2)_6 - \text{NH}_2 \quad \longrightarrow$$

adipoyl chloride   hexamethylenediamine

$$--- \overset{\overset{\text{O}}{\|}}{\text{C}} - (\text{CH}_2)_4 - \overset{\overset{\text{O}}{\|}}{\text{C}} \overset{\text{\textit{amide linkages}}}{-} \text{NH} - (\text{CH}_2)_6 - \text{NH} - \overset{\overset{\text{O}}{\|}}{\text{C}} - (\text{CH}_2)_4 - \overset{\overset{\text{O}}{\|}}{\text{C}} \overset{\text{\textit{amide linkage}}}{-} \text{NH} - (\text{CH}_2)_6 - \text{NH} --- \quad + \quad n\text{HCl}$$

polyamide nylon

(6.16)

Three billion pounds of nylon and related polyamides are produced annually. About 60% of that total goes to make the nylon fiber used in home furnishings such as carpet. The remainder is used largely as a textile fiber in clothing (shirts, dresses, stockings, underwear, etc.) and as tire cord. Minor but important uses include fasteners, rope, parachutes, paint brushes, and electrical parts (see ■ Figure 6.4). In medicine, nylon is used in specialized tubing. Nylon sutures were the first synthetic sutures and are still used.

Silk and wool are natural polyamides that we classify as proteins (discussed in detail in Chapter 9). For now, note the repeating units and amide linkages characteristic of proteins as shown in the following portion of a polyamide protein chain:

amide linkages

➤ 6.5 Amines as Neurotransmitters

Neurotransmitters, the chemical messengers of the nervous system, carry nerve impulses from one nerve cell (neuron) to another. A neuron consists of a bulbous body called the soma (or cell body) attached to a long stemlike projection called an axon (see ■ Figure 6.5). Numerous short extensions called dendrites are attached to the large bulbous end of the soma, and filaments called

Neurotransmitter
A substance that acts as a chemical bridge in nerve impulse transmission between nerve cells.

➤ **STUDY SKILLS 6.1**

A REACTION MAP FOR AMINES

We have discussed four different reactions of amines. When trying to decide what happens in a reaction, remember to focus on the functional groups. After identifying an amine as a starting material, next decide what other reagent is involved and which of the four pathways is appropriate.

Amine

H₂O → An ammonium ion (plus OH⁻)

HCl → Salt

Acid chloride → Amide (plus HCl)

Acid anhydride → Amide (plus a carboxylic acid)

➤ **FIGURE 6.5** A nerve cell and the transmission of a nerve signal.

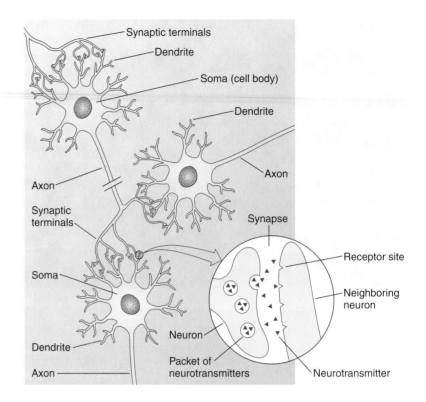

synaptic terminals are attached to the opposite end of the axon. At low magnification, the synaptic terminals of one neuron's axon appear to be attached to the dendrites of an adjacent neuron. However, upon further magnification, a small gap called a synapse is visible between the synaptic terminals of the neuron and the dendrites of the next neuron.

Molecules of a neurotransmitter are stored in small pockets in the neuron near the synapse. These pockets release neurotransmitter molecules into the synapse when an electrical current flows from the soma, along the axon to the synaptic terminals. The electrical current is generated by the exchange of positive and negative ions across membranes. The released neurotransmitter molecules diffuse across the synapse and bind to receptors on the dendrites of the next neuron. Once the neurotransmitter binds to the receptor, the message has been delivered. The receiving cell then sends an electrical current down its own axon until a synapse is reached, and a neurotransmitter delivers the message across the synapse to the next neuron. In the central nervous system—the brain, spinal cord, and nerves that radiate from the spinal cord—the most important neurotransmitters are acetylcholine and three other amines: norepinephrine, dopamine, and serotonin:

norepinephrine

dopamine

serotonin

acetylcholine cation

Neurotransmitters are not only chemical messengers for the nervous system, they may also be partly responsible for our moods. A simplified biochemical theory of mental illness is based on two amines found in the brain. The first is norepinephrine (NE). When an excess of norepinephrine is formed in the brain, the result is a feeling of elation. Extreme excesses of NE can even induce a manic state, while low NE levels may be a cause of depression.

At least six different receptors in the body can be activated by norepinephrine and related compounds. A class of drugs called beta blockers reduces the stimulant action of epinephrine and norepinephrine on various kinds of cells. Some beta blockers are used to treat cardiac arrhythmias, angina, and hypertension (high blood pressure). They function by slightly decreasing the force of each heartbeat. Depression is sometimes a side effect of using such drugs.

In a complex, multistep process, the amino acid tryrosine is used in the body to synthesize norepinephrine:

HO —⟨ ⟩— CH_2 — CH — NH_2 ⟶ HO —⟨ ⟩— CH_2 — CH — NH_2 ⟶
     | HO |
     COOH   COOH

    tyrosine   dopa

HO —⟨ ⟩— CH_2 — CH_2 — NH_2 ⟶ HO —⟨ ⟩— CH — CH_2 — NH_2
HO   HO |
                 OH

   dopamine   norepinephrine

Each step in the process is catalyzed by at least one enzyme, and each of the intermediates formed in the synthesis has physiological activity. Dopa is used as a treatment for Parkinson's disease, and dopamine has been used to treat low blood pressure. Tyrosine is an essential amino acid that must be obtained from the diet, so what we eat may make a direct contribution to our mental health.

The second amine often associated with the biochemical theory of mental illness is serotonin. Like norepinephrine, it also functions as a neurotransmitter. Serotonin is produced in the body from the amino acid tryptophan:

     O
     ‖
     C — OH
     |
  CH_2 — CH — NH_2 CH_2 — CH_2 — NH_2
            HO

  tryptophan   serotonin

Serotonin has been found to influence sleeping, the regulation of body temperature, and sensory perception, but its exact role in mental illness is not yet clear. Unusually low levels of 5-hydroxyindoleacetic acid, a product of serotonin utilization, are characteristically found in the spinal fluid of victims of violent suicide. Drugs that mimic serotonin are sometimes used to treat depression, anxiety, and obsessive-compulsive disorder. Serotonin blockers are used to treat migraine headaches and relieve the nausea that accompanies cancer chemotherapy. A better understanding of the biochemistry of the brain may lead to better medications for treating various forms of mental illness.

➤ 6.6 Other Biologically Important Amines

Epinephrine

Epinephrine, or adrenaline, has a molecular structure similar to that of norepinephrine. The difference is that epinephrine contains an *N*-methyl group:

epinephrine
(adrenaline)

CHEMISTRY AROUND US • 6.2

PARKINSON'S DISEASE

Parkinson's disease is a movement disorder characterized by tremors, spastic motion of the eyelids, rigidity of the limbs, poor balance, and difficulty initiating movement. Movement becomes slow, and there is great difficulty stopping one type of behavior and starting another. For example, once a patient begins walking, he or she will find it difficult to stop. Also, writing becomes labored and slow, and progressively smaller letters are formed. Posture is also affected; patients are often bent over and move with shuffling footsteps. When a healthy person is thrown off balance, he or she will quickly maneuver to restore equilibrium, whereas a person with Parkinson's disease will usually fall without making an attempt to break the fall.

The cause of Parkinson's disease is the degeneration of neurons that connect two parts of the brain's motor system. Dopamine functions as a neurotransmitter between these neurons, and in some patients, the receptors that receive the dopamine continue to function despite the neuron degeneration. In such patients, the administration of the drug L-dopa can relieve the symptoms of the disease. Dopamine itself has no effect when given directly to patients because it cannot penetrate the blood-brain barrier and therefore cannot reach the receptors in the brain. However, L-dopa (a precursor to dopamine) can be transported through arterial walls and enter the brain, where it is converted to dopamine.

L-dopa   dopamine

Unfortunately, this therapy is not effective in all Parkinson's patients. Occasionally a side effect of the L-dopa treatment is the onset of schizophrenic symptoms, which are thought to be caused by overactivity of the dopamine receptors. This side effect is usually eliminated by reducing the dosage of L-dopa.

Fortunately, a new drug has shown promise in reducing neuron degeneration. Deprenyl, a monoamine oxidase inhibitor (MAOI) seems to be very effective in reducing Parkinson's disease symptoms when administered along with L-dopa. Deprenyl appears to reduce neuron degeneration and also increases the level of dopamine in the brain by preventing it from being oxidized by MAOIs.

Epinephrine is more important as a hormone than as a neurotransmitter. It is synthesized in the adrenal gland and acts to increase the blood level of glucose, a key source of energy for body processes. The release of epinephrine—usually in response to pain, anger, or fear—provides glucose for a sudden burst of energy. For this reason, epinephrine has been called the fight-or-flight hormone.

Epinephrine also raises blood pressure. It does this by increasing the rate and force of heart contractions and by constricting the peripheral blood vessels. Injectable local anesthetics usually contain epinephrine because it constricts blood vessels in the vicinity of the injection. This prevents the blood from rapidly "washing" the anesthetic away and prolongs the anesthetic effect in the target tissue. Epinephrine is also used to reduce hemorrhage, treat asthma attacks, and combat anaphylactic shock (collapse of the circulatory system).

Amphetamines

Amphetamine (also known as Benzedrine®) is a powerful nervous system stimulant with an amine structure similar to that of epinephrine:

amphetamine
(Benzedrine®)

epinephrine
(adrenaline)

Notice that both compounds contain an aromatic ring attached to a substituted ethylamine. That structural arrangement, sometimes referred to as a phenethylamine, is apparently an essential feature of several physiologically active compounds. These structural similarities lead to similar behaviors in the body: Both compounds raise the glucose level in the blood, and increase pulse rate and blood pressure. Other closely related phenethylamine compounds, such as Methedrine®, also act as powerful nervous system stimulants. Compounds of this type can be thought of as being derived from amphetamine and as a result are sometimes collectively referred to as **amphetamines**:

N-methylamphetamine
(Methedrine, or "speed")

Amphetamines are used both legally (as prescribed by a physician) and illegally to elevate mood or reduce fatigue. In the drug culture, amphetamine tablets go by such names as bennies, pep pills, reds, red devils, speed, dexies, and uppers. Some amphetamines (known as STP, speed, and mescaline) cause hallucinations. The abuse of such drugs results in severe detrimental effects on both the body and the mind. They are addictive and concentrate in the brain and nervous system. Abusers often experience long periods of sleeplessness, weight loss, and paranoia. Abusers of amphetamines often compound their problems by taking other drugs in order to prevent the "crash," or deep depression, brought on by discontinuation of the drug.

Alkaloids

Plants are the sources of some of the most powerful drugs known. Numerous primitive tribes in the world possess knowledge about the physiological effects that result from eating or chewing the leaves, roots, or bark of certain plants.

Amphetamines
A class of drugs structurally similar to epinephrine, used to stimulate the central nervous system.

http://www.brookscole.com/chemistry/seager_chemistry
At Web site 6.4, you will find information on amphetamines from the Indiana Prevention Resource Center.

Alkaloids
A class of nitrogen-containing organic compounds obtained from plants.

The effects vary from plant species to plant species. Some cure diseases; others, such as opium, are addictive drugs. Still others are deadly poisons, such as the leaves of the belladonna plant and the hemlock herb. The substances responsible for these marked physiological effects are often nitrogen-containing compounds called **alkaloids**. The name, which means alkali-like, reflects the fact that these amine compounds are weakly basic.

The molecular structures of alkaloids vary from simple to complex. Two common and relatively simple alkaloids are nicotine, which is found in tobacco, and caffeine, which is present in coffee and cola drinks:

nicotine   caffeine

Many people absorb nicotine into their bodies as a result of smoking or chewing tobacco. In small doses, nicotine behaves like a stimulant and is not especially harmful. However, it is habit-forming, and habitual tobacco users are exposed to other harmful substances such as tars, carbon monoxide, and polycyclic carcinogens.

Besides coffee and cola drinks, other sources of caffeine are tea, chocolate, and cocoa. Caffeine is a mild stimulant of the respiratory and central nervous systems, the reason for its well-known side effects of nervousness and insomnia. These characteristics, together with its behavior as a mild diuretic, account for the use of caffeine in a wide variety of products, including pain relievers, cold remedies, diet pills, and "stay-awake" pills (No-Doz®). Because caffeine is considered to be a drug, pregnant women should be prudent about how much caffeine they consume. Like most other drugs, caffeine enters the bloodstream, crosses the placental barrier, and reaches the unborn child.

A number of alkaloids are used medicinally. Quinine, for example, is used to treat malaria. Atropine is used as a preoperative drug to relax muscles and reduce the secretion of saliva in surgical patients, and to dilate the pupil of the eye in patients undergoing eye examinations.

http://www.brookscole.com/chemistry/
seager_chemistry
At Web site 6.5, you will find additional information on caffeine.

quinine   atropine

Opium, the dried juice of the poppy plant, has been used for centuries as a painkilling drug. It contains numerous alkaloids, including morphine and its methyl ether, codeine. Both are central nervous system depressants, with codeine being much weaker. They exert a soothing effect on the body when administered in sufficient amounts. These properties make them useful as painkillers. Morphine is especially useful in this regard, being one of the most effective painkillers known. Codeine is used in some cough syrups to depress the action of the cough center in the brain. The major drawback to using these drugs is that they are all addictive. Heroin, a derivative of morphine, is one of the more destructive hard drugs used illegally in our society. A substantial number of addicts commit crimes to support their habits.

| morphine | codeine | heroin |

➤ 6.7 The Nomenclature of Amides

Simple amides are named after the corresponding carboxylic acids by changing the -ic ending (common names) or the -oic ending (IUPAC names) of the acid to -amide. Common names are used more often than IUPAC names. Some examples of the assignment of such names follow:

$$H-\overset{\overset{\displaystyle O}{\|}}{C}-OH \longrightarrow H-\overset{\overset{\displaystyle O}{\|}}{C}-NH_2$$

Common name	formic acid	formamide
IUPAC name	methanoic acid	methanamide

$$CH_3-\overset{\overset{\displaystyle O}{\|}}{C}-OH \longrightarrow CH_3-\overset{\overset{\displaystyle O}{\|}}{C}-NH_2$$

Common name	acetic acid	acetamide
IUPAC name	ethanoic acid	ethanamide

$$\text{C}_6\text{H}_5-\overset{\overset{\displaystyle O}{\|}}{C}-OH \longrightarrow \text{C}_6\text{H}_5-\overset{\overset{\displaystyle O}{\|}}{C}-NH_2$$

Common name	benzoic acid	benzamide
IUPAC name	benzoic acid	benzamide

a. Give the IUPAC name for the following simple amide:

$$CH_3CH_2CH_2-\overset{\overset{\displaystyle O}{\|}}{C}-NH_2$$

b. What is the IUPAC name for the simple amide derived from hexanoic acid?

Learning Check 6.9

If alkyl groups are attached to the nitrogen atom of amides (substituted amides), the name of the alkyl group precedes the name of the amide, and an italic capital *N* is used to indicate that the alkyl group is bonded to the nitrogen. See the examples at the top of the next page.

$$H-\overset{\overset{\displaystyle O}{\|}}{C}-NH-CH_3 \qquad CH_3-\overset{\overset{\displaystyle O}{\|}}{C}-\underset{\underset{\displaystyle CH_3}{|}}{N}-CH_3 \qquad \text{[benzene]}\overset{\overset{\displaystyle O}{\|}}{C}-\underset{\underset{\displaystyle CH_3}{|}}{N}-CH_2CH_3$$

Common name	N-methylformamide	N,N-dimethylacetamide	N-ethyl-N-methylbenzamide
IUPAC name	N-methylmethanamide	N,N-dimethylethanamide	N-ethyl-N-methylbenzamide

Learning Check 6.10

Give either a common name or the IUPAC name for each of the following amides:

a. $CH_3-\overset{\overset{\displaystyle O}{\|}}{C}-\underset{\underset{\displaystyle CH_2CH_3}{|}}{N}-CH_2CH_3$

b. $CH_3CH_2CH_2-\overset{\overset{\displaystyle O}{\|}}{C}-NH-CH_2\underset{\underset{\displaystyle CH_3}{|}}{CH}CH_3$

c. $\text{[benzene]}\overset{\overset{\displaystyle O}{\|}}{C}-\underset{\underset{\displaystyle CH_2CH_3}{|}}{N}-\text{[benzene]}$

d. $\text{[benzene]}\overset{\overset{\displaystyle O}{\|}}{C}-NH-CH_3$ with CH_3 substituent

➤ **FIGURE 6.6** Intermolecular hydrogen bonding in an unsubstituted amide.

➤ **FIGURE 6.7** Hydrogen bonding between water and (a) an unsubstituted amide and (b) a disubstituted amide.

➤ 6.8 Physical Properties of Amides

Formamide is a liquid at room temperature. It is the only unsubstituated amide that is a liquid; all others are solids.

Unsubstituted amides can form a complex network of intermolecular hydrogen bonds (see ■ Figure 6.6). It is this characteristic that causes melting points of these substances to be so high.

The substitution of alkyl or aromatic groups for hydrogen atoms on the nitrogen reduces the number of intermolecular hydrogen bonds that can form and causes the melting points to decrease. Thus, disubstituted amides often have lower melting and boiling points than monosubstituted and unsubstituted amides. We'll see in Chapter 9 that hydrogen bonding between amide groups is important in maintaining the shape of protein molecules.

Amides are rather water soluble, especially those containing fewer than six carbon atoms. This solubility results from the ability of amides to form hydrogen bonds with water. Even disubstituted amides do this because of the presence of the carbonyl oxygen (see ■ Figure 6.7).

(a)

(b)

Show how the amide below can form the following:

a. Intermolecular hydrogen bonds with other amide molecules
b. Hydrogen bonds with water molecules

$$CH_3-\overset{\overset{\displaystyle O}{\|}}{C}-\overset{\overset{}{\underset{\underset{\displaystyle H}{|}}{N}}}-CH_3$$

➤ 6.9 Chemical Properties of Amides

Neutrality

Because basicity is the most important property of amines, it is only natural to wonder if amides are also basic. The answer is no. Although they are formed from carboxylic acids and basic amines, amides are neither basic nor acidic; they are neutral. The carbonyl group bonded to the nitrogen has destroyed the basicity of the original amine, and the nitrogen of the amine has replaced the acidic —OH of the carboxylic acid.

Amide Hydrolysis

The most important reaction of amides is hydrolysis. This reaction corresponds to the reverse of amide formation; the amide is cleaved into two compounds, a carboxylic acid and an amine or ammonia:

$$\underset{\text{amide}}{R-\overset{\overset{\displaystyle O}{\|}}{C}-NH-R'} + H_2O \xrightarrow{\text{Acid or base}} \underset{\text{carboxylic acid}}{R-\overset{\overset{\displaystyle O}{\|}}{C}-OH} + \underset{\text{amine}}{R'-NH_2} \quad (6.17)$$

cleavage here

As with hydrolysis of carboxylic esters, amide hydrolysis requires the presence of a strong acid or base, and the nature of the final products depends on which of these catalysts is used. Notice in the following examples that under basic conditions, the salt of the carboxylic acid is formed along with the amine, and under acidic conditions, the salt of the amine is produced along with the carboxylic acid.

General reactions:

$$\underset{\text{amide}}{R-\overset{\overset{\displaystyle O}{\|}}{C}-NH-R'} + H_2O + HCl \longrightarrow \underset{\text{carboxylic acid}}{R-\overset{\overset{\displaystyle O}{\|}}{C}-OH} + \underset{\text{amine salt}}{R'-NH_3^+Cl^-} \quad (6.18)$$

$$\underset{\text{amide}}{R-\overset{\overset{\displaystyle O}{\|}}{C}-NH-R'} + NaOH \longrightarrow \underset{\text{carboxylate salt}}{R-\overset{\overset{\displaystyle O}{\|}}{C}-O^-Na^+} + \underset{\text{amine}}{R'-NH_2} \quad (6.19)$$

Specific examples:

$$CH_3-\overset{\overset{\displaystyle O}{\displaystyle \|}}{C}-NH_2 \;+\; H_2O \;+\; HCl \;\longrightarrow\; CH_3-\overset{\overset{\displaystyle O}{\displaystyle \|}}{C}-OH \;+\; NH_4^+Cl^- \qquad (6.20)$$

acetamide   acetic acid   ammonium chloride

$$CH_3-\overset{\overset{\displaystyle O}{\displaystyle \|}}{C}-NH-CH_3 \;+\; H_2O \;+\; HCl \;\longrightarrow\; CH_3-\overset{\overset{\displaystyle O}{\displaystyle \|}}{C}-OH \;+\; CH_3-NH_3^+Cl^- \qquad (6.21)$$

N-methylacetamide   acetic acid   methylammonium chloride

$$CH_3-\overset{\overset{\displaystyle O}{\displaystyle \|}}{C}-NH-CH_3 \;+\; NaOH \;\longrightarrow\; CH_3-\overset{\overset{\displaystyle O}{\displaystyle \|}}{C}-O^-Na^+ \;+\; CH_3-NH_2 \qquad (6.22)$$

N-methylacetamide   sodium acetate   methylamine

Because the form in which carboxylic acids and amines occur in solution depends on the solution pH, one of the hydrolysis products must always be in the form of a salt.

Learning Check 6.12

Complete the following hydrolysis reactions:

a.

$$\underset{\text{benzene ring}}{\bigcirc}\!-\overset{\overset{\displaystyle O}{\displaystyle \|}}{C}-NH-CH_2CH_3 \;+\; H_2O \;+\; HCl \;\longrightarrow$$

b.

$$\underset{\text{benzene ring}}{\bigcirc}\!-\overset{\overset{\displaystyle O}{\displaystyle \|}}{C}-NH-CH_2CH_3 \;+\; NaOH \;\longrightarrow$$

Amide hydrolysis, a very important reaction in biochemistry, is a central reaction in the digestion of proteins and the breakdown of proteins within cells. However, most amide hydrolysis in the body is catalyzed by enzymes rather than by strong acids or bases.

Because of their physiological properties, a number of amides play valuable roles in medicine. ■ Table 6.3 lists some of these important compounds.

TABLE 6.3 Some important amides in medicine

Structure	Generic name (trade name)	Use
	thiopental (Pentothal®)	Intravenous anesthesia
	amobarbital (Amytal®)	Treatment of insomnia
	diazepam (Valium®)	Tranquilizer
	ampicillin (Polycillin®)	Antibiotic

> ## FOR FUTURE REFERENCE VITAMIN B$_6$

Vitamin B$_6$ occurs in three chemical forms: pyridoxal, pyridoxine, and pyridoxamine. As the name indicates, the last form contains an amine functional group and demonstrates the chemical properties of amines (see Section 6.4). In the body, vitamin B$_6$ is involved in numerous important processes, including: the synthesis of certain amino acids, the metabolism of proteins and urea, the synthesis of a portion of the hemoglobin molecule, and the synthesis of nucleic acids such as DNA and RNA. Recent research results indicate that vitamin B$_6$ also influences cognitive performance, the immune reactions of the body and the activity of certain hormones. Unlike other water-soluble vitamins, a significant amount of the vitamin is stored in muscle tissue.

Many drugs interact with vitamin B$_6$ and cause it to be lost from the body. The most common of these is alcohol. Thus, vitamin B$_6$ deficiency is a potential health threat to individuals who use alcohol excessively. The best food sources of the vitamin are meats, fish, poultry, potatoes and certain other vegetables and fruits. The chart below lists some foods that provide a minimum of 0.25 mg of vitamin B$_6$ per serving. As the chart shows, multiple servings of even these relatively rich sources of the vitamin are needed daily to meet the RDA for adults.

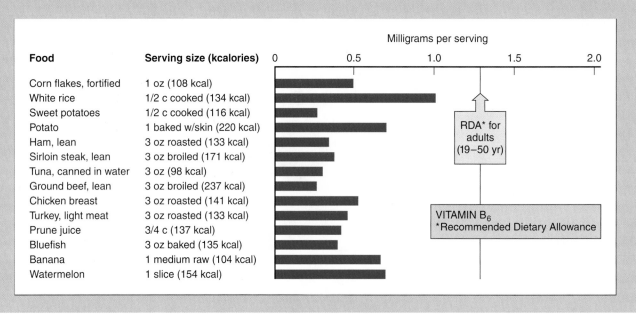

Food	Serving size (kcalories)
Corn flakes, fortified	1 oz (108 kcal)
White rice	1/2 c cooked (134 kcal)
Sweet potatoes	1/2 c cooked (116 kcal)
Potato	1 baked w/skin (220 kcal)
Ham, lean	3 oz roasted (133 kcal)
Sirloin steak, lean	3 oz broiled (171 kcal)
Tuna, canned in water	3 oz (98 kcal)
Ground beef, lean	3 oz broiled (237 kcal)
Chicken breast	3 oz roasted (141 kcal)
Turkey, light meat	3 oz roasted (133 kcal)
Prune juice	3/4 c (137 kcal)
Bluefish	3 oz baked (135 kcal)
Banana	1 medium raw (104 kcal)
Watermelon	1 slice (154 kcal)

SOURCE: Adapted from Whitney, E.N.; Rolfes, S.R. *Understanding Nutrition,* 8th ed. Belmont, CA: West/Wadsworth, 1999.

CONCEPT SUMMARY

Classification of Amines. Amines are organic derivatives of ammonia in which one or more of the ammonia hydrogens are replaced by alkyl or aromatic groups. Amines are classified as primary, secondary, or tertiary, depending on the number of groups (one, two, or three) attached to the nitrogen.

The Nomenclature of Amines. Common names are given to simple amines by adding the ending *-amine* to the names of the alkyl groups attached to the nitrogen. In IUPAC names, the —NH$_2$ group is treated as a chain substituent and given the name *amino.* Aromatic amines are named as derivatives of aniline.

Physical Properties of Amines. Primary and secondary amines have boiling points slightly lower than those of corresponding alcohols. Tertiary amines have boiling points similar to those of alkanes. Low molecular weight amines are water soluble.

Chemical Properties of Amines. Amines are weak bases. They react with water to liberate hydroxide ions, and they react with acids to form salts. Amines react with acid chlorides and acid anydrides to form amides.

Amines as Neurotransmitters. Neurotransmitters are the chemical messengers of the nervous system. They carry nerve impulses from one nerve cell (neuron) to another. The most important neurotransmitters are acetylcholine and three other amines: norepinephrine, dopamine, and serotonin.

Other Biologically Important Amines. Four neurotransmitters are acetylcholine, norepinephrine, dopamine, and serotonin. Epinephrine is also known as the fight-or-flight hormone. The amphetamines have structures similar to that of epinephrine. Alkaloids are nitrogen-containing compounds isolated from plants. They exhibit a variety of physiological effects on the body. Examples of alkaloids include nicotine, caffeine, quinine, atropine, morphine, and codeine.

The Nomenclature of Amides. Amides are named by changing the *-ic acid* or *-oic acid* ending of the carboxylic acid portion of the compound to *-amide*. Groups attached to the nitrogen of the amide are denoted by an italic capital N that precedes the name of the attached group.

Physical Properties of Amides. Low-molecular-weight amides are water soluble due to the formation of hydrogen bonds. The melting and boiling points of unsubstituted amides are higher than those of comparable substituted amides.

Chemical Properties of Amides. Amides undergo hydrolysis in acidic conditions to yield a carboxylic acid and an amine salt. Hydrolysis under basic conditions produces a carboxylate salt and an amine.

LEARNING OBJECTIVES

When you have completed your study of this chapter, you should be able to:

1. Given structural formulas, classify amines as primary, secondary, or tertiary.
2. Assign common and IUPAC names to simple amines and amides.
3. Discuss how hydrogen bonding influences the physical properties of amines and amides.
4. Write equations for the reactions of amines with water, the formation of amine salts, the reactions of amine salts with strong bases, and the reactions of amines with acid chlorides and anhydrides.
5. Give the products of acidic and basic hydrolysis of amides.
6. List uses for important amides and amines.

KEY TERMS AND CONCEPTS

Alkaloids (6.6)
Amide (6.4)
Amide linkage (6.4)
Amine (6.1)

Amphetamines (6.6)
Antihistamine (6.2)
Neurotransmitter (6.5)
Primary amine (6.1)

Quaternary ammonium salt (6.4)
Secondary amine (6.1)
Tertiary amine (6.1)

KEY REACTIONS

1. Reaction of amines with water (Section 6.4):

$$R-NH_2 + H_2O \rightleftharpoons R-NH_3^+ + OH^-$$ Reaction 6.3

2. Reaction of amines with acids (Section 6.4):

$$R-NH_2 + HCl \rightarrow R-NH_3^+Cl^-$$ Reaction 6.5

3. Conversion of amine salts back to amines (Section 6.4):

$$R-NH_3^+Cl^- + NaOH \rightarrow R-NH_2 + H_2O + NaCl$$ Reaction 6.9

4. Reaction of amines with acid chlorides to form amides (Section 6.4):

$$\underset{\underset{}{}}{R-\overset{\overset{O}{\|}}{C}-Cl} + \underset{\underset{H}{|}}{R'-N-H} \rightarrow \underset{\underset{H}{|}}{R-\overset{\overset{O}{\|}}{C}-N-R'} + HCl$$ Reaction 6.11

5. Reaction of amines with acid anhydrides to form amides (Section 6.4):

$$R-\overset{\overset{\displaystyle O}{\|}}{C}-O-\overset{\overset{\displaystyle O}{\|}}{C}-R + R'-\underset{\underset{\displaystyle H}{|}}{N}-H \rightarrow R-\overset{\overset{\displaystyle O}{\|}}{C}-\underset{\underset{\displaystyle H}{|}}{N}-R' + R-\overset{\overset{\displaystyle O}{\|}}{C}-OH \qquad \text{Reaction 6.12}$$

6. Acid hydrolysis of amides (Section 6.9):

$$R-\overset{\overset{\displaystyle O}{\|}}{C}-NH-R' + H_2O + HCl \longrightarrow R-\overset{\overset{\displaystyle O}{\|}}{C}-OH + R'-NH_3^+Cl^- \qquad \text{Reaction 6.18}$$

7. Basic hydrolysis of amides (Section 6.9):

$$R-\overset{\overset{\displaystyle O}{\|}}{C}-NH-R' + NaOH \longrightarrow R-\overset{\overset{\displaystyle O}{\|}}{C}-O^-Na^+ + R'-NH_2 \qquad \text{Reaction 6.19}$$

EXERCISES

CLASSIFICATION OF AMINES (SECTION 6.1)

6.1 What is the difference between primary, secondary, and tertiary amines in terms of bonding to the nitrogen atom?

6.2 Classify each of the following as a primary, secondary, or tertiary amine:

a. $CH_3CH_2-\underset{\underset{\displaystyle CH_3}{|}}{N}-CH_2CH_3$

b. $CH_3\underset{\underset{\displaystyle CH_3}{|}}{\overset{\overset{\displaystyle CH_3}{|}}{C}}-NH_2$

c. ⬡—NH—CH₃

6.3 Classify each of the following as a primary, secondary, or tertiary amine:

a. $CH_3CH_2-\underset{\underset{\displaystyle CH_3}{|}}{N}-CH_3$

b. ⬡—NH₂

c. (cyclopentane)—NH—CH₃

d. ⬡—NH₂

6.4 Draw structural formulas for the four amines that have the molecular formula C_3H_9N. Label each one as primary, secondary, or tertiary.

THE NOMENCLATURE OF AMINES (SECTION 6.2)

6.5 Give each of the following amines a common name by adding the ending -amine to alkyl group names:

a. $CH_3CH_2-NH-CH_2CH_3$

b. ⬡—NH—CH₃

c. $CH_3CH_2CH_2-\underset{\underset{\displaystyle CH_3}{|}}{N}-CH_3$

6.6 Give each of the following amines a common name by adding the ending -amine to alkyl group names:

a. $CH_3\underset{\underset{\displaystyle NH_2}{|}}{C}HCH_3$

b. ⬡—NH—CH₂CH₃

c. $CH_3CH_2CH_2-\underset{\underset{\displaystyle CH_2CH_3}{|}}{N}-CH_3$

6.7 Give each of the following amines an IUPAC name by treating the amino group as a substituent:

a. $CH_3\underset{\underset{\displaystyle CH_3}{|}}{C}HCH_2\underset{\underset{\displaystyle CH_3}{|}}{C}H-NH_2$

b. $CH_3\underset{\underset{\displaystyle NH_2}{|}}{C}HCH_2CH_2-NH_2$

c. (cyclopentane with NH₂ groups)

6.8 Give each of the following amines an IUPAC name by treating the amino group as a substituent:

a.

b. NH$_2$ NH$_2$
　　CH$_2$CH$_2$CHCH$_3$

c. 　$\overset{\overset{\displaystyle O}{\|}}{CH_3C}CH_2CH_2-NH_2$

6.9 Name the following aromatic amines as derivatives of aniline:

a. NH$_2$ CH$_3$ CH$_3$

b. NH—CH$_2$CH$_3$ CH$_3$

6.10 Name the following aromatic amines as derivatives of aniline:

a. NH—CHCH$_3$ CH$_3$

b. CH$_3$ N—CH$_2$CH$_3$

6.11 Name each of the following amines by one of the methods used in Exercises 6.5, 6.7, and 6.9:

a. $CH_3-\overset{\overset{\displaystyle CH_3}{|}}{\underset{\underset{\displaystyle CH_3}{|}}{C}}-NH_2$

c. NH$_2$ Cl

b. CH$_3$CH$_2$—N—CH$_2$CH$_3$

d. CH$_3$ NH$_2$
　　CH$_3$CHCHCHCH$_3$
　　　　　　NH$_2$

6.12 Draw the structural formula for each of the following amines:
a. 3-amino-3-methyl-1-hexene
b. p-propylaniline
c. N, N-dimethylaniline

6.13 Draw the structural formula for each of the following amines:
a. *sec*-butylamine
b. diethylmethylamine
c. 2-aminobutanoic acid

PHYSICAL PROPERTIES OF AMINES (SECTION 6.3)

6.14 Explain why all classes of low-molecular-weight amines are water soluble.

6.15 Why are the boiling points of amines lower than those of corresponding alcohols?

6.16 Why are the boiling points of tertiary amines lower than those of corresponding primary and secondary amines?

6.17 Draw diagrams similar to Figure 6.1 to illustrate hydrogen bonding between the following compounds:

a. $\overset{\overset{\displaystyle H}{|}}{CH_3CH_2-N-CH_3}$ and $\overset{\overset{\displaystyle H}{|}}{CH_3CH_2-N-CH_3}$

b. and H$_2$O

6.18 Draw diagrams similar to Figure 6.1 to illustrate hydrogen bonding between the following compounds:

a. CH$_3$CH$_2$—NH—CH$_3$ and H$_2$O

b. and

6.19 Arrange the following compounds in order of increasing boiling point:
a. CH$_3$CH$_2$CH$_2$—NH$_2$
b. CH$_3$—N—CH$_3$
　　　　　CH$_3$
c. CH$_3$CH$_2$CH$_2$—OH

CHEMICAL PROPERTIES OF AMINES (SECTION 6.4)

6.20 When diethylamine is dissolved in water, the solution becomes basic. Write an equation to account for this observation.

6.21 Complete the following equations. If no reaction occurs, write "no reaction."

a. $\overset{\overset{\displaystyle O}{\|}}{C}-Cl$ + NH$_3$ ⟶

b. —N—CH$_3$ + $\overset{\overset{\displaystyle O}{\|}}{CH_3C}$—OH ⟶
　　　CH$_3$

c. CH$_3$CH$_2$—N—CH$_3$ + $CH_3-\overset{\overset{\displaystyle O}{\|}}{C}-Cl$ ⟶
　　　　　　CH$_3$

d. N—H + HCl ⟶

e. N—H + H$_2$O ⇌

f. CH$_3$CH$_2$—NH$_3$$^+Cl^-$ + NaOH ⟶
　　　　　CH$_3$

6.22 Complete the following equations. If no reaction occurs, write "no reaction."

a. $CH_3CH_2CH_2CH_2-NH_2 + HCl \rightarrow$

b. $+ H_2O \rightleftharpoons$

c. $+ CH_3\overset{O}{\overset{\|}{C}}-OH \longrightarrow$

d. $CH_3\overset{\overset{\displaystyle NH_3{}^+Cl^-}{|}}{CH}CH_3 + NaOH \longrightarrow$

e. $CH_3\overset{\overset{\displaystyle}{|}}{\underset{\underset{\displaystyle CH_3}{|}}{CH}}-\overset{O}{\overset{\|}{C}}-Cl + CH_3-NH_2 \longrightarrow$

f. $-\overset{O}{\overset{\|}{C}}-O-\overset{O}{\overset{\|}{C}}-$$ + NH_3 \longrightarrow$

6.23 How does the structure of a quaternary ammonium salt differ from the structure of a salt of a tertiary amine?

6.24 Why are amine drugs commonly administered in the form of their salts?

6.25 Write an equation to show how the drug dextromethorphan, a cough suppressant, could be made more water soluble.

dextromethorphan

6.26 Write equations for two different methods of synthesizing the following amide:

$$CH_3-\overset{O}{\overset{\|}{C}}-\overset{\overset{\displaystyle}{|}}{\underset{\underset{\displaystyle CH_3}{|}}{N}}-CH_3$$

6.27 Write structures for the chemicals needed to carry out the following conversions:

a. $\longrightarrow$

b. $CH_3-NH_2 \longrightarrow$ $-\overset{O}{\overset{\|}{C}}-NH-CH_3$

c. $CH_3\overset{\overset{\displaystyle CH_3}{|}}{CH}-NH_2 \rightarrow CH_3\overset{\overset{\displaystyle CH_3}{|}}{CH}-NH_3{}^+ + OH^-$

6.28 Write structures for the chemicals needed to carry out the following conversions:

a. $-NH_3{}^+Cl^- \longrightarrow$ $-NH_2$

b. $CH_3CH_2-\overset{O}{\overset{\|}{C}}-Cl \longrightarrow CH_3CH_2-\overset{O}{\overset{\|}{C}}-NH_2$

c. $CH_3-\overset{\overset{\displaystyle}{|}}{\underset{\underset{\displaystyle CH_3}{|}}{NH}} \longrightarrow CH_3-\overset{\overset{\displaystyle}{|}}{\underset{\underset{\displaystyle CH_3}{|}}{NH_2{}^+}}Cl^-$

AMINES AS NEUROTRANSMITTERS (SECTION 6.5)

6.29 Describe the general structure of a neuron.

6.30 What term is used to describe the gap between one neuron and the next?

6.31 Name the two amino acids which are starting materials for the synthesis of neurotransmitters.

6.32 Name the two amines often associated with the biochemical theory of mental illness.

OTHER BIOLOGICALLY IMPORTANT AMINES (SECTION 6.6)

6.33 What role do neurotransmitters play in nerve impulse transmission?

6.34 List two neurotransmitters classified as catecholamines.

6.35 Why is epinephrine called the fight-or-flight hormone?

6.36 Describe one clinical use of epinephrine.

6.37 What are the physiological effects of amphetamines on the body?

6.38 What is the source of alkaloids?

6.39 Why are alkaloids mildly basic?

6.40 Give the name of an alkaloid for the following:
   a. Found in cola drinks
   b. Used to reduce saliva flow during surgery
   c. Present in tobacco
   d. A cough suppressant
   e. Used to treat malaria
   f. An effective pain killer

THE NOMENCLATURE OF AMIDES (SECTION 6.7)

6.41 Assign IUPAC names to the following amides:

a. $CH_3CH_2-\overset{O}{\overset{\|}{C}}-NH_2$

b. $CH_3CH_2-\overset{O}{\overset{\|}{C}}-NH-CH_3$

c.
$$
\underset{\text{(benzene ring)}}{\bigcirc}-\overset{\displaystyle O}{\overset{\|}{C}}-NH-CH_2CH_3
$$

6.42 Assign IUPAC names to the following amides:

a.
$$
CH_3CH_2CH-\overset{\displaystyle O}{\overset{\|}{C}}-NH_2 \\ \quad\ \underset{CH_2CH_3}{|}
$$

b.
$$
CH_3CH_2CH_2-\overset{\displaystyle O}{\overset{\|}{C}}-NH-\bigcirc
$$

c.
$$
\bigcirc-\overset{\displaystyle O}{\overset{\|}{C}}-NH-CHCH_3 \\ \qquad\qquad\quad \underset{CH_3}{|}
$$

d.
$$
CH_3CH_2-\overset{\displaystyle O}{\overset{\|}{C}}-\underset{\underset{CH_3}{|}}{N}-CH_3
$$

6.43 Draw structural formulas for the following amides:
 a. butanamide
 b. *N*-phenylbenzamide
 c. 4-methylbenzamide

6.44 Draw structural formulas for the following amides:
 a. benzamide
 b. *N*-methylethanamide
 c. *N*-methyl-*N*-phenylbutanamide

PHYSICAL PROPERTIES OF AMIDES (SECTION 6.8)

6.45 Draw diagrams similar to those in Figures 6.4 and 6.5 to illustrate hydrogen bonding between the following molecules:

a.
$$
CH_3-\overset{\displaystyle O}{\overset{\|}{C}}-\underset{\underset{CH_3}{|}}{N}-CH_3 \ \text{ and } \ H_2O
$$

b.
$$
CH_3-\overset{\displaystyle O}{\overset{\|}{C}}-NH_2 \ \text{ and } \ CH_3-\overset{\displaystyle O}{\overset{\|}{C}}-NH_2
$$

6.46 Draw diagrams similar to those in Figures 6.4 and 6.5 to illustrate hydrogen bonding between the following molecules:

a.
$$
\underset{\text{(cyclopentane ring)}}{\bigcirc}-\overset{\displaystyle O}{\overset{\|}{C}}-NH_2 \ \text{ and } \ H_2O
$$

b.
$$
CH_3CH-\overset{\displaystyle O}{\overset{\|}{C}}-NH_2 \ \text{ and } \ CH_3CH-\overset{\displaystyle O}{\overset{\|}{C}}-NH_2 \\ \quad\ \underset{CH_3}{|} \qquad\qquad\qquad\ \underset{CH_3}{|}
$$

6.47 Explain why the boiling points of disubstituted amides are often lower than those of unsubstituted amides.

CHEMICAL PROPERTIES OF AMIDES (SECTION 6.9)

6.48 Complete the following reactions:

a.
$$
CH_3CH_2-\overset{\displaystyle O}{\overset{\|}{C}}-NH-CH_3 \ + \ NaOH \ \longrightarrow
$$

b.
$$
CH_3CH_2-\overset{\displaystyle O}{\overset{\|}{C}}-NH-CH_3 \ + \ H_2O \ + \ HCl \ \longrightarrow
$$

6.49 Complete the following reactions:

a.
$$
\bigcirc-\overset{\displaystyle O}{\overset{\|}{C}}-\underset{\underset{CH_3}{|}}{N}-CH_3 \ + \ NaOH \ \longrightarrow
$$

b.
$$
\bigcirc-\overset{\displaystyle O}{\overset{\|}{C}}-\underset{\underset{CH_3}{|}}{N}-CH_3 \ + \ H_2O \ + \ HCl \ \longrightarrow
$$

6.50 One of the most successful mosquito repellants has the following structure and name. What are the products of the basic hydrolysis of *N, N*-diethyl-*m*-toluamide?

$$
\underset{CH_3}{\bigcirc}-\overset{\displaystyle O}{\overset{\|}{C}}-\underset{\underset{CH_2CH_3}{|}}{N}-CH_2CH_3
$$

N,N-diethyl-*m*-toluamide

6.51 What are the products of the acid hydrolysis of the local anesthetic lidocaine?

$$
CH_3CH_2-\underset{\underset{CH_2CH_3}{|}}{N}-CH_2-\overset{\displaystyle O}{\overset{\|}{C}}-NH-\underset{CH_3}{\overset{CH_3}{\bigcirc}}
$$

lidocaine (Xylocaine)

CHEMISTRY AROUND US AND OVER THE COUNTER

6.52 What chemical substance is responsible for watery eyes and other symptoms of hay fever?

6.53 Why should driving a car be avoided after taking some antihistamines?

6.54 What are the three categories of products that are available to combat the symptoms of hay fever?

6.55 Why is there a warning against the use of aspirin to treat children feverish with flu?

6.56 List the advantages and disadvantages of acetaminophen as an aspirin substitute.

6.57 List the advantages of ibuprofen as an aspirin substitute.

6.58 What compound can relieve the symptoms of Parkinson's disease in some patients?

► POINTS TO PONDER AND PURSUE

6.1 Why might it be more practical for chemical suppliers to ship amines as amine salts?

6.2 Give structural formulas for the four-carbon diamine and the four-carbon diacid chloride that would react to form nylon.

6.3 Why might it be dangerous on a camping trip to prepare a salad using an unknown plant?

6.4 One of the active ingredients in Triaminic® cold syrup is listed as chlorpheniramine maleate. What does the word *maleate* indicate about the structure of this active ingredient?

6.5 Write reactions to show how the following conversion can be achieved. An amine reagent is required, and reactions from earlier chapters are necessary.

$$CH_3CH_2-OH \longrightarrow CH_3-NH_3^+CH_3-\overset{\displaystyle O}{\overset{\displaystyle \|}{C}}-O^-$$

6.6 In Figure 6.2, several products are shown that contain amine salts. Sometimes the active ingredients are given common names ending in hydrochloride. What acid do you think is used to prepare an amine hydrochloride salt?

6.7 Using Web site 6.2, describe the symptoms of rhinitis.

6.8 According to Web site 6.4, what is the composition and physical nature of the street drug "ice"?

6.9 According to Web site 6.5, what is the most widely used drug in the world?

http://www.brookscole.com/chemistry/seager_chemistry

Carbohydrates

The nursing profession provides a broad range of services in a large number of different settings. This is possible because nurses specialize in one or more of a variety of patient care areas, such as emergency room, operating room, intensive care, pediatrics, and obstetrics. Here, a nurse adjusts a device that provides a patient with an intravenous infusion of a solution of glucose, a carbohydrate. In this chapter, you will learn many of the important characteristics of glucose and other carbohydrates.

http://www.brookscole.com/chemistry/seager_chemistry
At Web site 7.1, you will find further information on a career as a licensed practical nurse.

arbohydrates are compounds of tremendous biological and commercial importance. Widely distributed in nature, they include such familiar substances as cellulose, table sugar, and starch. Carbohydrates have four important functions in living organisms:

- To provide energy through their oxidation.
- To supply carbon for the synthesis of cell components.
- To serve as a stored form of chemical energy.
- To form a part of the structural elements of some cells and tissues.

With carbohydrates, we begin our study of **biomolecules,** substances closely associated with living organisms. All life processes, as far as we know, involve carbohydrates and other biomolecules, including lipids, proteins, and nucleic acids (see ■ Figure 7.1). Biomolecules are organic because they are compounds of carbon, but they are also a starting point for **biochemistry,** the chemistry of living organisms. Thus, these chapters on biomolecules represent an overlap of two areas of chemistry: organic and biochemistry.

Biomolecule

A general term referring to organic compounds essential to life.

Biochemistry

A study of the compounds and processes associated with living organisms.

➤ 7.1 Classes of Carbohydrates

The name *carbohydrate* comes from an early observation that heating these compounds produced water and a black residue of carbon, which was incorrectly interpreted to mean that they were hydrates of carbon.

The most striking chemical characteristic of carbohydrates is the large number of functional groups they have. For example, ribose has a functional group on every carbon atom of its five-carbon chain:

$$
\begin{array}{c}
H \quad\quad O \\
\diagdown \; \diagup \\
C \\
| \\
H - C - OH \\
| \\
H - C - OH \\
| \\
H - C - OH \\
| \\
CH_2OH
\end{array}
$$
ribose

➤ **FIGURE 7.1** Life in all its variety depends on biomolecules. Carbohydrates, lipids, proteins, and nucleic acids are essential components of every form of life.

monosaccharide

disaccharide

oligosaccharide

(chain containing
3–10 units)

polysaccharide

(long chain with possibly hundreds
or thousands of units)

➤ **FIGURE 7.2** Carbohydrate classification.

In addition to —OH groups, most carbohydrates have as part of their molecular structure a carbonyl group—either an aldehyde (sometimes shown as —CHO) or a ketone functional group. This is the origin of the modern definition: **Carbohydrates** are polyhydroxy aldehydes or ketones, or substances that yield such compounds on hydrolysis.

Carbohydrates can be classified according to the size of the molecules. **Monosaccharides** consist of a single polyhydroxy aldehyde or ketone unit. **Disaccharides** are carbohydrates composed of two monosaccharide units linked together chemically. Oligosaccharides (less common and of minor importance) contain from three to ten units. **Polysaccharides** consist of very long chains of linked monosaccharide units (see ■ Figure 7.2).

➤ 7.2 The Stereochemistry of Carbohydrates

Glyceraldehyde, the simplest carbohydrate, exhibits a fascinating stereoisomerism shown by many organic compounds but especially characteristic of carbohydrates and many other natural substances. Glyceraldehyde can exist in two isomeric forms that are mirror images of each other (see ■ Figure 7.3). Mirror-image isomers, called **enantiomers,** have the same molecular and structural formulas but different spatial arrangements of their atoms. No amount of rotation can convert one of these structures into the other. The molecules can be compared to right and left hands, which are also mirror images of each other (see ■ Figure 7.4). Test this yourself. Put your right hand up to a mirror, and verify that its mirror image is the same as the direct image of your left hand.

Carbohydrate
A polyhydroxy aldehyde or ketone, or substance that yields such compounds upon hydrolysis.

Monosaccharide
A simple carbohydrate most commonly consisting of three to six carbon atoms.

Disaccharide
A carbohydrate formed by the combination of two monosaccharides.

Polysaccharide
A carbohydrate formed by the combination of many monosaccharide units.

Enantiomers
Stereoisomers that are mirror images.

➤ **FIGURE 7.3** Enantiomers of glyceraldehyde. The designations L and D are discussed in Section 7.3.

➤ **FIGURE 7.4** The reflection of a right hand is a left hand.

➤ **FIGURE 7.5** Two right hands can be superimposed; a right hand and a left hand cannot.

Chiral

A descriptive term for compounds or objects that cannot be superimposed on their mirror image.

Chiral carbon

A carbon atom with four different groups attached.

Compare your right hand with the right hand of a friend. Note that they can be overlapped or superimposed (see ■ Figure 7.5). Now try to rotate or twist your right and left hands so that one is superimposed on the other and turned the exact same way. It cannot be done. Even if you place your hands palm-to-palm so the profiles superimpose, the nails are on the wrong sides. The same thing happens when you try to put a left-hand glove on your right hand.

When an object cannot be superimposed on its mirror image, the object is said to be **chiral.** Thus, a hand, a glove, and a shoe are chiral objects; they cannot be superimposed on their mirror images. However, a drinking glass, a sphere, and a cube are achiral objects because they can be superimposed on their mirror images.

Many molecules in addition to glyceraldehyde are chiral, but fortunately it is not necessary to try to superimpose them on their mirror images in order to determine whether or not they are chiral. Generally, an organic molecule that contains a carbon atom attached to four different groups will be a chiral and thus will have two nonsuperimposable mirror images. A carbon atom with four different groups attached is called a **chiral carbon.**

The center carbon of glyceraldehyde

$$\boxed{H}-C\,\substack{\boxed{CHO}\\ \\ \boxed{OH}\\ \\ \boxed{CH_2OH}}$$

is chiral because it is attached to CHO, H, OH, and CH_2OH groups. Like your hands, the two enantiomers of glyceraldehyde are nonsuperimposable mirror images. No matter how hard we try, there is no way, short of breaking bonds, to superimpose the four groups so that they coincide.

Because the presence of a single chiral carbon atom gives rise to stereoisomerism, it is important to be able to recognize one when you see it. It is really not difficult. If a carbon atom is connected to four different groups, no matter how slight the differences, it is a chiral carbon. If any two groups are identical, it is not a chiral carbon.

➤ **Example 7.1**

....

Which of the following compounds contain a chiral carbon?

a. CH_3CHCH_3
$\quad\quad\;\;|$
$\quad\quad\;OH$

b. $CH_3CHCH_2CH_3$
$\quad\quad\;\;|$
$\quad\quad\;OH$

c. $CH_3CCH_2CH_3$
$\quad\quad\;\; \overset{\|}{O}$

d. $\quad$ OH
$\quad\quad\;\;\; |$
$\quad\quad\; CH-CH_2$
$\quad\quad / \quad\quad\quad \backslash$
$\quad CH_2 \quad\quad\quad O$
$\quad\quad \backslash \quad\quad\quad /$
$\quad\quad\; CH_2-CH_2$

Solution

It may help to draw the structures in a more complete form.

a. The central carbon atom has one H, one OH, but two CH_3 groups attached. It is not chiral. Neither of the other two carbons has four different groups.

$$\begin{array}{c} H \\ | \\ CH_3 - C - CH_3 \\ | \\ OH \end{array}$$

b. The central carbon atom has one H, one OH, one CH_3, and one CH_2CH_3 group attached. Both CH_3 and CH_2CH_3 are bonded to the central carbon atom through a carbon atom. However, when analyzing for chiral carbons, we look at the entire collection of atoms in a group and not just at the four atoms directly bonded. Thus, CH_3 and CH_2CH_3 groups are different (as are H and OH), so the central carbon (identified by an asterisk) is chiral. None of the other carbons is chiral.

$$\begin{array}{c} H \\ | \\ CH_3 - \overset{*}{C} - CH_2CH_3 \\ | \\ OH \end{array}$$

c. The carbonyl carbon is attached to only three groups, so it is not chiral, even though the three groups are different. None of the other carbons have four different groups attached.

$$\begin{array}{c} CH_3 - C - CH_2 - CH_3 \\ \| \\ O \end{array}$$

d. The rules are no different for carbons in rings. There is a chiral carbon marked with an asterisk. H and OH are two of the groups. As we proceed around the ring, $-CH_2-O-$ (the third group) is encountered in the clockwise direction and $-CH_2-CH_2-$ (the fourth group) in the counterclockwise direction.

$\quad\quad\quad\quad$ OH
$\quad\quad\quad\quad |$
$\quad\quad\quad \overset{*}{C}H-CH_2$
$\quad\quad\; / \quad\quad\quad \backslash$
$\quad CH_2 \quad\quad\quad O$
$\quad\quad \backslash \quad\quad\quad /$
$\quad\quad\; CH_2-CH_2$

Which of the carbon atoms shown in color is chiral?

Learning Check 7.1

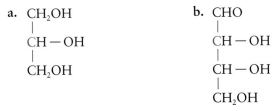

c.

$$
\begin{array}{c}
CH_2OH \\
| \\
C=O \\
| \\
CH-OH \\
| \\
CH_2OH
\end{array}
$$

d.

$$
\begin{array}{c}
CH_2OH \\
| \\
CH-O \\
\diagup \quad \diagdown \\
CH_2 \qquad CH_2 \\
CH_2-CH_2
\end{array}
$$

Organic molecules (and especially carbohydrates) may contain more than one chiral carbon. Erythrose, for example, has two chiral carbon atoms:

$$
\begin{array}{ccc}
OH & OH & OH \\
| & | & | \\
CH_2-CH & -CH & -CHO
\end{array}
$$

erythrose

➤ Example 7.2

Identify the chiral carbons in glucose:

$$
\begin{array}{c}
H \diagdown \quad \diagup O \\
{}^1C \\
| \\
H-{}^2C-OH \\
| \\
HO-{}^3C-H \\
| \\
H-{}^4C-OH \\
| \\
H-{}^5C-OH \\
| \\
{}^6CH_2OH
\end{array}
$$

Solution

Carbon number 1 is not chiral, because it has only three groups attached to it:

$$
\begin{array}{c}
\boxed{H} \diagdown \quad \diagup \boxed{O} \\
{}^1C \\
| \\
\begin{array}{|c|}
\hline
H-C-OH \\
| \\
HO-C-H \\
| \\
H-C-OH \\
| \\
H-C-OH \\
| \\
CH_2OH \\
\hline
\end{array}
\end{array}
$$

Carbon atoms 2, 3, 4, and 5 are chiral because they each have four different groups attached:

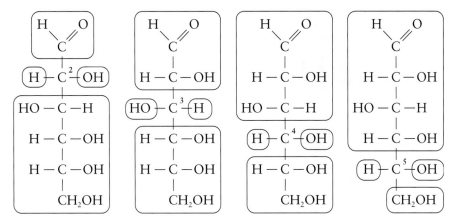

Carbon number 6 is not chiral because two of the four attached groups are the same.

Learning Check 7.2

Identify the chiral carbon atoms in the following:

a.

$$CH_2OH$$
$$|$$
$$C=O$$
$$|$$
$$HO-C-H$$
$$|$$
$$H-C-OH$$
$$|$$
$$CH_2OH$$

b.

$$CHO$$
$$|$$
$$CH-OH$$
$$|$$
$$CH-OH$$
$$|$$
$$CH-OH$$
$$|$$
$$CH_2OH$$

c.

HOCH₂ O OH
H   H
H   H
OH OH

When a molecule contains more than one chiral carbon, the possibility exists for two arrangements of attached groups at each chiral carbon atom. Thus, nearly all molecules having two chiral carbon atoms exist as two pairs of enantiomers, for a total of four stereoisomers. In a similar manner, when there are three chiral carbons present, there are eight stereoisomers (four pairs of enantiomers). The general formula is

$$\text{Maximum number of possible stereoisomers} = 2^n$$

where n is the number of chiral carbon atoms.

➤ Example 7.3

How many stereoisomers are possible for a molecule that contains four chiral carbons such as glucose?

Solution
Stereoisomers $= 2^n$

$$n = 4 \text{ chiral carbons}$$

$$2^4 = 2 \times 2 \times 2 \times 2 = 16$$

Learning Check 7.3

How many stereoisomers are possible for the following?

$$\text{OH} \quad \text{OH} \quad \text{OH} \quad \text{OH}$$
$$| \qquad | \qquad | \qquad |$$
$$CH_2-CH-CH-CH-CHO$$

➤ **FIGURE 7.6** Ball-and-stick models and Fischer projections of the two enantiomers of glyceraldehyde.

L-glyceraldehyde   **Mirror**   D-glyceraldehyde

➤ 7.3 Fischer Projections

It is time-consuming to draw molecules in the three-dimensional shapes shown for the two enantiomers of glyceraldehyde in Figure 7.3, but there is a way to represent these mirror images in two dimensions. Emil Fischer, a German chemist known for pioneering work in carbohydrate chemistry, introduced a method late in the nineteenth century. His two-dimensional structures, called **Fischer projections**, are illustrated in ■ Figure 7.6 for glyceraldehyde.

In Fischer projections, the chiral carbon is represented by the intersection of two lines. If the compound is a carbohydrate, the carbonyl group is placed at or near the top. The molecule is also positioned so that the two bonds coming toward you out of the plane of the paper are drawn horizontally in the Fischer projection. The two bonds projecting away from you into the plane of the paper are drawn vertically. Thus, when you see a Fischer projection, you must realize that the molecule has a three-dimensional shape with horizontal bonds coming toward you and vertical bonds going away from you.

Fischer designated the two enantiomers of glyceraldehyde as L and D compounds. Using this system, we can represent the isomers of similar compounds by the direction of bonds about the chiral carbon. A small capital L is used to indicate that an —OH group (or another functional group) is on the left of the chiral carbon when the carbonyl is at the top. A small capital D means the —OH is on the right of the chiral carbon.

Fischer projection
A method of depicting three-dimensional shapes for chiral molecules.

➤ Example 7.4

Draw Fischer projections for the D and L forms of the following:

a.

$$\begin{array}{c} OH \\ | \\ CH_3 - CH - COOH \end{array}$$
lactic acid

b.

$$\begin{array}{c} NH_2 \\ | \\ CH_3 - CH - COOH \end{array}$$
alanine

Solution

a. The second carbon of lactic acid is chiral with four different groups attached: H, OH, CH_3, and COOH. The chiral carbon is placed at the intersection of two lines. Lactic acid is not an aldehyde, but it does contain a carbon–oxygen double bond (in the carboxyl group). This carboxyl group is placed at the top of the vertical line. The direction of the —OH groups on the chiral carbons determines the D and L notations:

$$
\begin{array}{ccc}
\text{COOH} & \bigg| & \text{COOH} \\
\text{HO}\!-\!\!\!\!-\!\text{H} & \bigg| & \text{H}\!-\!\!\!\!-\!\text{OH} \\
\text{CH}_3 & \bigg| & \text{CH}_3
\end{array}
$$

Mirror

L-lactic acid   D-lactic acid

b. Similarly, the amino acid alanine has a chiral carbon and a carboxyl group. The direction of the —NH$_2$ group determines D and L notations:

$$
\begin{array}{ccc}
\text{COOH} & \bigg| & \text{COOH} \\
\text{H}_2\text{N}\!-\!\!\!\!-\!\text{H} & \bigg| & \text{H}\!-\!\!\!\!-\!\text{NH}_2 \\
\text{CH}_3 & \bigg| & \text{CH}_3
\end{array}
$$

Mirror

L-alanine   D-alanine

Draw Fischer projections for the D and L forms of the following:

$$
\begin{array}{cc}
\text{CH}_3 & \text{NH}_2 \\
| & | \\
\end{array}
$$
$$\text{CH}_3\text{CH}\!-\!\text{CH}\!-\!\text{COOH}$$

Learning Check 7.4

The existence of more than one chiral carbon in a carbohydrate molecule could lead to confusion of D and L designations. This is avoided by focusing on the hydroxy group attached to the chiral carbon farthest from the carbonyl group. By convention, the D family of compounds is that in which this hydroxy group projects to the right when the carbonyl is "up." In the L family of compounds, it projects to the left:

$$
\begin{array}{cccc}
& & \text{CHO} & \text{CHO} \\
\text{CHO} & \text{CHO} & \text{H}\!-\!\text{OH}\;\;\text{HO}\!-\!\text{H} & \\
\text{H}\!-\!\text{OH}\;\;\text{HO}\!-\!\text{H} & \text{HO}\!-\!\text{H}\;\;\;\text{H}\!-\!\text{OH} & \\
\text{H}\!-\!\text{OH}\;\;\text{HO}\!-\!\text{H} & \text{H}\!-\!\text{OH}\;\;\text{HO}\!-\!\text{H} & \\
\text{CH}_2\text{OH}\;\;\;\text{CH}_2\text{OH} & \text{H}\!-\!\text{OH}\;\;\text{HO}\!-\!\text{H} & \\
& & \text{CH}_2\text{OH}\;\;\;\text{CH}_2\text{OH} &
\end{array}
$$

D-erythrose   L-erythrose   D-glucose   L-glucose

Notice that D and L compounds are mirror images of each other (enantiomers).

Learning Check 7.5

Identify each structure as D or L:

a.

```
        CHO
        |
HO ─────┼───── H
        |
HO ─────┼───── H
        |
HO ─────┼───── H
        |
       CH₂OH
```

b.

```
       CH₂OH
        |
        C═O
        |
HO ─────┼───── H
        |
 H ─────┼───── OH
        |
 H ─────┼───── OH
        |
       CH₂OH
```

c.

```
        CHO
        |
 H ─────┼───── OH
        |
HO ─────┼───── H
        |
       CH₂OH
```

The physical properties of the two compounds that make up a pair of D and L isomers (enantiomers) are generally the same. The only exception is the way solutions of the two compounds affect polarized light that is passed through them. Light is polarized by passing it through a special polaroid lens, such as those found in polarized sunglasses. When polarized light is passed through a solution of one enantiomer, the plane of polarization of the light is rotated to either the right or the left when viewed by looking toward the source of the light (see ■ Figure 7.7). The enantiomer that rotates it to the left is called the **levorotatory** (to the left) or (−) enantiomer. The one that rotates it to the right is the **dextrorotatory** (to the right) or (+) enantiomer.

The D and L designations do not represent the words *dextrorotatory* and *levorotatory* but only the spatial relationships of a Fischer projection. Thus, for example, some D compounds rotate polarized light to the right (dextrorotatory), and some rotate it to the left (levorotatory). The spatial relationships D and L and the rotation of polarized light are entirely different concepts and should not be confused with each other.

The property of rotating the plane of polarized light is called optical activity, and molecules with this property are said to be **optically active.** Measurements of optical activity are useful for differentiating between enantiomers.

Why is it so important to be able to describe stereoisomerism and recognize it in molecules? The reason stems from the fact that living organisms, both plant and animal, consist largely of chiral substances. Most of the

Levorotatory
Rotates plane-polarized light to the left.

Dextrorotatory
Rotates plane-polarized light to the right.

Optically active molecule
A molecule that rotates the plane of polarized light.

➤ **FIGURE 7.7** The rotation of plane-polarized light by the solution of an enantiomer.

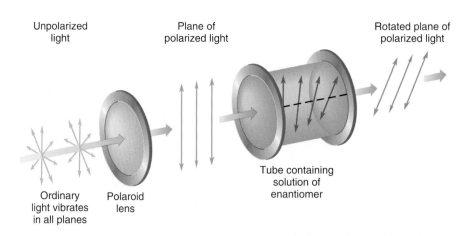

compounds in natural systems—including biomolecules such as carbohydrates, lipids, proteins, and nucleic acids—are chiral. Although these molecules can in principle exist as a mixture of stereoisomers, quite often only one stereoisomer is found in nature.

In some instances, two enantiomers are found in nature, but they are not found together in the same biological system. For example, lactic acid occurs naturally in both the forms. The L-lactic acid is found in living muscle, whereas the D-lactic acid is present in sour milk. When we realize that only one enantiomer is found in a given biological system, it is not too surprising to find that the system can usually use or assimilate only one enantiomer. Humans utilize the D-isomers of monosaccharides and are unable to metabolize the L-isomers. The D form of glucose tastes sweet, is nutritious, and is an important component of our diets. The L-isomer, on the other hand, is tasteless, and the body cannot use it. Yeast can ferment only D-glucose to produce alcohol, and most animals are able to utilize only L-amino acids in the synthesis of proteins.

TABLE 7.1 Monosaccharide classification based on the number of carbons in their chains

Number of carbon atoms	Sugar class
3	Triose
4	Tetrose
5	Pentose
6	Hexose

➤ 7.4 Monosaccharides

The simplest carbohydrates are the monosaccharides, consisting of a single polyhydroxy aldehyde or ketone unit. Monosaccharides are further classified according to the number of carbon atoms they contain (see ■ Table 7.1). Thus, simple sugars containing three, four, five, and six carbon atoms are called, respectively, trioses, tetroses, pentoses, and hexoses. The presence of an aldehyde group in a monosaccharide is indicated by the prefix *aldo-*. Similarly, a ketone group is denoted by the prefix *keto-*. Thus, glucose is an aldohexose, and ribulose is a ketopentose:

$$
\begin{array}{c}
\text{CHO} \\
| \\
\text{H}-\text{C}-\text{OH} \\
| \\
\text{HO}-\text{C}-\text{H} \\
| \\
\text{H}-\text{C}-\text{OH} \\
| \\
\text{H}-\text{C}-\text{OH} \\
| \\
\text{CH}_2\text{OH}
\end{array}
\qquad
\begin{array}{c}
\text{CH}_2\text{OH} \\
| \\
\text{C}=\text{O} \\
| \\
\text{H}-\text{C}-\text{OH} \\
| \\
\text{H}-\text{C}-\text{OH} \\
| \\
\text{CH}_2\text{OH}
\end{array}
$$

glucose, an aldohexose   ribulose, a ketopentose

Learning Check 7.6

Classify each of the following monosaccharides by combining the aldehyde–ketone designation with terminology indicating the number of carbon atoms:

a.
$$
\begin{array}{c}
\text{CHO} \\
| \\
\text{H}-\text{C}-\text{OH} \\
| \\
\text{HO}-\text{C}-\text{H} \\
| \\
\text{H}-\text{C}-\text{OH} \\
| \\
\text{CH}_2\text{OH}
\end{array}
$$

b.
$$
\begin{array}{c}
\text{CH}_2\text{OH} \\
| \\
\text{C}=\text{O} \\
| \\
\text{H}-\text{C}-\text{OH} \\
| \\
\text{CH}_2\text{OH}
\end{array}
$$

c.
$$
\begin{array}{c}
\text{CHO} \\
| \\
\text{HO}-\text{C}-\text{H} \\
| \\
\text{H}-\text{C}-\text{OH} \\
| \\
\text{CH}_2\text{OH}
\end{array}
$$

CHO
H —— OH
CH$_2$OH

D-glyceraldehyde

Aldotriose

CHO
H —— OH
H —— OH
CH$_2$OH

D-erythrose

CHO
HO —— H
H —— OH
CH$_2$OH

D-threose

Aldotetroses

CHO
H —— OH
H —— OH
H —— OH
CH$_2$OH

D-ribose

CHO
HO —— H
H —— OH
H —— OH
CH$_2$OH

D-arabinose

CHO
H —— OH
HO —— H
H —— OH
CH$_2$OH

D-xylose

CHO
HO —— H
HO —— H
H —— OH
CH$_2$OH

D-lyxose

Aldopentoses

CHO
H —— OH
H —— OH
H —— OH
H —— OH
CH$_2$OH

D-allose

CHO
HO —— H
H —— OH
H —— OH
H —— OH
CH$_2$OH

D-altrose

CHO
H —— OH
HO —— H
H —— OH
H —— OH
CH$_2$OH

D-glucose

CHO
HO —— H
HO —— H
H —— OH
H —— OH
CH$_2$OH

D-mannose

CHO
H —— OH
H —— OH
HO —— H
H —— OH
CH$_2$OH

D-gulose

CHO
HO —— H
H —— OH
HO —— H
H —— OH
CH$_2$OH

D-idose

CHO
H —— OH
HO —— H
HO —— H
H —— OH
CH$_2$OH

D-galactose

CHO
HO —— H
HO —— H
HO —— H
H —— OH
CH$_2$OH

D-talose

Aldohexoses

➤ **FIGURE 7.8** The family of D aldoses, shown in Fischer projections.

Most monosaccharides are aldoses, and almost all natual monosaccharides belong to the D series. The family of D aldoses is shown in ■ Figure 7.8. D-glyceraldehyde, the smallest monosaccharide with a chiral carbon, is the standard on which the whole series is based. Notice that the bottom chiral carbon in each compound is directed to the right. The 2^n formula tells us there must be 2 trioses, 4 tetroses, 8 pentoses, and 16 hexoses. Half of those are the D compounds shown in Figure 7.8. The other half (not shown) are the enantiomers or L compounds.

➤ **7.5 Properties of Monosaccharides**

Physical Properties

Most monosaccharides (and disaccharides) are called sugars because they taste sweet. The degree of sweetness varies, as shown in ■ Table 7.2. Fructose, the sweetest of the common sugars, is about 73% sweeter than sucrose, ordinary table sugar.

All carbohydrates are solids at room temperature, and because of the many —OH groups present, monosaccharide carbohydrates are extremely soluble in water. The —OH groups form numerous hydrogen bonds with the surrounding water molecules. In the body, this solubility allows carbohydrates to be transported rapidly in the circulatory system.

TABLE 7.2	The relative sweetness of sugars (sucrose = 1.00)	
Sugar	**Relative sweetness**	**Type**
Lactose	0.16	Disaccharide
Galactose	0.22	Monosaccharide
Maltose	0.32	Disaccharide
Xylose	0.40	Monosaccharide
Glucose	0.74	Monosaccharide
Sucrose	1.00	Disaccharide
Invert sugar	1.30	Mixture of glucose and fructose
Fructose	1.73	Monosaccharide

Chemical Properties, Cyclic Forms

So far we have represented the monosaccharides as open-chain polyhydroxy-aldehydes and ketones. These representations are useful in discussing the structural features and stereochemistry of monosaccharides. However, we learned in Section 4.3 that aldehydes and ketones react with alcohols to form hemiacetals and hemiketals. When these functional groups are present in the same molecule, we noted that the result of their reaction is a stable cyclic structure. You should not be surprised then to find that all monosaccharides with at least five carbon atoms exist predominantly as cyclic hemiacetals and hemiketals.

To help depict the cyclization of glucose in Reaction 7.1, the open-chain structure has been bent around to position the functional groups in closer proximity. Notice the numbering of the carbon atoms that begins at the end of the chain, giving the lowest number to the carbonyl group carbon:

(7.1)

β-D-glucose

α-D-glucose

Pyranose ring
A six-membered sugar ring system containing an oxygen atom.

Haworth structure
A method of depicting three-dimensional carbohydrate structures.

Anomeric carbon
An acetal, ketal, hemiacetal, or hemiketal carbon atom giving rise to two stereoismers.

Anomers
Stereoisomers that differ in the three-dimensional arrangement of groups at the carbon of an acetal, ketal, hemiacetal, or hemiketal group.

In the reaction, the alcohol group on carbon 5 adds to the aldehyde group on carbon 1. The result is a **pyranose ring,** a six-membered ring containing an oxygen atom. The attached groups have been drawn above or below the plane of the ring. This kind of drawing, called a **Haworth structure,** is used extensively in carbohydrate chemistry. Note that as the reaction occurs, a new chiral carbon is produced at position 1. Thus, two steroisomers are possible: one with the —OH group pointing down (the α form), and the other with the —OH group pointing up (the β form). The C-1 carbon is called an **anomeric carbon,** and the α and β forms are called **anomers.** Convenient condensed structures for the cyclic compounds omit the carbon atoms in the ring and the hydrogen atoms attached to the ring carbons, as shown in Reaction 7.2:

$$\text{(7.2)}$$

| α-D-glucose (36%) | D-glucose (0.02%) | β-D-glucose (64%) |

Thus, there are three forms of D-glucose: an open-chain structure and two ring forms. Because the reaction to form a cyclic hemiacetal is reversible, the three isomers of D-glucose are interconvertible. Studies indicate that the equilibrium distribution in water solutions is approximately 36% α-glucose, 0.02% open-chain form, and 64% β-glucose.

Other monosaccharides also form cyclic structures. However, some form five-membered rings, called **furanose rings,** rather than the six-membered kind. An example is D-fructose.

The five-membered ring cyclization occurs like the ring formation in glucose. An alcohol adds across the carbonyl double bond (in this case, a ketone). The orientation of the —OH group at position 2 determines whether fructose is in the α or β form:

Furanose ring
A five-membered sugar ring system containing an oxygen atom.

D-fructose

$$\text{(7.3)}$$

α-D-fructose

β-D-fructose

CHEMISTRY AROUND US • 7.1

SUGAR-FREE FOODS AND DIABETES

In the past, ice cream, cookies, and other sweets were considered to be off-limits to people with diabetes mellitus (Chemistry Around Us 13.3). However, advances in food science and a better understanding of the relationship between different foods and blood glucose levels have changed the rules, and sweet foods are now acceptable as part of the meal plan for diabetics. Diabetes is characterized by an impairment of the body's ability to utilize blood glucose.

Among the food science developments helpful to diabetics has been the discovery and commercial distribution of several synthetic noncarbohydrate sweetening agents. Three that have been approved by the American Diabetes Association are saccharin, acesulfame-K, and aspartame:

saccharin   aspartame

acesulfame-K

When compared with an equal weight of sucrose (table sugar), saccharin is about 450 times sweeter, acesulfame-K is about 200 times sweeter, and aspartame is about 180 times sweeter. Saccharin and acesulfame-K are noncaloric, whereas aspartame contributes very few calories when used in the quantities necessary to provide sweetness. Thus, these substances are useful dietary substitutes for sucrose or other calorie-laden sweeteners for people who wish to control calorie intake or for those who must avoid sugar, such as diabetics.

Until the approval of aspartame (trade name Nutrasweet®) in 1981, people with diabetes shopped for "dietetic" foods primarily in a special section of the supermarket. Aspartame vastly expanded the range of acceptable foods. Studies have verified its general safety when used in moderate amounts. The one exception is that it should not be used by individuals suffering from phenylketonuria (PKU), an inherited inability to metabolize phenylalanine properly. This condition is easily detected in newborn babies and infants, so routine screening is done shortly after birth.

In 1988, after 6 years of review, the FDA approved acesulfame-K under the trade name Sunette®. Pepsi-Cola® is using this artificial sweetener in a soft drink called Pepsi One®. Unlike aspartame, acesulfame-K is heat-stable and can survive the high temperatures of cooking processes.

Other sweeteners that have expanded the food options for diabetics are known as sugar alcohols. These are carbohydrate derivatives, such as sorbitol, in which the carbon–oxygen double bond of the aldehyde or ketone has been converted to an alcohol:

sorbitol

Sugar alcohols are incompletely absorbed during digestion and consequently contribute fewer calories than other carbohydrates. Sugar alcohols are found naturally in fruits. They are also produced commercially from carbohydrates for use in sugar-free candies, cookies, and chewing gum.

http://www.brookscole.com/chemistry/seager_chemistry
Web site 7.2 provides information on sugars and artificial sweeteners from the American Diabetes Association.

Some common artificial sweeteners.

In drawing cyclic Haworth structures of monosaccharides, certain rules should be followed so that all our structures will be consistent. First we must draw the ring with its oxygen to the back:

furanose
ring

pyranose
ring

We must also put the anomeric carbon on the right side of the ring. Then we envision the ring as planar with groups pointing up or down. The terminal —CH₂OH group (position 6) is always shown above the ring for D-monosaccharides.

➤ **Example 7.5**

The aldohexose D-galactose exists predominantly in cyclic forms. Given the structure below, draw the Haworth structure for the anomer. Label the new compound as α or β.

Solution

Draw the pyranose ring with the oxygen atom to the back. Number the ring starting at the right side. Position number 1 will be the anomeric carbon. Place the —OH group at position 1 in the up direction (β form) so that it is the anomer of the given compound. Place groups at the other positions exactly as they are in the given compound. Remember that anomers differ only in the position of the OH attached to the anomeric carbon.

β-D-galactose

Learning Check 7.7

Draw the Haworth structure for the anomer of D-ribose. Label the new compound as α or β.

D-ribose

Oxidation

We learned in Section 4.3 that aldehydes and those ketones that contain an adjacent —OH group are readily oxidized by an alkaline solution of Cu^{2+} (Benedict's reagent). Open-chain forms of monosaccharides exist as aldehydes or hydroxyketones and are readily oxidized. As the oxidation proceeds, the cyclic forms in equilibrium are converted to open-chain forms (Le Châtelier's principle, Section 8.8) and also react.

Sugars that can be oxidized by weak oxidizing agents are called **reducing sugars.** Thus, all monosaccharides are reducing sugars.

Reducing sugar
A sugar that can be oxidized by Cu^{2+} solutions.

General reaction:

$$\text{Reducing sugar} \quad + \quad Cu^{2+} \quad \rightarrow \quad \text{oxidized compound} \quad + \quad Cu_2O \qquad (7.4)$$

(complex)
deep blue solution

red-orange
precipitate

Specific example:

$$
\begin{array}{c}
\text{CHO} \\
| \\
\text{H}-\text{C}-\text{OH} \\
| \\
\text{HO}-\text{C}-\text{H} \\
| \\
\text{H}-\text{C}-\text{OH} \\
| \\
\text{H}-\text{C}-\text{OH} \\
| \\
\text{CH}_2\text{OH} \\
\text{D-glucose}
\end{array}
\quad + Cu^{2+} \rightarrow \quad
\begin{array}{c}
\text{COOH} \\
| \\
\text{H}-\text{C}-\text{OH} \\
| \\
\text{HO}-\text{C}-\text{H} \\
| \\
\text{H}-\text{C}-\text{OH} \\
| \\
\text{H}-\text{C}-\text{OH} \\
| \\
\text{CH}_2\text{OH} \\
\text{D-gluconic} \\
\text{acid}
\end{array}
\quad + Cu_2O
\qquad (7.5)
$$

Benedict's reagent is deep blue in color. As the open-chain form of a monosaccharide is oxidized, Cu^{2+} is reduced and precipitated as Cu_2O, a red-orange solid.

When we study energy transformations in cells (Chapters 12 and 13), we will encounter many reactions that involve the stepwise oxidation of carbohydrates to CO_2 and H_2O. The controlled oxidation of carbohydrates in the body serves as an important source of heat and other forms of energy.

Phosphate Esters

The hydroxy groups of monosaccharides can behave as alcohols and react with acids to form esters. Esters formed from phosphoric acid and various monosaccharides are found in all cells, and some serve as important intermediates in carbohydrate metabolism. The structures of two representative phosphate esters are shown below:

glucose 6-phosphate

fructose 6-phosphate

Glycoside Formation

Glycoside
Another name for a carbohydrate containing an acetal or ketal group.

In Section 4.3, we learned that hemiacetals and hemiketals can react with alcohols in acid solutions to yield acetals and ketals, respectively. Thus, cyclic monosaccharides (hemiacetals and hemiketals) readily react with alcohols in the presence of acid to form acetals and ketals. The general name for these carbohydrate products is **glycosides**. The reaction of α-D-glucose with methanol is shown in Reaction 7.6.

α-D-glucose   methyl α-D-glucopyranoside   methyl β-D-glucopyranoside

glycosides

(7.6)

The hemiacetal in glucose is located at position 1, and the acetal thus forms at that same position. All the other —OH groups in glucose are ordinary alcohol groups and do not react under these conditions. As the glycoside reaction takes place, a new bond is established between the pyranose ring and the —OCH$_3$ groups. The new group may point up or down from the ring. Thus, a mixture of α- and β-glycosides is formed. The new carbon–oxygen–carbon linkage that joins the components of the glycoside is called a **glycosidic linkage**.

Glycosidic linkage
The carbon–oxygen–carbon linkage that joins the components of a glycoside to the ring.

Although carbohydrate cyclic hemiacetals and hemiketals are in equilibrium with open-chain forms of the monosaccharides, the glycosides (acetals and ketals) are much more stable and do not exhibit open-chain forms. Therefore, glycosides of monosaccharides are not reducing sugars.

As we'll see later in this chapter, both disaccharides and polysaccharides are examples of glycosides in which monosaccharide units are joined together by acetal (glycosidic) linkages.

Learning Check 7.8

Two glycosides are shown below. Circle any acetal or ketal groups and use an arrow to identify the glycosidic linkages.

a.

b.

►7.6 Important Monosaccharides

Ribose and Deoxyribose

Two pentoses, ribose and deoxyribose, are extremely important because they are used in the synthesis of nucleic acids (DNA and RNA), substances essential in protein synthesis and the transfer of genetic material (Chapter 11). Ribose (along with phosphate groups) forms the long chains that make up ribonucleic acid (RNA). Deoxyribose (along with phosphate groups) forms the long chains of deoxyribonucleic acid (DNA). Deoxyribose differs from ribose in that the OH group on carbon 2 has been replaced by a hydrogen atom (deoxy form).

β-D-ribose   β-D-deoxyribose
                two hydrogens at this position

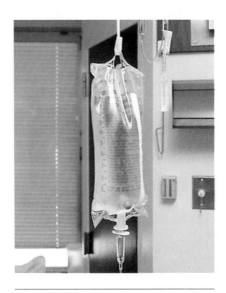

➤ **FIGURE 7.9** A glucose solution prepared for intravenous use.

Glucose

Of the monosaccharides, the hexose glucose is the most important nutritionally and the most abundant in nature. Glucose is present in honey and fruits such as grapes, figs, and dates. Ripe grapes, for example, contain 20–30% glucose. Glucose is sometimes called dextrose. Glucose is also known as blood sugar because it is the sugar transported by the blood to body tissues to satisfy energy requirements (see ■ Figure 7.9). Other sugars absorbed into the body must be converted to glucose by the liver. Glucose is commonly used as a sweetener in confections and other foods, including some baby foods.

β-D-glucose

Galactose

Galactose is a hexose with a structure very similar to that of glucose. The only difference between the two is the orientation of the hydroxy group attached to carbon 4. Like glucose, galactose can exist in α, β, or open-chain forms; the β form is shown below:

—OH group is down in glucose

β-D-galactose

Galactose is synthesized in the mammary glands, where it is incorporated into lactose, the sugar found in milk. It is also a component of substances present in nerve tissue.

Fructose

Fructose (Section 7.5) is the most important ketohexose. It is also known as levulose and, because of its presence in many fruits, fruit sugar. It is present in honey in a 1:1 ratio with glucose and is abundant in corn syrup. This sweetest of the common sugars is important as a food sweetener because less fructose is needed than other sugars to achieve the same degree of sweetness.

➤ 7.7 Disaccharides

Disaccharides are sugars composed of two monosaccharide units linked together by the acetal or ketal linkages described earlier. They can be hydrolyzed to yield their monosaccharide building blocks by boiling with dilute acid or reacting them with appropriate enzymes. Nutritionally, the most important members of this group are maltose, lactose, and sucrose.

Maltose

Maltose, also called malt sugar, contains two glucose units joined by a glycosidic linkage between carbon 1 of the first glucose unit and carbon 4 of the second unit (Reaction 7.7). The configuration of the linkage on carbon 1 in the glycosidic linkage between glucose units is α, and the linkage is symbolized α(1 → 4) to indicate this fact.

$$\text{α-D-glucose} + \text{α-D-glucose} \longrightarrow \text{maltose} + H_2O \tag{7.7}$$

Notice that maltose contains both an acetal carbon (left glucose unit, position 1) and a hemiacetal carbon (right glucose unit, position 1). The —OH group at the hemiacetal position can point either up or down (the α is shown). The presence of a hemiacetal group also means that the right glucose ring can

STUDY SKILLS 7.1

BIOMOLECULES: A NEW FOCUS

The chapters on organic chemistry (Chapters 1–6) were organized around the functional group concept. Each chapter dealt with a particular functional group or related groups. The nomenclature, properties, and uses of the compounds were discussed. The focus is different in this and the next four chapters (lipids, proteins, enzymes, and nucleic acids). Each of these chapters is devoted to a particular class of biomolecules—substances closely associated with life. Reactions and nomenclature receive much less emphasis; more time is spent describing structures and how they contribute to the processes of living organisms. Another key difference between these and earlier chapters is the reduced emphasis on mastering specific skills, such as solving numerical problems, balancing equations, and naming compounds. More attention should now be given to recognizing structures and understanding key terms and processes. It is a good time to increase your efforts to highlight important concepts in both the text and your lecture notes.

open up to expose an aldehyde group. Thus, maltose is a reducing sugar (Reaction 7.8).

$$hemiacetal\ group$$

(7.8)

$$aldehyde\ group$$

maltose

Maltose, which is found in germinating grain, is formed during the digestion (hydrolysis) of starch to glucose. Its name is derived from the fact that during the germinating (or malting) of barley, starch is hydrolyzed, and the disaccharide is formed. On hydrolysis, maltose forms two molecules of D-glucose:

$$\text{maltose} + H_2O \xrightarrow{H^+} \text{D-glucose} + \text{D-glucose} \qquad (7.9)$$

Lactose

Lactose, or milk sugar, constitutes 5% of cow's milk and 7% of human milk by weight (see ■ Figure 7.10). Pure lactose is obtained from whey, the watery by-product of cheese production. Lactose is composed of one molecule of D-galactose and one of D-glucose. The linkage between the two sugar units is $\beta(1 \rightarrow 4)$:

$$hemiacetal\ of\ \alpha\text{-}D\text{-}glucose\ unit$$

$$\beta\text{-}D\text{-}galactose\ unit$$

lactose

The presence of a hemiacetal group in the glucose unit makes lactose a reducing sugar.

http://www.brookscole.com/chemistry/seager_chemistry
Web site 7.3 provides information on the benefits of breastfeeding from the University of Washington Health Sciences Center.

> **FIGURE 7.10** Human milk is about 7% lactose. What two products form when the lactose is hydrolyzed in the infant's digestive system?

CHEMISTRY AROUND US • 7.2

SWEETNESS AND "LITE"

Over the years, sugar has gained a bad reputation that is not entirely deserved. Despite popular opinion, high-sugar diets are not a cause of hyperactivity or diabetes. In fact, the American Diabetes Association now includes sugar as part of a recommended diet for diabetics. A problem with sugar-laden foods like candy, soft drinks, syrups, and sherbet is that they contain a lot of simple sugars that provide plenty of calories (referred to as "empty calories") but few other beneficial nutrients.

For example, if you have 200 calories to "spend" on something to eat and you spend them on pure sugar, you get nothing of value except a source of energy. Of course, energy is required by the body, but so are many important nutrients. If you spent your 200 calories on three slices of whole wheat bread, in addition to the energy your body needs, you would get 14% of the protein, 24% of the thiamin, and 17% of the niacin of your body's requirement for the day, as well as comparable amounts of other nutrients. These percentages are according to the RDI (reference daily intake; see Chapter 12). It is easy to see that an inappropriate amount of dietary sugar could lead to malnutrition, because nutrients would be displaced despite a plentiful supply of energy.

Sugar is sometimes targeted as a cause of obesity. Overeating any type of food, including foods rich in sugar, can lead to weight problems, but fat in the diet is a more likely culprit be-

cause it contains more than twice the calories per gram than sugar. However, people who are overweight or have low caloric needs should still use simple sugars sparingly to ensure that nutrients—in addition to calories—are contained in the foods they eat.

Dental experts agree that sugar-containing foods promote tooth decay, but they also point out that carbohydrates in the form of starchy foods also contribute to the problem. A thorough brushing after every meal is recommended to minimize the problem.

The Dietary Guidelines for Americans recommend moderate amounts of sugar for individuals in good health. If people wish to reduce the amount of sugar in their diet, some steps are obvious but require a constant effort and awareness. They should avoid adding sugar to food, and should also pay attention to the sugar content of foods by reading labels and comparing similar foods. Foods with high sugar content should be eaten less frequently and in smaller amounts. For example, orange juice is available with and without added sugar, as indicated on the product label.

http://www.brookscole.com/chemistry/seager_chemistry
Additional information on diabetes is available at Web site 7.4.

Sucrose

The disaccharide sucrose, common househould sugar, is extremely abundant in the plant world. It occurs in many fruits, in the nectar of flowers (see ■ Figure 7.11), and in the juices of many plants, especially sugar cane and sugar beets, the commercial sources. Sucrose contains two monosaccharides, glucose and fructose, joined together by a linkage that is α from carbon 1 of glucose and β from carbon 2 of fructose:

α-D-glucose unit

β-D-fructose unit

sucrose

> **FIGURE 7.11** Hummingbirds depend on the sucrose and other carbohydrates of nectar for their energy needs.

Neither ring in the sucrose molecule contains a hemiacetal or hemiketal group necessary for ring opening. This is true because the anomeric positions in both glucose and fructose are part of the glycosidic linkage that connects

From left to right, four test tubes containing Benedict's reagent, 2% maltose solution, 2% sucrose solution, and 2% lactose solution.

Both maltose and lactose are reducing sugars. The sucrose has remained unreacted.

➤ **FIGURE 7.12** Benedict's test on disaccharides.

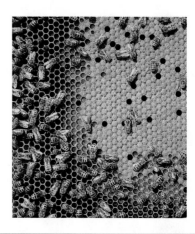

➤ **FIGURE 7.13** Sucrose is hydrolyzed by bees in making honey. Why is honey sweeter than the starting sucrose?

the rings to each other. Thus, in contrast to maltose and lactose, both rings of sucrose are locked in the cyclic form, and sucrose is therefore not a reducing sugar (see ■ Figure 7.12).

The hydrolysis of sucrose (see ■ Figure 7.13) produces a 1:1 mixture of D-glucose and D-fructose called **invert sugar** (Reaction 7.10):

$$\text{sucrose} + H_2O \xrightarrow{H^+} \text{D-glucose} + \text{D-fructose} \qquad (7.10)$$

This mixture is sweeter than the original sucrose (see Table 7.2).

When sucrose is cooked with acid-containing fruits or berries, partial hydrolysis takes place, and some invert sugar is formed. The jams and jellies prepared in this manner are actually sweeter than the pure sugar originally put in them. ■ Table 7.3 summarizes some features of the three disaccharides we have discussed.

Invert sugar
A mixture of equal amounts of glucose and fructose.

➤ 7.8 Polysaccharides

Just as two sugar units are linked together to form a dissacharide, additional units can be added to form higher saccharides. Nature does just this in forming polysaccharides, which are condensation polymers containing thousands of units. Because of their size, polysaccharides are not water soluble, but their many hydroxy groups become hydrated individually when exposed to water, and some polysaccharides form thick colloidal dispersions when heated in water. Thus, the polysaccharide known as starch can be used as a thickener

TABLE 7.3 Some important disaccharides

Name	Monosaccharide constituents	Glycoside linkage	Source
Maltose	Two glucose units	α(1 → 4)	Hydrolysis of starch
Lactose	Galactose and glucose	β(1 → 4)	Mammalian milk
Sucrose	Glucose and fructose	α-1 → β-2	Sugar cane and sugar beet juices

TABLE 7.4 Properties of polysaccharides compared with those of monosaccharides and disaccharides

Property	Monosaccharides and disaccharides	Polysaccharides
Molecular weight	Low	Very high
Taste	Sweet	Tasteless
Solubility in water	Soluble	Insoluble or form colloidal dispersions
Size of particles	Pass through a membrane	Do not pass through a membrane
Test with Cu^{2+} for reducing sugars	Positive (except for sucrose)	Negative

in sauces, gravies, pie fillings, and other food preparations. As shown in ■ Table 7.4, the properties of polysaccharides differ markedly from those of monosaccharides and disaccharides.

Starch

Starch is a polymer consisting entirely of D-glucose units. It is the major storage form of D-glucose in plants. Two starch fractions, amylose (10–20%) and amylopectin (80–90%), can usually be isolated from plants. Amylose is made up of long unbranched chains of glucose units connected by $\alpha(1 \rightarrow 4)$ linkages (see ■ Figure 7.14). This is the same type of linkage found in maltose. The long chain, often containing between 1000 and 2000 glucose units, is flexible enough to allow the molecules to twist into the shape of a helix (see ■ Figure 7.15a).

An important test for the presence of starch is the reaction that occurs between iodine, I_2, and the coiled form of amylose. The product of the reaction is deep blue in color (see ■ Figure 7.16) and is thought to consist of the amylose helix filled with iodine molecules (see Figure 7.15b). This same iodine reaction is also widely used to monitor the hydrolysis of starch. The color gradually fades and finally disappears, as starch is hydrolyzed by either acid or enzymes to form dextrins (smaller polysaccharides), then maltose, and finally glucose. The disappearance of the blue iodine color is thought to be the result of the breakdown of the starch helix.

Amylopectin, the second component of starch, is not a straight-chain molecule like amylose but contains random branches. The branching point is the

$\alpha\ (1 \rightarrow 4)\ linkage$

➤ **FIGURE 7.14** The structure of amylose.

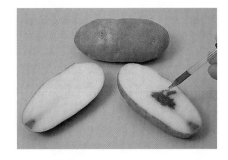

➤ **FIGURE 7.16** A dark blue color is the characteristic result when a solution of iodine encounters the starch in potatoes. What other foods would you expect to give a positive starch test?

➤ **FIGURE 7.15** The molecular conformation of starch and the starch–iodine complex: (a) the helical conformation of the amylose chain and (b) the starch–iodine complex.

α(1 → 6) glycosidic linkage (see ■ Figure 7.17). There are usually 24 to 30 D-glucose units, all connected by α(1 → 4) linkages, between each branch point of amylopectin. These branch points give amylopectin the appearance of a bushy molecule (Figure 7.17). Amylopectin contains as many as 10^5 glucose units in one gigantic molecule.

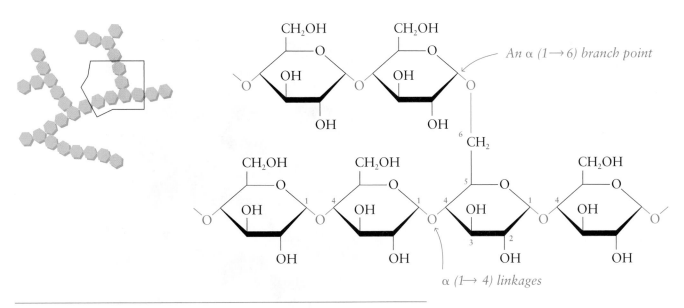

➤ **FIGURE 7.17** The partial structure of an amylopectin molecule. Glycogen has a similar structure.

➤ **FIGURE 7.18** A simplified representation of the branched polysaccharide glycogen (branches every 8–12 glucose units). Amylopectin is much less densely branched (branches every 24–30 glucose units). Each small circle represents a single glucose unit.

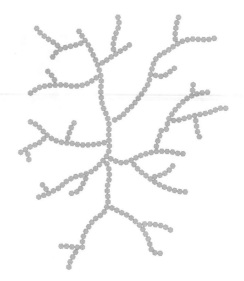

Glycogen

The polysaccharide glycogen is sometimes called animal starch because it is a storage carbohydrate for animals that is analogous to the starch of plants. It is especially abundant in the liver and muscles, where much of the excess glucose taken in by an animal is stored for future use. On hydrolysis, glycogen forms D-glucose, which helps maintain the normal blood sugar level and provide the muscles with energy. Structurally, glycogen is very similar to amylopectin, containing both $\alpha(1 \rightarrow 4)$ and $\alpha(1 \rightarrow 6)$ linkages between glucose units (see ■ Figure 7.18). The main difference between amylopectin and glycogen is that glycogen is even more highly branched; there are only 8 to 12 D-glucose units between branch points.

Cellulose

Cellulose is the most important structural polysaccharide and is the single most abundant organic compound on earth (see ■ Figure 7.19). It is the material in plant cell walls that provides strength and rigidity. Wood is about 50%

➤ **FIGURE 7.19** Cellulose and the plant kingdom are responsible for much of the world's natural beauty.

OVER THE COUNTER • 7.1

DIETARY FIBER

A large amount of research has been conducted on the benefits of **fiber** in the diet. Reports of the results in the mass media have often created conflicts in the minds of consumers. A few years ago, oatmeal was specifically reported to significantly reduce the chances of contracting a number of serious ailments. As a result, grocery stores had trouble keeping oatmeal on the shelves. But later reports of research findings moderated the beneficial role of oatmeal, and sales returned to normal.

For years, dietary fiber in general has been credited with helping prevent colon cancer and several other health problems. However, a prestigious new study involving more than 88,000 women has found no evidence that consuming large amounts of high-fiber food helps lower the risk of colon cancer. No research has yet disproved the results of previous studies indicating that dietary fiber may help prevent digestive disorders, some types of diabetes, heart disease, high blood pressure, and obesity.

Dietary fiber comes from plants; it consists of complex carbohydrates, such as cellulose, and other substances that make up the cell walls and structural parts of plants. Some examples of good sources of dietary fiber are cereal grains (including oatmeal), fresh fruits and vegetables, and grain products such as breads.

Dietary fiber is classified as soluble or insoluble on the basis of its solubility in water. Soluble fiber, such as pectin, has a lower molecular weight and structural features that make it more water soluble than insoluble fiber, such as cellulose. Perhaps the most important function of insoluble fiber in the diet is providing bulk to the stool. This helps the body eliminate solid wastes and in particular helps reduce the risk of *diverticulosis,* a condition in which small pouches form in the colon wall. Increased fiber consumption also helps alleviate the symptoms of diarrhea, abdominal pain, and flatulence. Also, because insoluble fiber is indigestible and passes through the body without providing many calories, it causes people to feel full and therefore eat less, thereby helping them avoid weight gain. Insoluble fiber may also contribute to weight control by inhibiting the absorption of some calorie-dense dietary fats from the digestive system.

Soluble fiber traps carbohydrates and thereby slows their digestion and absorption. This has the effect of leveling out blood sugar levels during the day. Studies indicate that a diet high in sugar and low in fiber more than doubles the chances for a woman to develop Type II diabetes. Clinical studies have also shown that a diet low in saturated fat and cholesterol, and high in fruits, vegetables, and grain products that contain soluble fiber, can lower blood cholesterol levels. Recently, the FDA approved specific labels for breakfast cereal and other foods that contain soluble fiber from psyllium seed husk. The labels state that the food (especially oats) may reduce the risk of coronary heart disease if the food is included as part of a diet that is also low in saturated fat and cholesterol. Maybe the sales of oatmeal will again increase. The soluble fiber lowers blood cholesterol levels by binding to dietary cholesterol and thus helping the body eliminate it.

If dietary fiber looks good to you and you want to include more in your daily diet, start by reading food labels and choosing those that contain more fiber. However, if you follow the current dietary recommendations and each day consume 2–4 servings of fruit, 3–5 servings of vegetables, and 6–11 servings of cereal and grain foods (including oatmeal), you will get enough fiber. Here are some suggestions to help you increase your fiber intake:

• Start each day with a whole-grain cereal.
• Try eating vegetables raw, instead of cooked.
• Don't peel fruits and vegetables before eating them.
• Add beans to soups, stews, and salads as a meat replacement.

As you make the transition to a high-fiber diet, there are some precautions to consider. First, fiber supplements have not been shown to be safe or effective for weight loss. A low-fat diet, combined with an increase in exercise, is still the most effective method. Second, don't increase the fiber content of your diet too rapidly. A gradual increase of a few grams each day will allow your digestive system to adjust. The unpleasant, common results of a sudden dietary fiber increase include abdominal cramps, gas, bloating, and diarrhea or constipation. Third, drink at least 8 cups of liquid (preferably water) daily. Finally, the use of an enzyme product, such as Beano®, will help minimize unpleasant symptoms that may accompany your change in diet.

http://www.brookscole.com/chemistry/seager_chemistry
Web site 7.5 offers information on the nature of fiber from the American Dietetic Association.

A variety of high-fiber foods.

Fiber
Indigestible plant material composed primarily of cellulose.

cellulose. Like amylose, cellulose is a linear polymer consisting of D-glucose units joined by the 1 → 4 linkage. It may contain from 300 to 3000 glucose units in one molecule. The main structural difference between amylose and cellulose is that all the 1 → 4 glycosidic linkages in cellulose are β instead of α (see ■ Figure 7.20). This seemingly small difference causes tremendous differences between amylose and cellulose. The shapes of the molecules are quite different. The α(1 → 4)-linked amylose tends to form loose spiral structures (Figure 7.15), whereas β(1 → 4)-linked cellulose tends to form extended straight chains. These chains become aligned side-by-side to form well-organized, water-insoluble fibers in which the hydroxy groups form numerous hydrogen bonds with the neighboring chains. These parallel chains of cellulose confer rigidity and strength to the molecules.

Second, although starch with its α linkages is readily digestible, humans and most other animals lack the necessary enzymes to hydrolyze the β linkages of cellulose. Thus, cellulose passes unchanged through the digestive tract and does not contribute to the caloric value of food. However, it still serves a useful purpose in digestion. Cellulose, a common constituent of dietary fiber, is the roughage that provides bulk, stimulates contraction of the intestines, and aids the passage of food through the digestive system.

Animals that use cellulose as a food do so only with the help of bacteria that possess the necessary enzymes. Herbivores such as cows, sheep, and horses are animals in this category. Each has a colony of such bacteria somewhere in the digestive system and uses the simple carbohydrates resulting from the bacterial hydrolysis of cellulose. Fortunately, the soil also contains organisms with appropriate enzymes. Otherwise, debris from dead plants would accumulate rather than disappear through biodegradation.

➤ **FIGURE 7.20** The structure of cellulose.

> **FOR FUTURE REFERENCE** THIAMIN IN SELECTED FOODS

Thiamin is a water-soluble B vitamin that plays an important role in the processes of the body that convert food into energy. Thiamin is also involved in the proper function of nerves and their responding tissues such as muscles. Significant amounts of thiamin are found in many foods, but few foods provide large quantities. People who eat enough nutritious and varied foods to meet their energy needs will usually also satisfy their thiamin requirements. Thiamin deficiency may arise in those situations where people obtain most of their energy from empty-calorie items such as alcohol or foods rich in refined simple sugars. The chart below contains a list of foods that provide a minimum of 0.15 mg of thiamin per serving.

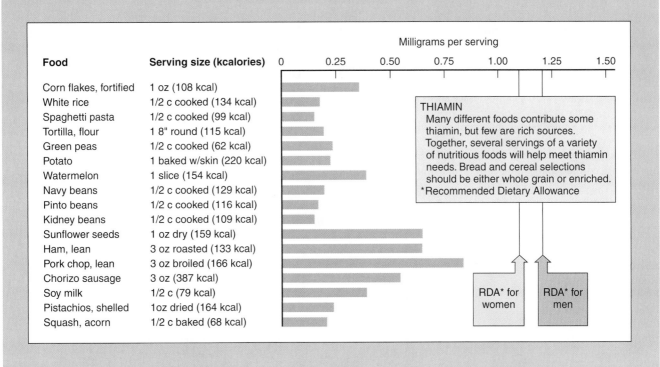

SOURCE: Adapted from Whitney, E.N.; Rolfes, S.R. *Understanding Nutrititon,* 8th ed. Belmont, CA: West/Wadsworth, 1999.

.... **CONCEPT SUMMARY** ....

Classes of Carbohydrates. Carbohydrates are polyhydroxy aldehydes or ketones, or substances that yield such compounds on hydrolysis. Carbohydrates can exist either as single units (monosaccharides) or joined together in molecules ranging from two units (disaccharides) to hundreds of units (polysaccharides).

The Stereochemistry of Carbohydrates. Carbohydrates, along with many other natural substances, exhibit a type of isomerism in which two isomers are mirror images of each other. The two isomers are called enantiomers. When a molecule has more than one chiral carbon, the maximum number of stereoisomers possible is 2^n, where n is the number of chiral carbons.

Fischer Projections. A useful way of depicting the structure of chiral molecules employs crossed lines (Fischer projections) to represent chiral carbon atoms. The prefixes D- and L- are used to distinguish between enantiomers. Signs indicating the rotation of plane-polarized light to the right (+) or to the left (−) may also be used to designate enantiomers.

Monosaccharides. Monosaccharides that contain an aldehyde group are called aldoses, whereas those containing a ketone group are ketoses. Monosaccharides are also classified by the number of carbon atoms as trioses, tetroses, etc. Most natural monosaccharides belong to the D family.

Properties of Monosaccharides. Monosaccharides are sweet-tasting solids that are very soluble in water. Noncarbohydrate low-calorie sweeteners such as aspartame have been developed as sugar substitutes. Pentoses and hexoses form cyclic hemiacetals or hemiketals whose structures can be represented by Haworth structures. Two isomers referred to as anomers (the α and β forms) are produced in the cyclization reaction. All monosaccharides are oxidized by Benedict's reagent and are called reducing sugars. Monosaccharides can react with alcohols to produce acetals or ketals that are called glycosides.

Important Monosaccharides. Ribose and deoxyribose are important as components of nucleic acids. The hexoses glucose, galactose, and fructose are the most important nutritionally and the most abundant in nature. Glucose, also known as blood sugar, is transported within the bloodstream to body tissues, where it supplies energy.

Disaccharides. Glycosidic linkages join monosaccharide units together to form disaccharides. Three important disaccharides are maltose (two glucose units $\alpha(1 \rightarrow 4)$-linked), lactose (a galactose linked to glucose by a $\beta(1 \rightarrow 4)$ glycosidic linkage), and sucrose (α-glucose joined to β-fructose).

Polysaccharides. Cellulose, starch, and glycogen are three important polysaccharides. Starch is the major storage form of glucose in plants, whereas glycogen is the storage form of glucose in animals. Cellulose is the structural material of plants.

LEARNING OBJECTIVES

When you have completed your study of this chapter, you should be able to:

1. Describe the four major functions of carbohydrates in living organisms.
2. Classify carbohydrates as monosaccharides, disaccharides, or polysaccharides.
3. Identify molecules possessing chiral carbon atoms and use Fischer projections to represent D and L compounds.
4. Classify monosaccharides as aldoses or ketoses, and classify them according to the number of carbon atoms they contain.
5. Write reactions for monosaccharide oxidation and glycoside formation.
6. Describe uses for important monosaccharides.
7. Draw the structures and list sources and uses for important disaccharides.
8. Write reactions for the hydrolysis of disaccharides.
9. Determine, on the basis of structure, whether a disaccharide is a reducing sugar.
10. Describe the structures and list sources and uses for important polysaccharides.

KEY TERMS AND CONCEPTS

Anomeric carbon (7.5)
Anomers (7.5)
Biochemistry (Introduction)
Biomolecule (Introduction)
Carbohydrate (7.1)
Chiral (7.2)
Chiral carbon (7.2)
Dextrorotatory (7.3)

Disaccharide (7.1)
Enantiomers (7.2)
Fiber (7.8)
Fischer projection (7.3)
Furanose ring (7.5)
Glycoside (7.5)
Glycosidic linkage (7.5)
Haworth structure (7.5)

Invert sugar (7.7)
Levoratatory (7.3)
Monosaccharide (7.1)
Optically active molecule (7.3)
Polysaccharide (7.1)
Pyranose ring (7.5)
Reducing sugar (7.5)

KEY REACTIONS

1. Oxidation of a sugar (Section 7.5):

$$\text{reducing sugar} + Cu^{2+} \rightarrow \text{oxidized compound} + Cu_2O$$

Reaction 7.4

2. Glycoside formation (Section 7.5):

$$\text{monosaccharide} + \text{alcohol} \xrightarrow{H^+} \text{acetal or ketal}$$

3. Hydrolysis of disaccharides (Section 7.7):

$$\text{disaccharide} + H_2O \xrightarrow[\substack{\text{or} \\ \text{enzymes}}]{H^+} \text{two monosaccharides}$$

4. Hydrolysis of polysaccharides (Section 7.8):

$$\text{polysaccharide} + H_2O \xrightarrow[\substack{\text{or} \\ \text{enzymes}}]{H^+} \text{monosaccharides}$$

CLASSES OF CARBOHYDRATES (SECTION 7.1)

7.1 What are the four important roles of carbohydrates in living organisms?

7.2 Describe whether each of the following substances serves primarily as an energy source, a form of stored energy, or a structural material (some serve as more than one):
   a. cellulose   **c.** glycogen
   b. sucrose, table sugar   **d.** starch

7.3 What are the structural differences between monosaccharides, disaccharides, and polysaccharides?

7.4 Define *carbohydrate* in terms of the functional groups present.

7.5 Match the terms *carbohydrate, monosaccharide, disaccharide,* and *polysaccharide* to each of the following (more than one term may fit):
   a. table sugar

   b.

   c. starch
   d. fructose
   e. cellulose
   f.

   g. glycogen
   h. amylose

THE STEREOCHEMISTRY OF CARBOHYDRATES (SECTION 7.2)

7.6 Why are carbon atoms 1 and 3 of glyceraldehyde not considered chiral?

7.7 Locate the chiral carbon in amphetamine and identify the four different groups attached to it.

7.8 Which of the following molecules can have enantiomers? Identify any chiral carbon atoms.

   a. $CH_3CH_2 - CH - CH_2CH_3$
                |
                OH

   b.

   c.

7.9 Which of the following molecules can have enantiomers? Identify any chiral carbon atoms.

   a.

   b.

   c.

FISCHER PROJECTIONS (SECTION 7.3)

7.10 Explain what the following Fischer projection denotes about the three-dimensional structure of the compound:

7.11 Identify each of the following as a D or an L form and draw the structural formula of the enantiomer:

   a.

   b.

7.12 Identify each of the following as a D or an L form and draw the structural formula of the enantiomer:

a.

```
        CH₂OH
         |
        C=O
         |
   H ——— OH
         |
   H ——— OH
         |
        CH₂OH
```

b.

```
        CHO
         |
   H ——— OH
         |
   H ——— OH
         |
  HO ——— H
         |
        CH₂OH
```

7.13 Draw Fischer projections for both the D and L isomers of the following:
a. 2,3-dihydroxypropanoic acid
b. 2-amino-3-hydroxypropanoic acid

7.14 How many chiral carbon atoms are there in each of the following sugars? How many stereoisomers exist for each compound?

```
      OH   OH   OH   O
       |    |    |    ‖
a. CH₂ — CH — CH — C — H

      OH   O    OH   OH   OH
       |    ‖    |    |    |
b. CH₂ — C — CH — CH — CH — CH₃
```

7.15 How many chiral carbon atoms are there in each of the following sugars? How many stereoisomers exist for each compound?

```
      OH   OH   OH   O
       |    |    |    ‖
a. CH₂ — CH — CH — C — CH₂ — OH

      OH   OH   OH
       |    |    |
b. CH₂ — CH — CH — CHO
```

7.16 How many aldopentoses are possible? How many are the D form and how many the L form?

7.17 Why is the study of chiral molecules important in biochemistry?

7.18 What physical property is characteristic of optically active molecules?

MONOSACCHARIDES (SECTION 7.4)

7.19 Classify each of the following monosaccharides as an aldo- or keto- triose, tetrose, pentose, or hexose:

a.

```
        CHO
         |
  HO ——— H
         |
        CH₂OH
```

b.

```
        CH₂OH
         |
        C=O
         |
   H ——— OH
         |
        CH₂OH
```

7.20 Classify each of the following monosaccharides as an aldo- or keto- triose, tetrose, pentose, or hexose:

a.

```
        CHO
         |
   H ——— OH
         |
   H ——— OH
         |
  HO ——— H
         |
  HO ——— H
         |
        CH₂OH
```

b.

```
        CH₂OH
         |
        C=O
         |
   H ——— OH
         |
   H ——— OH
         |
        CH₂OH
```

7.21 Draw Fischer projections of any aldotetrose and any ketopentose.

PROPERTIES OF MONOSACCHARIDES (SECTION 7.5)

7.22 Explain why certain carbohydrates are called sugars.

7.23 Explain why monosaccharides are soluble in water.

7.24 Identify each of the following as an α or β form, and draw the structural formula of the other anomer:

7.25 Identify each of the following as an α or β form, and draw the structural formula of the other anomer:

a.

```
        CH₂OH
       � O
      |     |
      | OH  |
  HO  |_____| OH
        |
        OH
```

b.

```
   HOCH₂   O   OH
      \    |   /
       \___|__/  CH₂OH
       OH  OH
```

7.26 The structure of talose differs from galactose only in the direction of the —OH group at position 2. Draw and label Haworth structures for α- and β-talose.

7.27 What is the difference between pyranose and furanose rings?

7.28 An unknown sugar failed to react with a solution of Cu^{2+}. Classify the compound as a reducing or nonreducing sugar.

7.29 Explain how the cyclic compound β-D-galactose can react with Cu^{2+} and be classified as a reducing sugar.

7.30 Complete the following reactions:

a.

```
        CHO
         |
  HO ——— H
         |          + Cu²⁺ ⟶
  HO ——— H
         |
        CH₂OH
```

b.

```
  CH₂OH   O   CH₂OH
      \   |   /
       \__|__/          + CH₃CH₂ — OH  →ᴴ⁺
      HO  | 
          OH
        OH
```

c.

```
        CH₂OH
      ⌐——— O   OH
      |        /
      | OH    /          + CH₃ — OH  →ᴴ⁺
  HO  |_____|
        |
        OH
```

7.31 Use an arrow to identify the glycosidic linkage in each of the following:

a.

b.

IMPORTANT MONOSACCHARIDES (SECTION 7.6)

7.32 Explain why D-glucose can be injected directly into the bloodstream to serve as an energy source.

7.33 Give two other names for D-glucose.

7.34 How do the hexoses glucose and galactose differ structurally?

7.35 Give the natural sources for each of the following:
a. glucose
b. fructose
c. galactose

7.36 Explain why fructose can be used as a low-calorie sweetener.

DISACCHARIDES (SECTION 7.7)

7.37 What one monosaccharide is a component of all three of the disaccharides sucrose, maltose, and lactose?

7.38 Identify a disaccharide that fits each of the following:
a. The most common household sugar
b. Converted to lactic acid when milk sours
c. An ingredient of synthetic mother's milk
d. Formed in germinating grain
e. Hydrolyzes when cooked with acidic foods to give invert sugar
f. Found in high concentrations in sugar cane

7.39 What type of linkage is broken when disaccharides are hydrolyzed to monosaccharides?

7.40 Write word equations for the hydrolysis of the following disaccharides to monosaccharides. Name each of the resulting products.
a. sucrose
b. maltose
c. lactose
d. cellobiose, composed of two glucose units linked β(1 → 4)

7.41 Why is invert sugar sweeter than sucrose?

7.42 Sucrose and honey are commonly used sweeteners. Suppose you had a sweet-tasting water solution that contained either honey or sucrose. How would you chemically determine which sweetener was present?

7.43 Draw Haworth projection formulas of the disaccharides maltose and sucrose. Label the hemiacetal, hemiketal, acetal, or ketal carbons.

7.44 Using structures, show why
a. lactose is a reducing sugar.
b. sucrose is not a reducing sugar.

7.45 What type of linkage, for example, α(1 → 4), is present in the following?
a. lactose
b. maltose
c. sucrose

7.46 Which of the following disaccharides are reducing sugars? Explain.

a.

b.

c.

7.47 The disaccharide melibiose is present in some plant juices.

a. What two monosaccharides are formed on hydrolysis of melibiose?
b. Is melibiose a reducing sugar? Explain.
c. Describe the glycosidic linkage between the two monosaccharide units.

7.48 Raffinose is a trisaccharide found in some plants.
  a. What are the three monosaccharides formed on hydrolysis of raffinose?

  b. Is raffinose a reducing sugar? Explain.
  c. What types of glycosidic linkages exist between the three monosaccharide units?

POLYSACCHARIDES (SECTION 7.8)

7.49 Match the following structural characteristics to the polysaccharides amylose, amylopectin, glycogen, and cellulose (a characteristic may fit more than one):

  a. Contains both $\alpha(1 \rightarrow 4)$ and $\alpha(1 \rightarrow 6)$ glycosidic linkages
  b. Is composed of glucose monosaccharide units
  c. Contains acetal linkages between monosaccharide units
  d. Is composed of highly branched molecular chains
  e. Is composed of unbranched molecular chains
  f. Contains only $\alpha(1 \rightarrow 4)$ glycosidic linkages
  g. Contains only $\beta(1 \rightarrow 4)$ glycosidic linkages

7.50 Name a polysaccharide that fits each of the following:
  a. The unbranched polysaccharide in starch
  b. A polysaccharide widely used as a textile fiber
  c. The most abundant polysaccharide in starch
  d. The primary constituent of paper
  e. A storage form of carbohydrates in animals

7.51 Polysaccharides are abundant in celery, and yet celery is a good snack for people on a diet. Explain why.

CHEMISTRY AROUND US AND OVER THE COUNTER

7.52 Name three chemical substances used as artificial sweeteners in the United States.

7.53 Food products containing aspartame (Nutrasweet®) carry a precautionary label. What group of people should avoid using aspartame?

7.54 Explain how eating an abundance of heavily sugared foods might possibly contribute to malnutrition.

7.55 Why is dietary fiber essential to the process of digestion?

7.56 Describe how dietary fiber might play a role in weight control.

> POINTS TO PONDER AND PURSUE

7.1 Suppose human intestinal bacteria were genetically altered so they could hydrolyze the β linkages of cellulose. Would the results be beneficial? Explain.

7.2 A sample of starch is found to have a molecular weight of 2.80×10^5 u. How many glucose units are present in a molecule of the starch?

7.3 The open-chain form of glucose constitutes only a small fraction of any glucose sample. Yet, when Cu^{2+} is used to oxidize the open-chain form, nearly all the glucose in a sample reacts. Use Le Châtelier's principle to explain this observation.

7.4 Aspartame (Nutrasweet®) contains calories and yet is used in diet drinks. Explain how a drink can contain aspartame as a sweetener and yet be low in calories.

7.5 From Figure 7.10, what two products form when the lactose in human milk is hydrolyzed in an infant's digestive system?

7.6 Answer the question raised in Figure 7.16. What foods in addition to potatoes would you expect to give a positive starch test?

7.7 In Web site 7.2, what is meant by the expression "sugar is not free food"?

7.8 According to Web site 7.3, how does a mother benefit from breastfeeding?

7.9 Based upon information at Web site 7.5, why might people with diabetes mellitus be interested in the fiber content of food?

http://www.brookscole.com/chemistry/seager_chemistry

Lipids

A certified surgical technologist (CST), a member of the operating room surgical team, is responsible for the preparation of supplies and equipment used in surgery. In addition, a CST usually scrubs the patients in preparation for surgery and may also assist the surgeon in other ways. Thus, a CST must be familiar with basic surgical procedures and cleansing techniques. One of the topics in this chapter on lipids describes the structural characteristics of soaps.

http://www.brookscole.com/chemistry/ seager_chemistry
At Web site 8.1, you will find information on a career as a surgical technologist.

CHAPTER OUTLINE

- ➤ Classification of Lipids
- ➤ Fatty Acids
- ➤ The Structure of Fats and Oils
- ➤ Chemical Properties of Fats and Oils
- ➤ Waxes
- ➤ Phosphoglycerides
- ➤ Sphingolipids
- ➤ Biological Membranes
- ➤ Steroids
- ➤ Steroid Hormones
- ➤ Prostaglandins

> **FIGURE 8.1** Waxes form a protective coating on these leaves.

Lipid
A biological compound that is soluble only in nonpolar solvents.

Simple lipid
An ester-containing lipid with just two types of components: an alcohol and one or more fatty acids.

Complex lipid
An ester-containing lipid with more than two types of components: an alcohol, fatty acids—plus others.

he group of compounds called lipids is made up of substances with widely different compositions and structures. Unlike the carbohydrates that are defined in terms of structure, lipids are defined in terms of a physical property—solubility. Lipids are biological molecules that are insoluble in water but soluble in nonpolar solvents. Lipids are the waxy, greasy, or oily compounds found in plants and animals (see ■ Figure 8.1 and ■ Figure 8.2). Lipids repel water, a useful characteristic of protective wax coatings found on some plants. Fats and oils are energy-rich and have relatively low densities. These properties account for their use as storage forms of energy in plants and animals. Still other lipids are used as structural components, especially in the formation of cellular membranes.

➤ 8.1 Classification of Lipids

Lipids are grouped into two main classes: saponifiable lipids and non-saponifiable lipids. Saponification refers to the process in which esters are hydrolyzed under basic conditions (Section 5.7). Triglycerides, waxes, phospholipids, and sphingolipids are esters and thus all belong to the first class. Nonsaponifiable lipids are not esters and cannot be hydrolyzed. Steroids and prostaglandins belong to this class. ■ Figure 8.3 summarizes the classification of different types of lipids. Notice from the figure that saponifiable lipids are further classified into categories of simple and complex, based on the number of components in their structure. **Simple lipids** contain just two types of components (fatty acids and an alcohol), whereas **complex lipids** contain more than two (fatty acids, an alcohol, plus other components).

➤ 8.2 Fatty Acids

Prior to discussing the chemical structures and properties of the saponifiable lipids, it is useful to first describe the chemistry of fatty acids, the fundamental building blocks of many lipids. Fatty acids (defined in Section 5.1) are long-

> **FIGURE 8.2** Thick layers of fat help insulate penguins against low temperatures.

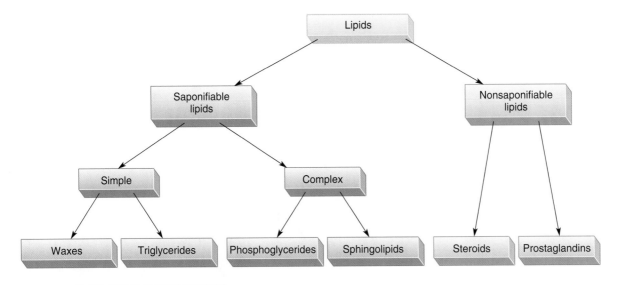

➤ **FIGURE 8.3** The major types of lipids.

chain carboxylic acids, as shown in ■ Figure 8.4. It is the long nonpolar tails of fatty acids that are responsible for most of the fatty or oily characteristics of fats. The carboxyl group, or polar head of fatty acids, is very hydrophilic under conditions of physiological pH, where it exists as the carboxylate anion —COO$^-$.

In aqueous solution, the ions of fatty acids associate with one another in a rather unique way. The ions form spherical clusters, called **micelles,** in which the nonpolar chains extend toward the interior of the structure away from water, and the polar carboxylate groups face outward in contact with the water. Some micelles are large on a molecular scale and contain hundreds or even thousands of fatty acid molecules. The nonpolar chains in a micelle are held together by weak dispersion forces. The structure of a micelle in a radial cross section is shown in ■ Figure 8.5. Micelle formation and structure are important in a number of biological functions, such as the transport of insoluble lipids in the blood (Section 14.1).

The fatty acids found in natural lipids have several characteristics in common:

1. They are usually straight-chain carboxylic acids (no branching).
2. The sizes of most common fatty acids range from 10 to 20 carbons.

Micelle
A spherical cluster of molecules in which the polar portions of the molecules are on the surface and the nonpolar portions are located in the interior.

$$CH_3—CH_2—CH_2—CH_2—CH_2—CH_2—CH_2—CH_2—CH_2—CH_2—CH_2—\overset{\displaystyle O}{\overset{\displaystyle \|}{C}}—OH$$

Nonpolar, hydrophobic tail
(water insoluble)

Polar, hydrophilic head
(water soluble)

(a)

Nonpolar tail   COOH   Polar head

(b)

➤ **FIGURE 8.4** The molecular structure of fatty acids: (a) lauric acid and (b) a simplified diagram of a fatty acid with a nonpolar tail and a polar head.

➤ **FIGURE 8.5** A cross section of a fatty acid micelle.

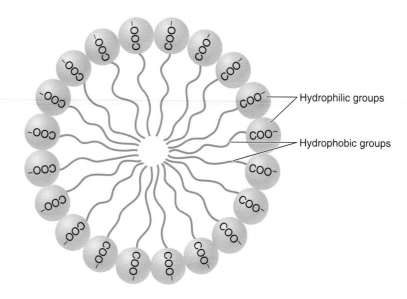

Hydrophilic groups

Hydrophobic groups

3. Fatty acids usually have an even number of carbon atoms (including the carboxyl group carbon).
4. Fatty acids can be saturated (containing no double bonds between carbons) or unsaturated (containing one or more double bonds between carbons). Apart from the carboxyl group and double bonds, there are usually no other functional groups present.

■ Table 8.1 lists some important fatty acids, along with their formulas and melting points.

Unsaturated fatty acids usually contain double bonds in the cis configuration.

$$\begin{array}{ccc} \text{H} & & \text{H} \\ \diagdown & & \diagup \\ \text{C} & = & \text{C} \\ \diagup & & \diagdown \\ \text{long chain} & & \text{long chain} \end{array}$$

This configuration creates a long characteristic bend, or kink, in the fatty acid chain that is not found in saturated fatty acids.

These kinks prevent unsaturated fatty acid chains from packing together as closely as do the chains of saturated acids (see ■ Figure 8.6). As a result, the intermolecular forces are weaker, and unsaturated fatty acids have lower melting points and are usually liquids at room temperature. For example, the melting point of stearic acid (18 carbons) is 71°C, whereas that of oleic acid (18 carbons with one cis double bond) is 13°C. The melting point of linoleic acid (18 carbons and two double bonds) is even lower (−5°C). Chain length also affects the melting point, as illustrated by the fact that palmitic acid (16 carbons) melts at 7°C lower than stearic acid (18 carbons). The presence of double bonds and the length of fatty acid chains in membrane lipids partly explain the fluidity of biological membranes, an important feature discussed in Section 8.8.

The human body can synthesize all except two of the fatty acids it needs. These two, linoleic acid and linolenic acid, are polyunsaturated fatty acids that contain 18 carbon atoms (Table 8.1). Because they are not synthesized within the body and must be obtained from the diet, they are called **essential fatty acids.** Both are widely distributed in plant and fish oils. In the body, both acids are used to produce hormonelike substances that regulate a wide range of functions and characteristics, including blood pressure, blood clotting, blood lipid

Essential fatty acid
A fatty acid needed by the body but not synthesized within the body.

TABLE 8.1	Some important fatty acids				
Compound type and number of carbons	**Name**	**Formula**		**Melting point (°C)**	**Common sources**
Saturated					
14	Myristic acid	$CH_3(CH_2)_{12}-COOH$		54	Butterfat, coconut oil, nutmeg oil
16	Palmitic acid	$CH_3(CH_2)_{14}-COOH$		63	Lard, beef fat, butterfat, cottonseed oil
18	Stearic acid	$CH_3(CH_2)_{16}-COOH$		70	Lard, beef fat, butterfat, cottonseed oil
20	Arachidic acid	$CH_3(CH_2)_{18}-COOH$		76	Peanut oil
Monounsaturated					
16	Palmitoleic acid	$CH_3(CH_2)_5CH=CH(CH_2)_7-COOH$		−1	Cod liver oil, butterfat
18	Oleic acid	$CH_3(CH_2)_7CH=CH(CH_2)_7-COOH$		13	Lard, beef fat, olive oil, peanut oil
Polyunsaturated					
18	Linoleic acid[a]	$CH_3(CH_2)_4(CH=CHCH_2)_2(CH_2)_6-COOH$		−5	Cottonseed oil, soybean oil, corn oil, linseed oil
18	Linolenic acid[b]	$CH_3CH_2(CH=CHCH_2)_3(CH_2)_6-COOH$		−11	Linseed oil, corn oil
20	Arachidonic acid[a]	$CH_3(CH_2)_4(CH=CHCH_2)_4(CH_2)_2-COOH$		−50	Corn oil, linseed oil, animal tissues
20	Eicosapentaenoic acid[b]	$CH_3CH_2(CH=CHCH_2)_5(CH_2)_2-COOH$			Fish oil, seafoods
22	Docosahexaenoic acid[b]	$CH_3CH_2(CH=CHCH_2)_6CH_2-COOH$			Fish oil, seafoods

[a]Omega-6 fatty acid.
[b]Omega-3 fatty acid.

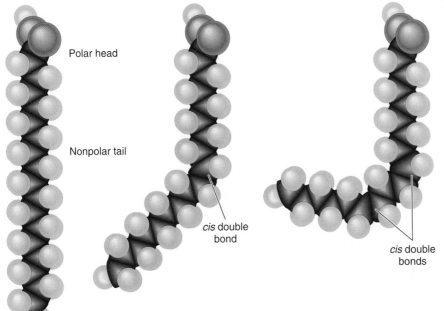

Polar head

Nonpolar tail

cis double bond

cis double bonds

stearic acid
(melting point 71°C)

oleic acid
(melting point 13°C)

linoleic acid
(melting point −5°C)

➤ **FIGURE 8.6** Space-filling models of fatty acids. Unsaturated fatty acids do not pack as tightly as saturated acids, and their melting points are lower than those of saturated acids.

levels, the immune response, and the inflammation response to injury and infection.

Linolenic acid is called an omega-3 fatty acid, which means that the endmost double bond is three carbons from the methyl end of the chain:

$$CH_3-CH_2-(CH=CHCH_2)_3-(CH_2)_6-COOH$$
linolenic acid

In linoleic acid, an omega-6 fatty acid, the endmost double bond is located six carbons from the methyl end of the chain. Both linoleic and linolenic acids can be converted to other omega-3 and omega-6 fatty acids.

Omega-3 fatty acids have been a topic of interest in medicine since 1985. In that year, researchers reported the results of a study involving natives of Greenland. These individuals have a very low death rate from heart disease, even though their diet is very high in fat. Studies led researchers to conclude that the abundance of fish in the diet of the Greenland natives was involved. Continuing studies led to a possible involvement of the omega-3 fatty acids in the oil of the fish.

A recent Harvard University study indicated that male adults who ate just a couple of servings of fish per month were 25% less likely to die of heart disease than those who ate it less than once per month or not at all.

➤ 8.3 The Structure of Fats and Oils

Animal fats and vegetable oils are the most widely occurring lipids. Chemically, fats and oils are both esters, and as we learned in Chapter 5, esters consist of an alcohol portion and an acid portion. In fats and oils, the alcohol portion is always derived from glycerol, and the acid portion is furnished by fatty acids. Because glycerol has three —OH groups, a single molecule of glycerol can be attached to three different acid molecules. An example is the esterification reaction of stearic acid and glycerol:

The diagram below may help you remember the components of a triglyceride.

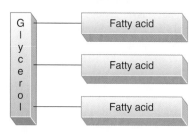

Triglyceride or triacylglycerol
A triester of glycerol in which all three alcohol groups are esterified.

glycerol   stearic acid   glyceryl tristearate
(1 molecule)   (3 molecules)   (a triglyceride—
                       1 molecule)

(8.1)

The resulting esters are called **triglycerides** or **triacylglycerols**.

The fatty acid components in naturally occurring triglyceride molecules are rarely identical (as in the case of glyceryl tristearate). In addition, natural triglycerides (fats and oils) are usually mixtures of different triglyceride molecules. Butterfat, for example, contains at least 14 different fatty acid components.

Learning Check 8.1

Write one structure (several possibilities exist) for a triglyceride derived from stearic acid, oleic acid, and palmitic acid.

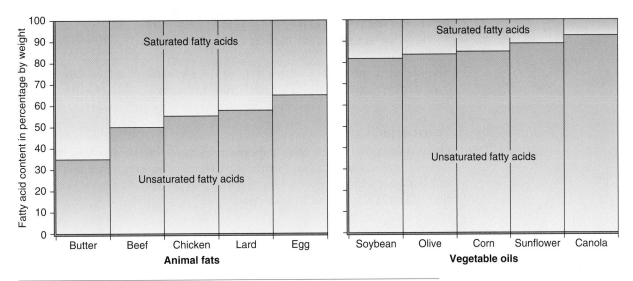

➤ **FIGURE 8.7** A comparison of saturated and unsaturated fatty acids in some foods.

With some exceptions, triglycerides that come from animals other than fish are solids at room temperature and are called **fats**. Triglycerides from plants or fish are liquids at room temperature and are usually referred to as **oils**. Although both fats and oils are triglycerides, they differ structurally in one important respect—the degree of unsaturation. As shown in ■ Figure 8.7, animal fats contain primarily triglycerides of long-chain saturated fatty acids (higher melting points). In contrast, vegetable oils such as corn oil and sunflower oil, consist of triglycerides containing unsaturated fatty acids (lower melting points) (see ■ Figure 8.8). Thus, we see that the structural and physical properties of fatty acids determine the properties of the triglycerides derived from them.

Excessive fat in the diet is recognized as one risk factor influencing the development of chronic disease. The main concern about excessive dietary fat, especially saturated fat, centers on its role in raising blood cholesterol levels (Section 8.9). High blood cholesterol is a recognized risk factor in the development of coronary heart disease, the leading cause of death in Americans every year.

Fat
A triglyceride that is a solid at room temperature.

Oil
A triglyceride that is a liquid at room temperature.

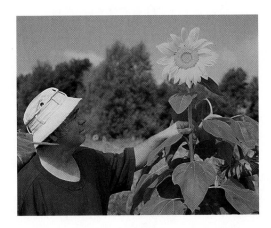

➤ **FIGURE 8.8** Sunflowers are an important crop in some regions of the country because the seeds are a good source of unsaturated oils. Why do you think seeds are rich in oils?

➤ 8.4 Chemical Properties of Fats and Oils

The chemical properties of triglycerides are typical of esters and alkenes because these are the two functional groups present.

Hydrolysis

Hydrolysis is one of the most important reactions of fats and oils, just as it was for carbohydrates. The treatment of fats or oils with water and an acid catalyst causes them to hydrolyze to form glycerol and fatty acids. This reaction is simply the reverse of ester formation:

$$
\begin{array}{l}
CH_2-O-\overset{\overset{\displaystyle O}{\|}}{C}-(CH_2)_{14}CH_3 \\[2mm]
CH-O-\overset{\overset{\displaystyle O}{\|}}{C}-(CH_2)_7CH=CH(CH_2)_7CH_3 \quad\quad + 3H_2O \quad \xrightarrow[\text{or lipase}]{H^+} \\[2mm]
CH_2-O-\overset{\overset{\displaystyle O}{\|}}{C}-(CH_2)_6(CH_2CH=CH)_2(CH_2)_4CH_3
\end{array}
$$

(8.2)

$$
\begin{array}{l}
CH_2-OH \\
CH-OH \quad + \\
CH_2-OH
\end{array}
\left\{
\begin{array}{l}
CH_3(CH_2)_{14}-\overset{\overset{\displaystyle O}{\|}}{C}-OH \\
\text{palmitic acid} \\[2mm]
CH_3(CH_2)_7CH=CH(CH_2)_7-\overset{\overset{\displaystyle O}{\|}}{C}-OH \\
\text{oleic acid} \\[2mm]
CH_3(CH_2)_4(CH=CHCH_2)_2(CH_2)_6-\overset{\overset{\displaystyle O}{\|}}{C}-OH \\
\text{linoleic acid}
\end{array}
\right.
$$

glycerol

Enzymes (lipases) of the digestive system also catalyze the hydrolysis process. This reaction represents the only important change that takes place in fats and oils during digestion. The breakdown of cellular fat deposits to supply energy also begins with the lipase-catalyzed hydrolysis reaction.

Learning Check 8.2

Write the structures of the hydrolysis products of the following reaction:

$$
\begin{array}{l}
CH_2-O-\overset{\overset{\displaystyle O}{\|}}{C}-(CH_2)_7CH=CH(CH_2)_7CH_3 \\[2mm]
CH-O-\overset{\overset{\displaystyle O}{\|}}{C}-(CH_2)_{14}CH_3 \quad\quad\quad + 3H_2O \xrightarrow{H^+} \\[2mm]
CH_2-O-\overset{\overset{\displaystyle O}{\|}}{C}-(CH_2)_{16}CH_3
\end{array}
$$

Saponification

When triglycerides are reacted with a strong base, the process of saponification (soap making) occurs. In this commercially important reaction, the products are glycerol and the salts of fatty acids, which are also called **soaps** (Reaction 8.3).

Soap
A salt of a fatty acid often used as a cleaning agent.

$$
\begin{array}{l}
CH_2-O-\overset{\overset{\displaystyle O}{\|}}{C}-(CH_2)_{14}CH_3 \\[2pt]
CH-O-\overset{\overset{\displaystyle O}{\|}}{C}-(CH_2)_7CH=CH(CH_2)_7CH_3 \; + \; 3NaOH \; \longrightarrow \\[2pt]
\qquad\qquad\qquad\qquad\qquad\qquad\qquad\quad \text{strong base} \\[2pt]
CH_2-O-\overset{\overset{\displaystyle O}{\|}}{C}-(CH_2)_6(CH_2CH=CH)_2(CH_2)_4CH_3
\end{array}
$$

(8.3)

$$
\begin{array}{l}
CH_2-OH \\
CH-OH \\
CH_2-OH \\
\text{glycerol}
\end{array}
\quad + \quad
\left\{
\begin{array}{l}
CH_3(CH_2)_{14}-\overset{\overset{\displaystyle O}{\|}}{C}-O^-Na^+ \\
\quad \text{sodium palmitate} \\
CH_3(CH_2)_7CH=CH(CH_2)_7-\overset{\overset{\displaystyle O}{\|}}{C}-O^-Na^+ \\
\quad \text{sodium oleate} \\
CH_3(CH_2)_4(CH=CHCH_2)_2(CH_2)_6-\overset{\overset{\displaystyle O}{\|}}{C}-O^-Na^+ \\
\quad \text{sodium linoleate}
\end{array}
\right.
$$

soaps

The salts obtained from saponification depend on the base used. Sodium salts, known as hard salts, are found in most cake soap used in the home. Potassium salts, soft soaps, are used in some shaving creams and liquid soap preparations.

In traditional soap making, animal fat is the source of triglycerides, and lye (crude NaOH) or an aqueous extract of wood ashes is the source of the base. The importance of soap making was amply proved after the soap maker's art was lost with the fall of the Roman Empire. The soapless centuries between A.D. 500 and 1500 were notorious for the devastating plagues that nearly depopulated an unsanitary western Europe.

Write structures for the products formed when the following triglyceride is saponified using NaOH:

Learning Check 8.3

$$
\begin{array}{l}
CH_2-O-\overset{\overset{\displaystyle O}{\|}}{C}-(CH_2)_{16}CH_3 \\[4pt]
CH-O-\overset{\overset{\displaystyle O}{\|}}{C}-(CH_2)_7CH=CH(CH_2)_7CH_3 \\[4pt]
CH_2-O-\overset{\overset{\displaystyle O}{\|}}{C}-(CH_2)_7CH=CH(CH_2)_7CH_3
\end{array}
$$

Hydrogenation

In Chapter 2, we learned that double bonds can be reduced to single bonds by treatment with hydrogen (H_2) in the presence of a catalyst. An important commercial reaction of fats and oils uses this same hydrogenation process, in which some of the fatty acid double bonds are converted to single bonds.

> **FIGURE 8.9** Hydrogenation of vegetable oils is used to prepare both shortenings and margarines. How do oils used in automobiles differ chemically from these vegetable oils?

The result is a decrease in the degree of unsaturation and a corresponding increase in the melting point of the fat or oil (Figure 8.6). The peanut oil in many popular brands of peanut butter has been partially hydrogenated to convert the oil to a semisolid that does not separate. Hydrogenation is most often used in the production of semisolid cooking shortenings (such as Crisco®, Spry®, and Fluffo®) or margarines from liquid vegetable oils (see ■ Figure 8.9).

Partially hydrogenated vegetable oils were developed in part to help displace highly saturated animal fats used in frying, baking, and spreads. The use of partially hydrogenated products provides a more healthful alternative to butter, beef tallow, and lard because the former are lower in saturated fatty acids, which are implicated in higher blood cholesterol levels (Section 8.9). It is important not to complete the reaction and totally saturate all the double bonds. If this is done, the product is hard and waxy—not the smooth, creamy product desired by consumers. Equation 8.4 gives an example in which some of the double bonds react, showing one of the possible products.

$$
\begin{array}{l}
CH_2-O-\overset{\displaystyle O}{\overset{\|}{C}}-(CH_2)_{14}CH_3 \\[4pt]
CH-O-\overset{\displaystyle O}{\overset{\|}{C}}-(CH_2)_7CH=CH(CH_2)_7CH_3 \quad +\ 2H_2 \\[4pt]
CH_2-O-\overset{\displaystyle O}{\overset{\|}{C}}-(CH_2)_6(CH_2CH=CH)_2(CH_2)_4CH_3
\end{array}
\xrightarrow{Ni}
\begin{array}{l}
CH_2-O-\overset{\displaystyle O}{\overset{\|}{C}}-(CH_2)_{14}CH_3 \\[4pt]
CH-O-\overset{\displaystyle O}{\overset{\|}{C}}-(CH_2)_7CH=CH(CH_2)_7CH_3 \\[4pt]
CH_2-O-\overset{\displaystyle O}{\overset{\|}{C}}-(CH_2)_6(CH_2CH_2CH_2)_2(CH_2)_4CH_3
\end{array}
\quad (8.4)
$$

Learning Check 8.4

If an oil with the following structure is completely hydrogenated, what is the structure of the product?

$$
\begin{array}{l}
CH_2-O-\overset{\displaystyle O}{\overset{\|}{C}}-(CH_2)_7CH=CH(CH_2)_7CH_3 \\[6pt]
CH-O-\overset{\displaystyle O}{\overset{\|}{C}}-(CH_2)_7CH=CHCH_2CH=CHCH_2CH=CHCH_2CH_3 \\[6pt]
CH_2-O-\overset{\displaystyle O}{\overset{\|}{C}}-(CH_2)_7CH=CHCH_2CH=CH(CH_2)_4CH_3
\end{array}
$$

During hydrogenation, some fatty acid molecules in the common *cis* configuration (Section 8.2) are isomerized into *trans* fatty acids. The main sources of *trans* fatty acids in the diet are stick margarine, shortening, and high-fat baked goods. Some clinical studies have shown that blood cholesterol levels may be raised somewhat by the consumption of *trans* fatty acids. Because *trans* fatty acids account for only a small component of total fat intake, current dietary advice to reduce the consumption of saturated fatty acids far outweighs any cause for alarm about *trans* fatty acids in foods.

➤ 8.5 Waxes

Wax
An ester of a long-chain fatty acid and a long-chain alcohol.

Waxes represent a second group of simple lipids that, like fats and oils, are esters of fatty acids. However, the alcohol portion of waxes is derived from long-chain alcohols (12–32 carbons) rather than glycerol. Beeswax, for example, contains a wax with the following structure:

A REACTION MAP FOR TRIGLYCERIDES

We've discussed three different reactions of triglycerides. Remember, when you're trying to decide what happens in a reaction to focus attention on the functional groups. After identifying a triglyceride as a starting material, next decide what other reagent is involved and which of the three pathways is appropriate.

A wax.

$$CH_3(CH_2)_{14} - \overset{\overset{\textstyle O}{\|}}{C} - O - (CH_2)_{29}CH_3$$

palmitic acid portion   long-chain alcohol portion

Waxes are water-insoluble and not as easily hydrolyzed as the fats and oils; consequently, they often occur in nature as protective coatings on feathers, fur, skin, leaves, and fruits. Sebum, a secretion of the sebaceous glands of the skin, contains many different waxes. It keeps the skin soft and prevents dehydration. Waxes are used commercially to make cosmetics, candles, ointments, and protective polishes.

➤ 8.6 Phosphoglycerides

Phosphoglycerides are complex lipids that serve as the major components of cell membranes. Phosphoglycerides and related compounds are also referred to as **phospholipids**. In these compounds, one of the —OH groups of glycerol is joined by an ester linkage to phosphoric acid, which in turn is linked to another alcohol (usually an amino alcohol). The other two —OH groups of glycerol are linked to fatty acids, resulting in the following general structure:

Phosphoglyceride
A complex lipid containing glycerol, fatty acids, phosphoric acid, and an alcohol component.

Phospholipid
A phosphorus-containing lipid.

A phosphoglyceride.

a phosphoglyceride

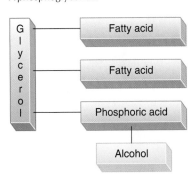

CHEMISTRY AROUND US • 8.1

OLESTRA: FAKE FAT'S BIG TEST

Excess weight is often implicated as a contributing factor in conditions and diseases such as diabetes, high blood pressure, some types of cancer, and atherosclerosis. Dietary guidelines now rank weight control as a more important factor for maintaining good health and avoiding heart attacks than a diet low in fat, saturated fat, and cholesterol.

Fat substitutes offer these benefits without compromising the desirable flavor and texture that fats and oils give foods. The substitutes also contain fewer calories than fats and thus help control body weight. The search for suitable artificial fats has occupied food chemists since the 1960s, but commercial products have only recently become available.

Olestra is a synthetic polyester of sucrose and fatty acids, but unlike either of these constituents, it is indigestible and passes through the digestive tract unabsorbed. Marketed under the trade name Olean®, olestra has the appearance, texture, and taste of natural dietary lipids and can be used in place of natural fats and oils in the diet without providing any calories or significantly changing the flavor of foods.

Olestra is noteworthy because it is the first heat-stable fat substitute and can be used for frying foods. For those who enjoy snacks such as potato chips, the fat and calorie savings are considerable. An ounce of regular potato chips (a vending-machine-sized bagful) contains 10 g of fat and 150 kcal. Olestra-fried chips contain 0 g of fat and 70 kcal, and they taste virtually the same. Olestra may cause gastrointestinal side effects such as diarrhea, and it can rob the body of nutrients, including the fat-soluble vitamins A, D, E, and K, and beta-carotene, which are absorbed as the olestra passes through the intestines and are then excreted. However, the absorption problem has been solved with vitamin supplementation included in Olean.®

http://www.brookscole.com/chemistry/seager_chemistry
At Web site 8.2, you will find a presentation on the properties and use of olestra.

Fat-free foods have become very popular with consumers.

The most abundant phosphoglycerides have one of the alcohols choline, ethanolamine, or serine attached to the phosphate group. These amino alcohols are shown below in charged forms:

$$HO-CH_2CH_2-\overset{+}{N}(CH_3)_3 \qquad HO-CH_2CH_2-\overset{+}{N}H_3 \qquad HO-CH_2CH-\overset{+}{N}H_3$$
$$\qquad\qquad\qquad\qquad\qquad\qquad\qquad\qquad\qquad\qquad\qquad | $$
$$\qquad\qquad\qquad\qquad\qquad\qquad\qquad\qquad\qquad\qquad COO^-$$

| choline (a quaternary ammonium cation) | ethanolamine (cation form) | serine (two ionic groups present) |

Lecithin
A phosphoglyceride containing choline.

A typical phosphoglyceride is phosphatidyl choline, which is commonly called **lecithin:**

$$^1CH_2-O-\overset{O}{\overset{\|}{C}}-(CH_2)_{16}CH_3$$
$$|$$
$$^2CH-O-\overset{O}{\overset{\|}{C}}-(CH_2)_7CH=CHCH_2CH=CH(CH_2)_4CH_3$$
$$|$$
$$^3CH_2-O-\overset{O}{\overset{\|}{P}}-O-CH_2CH_2-\overset{+}{N}(CH_3)_3$$
$$|$$
$$O^-$$

phosphatidylcholine

All phosphoglycerides that contain choline are classified as lecithins. Because different fatty acids may be bonded at positions 1 and 2 of the glycerol, a number of different lecithins are possible. Note that the lecithin shown above contains a negatively charged phosphate group and a positively charged quaternary nitrogen. These charges make that end of the molecule strongly hydrophilic, whereas the rest of the molecule is hydrophobic. This structure of lecithins enables them to function as very important structural components of most cell membranes (Section 8.8). This characteristic also allows lecithins to function as emulsifying and micelle-forming agents. Such micelles play an important role in the transport of lipids in the blood (Chapter 14). Lecithin (phosphatidylcholine) is commercially extracted from soybeans for use as an emulsifying agent in food. It gives a smooth texture to such products as margarine and chocolate candies (see ■ Figure 8.10).

Phosphoglycerides in which the alcohol is ethanolamine or serine, rather than choline, are called **cephalins**. They are found in most cell membranes and are particularly abundant in brain tissue. Cephalins are also found in blood platelets, where they play an important role in the blood-clotting process. A typical cephalin is represented as

Cephalin
A phosphoglyceride containing ethanolamine or serine.

$$
\begin{array}{l}
\text{CH}_2-\text{O}-\overset{\displaystyle \overset{\text{O}}{\|}}{\text{C}}-(\text{CH}_2)_{14}\text{CH}_3 \\[2em]
\text{CH}-\text{O}-\overset{\displaystyle \overset{\text{O}}{\|}}{\text{C}}-(\text{CH}_2)_7\text{CH}=\text{CH}(\text{CH}_2)_7\text{CH}_3 \\[2em]
\text{CH}_2-\text{O}-\overset{\displaystyle \overset{\text{O}}{\|}}{\underset{\displaystyle \text{O}^-}{\text{P}}}-\text{O}-\text{CH}_2\overset{\displaystyle}{\underset{\displaystyle \text{COO}^-}{\text{CH}}}-\overset{+}{\text{NH}_3}
\end{array}
$$

Note the presence of the negatively charged phosphate and carboxyl groups and positively charged nitrogen. This characteristic allows cephalins, like the lecithins, to function as emulsifiers.

Draw a typical structure for a cephalin containing the cation of ethanolamine:

$$\text{HO}-\text{CH}_2\text{CH}_2-\text{NH}_3$$

Learning Check 8.5

➤ 8.7 **Sphingolipids**

Sphingolipid
A complex lipid containing sphingosine, a fatty acid, phosphoric acid, and choline.

Sphingolipids, a second type of complex lipid found in cell membranes, do not contain glycerol. Instead, they contain sphingosine, a long-chain unsaturated amino alcohol. This substance and a typical sphingolipid, called a sphingomyelin, are represented below:

$$CH_3(CH_2)_{12}CH = CH - CH - OH$$
$$| $$
$$CH - NH_2$$
$$|$$
$$CH_2OH$$

sphingosine

A sphingomyelin.

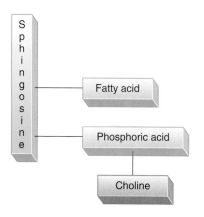

$$CH_3(CH_2)_{12}CH = CH - CH - OH \overbrace{}^{fatty\ acid}$$
$$|\qquad\qquad O$$
$$\qquad\qquad\qquad \|$$
$$CH - NH - C - (CH_2)_7 CH = CH(CH_2)_7 CH_3$$
$$|\qquad\qquad O$$
$$\qquad\qquad \|$$
$$CH_2 - O - P - O - CH_2 CH_2 \overset{+}{N}(CH_3)_3$$
$$phosphate\quad O^-\qquad\qquad choline$$

sphingomyelin

In sphingomyelin, choline is attached to sphingosine through a phosphate group. A single fatty acid is also attached to the sphingosine, but an amide linkage is involved instead of an ester linkage. A number of sphingomyelins are known; they differ only in the fatty acid component. Large amounts of sphingomyelins are found in brain and nerve tissue and in the protective myelin sheath that surrounds nerves.

Glycolipid
A complex lipid containing a sphingosine, a fatty acid, and a carbohydrate.

Glycolipids are another type of sphingolipid. Unlike sphingomyelins, however, these complex lipids contain carbohydrates (usually monosaccharides such as glucose or galactose). Glycolipids are often called cerebrosides because of their abundance in brain tissue. The structure of a typical cerebroside is

A glycolipid.

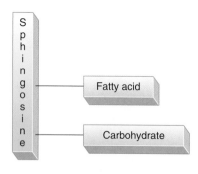

$$CH_3(CH_2)_{12}CH = CH - CH - OH \overbrace{}^{fatty\ acid}$$
$$|\qquad\qquad O$$
$$\qquad\qquad\qquad \|$$
$$CH - NH - C - (CH_2)_7 CH = CH(CH_2)_7 CH_3$$
$$|$$
$$CH_2$$

D-galactose

a cerebroside

Notice that, unlike the sphingomyelins, cerebrosides do not contain phosphate.

Several human diseases are known to result from an abnormal accumulation of sphingomyelins and glycolipids in the body (see ■ Table 8.2). Research has shown that each of these diseases is the result of an inherited absence of an enzyme needed to break down these complex lipids.

TABLE 8.2 Diseases originating from abnormal metabolism and accumulation of glycolipids and sphingomyelins

Disease	Organs(s) affected	Type of lipid accumulated
Tay-Sachs	Brain	Glycolipid (ganglioside G_{M2})
Gaucher's	Spleen, liver	Cerebrosides containing glucose
Niemann-Pick	Several, particularly liver and spleen	Sphingomyelins

➤ 8.8 Biological Membranes

Cell Structure

Two major cell types are found in living organisms: prokaryotic and eukaryotic. **Prokaryotic cells,** comprising bacteria and cyanobacteria (blue-green algae), are smaller and less complex than eukaryotic cells. The more complex **eukaryotic cells** make up the tissues of all other organisms, including humans. Both types of cells are essentially tiny membrane-enclosed units of fluid containing the chemicals involved in life processes. In addition, eukaryotic cells contain small bodies called **organelles,** which are enclosed in membranes and are suspended in the cellular fluid, or cytoplasm. Organelles are the sites of many specialized functions in eukaryotes. The most prominent organelles and their functions are listed in ■ Table 8.3. Membranes perform two vital functions in living organisms: The external cell membrane acts as a selective barrier between the living cell and its environment, and internal membranes surround some organelles, creating cellular compartments that have separate organization and functions.

Prokaryotic cell
A simple unicellular organism that contains no nucleus and no membrane-enclosed organelles.

Eukaryotic cells
A cell containing membrane-enclosed organelles, particularly a nucleus.

Organelle
A membrane-encased body in the interior of a cell.

Membrane Structure

Most cell membranes contain about 60% lipid and 40% protein. Phosphoglycerides (such as lecithin and cephalin), sphingomyelin, and cholesterol are the predominant types of lipids found in most membranes. Precisely how the

TABLE 8.3 The functions of some cellular organelles

Organelle	Function
Endoplasmic reticulum	Synthesis of proteins, lipids, and other substances
Lysosome	Digestion of substances taken into cells
Microfilaments and microtubules	Cellular movements
Mitochondrion	Cellular respiration and energy production
Nucleus	Contains hereditary material (DNA), which directs protein synthesis
Plastids	Contains plant pigments such as chlorophyll (photosynthesis)
Ribosome	Protein synthesis

➤ **FIGURE 8.11** The fluid-mosaic model of membrane structure. Phosphoglycerides are the chief lipid component. They are arranged in a bilayer. Proteins float like icebergs in a sea of lipid.

Fluid-mosaic model

A model of membrane structure in which proteins are embedded in a flexible lipid bilayer.

Lipid bilayer

A structure found in membranes, consisting of two sheets of lipid molecules arranged so that the hydrophobic portions are facing each other.

lipids and proteins are organized to form membranes has been the subject of a great deal of research. A widely accepted model called the **fluid-mosaic model** is diagrammed in ■ Figure 8.11. The lipids are organized in a **bilayer** in which the hydrophobic chains extend toward the inside of the bilayer, and the hydrophilic groups (the phosphate groups and other polar groups) are oriented toward the outside, where they come in contact with water (see ■ Figure 8.12). Like the micelle structure discussed in Section 8.2, a lipid bilayer is

➤ **FIGURE 8.12** The lipid bilayer model. Circles represent the polar portion of phosphoglycerides. The hydrocarbon "tails" are within the interior of the bilayer.

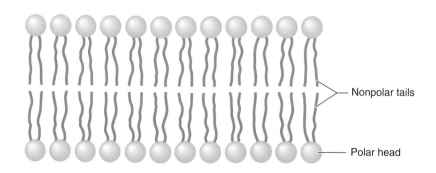

Nonpolar tails

Polar head

a very stable arrangement where "like" associates with "like." The hydrophobic tails of the lipids are protected from water, and the hydrophilic heads are in a position to interact with water. When a lipid bilayer is broken and the interior hydrocarbon tails are exposed to water, the resulting repulsion causes the bilayer to re-form, and the break seals spontaneously.

Membrane lipids usually contain unsaturated fatty acid chains that fit into bilayers more loosely than do saturated fatty acids (Section 8.2). This increases the flexibility or fluidity of the membrane. Some of the proteins in the membrane float in the lipid bilayer like icebergs in the sea, whereas others extend completely through the bilayer. The lipid molecules are free to move laterally within the bilayer like dancers on a crowded dance floor—hence the term "fluid mosaic."

➤ 8.9 Steroids

Steroids exhibit the distinguishing feature of other lipids: They are soluble in nonpolar solvents. Structurally, however, they are completely different from the lipids already discussed. **Steroids** have as their basic structure a set of three six-membered rings and a single five-membered ring fused together:

Steroid
A compound containing four rings fused in a particular pattern.

steroid ring system

Cholesterol

The most abundant steroid in the human body, and the most important, is cholesterol:

cholesterol

Cholesterol is an essential component of cell membranes and is a precursor for other important steroids, including bile salts, male and female sex hormones, vitamin D, and the adrenocorticoid hormones (Section 8.10). Cholesterol for use in these important functions is synthesized in the liver. Additional amounts of cholesterol are present in the foods we eat.

On the other side of the coin, cholesterol has received much attention because of a correlation between its levels in the blood and the disease known as atherosclerosis, or hardening of the arteries (Section 14.1). Although our knowledge of the role played by cholesterol in atherosclerosis is incomplete, it is now considered advisable to reduce the amount of cholesterol in the foods we eat. In addition, reducing the amount of saturated fatty acids in the diet appears to lower cholesterol production by the body.

CHEMISTRY AROUND US • 8.2

LOW CHOLESTEROL AND A HEALTHY HEART

Lowering high blood cholesterol levels in otherwise healthy people dramatically reduces the risk of heart disease and heart attack. Elevated blood cholesterol levels lead to the deposit of cholesterol on the walls of arteries and other blood vessels. In time, these deposits harden, leading to the disease *atherosclerosis*. The resulting narrowed blood vessels and blocked blood flow to the heart may result in heart attacks.

Three major factors that increase the risk of heart disease are high blood cholesterol, high blood pressure, and cigarette smoking. Blood cholesterol levels can be raised by cigarette smoking and two other risk factors: obesity and an inactive lifestyle. The National Cholesterol Education Program has set the following total blood cholesterol risk level guidelines for adults age 20 and older: desirable, less than 200 mg/dL; borderline, 200–239 mg/dL; and high, 240 or more mg/dL.

Changing eating patterns is the best approach for controlling blood cholesterol levels. Eating patterns affect not only dietary cholesterol but also the extent of liver synthesis of cholesterol (Section 14.2). Research shows that an excessive intake of saturated fats tends to increase blood cholesterol levels. The most effective defense against high blood cholesterol is a diet that is low in total fat, saturated fatty acids, and dietary cholesterol. The diet should also be calorically balanced to control weight.

Measuring blood cholesterol levels is a simple procedure.

Bile Salts

Bile is a yellowish brown or green liver secretion that is stored and concentrated in the gallbladder. The entry of partially digested fatty food into the small intestine causes the gallbladder to contract and empty bile into the intestine. Bile does not catalyze any digestive reactions, but it is important in lipid digestion. Fats are more difficult to hydrolyze than starch and proteins because they are not soluble in water. In the watery medium of the digestive system, fats tend to form large globules with a limited surface area exposed to attack by the digestive juices. The chief constituents of bile are the bile salts and two waste components: cholesterol and bile pigments. One of the principal bile salts is sodium glycocholate:

sodium glycocholate

The bile salts act much like soaps; they emulsify the lipids and break the large globules into many smaller droplets. After such action, the total amount of lipid is still the same, but a much larger surface area is available for the hydrolysis reactions. Some researchers feel that bile salts also remove fatty coatings from particles of other types of food and thus aid in their digestion.

A second function of bile salts is to emulsify cholesterol found in the bile. Excess body cholesterol is concentrated in bile and is passed into the small intestine for excretion. If cholesterol levels become too high in the bile, or if the concentration of bile salts is too low, the cholesterol precipitates and forms gallstones (see ■ Figure 8.13). The most common gallstones contain about 80% cholesterol and are colored by entrapped bile pigments.

The passage of a gallstone from the gallbladder down the common bile duct to the intestine causes excruciating pain. Sometimes one or more stones become lodged in the duct and prevent bile from passing into the duodenum, and fats can no longer be digested normally. A person afflicted with this condition feels great pain and becomes quite nauseated and ill. Bile pigments are absorbed into the blood, the skin takes on a yellow coloration, and the stool becomes gray-colored because of lack of excreted bile pigments. If the condition becomes serious, both the gallbladder and the stones can be surgically removed.

➤ 8.10 Steroid Hormones

A number of steroids in the body serve important roles as **hormones**. The two major categories of steroid hormones are the adrenocorticoid hormones and the sex hormones.

Adrenocorticoid Hormones

Adrenal glands are small mounds of tissue located at the top of each kidney. An outer layer of the gland, the adrenal cortex, produces a number of potent steroid hormones, the adrenocorticoids. They are classified into two groups according to function: Mineralocorticoids regulate the concentration of ions (mainly Na^+) in body fluids, and glucocorticoids enhance carbohydrate metabolism.

Cortisol, the major glucocorticoid, functions to increase the glucose and glycogen concentrations in the body. This is accomplished by the conversion of lactate and amino acids from body proteins into glucose and glycogen. The reactions take place in the liver under the influence of cortisol.

cortisol

Cortisol and its ketone derivative, cortisone, exert powerful anti-inflammatory effects in the body. These, or similar synthetic derivatives such as prednisolone, are used to treat inflammatory diseases such as rheumatoid arthritis and bronchial asthma.

cortisone

prednisolone

➤ **FIGURE 8.13** Three gallstones (green) in a gallbladder (red).

http://www.brookscole.com/chemistry/seager_chemistry
At Web site 8.3, you can learn more about gallstones.

Hormone
A chemical messenger secreted by specific glands and carried by the blood to a target tissue, where it triggers a particular response.

OVER THE COUNTER • 8.1

MELATONIN AND DHEA: HORMONES AT YOUR OWN RISK

The search for the fountain of youth continues. Melatonin and dehydroepiandrosterone (DHEA) have both been given the image of wonder drugs with the ability to reduce the chances of contracting several diseases or conditions, including cancer, high blood pressure, Alzheimer's disease, AIDS, and coronary heart disease. In addition, these drugs are popularly believed to improve sleep, increase sexual vitality, and increase longevity. Most of these claims are based on animal studies involving mice, or on short-term human clinical studies. No results of long-term studies about the effects of these compounds are yet available. Research is progressing slowly because both compounds are natural substances that are classified as food supplements and are readily available over the counter. Since the effects or uses for natural substances cannot be patented, there is no economic incentive for expensive research.

The basis for many of the claims made for the compounds relies on the fact that as humans age, smaller amounts of both of them are found in the body. Therefore, some conclude that increasing the levels of melatonin and DHEA in older adults will restore many characteristics of their youthful bodies.

Melatonin is a hormone produced by the pea-sized pineal gland located in the center of the brain. The amount of melatonin released by the pineal gland depends on the time of day. During daylight hours, the rate of release is lower than during the night. Because of this characteristic, it is thought that melatonin levels may help the body determine the time of day. Melatonin influences biological rhythms, such as tiredness in the evening, hormone production, and internal clocks. These natural rhythms have been delayed by melatonin injections administered in the early morning or at noon. Injections given in the late afternoon or early evening have been found to speed up the onset of the rhythms. This effect accounts for the use of melatonin in some sleep-aid products, but it must be taken at the appropriate time of day in order to be effective.

$$CH_3O \quad \text{---} \quad CH_2CH_2 - NH - \overset{\overset{\textstyle O}{\|}}{C} - CH_3$$

melatonin

Melatonin is generally considered safe when taken in low doses for short periods of time. The FDA has not received many complaints about its effects, yet most of the previously mentioned claims have yet to be verified. Therefore, it would not be wise to take melatonin in large doses, or for long periods of time, until more is learned about its effects.

DHEA is not itself a hormone, but a precursor to both testosterone, the male sex hormone, and estrogen, the female sex hormone. The results of research on DHEA have thus far been contradictory. Some short-term studies have shown that DHEA has a tendency to increase resistance to infection, prevent or inhibit some types of cancerous growths, reverse the disease known as lupus, and provide benefits in the treatment of age-related diabetes. However, other studies have indicated that DHEA increases the risk of endometrial cancer in women, and may also increase the incidence of prostate cancer in men.

As yet, not enough long-term studies have been done to completely evaluate the safety and usefulness of DHEA. In fact, the FDA banned the use of DHEA in 1985 because of its side effects. However, the compound was put back on the market as a result of the 1994 Dietary Supplements Health and Education Act. Because of the contradictory nature of the available research results, and the lack of results from long-term studies, people who take either melatonin or DHEA should be aware that they are doing it at their own risk.

http://www.brookscole.com/chemistry/seager_chemistry
Information on the side effects of DHEA is found at Web site 8.4.

A number of companies market melatonin and DHEA.

By far the most important mineralcorticoid is aldosterone, which influences the absorption of Na^+ and Cl^- in kidney tubules. An increase in the level of the hormone results in a corresponding increase in absorption of Na^+ and Cl^-. Because the concentration of Na^+ in tissues influences the retention of water, aldosterone is involved in water balance in the body.

CH$_2$OH
O
‖
C=O
CH

HO
CH$_3$

O

aldosterone

Sex Hormones

The testes and ovaries produce steroids that function as sex hormones. Secondary sex characteristics that appear at puberty develop under the influence of these sex hormones, including a deep voice, beard, and increased muscular development in males, and a higher voice, increased breast size, and lack of facial hair in females.

The testes in males perform two functions: One is to produce sperm; the other is to produce male sex hormones (androgens), the most important of which is testosterone. Testosterone promotes the normal growth of the male genital organs and aids in the development of secondary sex characteristics.

CH$_3$ OH
CH$_3$

O

testosterone

OH
CH$_3$
CH$_3$
CH$_3$

O

methandrostenolone

Growth-promoting (anabolic) steroids, including the male hormone testosterone and its synthetic derivatives such as methandrostenolone (Dianabol), are among the most widely used drugs banned by sports-governing organizations because they are alleged to be dangerous and to confer an unfair advantage to users. These drugs, used by both male and female athletes, promote muscular development without excessive masculinization. Their use is particularly prevalent in sports where strength and muscle mass are advantageous, such as weight lifting, track-and-field events like the shot put and hammer throw, and in body building.

The side effects of anabolic steroids range from acne to deadly liver tumors. Effects on the male reproductive system include testicular atrophy, a decrease in sperm count, and, occasionally, temporary infertility.

The primary female sex hormones are the estrogens estradiol and estrone, and progesterone. Estrogens and progesterone are important in the reproductive process. Estrogens are involved in egg (ovum) development in the ovaries, and progesterone causes changes in the wall of the uterus to prepare it to accept a fertilized egg and maintain the resulting pregnancy.

CH$_3$ OH

HO

estradiol

CH$_3$ O

HO

estrone

CH$_3$
C=O
CH$_3$
CH$_3$

O

progesterone

CHEMISTRY AROUND US • 8.3

ESTROGEN: RISKS AND BENEFITS

The onset of menopause marks the end of a woman's child-bearing years and is caused by a decrease in the body's production of the hormones estrogen and progesterone. A number of physical changes accompany this biological shift, including irregular menstrual periods, forgetfulness, mood swings, hot flashes, sleep disruptions, and reduced sexual desire. After menopause, women also develop an increased risk of osteoporosis and heart disease as time goes on. Many doctors advise women to use *hormone replacement therapy (HRT)* to replace the hormones that are no longer being produced naturally in sufficient amounts. Estrogen is one of the hormones commonly used in HRT.

During a woman's childbearing years, estrogen plays its well-known role of maintaining fertility. In addition, it is a steroid hormone that affects tissues throughout the body. It is thought to decrease blood cholesterol levels and relax blood vessels, thus reducing the risk of heart disease. It also appears to inhibit the clotting of blood, thereby reducing the chance of heart attack caused by coronary artery blood clots. Estrogen also helps build bone and may contribute to the prevention of osteoporosis.

The effects of estrogen therapy are not all positive. Studies have shown that when estrogen is taken alone, the risk of developing endometrial cancer (cancer of the lining of the uterus) greatly increases. Because of this, a synthetic hormone called progestin is usually taken along with estrogen. Progestin helps protect against endometrial cancer by causing a monthly shedding of the uterine lining to take place.

Another drawback to estrogen use is its relationship to breast cancer. In one study, reported in 1977, women who took estrogen for more than 10 years were found to have a 43% greater chance of developing breast cancer than women who did not take it. Because breast cancer tumors need estrogen to grow, estrogen replacement could promote tumor growth. The risk is greatest after about 5 years of estrogen use. The anti-estrogen drug Tamoxifen® can be used to reduce the risk. It blocks the estrogen from reaching tumors, and helps prevent recurrences.

In summary, the short-term use of estrogen in HRT, lasting 2–5 years, relieves the symptoms of menopause without providing the benefits or creating the risks associated with long-term use. Long-term use reduces the risk of death from heart disease, but this life-saving effect begins to decrease at the same time the risk of breast cancer begins to increase. Ultimately, the decision of whether or not to use hormone replacement therapy is a difficult one, and should be considered carefully by each individual woman and her doctor.

http://www.brookscole.com/chemistry/seager_chemistry
Information on the benefits and dangers of estrogen is provided at Web site 8.5.

Products commonly used in hormone-replacement therapy.

➤ 8.11 Prostaglandins

This group of compounds was given its name because the first prostaglandins were identified from the secretions of the male prostate gland. Recent research has identified as many as 20 prostaglandins in a variety of tissues within both males and females.

Prostaglandins are cyclic compounds synthesized in the body from the 20-carbon unsaturated fatty acid arachidonic acid. Prostaglandins are designated by codes that refer to the ring substituents and the number of side-chain dou-

Prostaglandin
A substance derived from unsaturated fatty acids, with hormonelike effects on a number of body tissues.

ble bonds. For example, prostaglandin E_2 (PGE_2) has a carbonyl group on the ring and a side chain containing two double bonds:

arachidonic acid

Prostaglandins are similar to hormones in the sense that they are intimately involved in a host of body processes. Clinically, it has been found that prostaglandins are involved in almost every phase of the reproductive process; they can act to regulate menstruation, prevent conception, and induce uterine contractions during childbirth. Certain prostaglandins stimulate blood clotting. They also lead to inflammation and fever. Recently, it has been found that aspirin inhibits prostaglandin production. This explains in part why aspirin is such a powerful drug in the treatment of inflammatory diseases such as arthritis.

From a medical viewpoint, prostaglandins have enormous therapeutic potential. For example, PGE_2 and PGF_2 induce labor and are used for therapeutic abortion in early pregnancy. PGE_2 in aerosol form is used to treat asthma; it opens up the bronchial tubes by relaxing the surrounding muscles. Other prostaglandins inhibit gastric secretions and are used in treating peptic ulcers. Many researchers believe that when prostaglandins are fully understood, they will be found useful for treating a much wider variety of ailments.

FOR FUTURE REFERENCE   CHOLESTEROL CONTENT OF SELECTED FOODS

Cholesterol, the most abundant steroid in the body, is synthesized in the liver, and is used in the body as an essential component of cell membranes as well as the starting material for the synthesis of numerous important compounds (see Section 8.9). Despite these important functions, cholesterol is regarded by some people as an undesirable component of some foods that should be avoided. High blood cholesterol levels have been identified as a major factor in an increased risk of heart disease (see Chemistry Around Us 8.3). However, most cholesterol circulating in the blood does not come from cholesterol-containing foods; it is synthesized in the liver. Nevertheless, the best defense against high blood cholesterol levels and their associated risk of heart disease includes a diet that is low in cholesterol (see Chemistry Around Us 8.3). Because cholesterol is produced only by animals, foods of animal origin are the only dietary sources. This is illustrated in the chart below, which lists common foods that contain cholesterol.

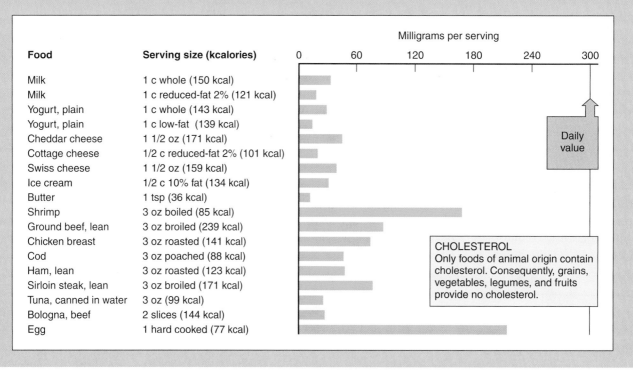

SOURCE: Adapted from Whitney, E. N., Rolfes, S. R. *Understanding Nutrition*, 8th ed. Belmont, CA: West/Wadsworth, 1999.

CONCEPT SUMMARY

Classification of Lipids. Lipids are a family of naturally occurring compounds grouped together on the basis of their relative insolubility in water and solubility in nonpolar solvents. Lipids are classified as saponifiable (ester-containing) or nonsaponifiable. Saponifiable lipids are further classified as simple or complex, depending on the number of structural components.

Fatty Acids. A fatty acid consists of a long nonpolar chain of carbon atoms, with a polar carboxylic acid group at one end. Most natural fatty acids contain an even number of carbon atoms. They may be saturated, unsaturated, or polyunsaturated (containing two or more double bonds).

The Structure of Fats and Oils. Triglycerides or triacylglycerols in the form of fats and oils are the most abundant lipids. Fats and oils are simple lipids that are esters of glycerol and fatty acids. The difference between fats and oils is the melting point, which is essentially a function of the fatty acids in the compound.

Chemical Properties of Fats and Oils. Fats and oils can be hydrolyzed in the presence of acid to produce glycerol and fatty acids. When the hydrolysis reaction is carried out in the presence of a strong base, salts of the fatty acids (soaps) are produced. During hydrogenation, some multiple bonds of unsaturated fatty acids contained in fats or oils are reacted with hydrogen and converted to single bonds.

Waxes. Waxes are simple lipids composed of a fatty acid esterified with a long-chain alcohol. Waxes are insoluble in water and serve as protective coatings in nature.

Phosphoglycerides. Phosphoglycerides consist of glycerol esterified to two fatty acids and phosphoric acid. The phosphoric acid is further esterified to choline (in the lecithins) and to ethanolamine or serine (in the cephalins). The phosphoglycerides are particularly important in membrane formation.

Sphingolipids. These complex lipids contain a backbone of sphingosine rather than glycerol and only one fatty acid component. They are abundant in brain and nerve tissue.

Biological Membranes. Membranes surround tissue cells as a selective barrier, and they encase the organelles found in eukaryotic cells. Membranes contain both proteins and lipids. According to the fluid-mosaic model, the lipids are arranged in a bilayer fashion with the hydrophobic protions on the inside of the bilayer. Proteins float in the bilayer.

Steroids. Steroids are compounds that have four rings fused together in a specific way. The most abundant steroid in humans is cholesterol, which serves as a starting material for other important steroids such as bile salts, adrenocorticoid hormones, and sex hormones.

Steroid Hormones. Hormones are chemical messengers, synthesized by specific glands, that affect various target tissues in the body. The adrenal cortex produces a number of steroid hormones that regulate carbohydrate utilization (the glucocorticoids) and electrolyte balance (the mineralocorticoids). The testes and ovaries produce steroid hormones that determine secondary sex characteristics and regulate the reproductive cycle in females.

Prostaglandins. These compounds are synthesized from the 20-carbon fatty acid arachiodonic acid. They exert many hormone-like effects on the body and are used therapeutically to induce labor, treat asthma, and control gastric secretions.

LEARNING OBJECTIVES

When you have completed your study of this chapter, you should be able to:

1. Classify lipids as saponifiable or nonsaponifiable, and list five major functions of lipids.
2. Describe four general characteristics of fatty acids.
3. Draw structural formulas of triglycerides given the formulas of the component parts.
4. Describe the structural similarities and differences of fats and oils.
5. Write key reactions for the following:

a. Acid- or enzyme-catalyzed hydrolysis of a triglyceride.
b. Saponification of a triglyceride.
c. Hydrogenation of an oil.
6. Draw structural formulas and describe uses for phosphoglycerides and sphingolipids.
7. Identify the structural characteristic typical of steroids, and list three important groups of steroids in the body.
8. Name the major categories of steroid hormones.
9. Describe the biological importance and therapeutic uses of the prostaglandins.

KEY TERMS AND CONCEPTS

Cephalin (8.6)
Complex lipid (8.1)
Essential fatty acid (8.2)
Eukaryotic cell (8.8)
Fat (8.3)
Fluid-mosaic model (8.8)
Glycolipid (8.7)
Hormone (8.10)

Lecithin (8.6)
Lipid (8.1)
Lipid bilayer (8.8)
Micelle (8.2)
Oil (8.3)
Organelle (8.8)
Phosphoglyceride (8.6)
Phospholipid (8.6)

Prokaryotic cell (8.8)
Prostaglandin (8.11)
Simple lipid (8.1)
Soap (8.4)
Sphingolipid (8.7)
Steroid (8.9)
Triglyceride or triacylglycerol (8.3)
Wax (8.5)

KEY REACTIONS

1. Hydrolysis of a triglyceride to glycerol and fatty acids—general reaction (Section 8.4):

$$
\begin{array}{l}
\text{CH}_2-\text{O}-\overset{\displaystyle \overset{\text{O}}{\|}}{\text{C}}-\text{R} \\[2ex]
\text{CH}-\text{O}-\overset{\displaystyle \overset{\text{O}}{\|}}{\text{C}}-\text{R} + 3\text{H}_2\text{O} \xrightarrow[\substack{\text{or} \\ \text{lipase}}]{\text{H}^+} \\[2ex]
\text{CH}_2-\text{O}-\overset{\displaystyle \overset{\text{O}}{\|}}{\text{C}}-\text{R}
\end{array}
\qquad
\begin{array}{l}
\text{CH}_2-\text{OH} \\[2ex]
\text{CH}-\text{OH} + 3\text{R}-\overset{\displaystyle \overset{\text{O}}{\|}}{\text{C}}-\text{OH} \\[2ex]
\text{CH}_2-\text{OH}
\end{array}
$$

2. Saponification of a triglyceride to glycerol and fatty acid salts—general reaction (Section 8.4):

$$
\begin{array}{l}
CH_2\!-\!O\!-\!\overset{\displaystyle O}{\overset{\|}{C}}\!-\!R \\[4pt]
CH\!-\!O\!-\!\overset{\displaystyle O}{\overset{\|}{C}}\!-\!R \;+\; 3NaOH \;\longrightarrow\; CH\!-\!OH \;+\; 3R\!-\!\overset{\displaystyle O}{\overset{\|}{C}}\!-\!O^-Na^+ \\[4pt]
CH_2\!-\!O\!-\!\overset{\displaystyle O}{\overset{\|}{C}}\!-\!R
\end{array}
$$

with glycerol products $CH_2\!-\!OH$, $CH\!-\!OH$, $CH_2\!-\!OH$

3. Hydrogenation of a triglyceride—general reaction (Section 8.4):

$$
\begin{array}{l}
CH_2\!-\!O\!-\!\overset{\displaystyle O}{\overset{\|}{C}}\!-\!R \\[4pt]
CH\!-\!O\!-\!\overset{\displaystyle O}{\overset{\|}{C}}\!-\!(CH_2)_7CH\!=\!CH(CH_2)_7CH_3 \;+\; H_2 \;\xrightarrow{\;Ni\;}\; \\[4pt]
CH_2\!-\!O\!-\!\overset{\displaystyle O}{\overset{\|}{C}}\!-\!R
\end{array}
\qquad
\begin{array}{l}
CH_2\!-\!O\!-\!\overset{\displaystyle O}{\overset{\|}{C}}\!-\!R \\[4pt]
CH\!-\!O\!-\!\overset{\displaystyle O}{\overset{\|}{C}}\!-\!(CH_2)_{16}CH_3 \\[4pt]
CH_2\!-\!O\!-\!\overset{\displaystyle O}{\overset{\|}{C}}\!-\!R
\end{array}
$$

EXERCISES

INTRODUCTION AND CLASSIFICATION OF LIPIDS (SECTION 8.1)

8.1 What is the basis for deciding if a substance is a lipid?

8.2 List two major functions of lipids in the human body.

8.3 What functional group is common to all saponifiable lipids?

8.4 Classify the following as saponifiable or nonsaponifiable lipids:
   a. A steroid   d. A phosphoglyceride
   b. A wax   e. A glycolipid
   c. A triglyceride   f. A prostaglandin

FATTY ACIDS (SECTION 8.2)

8.5 Draw the structure of a typical saturated fatty acid. Label the polar and nonpolar portions of the molecule. Which portion is hydrophobic? Hydrophilic?

8.6 Describe the structure of a micelle formed by the association of fatty acid molecules in water. What forces hold the micelle together?

8.7 Describe four structural characteristics exhibited by most fatty acids.

8.8 Name two essential fatty acids, and explain why they are called essential.

8.9 What structural feature of a fatty acid is responsible for the name omega-3 fatty acid?

THE STRUCTURE OF FATS AND OILS (SECTION 8.3)

8.10 How are fats and oils structurally similar? How are they different?

8.11 Draw the structure of a triglyceride that contains one palmitic acid, one oleic acid, and one stearic acid. Identify the ester bonds.

8.12 From Figure 8.7, arrange the following substances in order of increasing percentage of unsaturated fatty acids: eggs, canola oil, lard, olive oil, butter.

8.13 From what general source do triglycerides tend to have more saturated fatty acids? More unsaturated fatty acids?

8.14 The percentage of fatty acid composition of two triglycerides is reported below. Predict which triglyceride has the lower melting point.

	Palmitic acid	Stearic acid	Oleic acid	Linoleic acid
Triglyceride A	20.4	28.8	38.6	10.2
Triglyceride B	9.6	7.2	27.5	55.7

8.15 Why is the amount of saturated fat in the diet a health concern?

CHEMICAL PROPERTIES OF FATS AND OILS (SECTION 8.4)

8.16 In general terms, name the products of the reactions below:
   a. Acid hydrolysis of a fat
   b. Acid hydrolysis of an oil

c. Saponification of a fat

d. Saponification of an oil

8.17 Write equations for the following reactions using the oil shown below:

$$CH_2-O-\overset{\displaystyle O}{\overset{\|}{C}}-(CH_2)_7CH=CH(CH_2)_7CH_3$$

$$CH-O-\overset{\displaystyle O}{\overset{\|}{C}}-(CH_2)_7CH=CH(CH_2)_7CH_3$$

$$CH_2-O-\overset{\displaystyle O}{\overset{\|}{C}}-(CH_2)_{16}CH_3$$

a. Lipase-catalyzed hydrolysis

b. Saponification

c. Hydrogenation

8.18 What process is used to prepare a number of useful products such as margarines and cooking shortenings from vegetable oils?

8.19 Palmolive soap is mostly sodium palmitate. Write the structure of this compound.

8.20 Write reactions to show how each of the following products might be prepared from a typical triglyceride present in the source given. Use Table 8.1 as an aid:

a. Glycerol from beef fat

b. Stearic acid from beef fat

c. A margarine from corn oil

d. Soaps from lard

WAXES (SECTION 8.5)

8.21 Like fats, waxes are esters of long-chain fatty acids. What structural difference exists between them that warrants placing waxes in a separate category?

8.22 Draw the structure of a wax formed from myristic acid and cetyl alcohol.

8.23 What role do waxes play in nature?

PHOSPHOGLYCERIDES (SECTION 8.6)

8.24 Draw the general block diagram structure of a phosphoglyceride.

8.25 Draw the structure of a phosphoglyceride containing serine:

$$HO-CH_2-CH-\overset{+}{N}H_3$$
$$|$$
$$COO^-$$

Is it a lecithin or cephalin?

8.26 Describe two biological roles served by the lecithins.

8.27 Draw the structure of a lecithin. What structural features make lecithin an important commercial emulsifying agent in certain food products?

8.28 What is the structural difference between a lecithin and a cephalin?

8.29 Where are cephalins found in the human body?

SPHINGOLIPIDS (SECTION 8.7)

8.30 Draw the general block diagram structure of a sphingolipid.

8.31 Describe the structural similarities and differences between the sphingomyelins and the glycolipids.

8.32 Give another name for cerebrosides. In what tissues are they found?

BIOLOGICAL MEMBRANES (SECTION 8.8)

8.33 Where in the body are membranes found? What functions do they serve?

8.34 What three classes of lipids are found in membranes?

8.35 How does the polarity of the phosphoglycerides contribute to their function of forming cell membranes?

8.36 Describe the major features of the fluid-mosaic model of cell membrane structure.

STEROIDS (SECTION 8.9)

8.37 Draw the characteristic chemical structure that applies to all steroid molecules.

8.38 Why is it suggested that some people restrict cholesterol intake in their diet?

8.39 List three important groups of compounds the body synthesizes from cholesterol.

8.40 Explain how bile salts aid in the process of digestion.

8.41 What symptoms might indicate the presence of gallstones in the gallbladder?

8.42 What is the major component in gallstones?

STEROID HORMONES (SECTION 8.10)

8.43 What is a hormone? What are the two major categories of steroid hormones?

8.44 Name the two groups of adrenocorticoid hormones, give a specific example of each group, and explain the function of those compounds in the body.

8.45 How are testosterone and progesterone structurally similar? How are they different?

8.46 Name the primary male sex hormone and the three principal female sex hormones.

8.47 What side effects are associated with the use of anabolic steroids to increase athletic performance?

8.48 What role do the estrogens and progesterone serve in preparation for pregnancy?

PROSTAGLANDINS (SECTION 8.11)

8.49 What structural features characterize the prostaglandins?

8.50 What compound serves as a starting material for prostaglandin synthesis?

8.51 What body processes appear to be regulated in part by prostaglandins?

8.52 Name three therapeutic uses of prostaglandins.

CHEMISTRY AROUND US AND OVER THE COUNTER

8.53 Why is it believed that melatonin may help the body tell what time of day it is?

8.54 What hormones are formed from DHEA?

8.55 Why is estrogen therapy during menopause recommended by some doctors?

8.56 Why is the synthetic hormone progestin usually taken along with estrogen in hormone replacement therapy?

8.57 Why might the use of fat substitutes in certain foods be advantageous?

8.58 What side effects might possibly result from consuming excessive amounts of Olestra-containing foods?

8.59 What are the three major risk factors for heart disease?

8.60 What lifestyle factors result in an increased blood cholesterol level?

> **POINTS TO PONDER AND PURSUE**

8.1 Gasoline is soluble in nonpolar solvents. Why is gasoline not classified as a lipid?

8.2 The structure of cellular membranes is such that ruptures are closed naturally. Describe the molecular forces that cause the closing to occur.

8.3 In Figure 8.7, the five vegetable oils listed are derived from seeds. Why do you think seeds are rich in oils?

8.4 Answer the question asked in Figure 8.9. How do oils used in automobiles differ chemically from vegetable oils?

8.5 Why doesn't honey dissolve beeswax?

8.6 In Figure 8.5 (where the nonpolar tails are together), why is the structure more stable than a structure in which the polar heads are together?

8.7 Using Web site 8.2, explain why Olestra® is calorie free.

8.8 According to Web site 8.3, what segment of the population has the highest prevalence of gallstones?

8.9 According to Web site 8.4, what are common side effects in individuals taking DHEA?

http://www.brookscole.com/chemistry/seager_chemistry

Proteins

A physical therapist assesses the limitations, and potential for improvement, of individuals with physical disabilities, then selects or develops rehabilitation methods. Physical therapists use such methods as body massage, hydrotherapy, heat, light, and ultrasound to help people with sensory or motor deficiencies return to normal. Physical therapists need a basic knowledge of the protein structure of muscles and connective tissue to help them understand human physiology. Protein structure is one of the topics of this chapter.

http://www.brookscole.com/chemistry/seager_chemistry
Web site 9.1 provides information on a career in physical therapy.

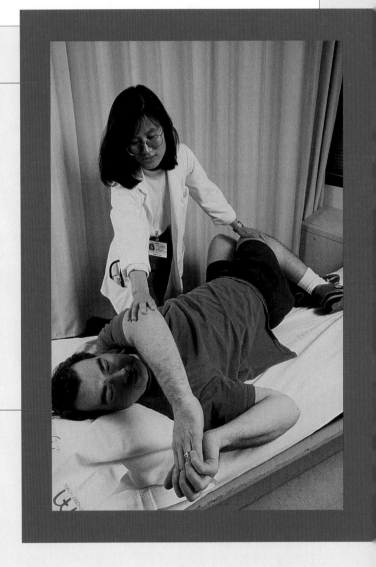

CHAPTER OUTLINE

The name *protein,* coined more than 100 years ago, is derived from the Greek *proteios,* which means "of first importance," and is appropriate for these most important of all biological compounds. Proteins are indispensable components of all living things, where they play crucial roles in all biological processes. In this chapter, we begin with a discussion of the structures and chemistry of the common amino acids, then progress to show how combinations of amino acids form large molecules called peptides and larger molecules called proteins.

➤ 9.1 The Amino Acids

Structure

All proteins, regardless of their origin or biological function, are structurally similar. They are polymers consisting of chains of amino acids chemically bound to each other. As the name implies, an amino acid is an organic compound that contains both an amino group and a carboxylate group in ionic form (Section 9.2). Hundreds of different amino acids, both synthetic and naturally occurring, are known, but only 20 are commonly found in natural proteins. The amino acids in proteins are called **alpha (α)-amino acids** because the amino group is attached to the α carbon (the carbon next to the carboxylate group), as shown in ■ Figure 9.1. Each amino acid has a characteristic side chain, or R group, that imparts chemical individuality to the molecule. These R side chains contain different structural features, such as aromatic rings, —OH groups, —NH$_3^+$ groups, and —COO$^-$ groups. This variety in side chains causes differences in the properties of the individual amino acids and the proteins containing different combinations of them.

The polarity of the R groups is an important characteristic and is the basis for classifying amino acids into the four groups shown in ■ Table 9.1: neutral and nonpolar side chains, neutral but polar side chains, basic side chains, and acidic side chains. Note that acidic side chains contain a carboxylate group and that basic side chains contain an additional amino group. The three-letter and one-letter abbreviations that are used to represent amino acids are given after the names:

Alpha-amino acid

An organic compound containing both an amino group and a carboxylate group, with the amino group attached to the carbon next to the carboxylate group.

The structure of proline differs slightly from the general formula because the amino group and R group are part of a ring.

cysteine (Cys) C

glutamine (Gln) Q

asparagine (Asn) N

➤ **FIGURE 9.1** The general structure of α-amino acids shown in ionic form.

TABLE 9.1 The common amino acids of proteins. Below each amino acid are its name, its three-letter abbreviation, and its one-letter abbreviation.

Neutral, nonpolar side chains

$$H_3\overset{+}{N}-CH-COO^-$$
$$|$$
$$H$$

Glycine (Gly) G

$$H_3\overset{+}{N}-CH-COO^-$$
$$|$$
$$CH_3$$

Alanine (Ala) A

$$H_3\overset{+}{N}-CH-COO^-$$
$$|$$
$$CH$$
$$\diagup \diagdown$$
$$CH_3 \quad CH_3$$

Valine (Val) V

$$H_3\overset{+}{N}-CH-COO^-$$
$$|$$
$$CH_2$$
$$|$$
$$CH$$
$$\diagup \diagdown$$
$$CH_3 \quad CH_3$$

Leucine (Leu) L

$$H_3\overset{+}{N}-CH-COO^-$$
$$|$$
$$CH-CH_3$$
$$|$$
$$CH_2$$
$$|$$
$$CH_3$$

Isoleucine (Ile) I

$$H_3\overset{+}{N}-CH-COO^-$$
$$|$$
$$CH_2$$
(benzene ring)

Phenylalanine (Phe) F

$$H_2\overset{+}{N}-CH-COO^-$$
$$|$$
$$CH_2 \quad CH_2$$
$$\diagdown \diagup$$
$$CH_2$$

Proline (Pro) P

$$H_3\overset{+}{N}-CH-COO^-$$
$$|$$
$$CH_2$$
$$|$$
$$CH_2$$
$$|$$
$$S-CH_3$$

Methionine (Met) M

Neutral, polar side chains

$$H_3\overset{+}{N}-CH-COO^-$$
$$|$$
$$CH_2$$
$$|$$
$$OH$$

Serine (Ser) S

$$H_3\overset{+}{N}-CH-COO^-$$
$$|$$
$$CH-CH_3$$
$$|$$
$$OH$$

Threonine (Thr) T

$$H_3\overset{+}{N}-CH-COO^-$$
$$|$$
$$CH_2$$
(benzene ring)
$$|$$
$$OH$$

Tyrosine (Tyr) Y

$$H_3\overset{+}{N}-CH-COO^-$$
$$|$$
$$CH_2$$
(indole ring)
$$N$$
$$|$$
$$H$$

Tryptophan (Trp) W

$$H_3\overset{+}{N}-CH-COO^-$$
$$|$$
$$CH_2$$
$$|$$
$$SH$$

Cysteine (Cys) C

$$H_3\overset{+}{N}-CH-COO^-$$
$$|$$
$$CH_2$$
$$|$$
$$C=O$$
$$|$$
$$NH_2$$

Asparagine (Asn) N

$$H_3\overset{+}{N}-CH-COO^-$$
$$|$$
$$CH_2$$
$$|$$
$$CH_2$$
$$|$$
$$C=O$$
$$|$$
$$NH_2$$

Glutamine (Gln) Q

Basic, polar side chains

$$H_3\overset{+}{N}-CH-COO^-$$
$$|$$
$$CH_2$$
(imidazole ring)
$$H\overset{+}{N} \quad NH$$

Histidine (His) H

$$H_3\overset{+}{N}-CH-COO^-$$
$$|$$
$$CH_2$$
$$|$$
$$CH_2$$
$$|$$
$$CH_2$$
$$|$$
$$CH_2$$
$$|$$
$$NH_3^+$$

Lysine (Lys) K

$$H_3\overset{+}{N}-CH-COO^-$$
$$|$$
$$CH_2$$
$$|$$
$$CH_2$$
$$|$$
$$CH_2$$
$$|$$
$$NH$$
$$|$$
$$C=NH_2^+$$
$$|$$
$$NH_2$$

Arginine (Arg) R

Acidic, polar side chains

$$H_3\overset{+}{N}-CH-COO^-$$
$$|$$
$$CH_2$$
$$|$$
$$COO^-$$

Aspartate (Asp) D

$$H_3\overset{+}{N}-CH-COO^-$$
$$|$$
$$CH_2$$
$$|$$
$$CH_2$$
$$|$$
$$COO^-$$

Glutamate (Glu) E

Stereochemistry

A glance at the general formula for the 20 amino acids

$$H_3\overset{+}{N}-\underset{\underset{R}{|}}{\overset{\overset{H}{|}}{C}}-COO^-$$

shows that with the exception of glycine, in which the R group is —H, there are four different groups attached to the α carbon. Thus, 19 of the 20 amino acids contain a chiral α carbon atom and can exist as two enantiomers; these are the L and D forms:

$$\overset{+}{N}H_3 \text{ on} \atop \text{the left} \searrow \quad H_3\overset{+}{N}-\underset{\underset{R}{|}}{\overset{\overset{COO^-}{|}}{C}}-H \qquad H-\underset{\underset{R}{|}}{\overset{\overset{COO^-}{|}}{C}}-\overset{+}{N}H_3 \quad \nearrow \overset{+}{N}H_3 \text{ on} \atop \text{the right}$$

an L-amino acid   a D-amino acid

As we saw when we studied carbohydrates, one of the two possible forms usually predominates in natural systems. With few exceptions, the amino acids found in living systems exist in the L form.

► 9.2 Zwitterions

The structural formulas used to this point have shown amino acids in an ionized form. The fact that amino acids are white crystalline solids with relatively high melting points and high water solubilities suggests that they exist in an ionic form. Studies of amino acids confirm that this is true both in the solid state and in solution. The presence of a carboxyl group and a basic amino group in the same molecule makes possible the transfer of a hydrogen ion in a kind of internal acid–base reaction (Reaction 9.1). The product of this reaction is a dipolar ion called a **zwitterion**.

Zwitterion
A dipolar ion that carries both a positive and a negative charge as a result of an internal acid–base reaction in an amino acid molecule.

$$H_2N-\underset{\underset{R}{|}}{\overset{\overset{O}{\|}}{CH}-C}-OH \rightarrow H_3\overset{+}{N}-\underset{\underset{R}{|}}{\overset{\overset{O}{\|}}{CH}-C}-O^- \qquad (9.1)$$

nonionized form   zwitterion
(does not exist)   (present in solids
                           and solutions)

The structure of an amino acid in solution varies with the pH of the solution. For example, if the pH of a solution is lowered by adding a source of H_3O^+, such as hydrochloric acid, the carboxylate group ($—COO^-$) of the zwitterion can pick up a proton to form —COOH (Reaction 9.2):

$$H_3\overset{+}{N}-\underset{\underset{R}{|}}{\overset{\overset{O}{\|}}{CH}-C}-O^- + H_3O^+ \rightarrow H_3\overset{+}{N}-\underset{\underset{R}{|}}{\overset{\overset{O}{\|}}{CH}-C}-OH + H_2O \qquad (9.2)$$

zwitterion
(0 net charge)   (positive net charge)

The zwitterion form has a net charge of 0, but the form in acid solution has a net positive charge.

When the pH of the solution is increased by adding OH^-, the $-NH_3^+$ of the zwitterion can lose a proton, and the zwitterion is converted into a negatively charged form (Reaction 9.3):

$$\overset{+}{H_3N}-CH-\overset{\overset{\displaystyle O}{\|}}{C}-O^- + OH^- \rightarrow H_2N-CH-\overset{\overset{\displaystyle O}{\|}}{C}-O^- + H_2O \qquad (9.3)$$

R	R
zwitterion (0 net charge)	(negative net charge)

Changes in pH also affect acidic and basic side chains and those proteins in which they occur. Thus, all amino acids exist in solution in ionic forms, but the actual form (and charge) is determined by the solution pH.

Because amino acids assume a positively charged form in acidic solutions and a negatively charged form in basic solutions, it seems reasonable to assume that at some pH between the acidic and basic extremes the amino acid will have no net charge (it will be a zwitterion). This assumption is correct, and the pH at which the zwitterion forms is called the **isoelectric point**. Each amino acid has a unique and characteristic isoelectric point; those with neutral R groups are all near a pH of 6 (that of glycine, e.g., is at pH 6.06), those with basic R groups have higher values (that of lysine is at pH 9.47), and the two acidic amino acids have lower values (that of aspartic acid is as pH 2.98). Proteins, which are composed of amino acids, also have characteristic isoelectric points.

The ability of amino acids to react with both H_3O^+ and OH^-, as shown by Reactions 9.2 and 9.3, allows solutions of amino acids or proteins to behave as buffers. The buffering action of blood proteins is one of their most important functions.

Isoelectric point
The characteristic solution pH at which an amino acid has a net charge of 0.

➤ Example 9.1

Draw the structure of the amino acid leucine

$$\overset{+}{H_3N}-CH-\overset{\overset{\displaystyle O}{\|}}{C}-O^-$$
$$|$$
$$CH_2$$
$$|$$
$$CH$$
$$\diagup \ \diagdown$$
$$CH_3 \quad CH_3$$

a. in acidic solution at a pH below the isoelectric point;
b. in basic solution at a pH above the isoelectric point.

Solution

In acidic solution, the carboxylate group of the zwitterion picks up a proton.

$$\overset{+}{H_3N}-CH-\overset{\overset{\displaystyle O}{\|}}{C}-O^- + H_3O^+ \rightarrow \overset{+}{H_3N}-CH-\overset{\overset{\displaystyle O}{\|}}{C}-OH$$
$$|\qquad\qquad\qquad\qquad\qquad |$$
$$CH_2\qquad\qquad\qquad\qquad CH_2$$
$$|\qquad\qquad\qquad\qquad\qquad |$$
$$CH\qquad\qquad\qquad\qquad CH$$
$$\diagup \ \diagdown\qquad\qquad\qquad \diagup \ \diagdown$$
$$CH_3 \quad CH_3\qquad\qquad CH_3 \quad CH_3$$

In basic solution, the $-\overset{+}{N}H_3$ group of the zwitterion loses a proton.

$$H_3\overset{+}{N}-\underset{\underset{\underset{\underset{CH_3 \quad CH_3}{\diagup \quad \diagdown}}{CH}}{\underset{|}{CH_2}}}{CH}-\overset{\overset{O}{\|}}{C}-O^- + OH^- \rightarrow H_2N-\underset{\underset{\underset{\underset{CH_3 \quad CH_3}{\diagup \quad \diagdown}}{CH}}{\underset{|}{CH_2}}}{CH}-\overset{\overset{O}{\|}}{C}-O^- + H_2O$$

Learning Check 9.1

Draw the structure of the amino acid serine

$$H_3\overset{+}{N}-\underset{\underset{|}{CH_2-OH}}{CH}-\overset{\overset{O}{\|}}{C}-O^-$$

a. in acidic solution at a pH below the isoelectric point;
b. in basic solution at a pH above the isoelectric point.

➤ 9.3 Reactions of Amino Acids

Amino acids can undergo reactions characteristic of any functional group in the molecule. This includes reactions of the carboxylate group, the amino group, and any other functional groups attached to the R side chain. However, two reactions are of special interest because of their influence in determining protein structure.

Oxidation of Cysteine

Cysteine, the only sulfhydryl (—SH)-containing amino acid of the 20 commonly found in proteins, has a chemical property not shared by the other 19. We learned in Section 3.9 that a sulfhydryl group can easily be oxidized to form a disulfide bond (—S—S—). Thus, two cysteine molecules react readily to form a disulfide compound called cystine. The disulfide, in turn, is easily converted to —SH groups by the action of reducing agents:

$$\underset{\text{cysteine}}{\begin{array}{c}H_3\overset{+}{N}-CH-\overset{\overset{O}{\|}}{C}-O^- \\ | \\ CH_2 \\ | \\ SH \\ SH \\ | \\ CH_2 \quad O \\ | \quad \| \\ H_3\overset{+}{N}-CH-C-O^-\end{array}} \begin{array}{c}\text{sulfhydryl} \\ \text{group}\end{array} + (O) \xrightarrow{\text{Oxidation}} \underset{\text{cystine}}{\begin{array}{c}H_3\overset{+}{N}-CH-\overset{\overset{O}{\|}}{C}-O^- \\ | \\ CH_2 \\ | \\ S \\ | \quad \text{disulfide} \\ S \quad \text{bond} \\ | \\ CH_2 \quad O \\ | \quad \| \\ H_3\overset{+}{N}-CH-C-O^-\end{array}} + H_2O \quad (9.4)$$

As will be seen, this reaction is important in establishing the structure of some proteins.

OVER THE COUNTER • 9.1

MEDICINES AND NURSING MOTHERS

A concern for all mothers who are breastfeeding a baby should be the effect on the baby of substances, including medicines, that are used by the mother. The effects on breast-fed infants of drugs taken by the mother have been determined for only a limited number of medications. However, because very few problems have been reported, most OTC and prescription drugs that are taken by nursing mothers, as needed and according to directions, are considered to be safe for their infants. For example, a nursing mother who takes daily medication for epilepsy, high blood pressure, or diabetes does not put her infant at risk. However, it is a good precaution to check with a pediatrician to determine whether a specific medication might cause problems. In order to minimize a baby's exposure to any medication taken by the mother, the drug should be taken just after the baby finishes nursing, or just before the baby goes to sleep for an extended time.

The following medications are generally considered to present minimal problems for nursing infants when taken by their mother:

- Acetaminophen
- Antibiotics
- Antiepileptics
  (use Primidone with caution)
- Antihistamines
- Antihypertensives
- Aspirin
  (use with caution)
- Codeine
- Decongestants
- Ibuprofen
- Insulin
- Quinine
- Thyroid medications

The drugs in the following list have been identified as hazardous to a nursing child, and should never be taken by a nursing mother:

- Bromocriptine (Parlodel): This is a drug for Parkinson's disease; it decreases a woman's milk supply.
- Cancer chemotherapy drugs: These drugs kill body cells in the mother and thus could harm the baby as well.
- Ergotamine (for migraine headaches): This drug causes vomiting, diarrhea, and convulsions in infants.
- Lithium (for manic depression): This substance is secreted in the mother's milk.
- Methotrexate (for arthritis): This drug can depress the baby's immune system.
- Tobacco smoke: Nicotine can cause vomiting, diarrhea, and restlessness in a baby, as well as decreased milk production in the mother. There is also an increased risk of sudden infant death syndrome (SIDS) and an increased frequency of respiratory and ear infections in nursing infants of smoking mothers.
- Illegal drugs: Some drugs, such as cocaine and PCP, intoxicate a baby when used by the mother. Others, such as amphetamines, heroin, and marijuana, cause a variety of symptoms in the baby, including irritability, tremors, vomiting, and poor sleeping patterns. Worst of all, the infants become addicted to the drugs.

Peptide Formation

In Section 6.4 we noted that amides could be thought of as being derived from a carboxylic acid and an amine (Reaction 9.5), although in actual practice the acid chlorides or anhydrides are generally used.

$$R-\overset{\overset{O}{\|}}{C}-OH + R'-NH_2 \xrightarrow{Heat} R-\overset{\overset{O}{\|}}{C}-NH-R' + H_2O \quad (9.5)$$

a carboxylic acid   an amine   an amide (*amide linkage*)

Peptide linkage or peptide bond
The amide linkage between amino acids that results when the amino group of one acid reacts with the carboxylate group of another.

In the same hypothetical way, it is possible to envision the carboxylate group of one amino acid reacting with the amino group of a second amino acid. For example, glycine could combine with alanine:

$$H_3\overset{+}{N}-CH_2-\overset{\overset{O}{\|}}{C}-O^- + H_3\overset{+}{N}-\underset{CH_3}{CH}-\overset{\overset{O}{\|}}{C}-O^- \longrightarrow H_3\overset{+}{N}-CH_2-\overset{\overset{O}{\|}}{C}-NH-\underset{CH_3}{CH}-\overset{\overset{O}{\|}}{C}-O^- + H_2O \quad (9.6)$$

glycine   alanine   glycylalanine (a dipeptide) (*peptide linkage*)

Compounds of this type made up of two amino acids are called **dipeptides**, and the amide linkage that holds them together is called a **peptide linkage** or **peptide bond.**

Dipeptide
A compound formed when two amino acids are bonded by an amide linkage.

It is important to realize that a different dipeptide could also form by linking glycine and alanine the other way (alanine on the left in this case):

$$\underset{\text{alanine}}{\overset{+}{H_3N}-\underset{\underset{CH_3}{|}}{CH}-\overset{\overset{O}{\|}}{C}-O^-} + \underset{\text{glycine}}{\overset{+}{H_3N}-CH_2-\overset{\overset{O}{\|}}{C}-O^-} \longrightarrow \underset{\text{alanylglycine}}{\overset{+}{H_3N}-\underset{\underset{CH_3}{|}}{CH}-\overset{\overset{O}{\|}}{C}\overset{\overset{\textit{peptide linkage}}{}}{-}NH-CH_2-\overset{\overset{O}{\|}}{C}-O^-} + H_2O \qquad (9.7)$$

Glycylalanine and alanylglycine are structural isomers of each other, and each has a unique set of properties. The reactions to form these and other dipeptides are more complex than Reactions 9.6 and 9.7 imply. In nature, these processes are carried out by a mechanism we shall examine in Section 11.8.

The presence of amino and carboxylate groups on the ends of dipeptides allows for the attachment of a third amino acid to form a tripeptide. For example, valine could be attached to alanylglycine:

two peptide linkages

$$\overset{+}{H_3N}-\underset{\underset{CH_3}{|}}{CH}-\overset{\overset{O}{\|}}{C}-NH-CH_2-\overset{\overset{O}{\|}}{C}-NH-\underset{\underset{\underset{CH_3}{|}}{CH-CH_3}}{CH}-\overset{\overset{O}{\|}}{C}-O^- + H_2O \qquad (9.8)$$

alanylglycylvaline, a tripeptide

More amino acids can react in the same way to form a tetrapeptide, a pentapeptide, and so on, until a chain of hundreds or even thousands of amino acids has formed. The compounds with shortest chains are often simply called **peptides**, those with longer chains are called **polypeptides**, and those with still longer chains are called **proteins**. Chemists differ about where to draw the lines in the use of these names, but generally polypeptide chains with more than 50 amino acids are called proteins. However, the terms *protein* and *polypeptide* are often used interchangeably.

Amino acids that have been incorporated into chains are called **amino acid residues**. The amino acid residue with an unreacted or free amino group at the end of the chain is designated the **N-terminal residue**, and, according to convention, it is written on the left end of the peptide chain. Similarly, the residue with a free carboxylate group at the other end of the chain is the **C-terminal residue**; it is written on the right end of the chain. Thus, alanine is the N-terminal residue, and valine is the C-terminal residue in the tripeptide alanylglycylvaline:

Peptide
An amino acid polymer of short chain length.

Polypeptide
An amino acid polymer of intermediate chain length containing up to 50 amino acid residues.

Protein
An amino acid polymer made up of more than 50 amino acids.

Amino acid residue
An amino acid that is a part of a peptide, polypeptide, or protein chain.

N-terminal residue
An amino acid on the end of a chain that has an unreacted or free amino group.

C-terminal residue
An amino acid on the end of a chain that has an unreacted or free carboxylate group.

alanine, the N-terminal residue
$$\overset{+}{H_3N}-\underset{\underset{CH_3}{|}}{CH}-\overset{\overset{O}{\|}}{C}-NH-CH_2-\overset{\overset{O}{\|}}{C}-NH-\underset{\underset{\underset{CH_3}{|}}{CH-CH_3}}{CH}-\overset{\overset{O}{\|}}{C}-O^-$$
valine, the C-terminal residue

alanylglycylvaline

Peptides are named by starting at the N-terminal end of the chain and listing the amino acid residues in order from left to right. Structural formulas and even full names for large peptides become very unwieldy and time-consuming to write. This is simplified by representing peptide and protein structures in terms of three-letter abbreviations for the amino acid residues with dashes to show peptide linkages. Thus, the structure of glycylalanine is represented by Gly-Ala. The isomeric dipeptide, alanylglycine, is Ala-Gly. In this case, alanine is the N-terminal residue, and glycine is the C-terminal residue.

The tripeptide alanylglycylvaline can be represented by Ala-Gly-Val. In addition to this sequence, five other arrangements of these three components are possible, and each one represents an isomeric tripeptide with different properties. These sequences are Ala-Val-Gly, Val-Ala-Gly, Val-Gly-Ala, Gly-Val-Ala, and Gly-Ala-Val.

Insulin, with 51 amino acids, has 1.55×10^{66} sequences possible! From these possibilities, the body reliably produces only one, which illustrates the remarkable precision of the life process. From the simplest bacterium to the human brain cell, only the amino acid sequences needed to form essential cell peptides are produced.

Use Table 9.1 to draw the full structure of the following tetrapeptide. Label the N-terminal and C-terminal residues.

Phe-Cys-Ser-Ile

Learning Check 9.2

➤ 9.4 Important Peptides

More than 200 peptides have been isolated and identified as essential to the proper functioning of the human body. The identity and sequence of amino acid residues in these peptides is extremely important to proper physiological function. Two well-known examples are vasopressin and oxytocin, which are hormones released by the pituitary gland. Each hormone is a nonapeptide (nine amino acid residues) with six of the residues held in the form of a loop by a disulfide bond (see ■ Figure 9.2). The disulfide bond in these hormones is formed by the oxidation of cysteine residues in the first and sixth positions (counting from the N-terminal end) of these peptide chains. In peptides and proteins, disulfide bonds such as these that draw a single peptide chain into a loop or hold two peptide chains together are called **disulfide bridges**.

Figure 9.2 shows that the structures of vasopressin and oxytocin differ only in the third and eighth amino acid residues of their peptide chains. The result of these variations is a significant difference in their biological functions.

Disulfide bridge
A bond produced by the oxidation of —SH groups on two cysteine residues. The bond loops or holds two peptide chains together.

➤ **FIGURE 9.2** The structures of vasopressin and oxytocin. Differences are in amino acid residues 3 and 8. The normal carboxylate groups of the C-terminal residues (Gly) have been replaced by amide groups.

CHEMISTRY AROUND US • 9.1

PRIONS AND MAD COW DISEASE

A central principle of biology states that agents capable of transmitting diseases, especially infectious ones, always contain some nucleic acid. However, this idea was challenged during the 1980s when a few fatal diseases did not seem to follow the pattern. Two such diseases are scrapies and kuru, both of which affect the central nervous system of mammals, turning their brains into pitted sponges and causing dementia. These diseases progress very slowly, leading some scientists to continue to blame the diseases on "slow viruses." However, most scientists today believe these diseases are caused by *prions*, an acronym derived from the words "protein infectious particle."

Prions are transmissible agents that do not lose their potency when subjected to materials and conditions that destroy nucleic acids, a fact that suggests prions may consist only of proteins. This suggestion is supported by the observation that materials and conditions that change the shape of only proteins decrease the potency of scrapies and some other neurological diseases. Other evidence suggests that the proteins that make up prions are relatively small. In 1998, Stanley Prusiner won the Nobel Prize in medicine for his work on prions.

Scrapies, a disorder of sheep and goats, causes severe enough irritation that the animals scrape off their wool or hair. Kuru is a fatal brain disease that was found among the Fore people of Papua New Guinea. This degenerative malady was caused by their practice of eating the brains of their dead. Victims of the disease lost their coordination and quickly became demented. When the brain-eating ritual was given up, kuru was essentially wiped out.

A prion has been identified as the cause of bovine spongiform encephalopathy (BSE), the so-called mad cow disease in cattle. Victims of the disease become agitated, lose motor coordination, and are seriously disoriented. In early 1996, British health authorities suspected BSE to be the cause of a new form of Creutzfeldt-Jacob disease that affected younger people who had unknowingly eaten beef obtained from cattle with mad cow disease. Suspicions arose because Creutzfeldt-Jacob disease appears to be inherited in some cases, and normally does not appear in people under age 60. When these suspicions were reported to the public, Europe went into a panic, and the sale of British beef was banned in many countries. British beef has not been imported into the United States since 1985, and no cases of BSE have been reported in the U.S.

How do prions function in the body to cause these diseases? The prion protein (PrP) is a variation of a protein made in the cells of many vertebrates and seems to be necessary for all organisms. The only difference between normal prion protein (PrP^C) and the protein of the abnormal prion that causes scrapies (PrP^{SC}) is the shape. The α-helices found in some regions of PrP^C are transformed into β-pleated sheets in those same regions of PrP^{SC}. This change in shape causes the PrP^{SC} protein to be resistant to protein-digesting enzymes, and therefore capable of surviving the digestive process when they are ingested. Some researchers believe the abnormal proteins get into the lymph system and are delivered to target tissues, where they change normal prion protein (PrP^C) into abnormal prion protein (PrP^{SC}). Thus, these proteins transmit diseases by changing the conformation of normal proteins.

Prion
A protein infectious particle capable of causing disease.

Vasopressin is known as antidiuretic hormone (ADH) because of its action in reducing the volume of urine formed, thus conserving the body's water. It also raises blood pressure. Oxytocin causes the smooth muscles of the uterus to contract and is often administered to induce labor. Oxytocin also acts on the smooth muscles of lactating mammary glands to stimulate milk ejection.

Another important peptide hormone is adrenocorticotropic hormone (ACTH). It is synthesized by the pituitary gland and consists of a single polypeptide chain of 39 amino acid residues with no disulfide bridges (see ■ Figure 9.3). The major function of ACTH is to regulate the production of steroid hormones in the cortex of the adrenal gland. ■ Table 9.2 lists these and some other important peptide or protein hormones.

N-terminal Ser Tyr Ser Met Glu His Phe Arg Trp Gly Lys Pro Val Gly Lys Lys Arg Arg Pro

Val

C-terminal Phe Glu Leu Pro Phe Ala Glu Ala Ser Gln Asp Glu Gly Ala Asp Pro Tyr Val Lys

➤ **FIGURE 9.3** The amino acid sequence of ACTH.

TABLE 9.2 Examples of peptide and protein hormones

Name	Origin	Action
Adrenocorticotropic hormone (ACTH)	Pituitary	Stimulates production of adrenal hormones
Angiotensin II	Blood plasma	Causes blood vessels to constrict
Follicle-stimulating hormone (FSH)	Pituitary	Stimulates sperm production and follicle maturation
Gastrin	Stomach	Stimulates stomach to secrete acid
Glucagon	Pancreas	Stimulates glycogen metabolism in liver
Human growth hormone (HGH)	Pituitary	General effects; bone growth
Insulin	Pancreas	Controls metabolism of carbohydrates
Oxytocin	Pituitary	Stimulates contraction of uterus and other smooth muscles
Prolactin	Pituitary	Stimulates lactation
Somatostatin	Hypothalamus	Inhibits production of HGH
Vasopressin	Pituitary	Decreases volume of urine excreted

➤ 9.5 Characteristics of Proteins

Size

Proteins are extremely large natural polymers of amino acids with molecular weights that vary from about 6000 to several million u. The immense size of proteins (in the molecular sense) can be appreciated by comparing glucose with hemoglobin, a relatively small protein. Glucose has a molecular weight of 180 u, whereas the molecular weight of hemoglobin is 65,000 u (see ■ Table 9.3). The molecular formula of glucose is $C_6H_{12}O_6$ and that of hemoglobin is $C_{2952}H_{4664}O_{832}N_{812}S_8Fe_4$.

TABLE 9.3 Molecular weights of some common proteins

Protein	Molecular weight (u)	Number of amino acid residues
Insulin	6,000	51
Cytochrome c	16,000	104
Growth hormone	49,000	191
Hemoglobin	65,000	574
Hexokinase	96,000	730
Gamma globulin	176,000	1320
Myosin	800,000	6100

Protein molecules are too large to pass through cell membranes and are contained inside the normal cells where they were formed. However, if cells become damaged by disease or trauma, the protein contents can leak out. Thus, persistent excessive amounts of proteins in the urine are indicative of damaged kidney cells. A routine urinalysis usually includes a test for protein. Similarly, a heart attack can be confirmed by the presence in the blood of certain proteins (enzymes) that are normally confined to cells in heart tissue (Section 10.9).

Acid–Base Properties

The 20 different R groups of amino acids are important factors in determining the physical and chemical properties of proteins that contain the amino acids. Acid–base behavior is one of the most important of these properties. Proteins, like amino acids, take the form of zwitterions. The R groups of glutamate and aspartate contain $—COO^-$ groups and provide H_3O^+ ions, whereas the R groups of arginine and histidine are basic (see Table 9.1). Thus, proteins, like amino acids, have characteristic isoelectric points and can behave as buffers in solutions.

The tendency for large molecules such as proteins to remain in solution or form a stable colloidal dispersion depends to a large extent on the repulsive forces acting between molecules with like charges on their surfaces. When protein molecules are at a pH where they take forms that have a net positive or negative charge, the presence of these like charges causes the molecules to repel one another and remain dispersed. These repulsive forces are smallest at the isoelectric point, when the net molecular charges are essentially zero. Thus, protein molecules tend to clump together and precipitate from solutions in which the pH is equal to or close to the isoelectric point.

Protein Function

A large part of the importance of proteins results from the crucial roles they serve in all biological processes. The various functions of proteins are discussed below and summarized in ■ Table 9.4.

1. **Catalytic function:** Nearly all the reactions that take place in living organisms are catalyzed by proteins functioning as enzymes. Without these catalytic proteins, biological reactions would take place too slowly to support life. Enzyme-catalyzed reactions range from relatively simple processes to the very complex, such as the duplication of hereditary material in cell nuclei. Enzymes are discussed in more detail in the next chapter.
2. **Structural function:** We saw in Section 7.9 that the main structural material for plants is cellulose. In the animal kingdom, structural materials other than the inorganic components of the skeleton are composed of protein. Collagen, a fiberlike protein, is responsible for the mechanical strength of skin and bone. Keratin, the chief constituent of hair, skin, and fingernails, is another example.
3. **Storage function:** Some proteins provide a way to store small molecules or ions. Ovalbumin, for example, is a stored form of amino acids that is used by embryos developing in bird eggs. Casein, a milk protein, and gliadin, in wheat seeds, are also stored forms of protein intended to nourish animals and plants, respectively. Ferritin, a liver protein, attaches to iron ions and forms a storage complex in humans and other animals.
4. **Protective function:** **Antibodies** are tremendously important proteins that protect the body from disease. These highly specific proteins combine with and help destroy viruses, bacteria, and other foreign substances, called **antigens,** that get into the blood or tissue of the body. Blood clotting,

Antibody
A substance that helps protect the body from invasion by foreign materials known as antigens.

Antigen
A foreign substance that can stimulate the production of antibodies.

TABLE 9.4   Biological functions of proteins

Function	Examples	Occurrence or role
Catalysis	lactate dehydrogenase	Oxidizes lactic acid
	cytochrome c	Transfers electrons
	DNA polymerase	Replicates and repairs DNA
Structure	viral-coat proteins	Sheath around nucleic acid of viruses
	glycoproteins	Cell coats and walls
	α-keratin	Skin, hair, feathers, nails, and hooves
	β-keratin	Silk of cocoons and spider webs
	collagen	Fibrous connective tissue
	elastin	Elastic connective tissue
Storage	ovalbumin	Egg-white protein
	casein	A milk protein
	ferritin	Stores iron in the spleen
	gliadin	Stores amino acids in wheat
	zein	Stores amino acids in corn
Protection	antibodies	Form complexes with foreign proteins
	complement	Complexes with some antigen-antibody systems
	fibrinogen	Involved in blood clotting
	thrombin	Involved in blood clotting
Regulation	insulin	Regulates glucose metabolism
	growth hormone	Stimulates growth of bone
Nerve impulse transmission	rhodopsin	Involved in vision
	acetylcholine receptor protein	Impulse transmission in nerve cells
Movement	myosin	Thick filaments in muscle fiber
	actin	Thin filaments in muscle fiber
	dynein	Movement of cilia and flagella
Transport	hemoglobin	Transports O_2 in blood
	myoglobin	Transports O_2 in muscle cells
	serum albumin	Transports fatty acids in blood
	transferrin	Transports iron in blood
	ceruloplasmin	Transports copper in blood

another protective process, is carried out by the proteins thrombin and fibrinogen. Without this process, even small wounds would result in life-threatening bleeding.

5. **Regulatory function:** Numerous body processes are regulated by hormones, many of which are proteins. Examples are growth hormone, which regulates the growth rate of young animals, and thyrotropin, which stimulates the activity of the thyroid gland.

6. **Nerve impulse transmission:** Some proteins behave as receptors of small molecules that pass between gaps (synapses) separating nerve cells. In this way, they transmit nerve impulses from one nerve to another. Rhodopsin, a protein found in the rod cells of the retina of the eye, functions this way in the vision process.

7. **Movement function:** Every time we climb stairs, push a button, or blink an eye, we use muscles that have proteins as their major components. The proteins actin and myosin are particularly important in processes involving

CHEMISTRY AROUND US • 9.2

ALZHEIMER'S DISEASE

Alzheimer's disease is a progressive, degenerative disease that attacks the brain, causing impaired thinking, behavior, and memory. As the disease progresses, nerve cells in the brain degenerate, and the size of the brain actually decreases. About 4 million Americans suffer from this malady; it afflicts 1 in 10 people over age 65, and nearly half of all people over age 85. The disease, first identified by Alois Alzheimer in 1906, is best described by the words of one of its victims: "I have lost myself."

Some warning signs of Alzheimer's disease include difficulty in performing familiar tasks, language problems, disorientation related to time or place, poor or decreased quality of judgment, memory loss affecting job skills, problems with abstract thinking, misplacing things or putting them in inappropriate places, passivity and loss of initiative, and changes in mood, personality, or behavior. If several of these symptoms are present in an older person, a medical evaluation for Alzheimer's disease is in order.

The causes of the disease are under debate. Some researchers feel it is caused by the formation in the brain of fibrils (strands) and plaque deposits consisting largely of amyloid beta-protein. Others think the formation of tangles in a microtubule-associated protein called Tau is the culprit. Still others think both of these explanations are correct, and that the formation of abnormal amyloid beta strands causes the abnormal killer forms of the Tau protein to form. This position is supported by a recent study of the brains of aging rhesus monkeys. This study showed that amyloid beta-protein caused the same damage to the monkey brains that was found in the brains of Alzheimer's

victims. In addition, the amyloid beta-protein caused changes in the Tau protein of the monkey brains that made it capable of forming tangles.

Currently, two drugs have been approved for use in treating Alzheimer's disease, Tacrine and Aricept, which treat some of the associated memory loss. Both drugs function by inhibiting the reuptake of acetylcholine in neural synapses. Unfortunately, they neither cure the disease nor slow its progress.

Current research is focused on preventing or reversing the process of changing normal amyloid protein into the deformed amyloid protein found in the brains of Alzheimer's patients. As is true of prion diseases (Chemistry Around Us 9.1), the α-helices of normal proteins is changed to β-pleated sheets in the abnormal proteins characteristic of the disease. In a recent promising study, an experimental peptide containing five amino acids was shown to mimic a region of amyloid beta-protein that regulates folding. However, the experimental peptide is so constructed that it prevents other amino acids of amyloid beta-protein from folding into the β-pleated sheet conformation necessary to form amyloid beta-protein fibrils. In cell culture studies, the experimental peptide inhibited the formation of amyloid fibrils, broke down existing fibrils, and prevented the death of neurons. This type of promising discovery may eventually lead to a drug that can stop the progress—or even reverse the effects—of Alzheimer's disease.

http://www.brookscole.com/chemistry/seager_chemistry
More information about Alzheimer's disease is found at Web site 9.2.

---

movement. They are long-filament proteins that slide along each other during muscle contraction.

8. **Transport function:** Numerous small molecules and ions are transported effectively through the body only after binding to proteins. For example, fatty acids are carried between fat (adipose) tissue and other tissues or organs by serum albumin, a blood protein. Hemoglobin, a well-known example, carries oxygen from the lungs to other body tissues, and transferrin is a carrier of iron in blood plasma.

Proteins perform other functions, but these are among the most important. It is easy to see that properly functioning cells of living organisms must contain many proteins. It has been estimated that a typical human cell contains 9000 different proteins, and that a human body contains about 100,000 different proteins.

Classification

Fibrous protein
A protein made up of long rod-shaped or stringlike molecules that intertwine to form fibers.

Proteins can be classified by the functions just discussed. They can also be classified into two major types, fibrous and globular, on the basis of their structural shape. **Fibrous proteins** are made up of long rod-shaped or stringlike molecules that can intertwine with one another and form strong fibers. They are water-insoluble and are usually found as major components of connective tissue, elastic tissue, hair, and skin. Examples are collagen, elastin, and keratin (see ■ Figure 9.4).

TABLE 9.5 Conjugated proteins

Class	Prosthetic groups	Examples
Nucleoproteins	nucleic acids (DNA, RNA)	Viruses
Lipoproteins	lipids	Fibrin in blood, serum lipoproteins
Glycoproteins	carbohydrates	Gamma globulin in blood, mucin in saliva
Phosphoproteins	phosphate groups	Casein in milk
Hemoproteins	heme	Hemoglobin, myoglobin, cytochromes
Metalloproteins	iron zinc	Ferritin, hemoglobin Alcohol dehydrogenase

▶ **FIGURE 9.4** Human hair, animal fur, and spiderwebs are constructed of the fibrous proteins α-keratin and β-keratin, respectively. What properties of fibrous proteins make hair, fur, and spiderwebs useful?

Globular proteins are more spherical and either dissolve in water or form stable suspensions in water. They are not found in structural tissue but are transport proteins, or proteins that must be moved easily through the body by the circulatory system. Examples of these are hemoglobin and enzymes.

Another classification scheme for proteins is based on composition; proteins are either simple or conjugated. **Simple proteins** are those that contain only amino acid residues. **Conjugated proteins** contain amino acid residues and other organic or inorganic components. The non–amino acid parts of conjugated proteins are called **prosthetic groups**. The type of prosthetic group present is often used to further classify conjugated proteins. For example, proteins containing lipids, carbohydrates, or metal ions are called lipoproteins, glycoproteins, and metalloproteins, respectively. Thus, the iron-containing proteins hemoglobin and myoglobin are metalloproteins. ■ Table 9.5 lists the common classes of conjugated proteins.

Globular protein
A spherical-shaped protein that usually forms stable suspensions in water or dissolves in water.

Simple protein
A protein made up entirely of amino acid residues.

Conjugated protein
A protein made up of amino acid residues and other organic or inorganic components.

Prosthetic group
The non–amino acid parts of conjugated proteins.

Protein Structure

As you might expect, the structures of extremely large molecules, such as proteins, are much more complex than those of simple organic compounds. Many protein molecules consist of a chain of amino acids twisted and folded into a complex three-dimensional structure. This structural complexity imparts unique features to proteins that allow them to function in the diverse ways required by biological systems. To better understand protein function, we must look at four levels of organization in their structure. These levels are referred to as primary, secondary, tertiary, and quaternary.

Primary protein structure
The linear sequence of amino acid residues in a protein chain.

▶ 9.6 The Primary Structure of Proteins

http://www.brookscole.com/chemistry/ seager_chemistry
At Web site 9.3, you can learn more about primary protein structure.

Every protein, be it hemoglobin, keratin, or insulin, has a similar backbone of carbon and nitrogen atoms held together by peptide bonds:

$$-NH-CH-\overset{\overset{\displaystyle O}{\|}}{C}-NH-CH-\overset{\overset{\displaystyle O}{\|}}{C}-NH-CH-\overset{\overset{\displaystyle O}{\|}}{C}-NH-CH-\overset{\overset{\displaystyle O}{\|}}{C}-NH-CH-\overset{\overset{\displaystyle O}{\|}}{C}-$$
$$\quad\ \ R \qquad\qquad R' \qquad\qquad R'' \qquad\qquad R''' \qquad\qquad R''''$$

protein backbone (in color)

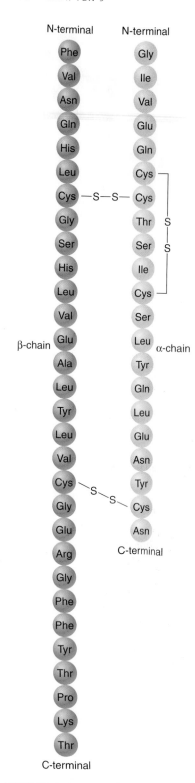

N-terminal

Phe
Val
Asn
Gln
His
Leu
Cys —S—S— Cys
Gly
Ser
His
Leu
Val
β-chain Glu
Ala
Leu
Tyr
Leu
Val
Cys —S—S—
Gly
Glu
Arg
Gly
Phe
Phe
Tyr
Thr
Pro
Lys
Thr

C-terminal

N-terminal

Gly
Ile
Val
Glu
Gln
Cys
Cys
Thr
Ser
Ile
Cys
Ser
Leu α-chain
Tyr
Gln
Leu
Glu
Asn
Tyr
Cys
Asn

C-terminal

> **FIGURE 9.5** The amino acid sequence (primary structure) of human insulin.

http://www.brookscole.com/chemistry/seager_chemistry
Web site 9.4 provides more information about Linus Pauling.

The difference in proteins is the length of the backbone and the sequence of the side chains (R groups) that are attached to the backbone. The order in which amino acid residues are linked together in a protein is called the **primary structure.** Each different protein in a biological organism has a unique sequence of amino acid residues. It is this sequence that causes a protein chain to fold and curl into the distinctive shape that, in turn, enables the protein to function properly. As we will see in Section 9.10, a protein molecule that loses its characteristic three-dimensional shape cannot function.

Biochemists have devised techniques for finding the order in which the residues are linked together in protein chains. A few of the proteins whose primary structures are known are listed in ■ Table 9.6.

■ Figure 9.5 shows the primary structure of human insulin. Notice that the individual molecules of this hormone consist of two chains having a total of 51 amino acid residues. In the molecule, two disulfide bridges hold the chains together, and there is one disulfide linkage within a chain. Insulin plays an essential role in regulating the use of glucose by cells. Inadequate production of insulin leads to diabetes mellitus, and people with severe diabetes must take insulin shots (see Chemistry Around Us 13.3).

As we saw in the case of vasopressin and oxytocin (Section 9.4), small changes in amino acid sequence can cause profound differences in the functioning of proteins. For example, a minor change in sequence in the blood protein hemoglobin causes the fatal sickle-cell disease.

➤ 9.7 The Secondary Structure of Proteins

If the only structural characteristics of proteins were their amino acid sequences (primary structures), all protein molecules would consist of long chains arranged in random fashion. However, protein chains fold and become aligned in such a way that certain orderly patterns result. These orderly patterns, referred to as **secondary structures,** result from hydrogen bonding and include the α-helix (alpha-helix) and the β-pleated (beta-pleated) sheet.

The α-Helix

In 1951, American chemists Linus Pauling and Robert Corey suggested that proteins could exist in the shape of an **α-helix,** a form in which a single protein chain twists so that it resembles a coiled helical spring (see ■ Figure 9.6). The chain is held in the helical shape by numerous intramolecular hydrogen bonds between carbonyl oxygens and amide hydrogens in adjacent turns of the helical backbone:

$$-C=O \ \cdots \ H-N-$$

The carbonyl group of each amino acid residue is hydrogen-bonded to the amide hydrogen of the amino acid four residues away in the chain; thus, all amide groups in the helix are hydrogen-bonded (Figure 9.6). The protein backbone forms the coil, and the side chains (R groups) extend outward from the coil.

Six years after Pauling and Corey proposed the α-helix, its presence in proteins was detected. Since that time it has been found that the amount of α-helical content is quite variable in proteins. In some it is the major struc-

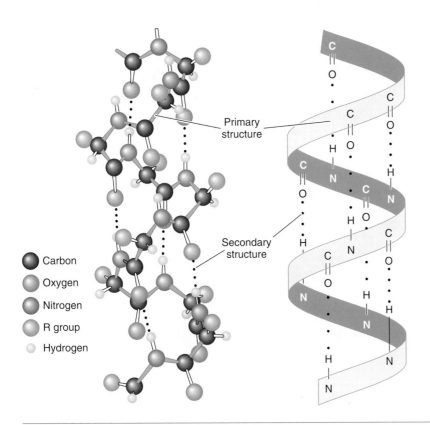

Primary structure

Secondary structure

- Carbon
- Oxygen
- Nitrogen
- R group
- Hydrogen

➤ **FIGURE 9.6** Two representations of the α-helix, showing hydrogen bonds between amide groups.

TABLE 9.6	Some proteins whose sequence of amino acids is known

Protein	Number of amino acid residues
Insulin	51
Hemoglobin	
α-chain	141
β-chain	146
Myoglobin	153
Human growth hormone	188
Trypsinogen	229
Carboxypeptidase A	307
Gamma globulin	1320

tural component, whereas in others a random coil predominates and there is little or no α-helix coiling.

In the proteins α-keratin (found in hair), myosin (found in muscle), epidermin (found in skin), and fibrin (found in blood clots), two or more helices interact (supracoiling) to form a cable (see ■ Figure 9.7). These cables make up bundles of fibers that lend strength to the tissue in which they are found.

The β-Pleated Sheet

A less common type of secondary structure is found in some proteins. This type, called a **β-pleated sheet**, results when several protein chains lie side-by-side and are held in position by hydrogen bonds between the amide carbonyl oxygens of one chain and the amide hydrogens of an adjacent chain (see ■ Figure 9.8).

The β-pleated sheet is found extensively only in the protein of silk. However, a number of proteins are known in which a single polypeptide chain forms sections of a β-pleated sheet structure by folding back on itself. ■ Figure 9.9 depicts a protein chain exhibiting the β-pleated sheet, the α-helix, and random structural conformations. Note the position of the hydrogen bonds in the two secondary structures.

Secondary protein structure
The arrangement of protein chains into patterns as a result of hydrogen bonds between amide groups of amino acid residues in the chain. The common secondary structures are the α-helix and the β-pleated sheet.

α-helix
The helical structure in proteins that is maintained by hydrogen bonds.

β-pleated sheet
A secondary protein structure in which protein chains are aligned side-by-side in a sheetlike array held together by hydrogen bonds.

➤ **FIGURE 9.7** The supracoiling of three α-helices in the keratins of hair and wool.

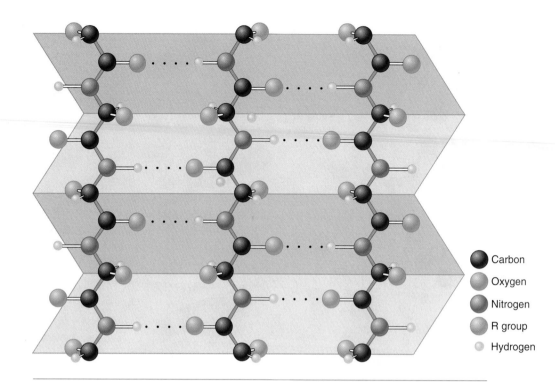

Carbon
Oxygen
Nitrogen
R group
Hydrogen

➤ **FIGURE 9.8** The β-pleated sheet. The four dots show hydrogen bonds between adjacent protein chains.

➤ 9.8 The Tertiary Structure of Proteins

Tertiary protein structure
A specific three-dimensional shape of a protein resulting from interactions between R groups of the amino acid residues in the protein.

Tertiary structure, the next higher level of complexity in protein structure, refers to the bending and folding of the protein into a specific three-dimensional shape. This bending and folding may seem rather disorganized, but nevertheless it results in a favored arrangement for a given protein.

The tertiary structure results from interactions between the R side chains of the amino acid residues. These R-group interactions are of four types:

1. **Disulfide bridges:** As in the structure of insulin (Figure 9.5), a disulfide linkage can form between two cysteine residues that are close to each other in the same chain or between cysteine residues in different chains. The existence and location of these disulfide bridges is a part of the ter-

➤ **FIGURE 9.9** A segment of a protein showing areas of α-helical, β-pleated sheet, and random coil molecular structure.

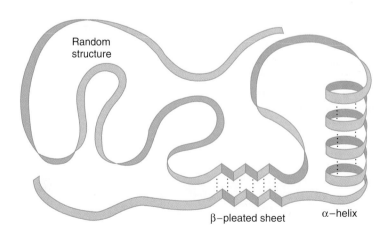

Random structure

β–pleated sheet

α–helix

CHEMISTRY AROUND US • 9.3

SICKLE-CELL DISEASE

Hemoglobin is composed of four polypeptide chains: two α-chains, each containing 141 amino acid residues, and two β-chains, each containing 146 amino acid residues (Figure 9.12). Hemoglobin is located in red blood cells and performs the life-sustaining function of transporting oxygen from the lungs to the cells and carbon dioxide from the cells to the lungs. In some individuals, the hemoglobin molecules have a slightly different sequence of amino acid residues. It is the so-called abnormal human hemoglobins that have attracted particular attention because of the diseases associated with them. One of the best-known abnormal hemoglobins (designated HbS) differs from the normal type (HbA) by only one amino acid residue in each of the two β-chains. In HbS, the glutamate in the sixth position of normal HbA is replaced by a valine residue:

	4	5	6	7	8	9
Normal HbA	—Thr—	Pro—	Glu—	Glu—	Lys—	Ala—
Sickle-cell HbS	—Thr—	Pro—	Val—	Glu—	Lys—	Ala—

This difference affects only two positions in a molecule containing 574 amino acid residues, and yet it is enough to result in a very serious disease, sickle-cell disease.

When combined with oxygen, the red blood cells of people with sickle-cell disease have the flat, disklike shape of normal red blood cells. However, after they give up their oxygen, red blood cells containing HbS hemoglobin become distorted into a characteristic sickle shape. These sickled cells tend to clump together and wedge in capillaries, particularly in the spleen, and cause excruciating pain. Cells blocking capillaries are rapidly destroyed, and the loss of red blood cells causes anemia. Children with sickle-cell disease have an 80% lower chance of surviving to adulthood than unafflicted children.

Sickle-cell anemia is a genetic disease that can be identified by screening a blood sample for abnormal hemoglobin. A person with an HbS gene from one parent and a normal HbA gene from another parent (a heterozygote) is said to have sickle-cell trait.

About 40% of the hemoglobin in these individuals is HbS, but there are generally no ill effects. A person with HbS genes from both parents (a homozygote) is said to have sickle-cell disease, and all of his or her hemoglobin is HbS. Laboratory screening tests to separate HbA and HbS proteins can easily determine whether prospective parents carry the sickle-cell trait.

There is an intriguing relationship between the sickle-cell trait and malaria resistance. In some parts of the world, up to 20% of the population has the sickle-cell trait. The trait produces an increased resistance to one type of malaria because the malarial parasite cannot feed on the hemoglobin in sickled red blood cells. Individuals with sickle-cell disease die young, while those without the sickle-cell trait also have a high probability of dying prematurely, but from malaria. Occupying the middle ground are those with the sickle-cell trait who do not suffer much from the effects of sickle-cell disease and also avoid the ravages of malaria.

http://www.brookscole.com/chemistry/seager_chemistry
Web site 9.5 has a public health information sheet on sickle-cell disease.

Scanning electron micrograph of normal red blood cells (round shape) in the midst of sickled cells (crescent shape).

tiary structure because these interactions hold the protein chain in a loop or some other three-dimensional shape.

2. **Salt bridges:** These interactions are a result of ionic bonds that form between the ionized side chain of an acidic amino acid ($-COO^-$) and the side chain of a basic amino acid ($-NH_3^+$).

3. **Hydrogen bonds:** Hydrogen bonds can form between a variety of side chains, especially those that possess the following functional groups:

$$-OH \qquad -NH_2 \qquad \overset{\displaystyle O}{\overset{\displaystyle \|}{-C}}-NH_2$$

One possible interaction, between the —OH groups of two serine residues, is shown in ■ Figure 9.10. We saw in Section 9.7 that hydrogen bonding also determines the secondary structure or proteins. The distinction

➤ **FIGURE 9.10** R-group interactions leading to tertiary protein structure.

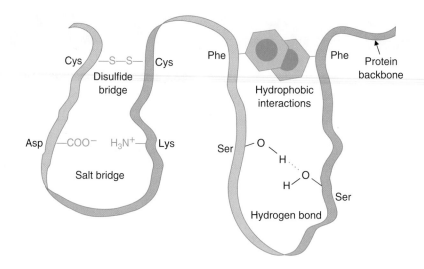

is that the tertiary hydrogen bonding occurs between R groups, whereas in secondary structures, the hydrogen bonding is between backbone —C=O and —N groups.

4. **Hydrophobic interactions:** These result when nonpolar groups are either attracted to one another or forced together by their mutual repulsion of aqueous solvent. Interactions of this type are common between R groups such as the nonpolar phenyl rings of phenylalanine residues (Figure 9.10). This is illustrated by the compact structure of globular proteins in aqueous solutions (see ■ Figure 9.11). Much like in the fatty acid micelles in Section 8.2, the globular shape results because polar groups are pointed outward toward the aqueous solvent, and nonpolar groups are pointed inward, away from the water molecules. This type of interaction is weaker than the other three, but it usually acts over large surface areas so that the net effect is an interaction strong enough to stabilize the tertiary structure.

The interactions leading to tertiary protein structures are summarized in Figure 9.10.

➤ **FIGURE 9.11** A globular protein with a hydrophobic region on the inside and polar groups on the outside extending into the aqueous surroundings.

VISUALIZING PROTEIN STRUCTURE

A visual aid to help you "see" protein structure and remember the difference between primary, secondary, and tertiary structure is the cord of a telephone receiver. The chainlike primary structure is represented by the long straight cord. The coiling of the cord into a helical arrangement represents the secondary structure, and the tangled arrangement the cord adopts after the receiver has been used several times is the tertiary structure.

Primary structure

Secondary structure

Tertiary structure

Three levels of structure of a telephone cord.

What kind of R-group interaction might be expected if the following side chains were in close proximity?

Learning Check 9.3

a.
$$
\begin{array}{cc}
\text{CH}_3 & \text{CH}_3 \\
| & | \\
-\text{CHCH}_2\text{CH}_3 \quad \text{and} \quad -\text{CHCH}_2\text{CH}_3
\end{array}
$$

b.
$$
-\overset{\overset{\textstyle O}{\|}}{\text{C}}-\text{NH}_2 \quad \text{and} \quad \langle\bigcirc\rangle-\text{OH}
$$

c.
$$
-\text{CH}_2-\overset{\overset{\textstyle O}{\|}}{\text{C}}-\text{O}^- \quad \text{and} \quad -\text{CH}_2\text{CH}_2\text{CH}_2\text{CH}_2-\text{NH}_3^+
$$

▶ 9.9 The Quaternary Structure of Proteins

Many functional proteins contain two or more polypeptide chains held together by forces such as ionic attractions, disulfide bridges, hydrogen bonds, and hydrophobic forces. Each of the polypeptide chains, called **subunits**, has its own primary, secondary, and tertiary structure. The arrangement of subunits to form the larger protein is called the **quaternary structure** of the protein.

Hemoglobin is a well-known example of a protein exhibiting quaternary structure. Hemoglobin is made of four chains (subunits): two identical chains (called alpha) containing 141 amino acid residues each and two other identical chains (beta) containing 146 residues each. Each of these four hemoglobin subunits contains a heme group (a planar ring structure centered around an iron atom, shown in ■ Figure 9.12b) located in crevices near the

Subunit

A polypeptide chain having primary, secondary, and tertiary structural features that are a part of a larger protein.

Quaternary protein structure

The arrangement of subunits that form a larger protein.

➤ **FIGURE 9.12** The structure of hemoglobin. (a) Hemoglobin exhibits quaternary structure with two α-chains and two β-chains. The purple disks are heme groups. (b) A heme group.

(a)

(b)

exterior of the molecule. The hemoglobin molecule is nearly spherical with the four subunits held together rather tightly by hydrophobic forces. A schematic drawing on the quaternary structure of hemoglobin is shown in Figure 9.12a.

➤ 9.10 Protein Hydrolysis and Denaturation

Hydrolysis

We learned in Section 6.8 that amides can be hydrolyzed by aqueous acid or base. The amide linkages of peptides and proteins also show this characteristic. As a result, the heating of a peptide or protein in the presence of acid or base causes it to break into smaller peptides, or even amino acids, depending on the hydrolysis time, temperature, and pH.

| Helical proteins in solution | Denatured proteins | Precipitated proteins |

➤ **FIGURE 9.13** Protein denaturation and coagulation of the proteins.

$$\text{protein} + H_2O \xrightarrow[OH^-]{H^+ \text{ or}} \text{smaller peptides} \xrightarrow[OH^-]{H^+ \text{ or}} \text{amino acids} \qquad (9.9)$$

The process of protein digestion (Section 12.6) involves hydrolysis reactions that are catalyzed by enzymes in the digestive tract. In Chapter 14, we'll also see that the hydrolysis of cellular proteins to amino acids is an ongoing process as the body resynthesizes needed molecules and tissues.

Denaturation

Proteins are maintained in their natural three-dimensional conformation, or **native state,** by stable secondary and tertiary structures and through the aggregation of subunits in quaternary structures. If proteins are exposed to physical or chemical conditions, such as extreme temperatures or pH values, that disrupt these stabilizing structures, the folded native structure breaks down and the protein takes on a random, disorganized conformation (see ■ Figure 9.13). This process, called **denaturation,** causes the protein to become inactive, and it may precipitate because hydrophobic residues are likely to be exposed to the aqueous surroundings. The change of egg white from a clear, jellylike material to a white solid when heated is an example of this process (see ■ Figure 9.14). Denaturation also explains why most proteins are biologically active only over a narrow temperature range, typically 0°C–40°C. Temperatures higher than those normal for a living organism can cause intramolecular vibrations to become so intense that hydrogen bonds or other stabilizing forces are disrupted and denaturation results.

The correlation between denaturation and loss of biological activity is clearly illustrated by the common methods used to kill microorganisms and deactivate their toxins. Most bacterial toxins, which are proteins, lose their ability to cause disease when they are heated. For example, the toxic protein excreted by *Clostridium botulinum* is responsible for botulism, an often fatal form of food poisoning. This dangerous substance loses its toxic properties after being heated at 100°C for a few minutes. Heating also deactivates the toxins of the bacteria that cause diphtheria and tetanus. Heat denaturation is used to prepare vaccines against these particular diseases. The deactivated (denatured) toxin can no longer cause the symptoms of the disease to appear,

Native state
The natural three-dimensional conformation of a functional protein.

Denaturation
The process by which a protein loses its characteristic native structure and function.

➤ **FIGURE 9.14** Denaturing the protein in eggs.

but they can stimulate the body to produce substances that induce immunity to the toxins. It is well known that high temperatures have been used for many years in hospitals and laboratories to sterilize surgical instruments and glassware that must be germ-free. The high temperatures denature and deactivate proteins needed within the germ cells for survival.

Similar changes in protein conformation resulting in denaturation can be brought about by heavy-metal ions such as Hg^{2+}, Ag^+, and Pb^{2+} that interact with —SH groups and carboxylate (—COO⁻) groups. Organic materials containing mercury (mercurochrome and merthiolate) were once commonly used as antiseptics in the home to treat minor skin abrasions because they denature proteins in bacteria. Mercury, silver, and lead salts are also poisonous to humans for the same reason: They deactivate critical proteins we need to live. Large doses of raw egg white and milk are often given to victims of heavy-metal poisoning. The proteins in the egg and milk bind to the metal ions and form a precipitate. The metal-containing precipitate is removed from the stomach by pumping or is ejected by induced vomiting. If the precipitated protein is not removed, it will be digested, and the toxic metal ions will be released again.

Some of the various substances and conditions that will denature proteins are summarized in ■ Table 9.7.

TABLE 9.7 Substances and conditions that denature proteins

Substance or condition	Effect on proteins
Heat and ultraviolet light	Disrupt hydrogen bonds and ionic attractions by making molecules vibrate too violently; produce coagulation, as in cooking an egg
Organic solvents (ethanol and others miscible with water)	Disrupt hydrogen bonds in proteins and probably form new ones with the proteins
Strong acids or bases	Disrupt hydrogen bonds and ionic attractions; prolonged exposure results in hydrolysis of protein
Detergents	Disrupt hydrogen bonds, hydrophobic interactions, and ionic attractions
Heavy metal ions (Hg^{2+}, Ag^+, Pb^{2+})	Form bonds to thiol groups and precipitate proteins as insoluble heavy-metal salts

Most of the iron in the body is found as a part of two proteins: hemoglobin in red blood cells, and myoglobin in muscle cells. In both of these forms, iron helps accept, transport, and then release oxygen to the cells of the body. Inside the cells, other iron-containing proteins are involved in the oxidation of nutrients by oxygen to yield the energy-rich compound ATP, which provides energy for use by the cells. Iron is also required by enzymes that catalyze the synthesis of amino acids, hormones, and neurotransmitters.

The typical American diet provides about 6–7 mg of iron per 1000 kcal. Most men eat more than 2000 kcal a day and obtain more than the RDA for men of 10 mg per day. However, most premenopausal women do not consume enough kilocalories to satisfy their RDA of 15 mg per day unless they include the most iron-rich foods in every food group at every meal. The chart below lists selected foods that contain a minimum of 1.5 mg of iron per serving.

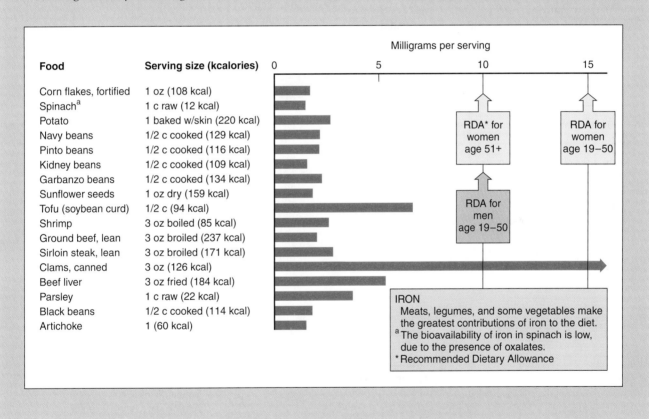

SOURCE: Adapted from Whitney, E. N.; Rolfes, S. R. *Understanding Nutrition*, 8th ed. Belmont, CA: West & Wadsworth, 1999.

CONCEPT SUMMARY

The Amino Acids. All proteins are polymers of amino acids, which are bifunctional organic compounds that contain both an amino group and a carboxylate group. Differences in the R groups of amino acids cause differences in the properties of amino acids and proteins.

Zwitterions. The presence of both amino groups and carboxyl groups in amino acids makes it possible for amino acids to exist in several ionic forms, including the form of a zwitterion. The zwitterion is a dipolar form in which the net charge on the ion is zero.

Reactions of Amino Acids. Amino acids can undergo reactions characteristic of any functional group in the molecule. Two important reactions are the reaction of two cysteine molecules to form a disulfide, and the reaction of amino groups and carboxylate groups of different molecules to form peptide (amide) linkages.

Important Peptides. More than 200 peptides have been shown to be essential to the proper functioning of the human body. Hormones and enkephalins are among the peptides for which functions have been identified.

Characteristics of Proteins. Proteins are large polymers of amino acids. Acidic and basic properties of proteins are determined by the acidic or basic character of the R groups of the amino acids constituting the protein. Proteins perform numerous important functions in the body. Proteins are classified structurally as fibrous or globular. They are classified on the basis of composition as simple or conjugated.

The Primary Structure of Proteins. The primary structure of a protein is the sequence of amino acids in the polymeric chain. This gives all proteins an identical backbone of carbon and nitrogen atoms held together by peptide linkages. The difference in proteins is the sequence of R groups attached to the backbone.

The Secondary Structure of Proteins. Protein chains are held in characteristic shapes called secondary structures by hydrogen bonds. Two specific structures that have been identified are the α-helix and the β-pleated sheet.

The Tertiary Structure of Proteins. A third level of complexity in protein structure results from interactions between the R groups of protein chains. These interactions include disulfide bridges, salt bridges, hydrogen bonds, and hydrophobic attractions.

The Quaternary Structure of Proteins. Some functional proteins consist of two or more polypeptide chains held together by forces such as ionic attractions, disulfide bridges, hydrogen bonds, and hydrophobic forces. The arrangement of these polypeptides to form the functional protein is called the quaternary structure of the protein.

Protein Hydrolysis and Denaturation. The peptide (amide) linkages of peptides and proteins can be hydrolyzed under appropriate conditions. This destroys the primary structure and produces smaller peptides or amino acids. The characteristic secondary, tertiary, and quaternary structures of proteins can also be disrupted by certain physical or chemical conditions such as extreme temperatures or pH values. The disruption of these structures is called denaturation and causes the protein to become nonfunctional and, in some cases, to precipitate.

LEARNING OBJECTIVES

When you have completed your study of this chapter, you should be able to:

1. Identify the characteristic parts of alpha-amino acids and classify the R groups according to polarity and acid–base character.
2. Draw structural formulas to illustrate the various ionic forms assumed by amino acids.
3. Write reactions to represent the oxidation of cysteine to give the disulfide cystine and the formation of peptides and proteins.
4. Correctly represent peptide and protein structures using three-letter abbreviations for amino acids.
5. Describe proteins in terms of the following characteristics: size, function, classification as fibrous or globular, and classification as simple or conjugated.
6. Explain what is meant by the primary, secondary, tertiary, and quaternary structure of proteins.
7. Describe the role of hydrogen bonding in the secondary structure of proteins and side-chain interactions in the tertiary structure of proteins.
8. Describe the conditions that can cause proteins to hydrolyze or become denatured.

KEY TERMS AND CONCEPTS

Alpha-amino acid (9.1)
α-helix (9.7)
Amino acid residue (9.3)
Antibody (9.5)
Antigen (9.5)
β-pleated sheet (9.7)
Conjugated protein (9.5)
C-terminal residue (9.3)
Denaturation (9.10)
Dipeptide (9.3)

Disulfide bridge (9.4)
Fibrous protein (9.5)
Globular protein (9.5)
Isoelectric point (9.2)
Native state (9.10)
N-terminal residue (9.3)
Peptide (9.3)
Peptide linkage (peptide bond) (9.3)
Polypeptide (9.3)
Primary protein structure (9.6)

Prion (9.4)
Prosthetic group (9.5)
Protein (9.3)
Quaternary protein structure (9.9)
Secondary protein structure (9.7)
Simple protein (9.5)
Subunit (9.9)
Tertiary protein structure (9.8)
Zwitterion (9.2)

1. Formation of a zwitterion (Section 9.2):

Reaction 9.1

2. Conversion of a zwitterion to a cation in an acidic solution (Section 9.2):

Reaction 9.2

3. Conversion of a zwitterion to an anion in a basic solution (Section 9.2):

Reaction 9.3

4. Oxidation of cysteine to cystine (Section 9.3):

Reaction 9.4

5. Formation of a peptide linkage—general reaction (Section 9.3):

Reaction 9.6

6. Hydrolysis of proteins in acid or base (Section 9.10):

$$\text{protein} + H_2O \xrightarrow[\text{OH}^-]{\text{H}^+ \text{ or}} \text{smaller peptides} \xrightarrow[\text{OH}^-]{\text{H}^+ \text{ or}} \text{amino acids}$$

Reaction 9.9

THE AMINO ACIDS (SECTION 9.1)

9.1 What functional groups are found in all amino acids?

9.2 Draw the structure of butanoic acid. Label the alpha-carbon.

9.3 Identify the R group of the side chain in the following amino acids that results in the side-chain classification indicated in parentheses (see Table 9.1):

a. aspartate (acidic)
b. cysteine (neutral, polar)
c. threonine (neutral, polar)
d. lysine (basic)
e. asparagine (neutral, polar)
f. valine (neutral, nonpolar)

9.4 Draw structural formulas for the following amino acids, identify the chiral carbon atom in each one, and circle the four different groups attached to the chiral carbon.
a. threonine
b. aspartate
c. serine
d. phenylalanine

9.5 Isoleucine contains two chiral carbon atoms. Draw the structural formula for isoleucine twice. In the first, identify one of the chiral carbons and circle the four groups attached to it. In the second, identify the other chiral carbon and circle the four groups attached to it.

9.6 Draw Fischer projections representing the D and L forms of the following:
a. cysteine
b. glutamate

ZWITTERIONS (SECTION 9.2)

9.7 What is meant by the term *zwitterion?*

9.8 What characteristics indicate that amino acids exist as zwitterions?

9.9 What is meant by the term *isoelectric point?*

9.10 Write structural formulas to show the form the following amino acids would have in a solution with a pH higher than the amino acid isoelectric point:
a. glutamine
b. proline

9.11 Write structural formulas to show the form the following amino acids would have in a solution with a pH lower than the amino acid isoelectric point:
a. valine
b. threonine

9.12 Write ionic equations to show how serine acts as a buffer against the following added ions:
a. OH^- b. H_3O^+

REACTIONS OF AMINO ACIDS (SECTION 9.3)

9.13 Write two reactions to represent the formation of the two dipeptides that form when valine and serine react.

9.14 Write a complete structural formula and an abbreviated formula for the tripeptide formed from aspartate, cysteine, and valine in which the C-terminal residue is cysteine and the N-terminal residue is valine.

9.15 Write abbreviated formulas for the six isomeric tripeptides of tryptophan, glycine, and methionine.

9.16 How many tripeptide isomers are possible that contain one residue each of valine, phenylalanine and lysine?

9.17 How many tripeptide isomers are possible that contain one glycine residue and two alanine residues?

IMPORTANT PEPTIDES (SECTION 9.4)

9.18 What special role does the amino acid cysteine have in the peptides vasopressin and oxytocin?

9.19 What is the difference between a peptide and a protein?

CHARACTERISTICS OF PROTEINS (SECTION 9.5)

9.20 Explain why the presence of certain proteins in body fluids such as urine or blood can indicate that cellular damage has occurred in the body.

9.21 Write equations to show how a protein containing both lysine and aspartate could act as a buffer against the following added ions:
a. OH^- b. H_3O^+

9.22 Explain why a protein is least soluble in an aqueous medium that has a pH equal to the isoelectric point of the protein.

9.23 Wool is a protein used as a textile fiber. Classify the protein of wool into a category of Table 9.4.

9.24 List the eight principal functions of proteins.

9.25 Classify each of the following proteins into one of the eight functional categories of proteins:
a. insulin
b. collagen
c. hemoglobin
d. cytochrome c
e. α-keratin
f. serum albumin

9.26 For each of the following two proteins listed in Table 9.4, predict whether it is more likely to be a globular or a fibrous protein. Explain your reasoning.
a. collagen
b. lactate dehydrogenase

9.27 For each of the following two proteins listed in Table 9.4, predict whether it is more likely to be a globular or a fibrous protein. Explain your reasoning.
a. elastin
b. ceruloplasmin

9.28 Differentiate between simple and conjugated proteins.

9.29 Give one example of a conjugated protein that contains the following prosthetic group:
a. iron
b. lipid
c. phosphate group
d. carbohydrate
e. heme

THE PRIMARY STRUCTURE OF PROTEINS (SECTION 9.6)

9.30 Describe what is meant by the term *primary structure of proteins.*

9.31 What type of bonding is present to account for the primary structure of proteins?

9.32 Write the structure for a protein backbone. Make the backbone long enough to attach four R groups symbolizing amino acid side chains.

THE SECONDARY STRUCTURE OF PROTEINS (SECTION 9.7)

9.33 What type of bonding between amino acid residues is most important in holding a protein or polypeptide in a specific secondary configuration?

9.34 Describe what is meant by "supracoiling" in proteins.

9.35 Describe the differences between alpha and beta secondary structures of proteins.

THE TERTIARY STRUCTURE OF PROTEINS (SECTION 9.8)

9.36 How do hydrogen bonds involved in tertiary protein structures differ from those involved in secondary structures?

9.37 Which amino acids have side-chain groups that can form salt bridges?

9.38 Refer to Table 9.1 and list the type of side-chain interaction expected between the side chains of the following pairs of amino acid residues:
a. tyrosine and glutamine
b. aspartate and lysine
c. leucine and isoleucine
d. phenylalanine and valine

9.39 Which amino acids have side-chain groups that can participate in hydrogen bonding?

9.40 A globular protein in aqueous surroundings contains the following amino acid residues: phenylalanine, methionine, glutamate, lysine, and alanine. Which amino acid side chains would be directed toward the inside of the protein and which would be directed toward the aqueous surroundings?

THE QUATERNARY STRUCTURE OF PROTEINS (SECTION 9.9)

9.41 What is meant by the term *quaternary protein structure?*

9.42 What types of forces give rise to quaternary structure?

9.43 Describe the quaternary protein structure of hemoglobin.

9.44 What is meant by the term *subunit?*

PROTEIN HYDROLYSIS AND DENATURATION (SECTION 9.10)

9.45 What is meant by the term *denaturation?*

9.46 Suppose a sample of a protein is completely hydrolyzed and another sample of the same protein is denatured. Compare the final products of each process.

9.47 As fish is cooked, the tissue changes from a soft consistency to a firm one. A similar change takes place when fish is pickled by soaking it in vinegar (acetic acid). In fact, some people prepare fish for eating by pickling instead of cooking. Propose an explanation why the two processes give somewhat similar results.

9.48 In what way is the protein in a raw egg the same as that in a cooked egg?

9.49 In addition to emulsifying greasy materials, what other useful function might be served by a detergent used to wash dishes?

9.50 Once cooked, egg whites remain in a solid form. However, egg whites that are beaten to form meringue will partially change back to a jellylike form if allowed to stand for a while. Explain these behaviors using the concept of reversible protein denaturation.

9.51 Lead is a toxic material. What form of the lead is actually the harmful substance?

9.52 Explain how egg white can serve as an emergency antidote for heavy metal poisoning.

OVER THE COUNTER AND CHEMISTRY AROUND US

9.53 When should medications be taken by a nursing mother to minimize a baby's exposure to the medication?

9.54 List three drugs that should never be taken by a nursing mother.

9.55 List three diseases believed to be caused by prions.

9.56 What aspect of Alzheimer's disease can be treated with some success?

9.57 What happens to the alpha helices of certain proteins in Alzheimer's patients?

9.58 What is the molecular basis of sickle-cell disease?

9.59 Why is the sickle-cell trait more prevalent in world regions where malaria is common?

▸ POINTS TO PONDER AND PURSUE

9.1 Some researchers feel that the "high" experienced by runners may be due to the brain's synthesis of enkephalins. Explain why this theory might be a reasonable explanation.

9.2 Suggest a reason why individuals with the sickle-cell trait might sometimes encounter breathing problems on high-altitude flights.

9.3 In Section 9.2, you learned that amino acids like alanine are crystalline solids with high melting points. However, the ethyl ester of alanine, which as a free NH_2 group, melts at only 87°C. Explain why the ethyl ester melts 200 degrees below the melting point of alanine.

9.4 Why do health care workers wipe a patient's skin with solutions of rubbing alcohol before giving injections?

9.5 Would you expect plasma proteins to be fibrous or globular? Explain.

9.6 Why must the protein drug insulin be given by injection rather than taken by mouth?

9.7 In Figure 9.4, it is noted that human hair, animal fur, and spiderwebs are fibrous proteins. What properties of fibrous proteins make hair, fur, and spiderwebs useful?

9.8 What protein's primary structure is found at Web site 9.3?

9.9 Using Web site 9.4, name the Nobel prizes that Linus Pauling received.

9.10 According to Web site 9.5, what special problems do babies with sickle-cell disease have?

http://www.brookscole.com/chemistry/seager_chemistry

Enzymes

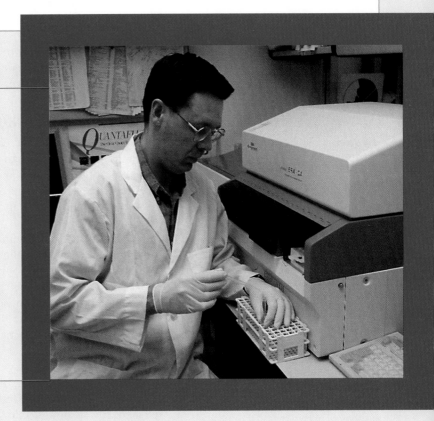

Clinical laboratory scientists often use sophisticated automated equipment to perform tests that can detect the failure or abnormal behavior of most organs of the body. The tests can identify such problems as diabetes, heart attacks, and liver and kidney damage. Many of the tests involve enzymes, the topic of this chapter.

http://www.brookscole.com/chemistry/ seager_chemistry
At Web site 10.1, you will find career information on clinical laboratory scientists.

CHAPTER OUTLINE

The catalytic behavior of proteins acting as enzymes is one of the most important functions performed by cellular proteins. Without catalysts, most cellular reactions would take place much too slowly to support life (see ■ Figure 10.1). With the exception of a few recently discovered RNA molecules (discussed in Chapter 11) that catalyze their own reactions, all enzymes are globular proteins. Enzymes are the most efficient of all known catalysts; some can increase reaction rates by 10^{20} times that of uncatalyzed reactions.

➤ 10.1 General Characteristics of Enzymes

Enzyme
A biomolecule that catalyzes chemical reactions.

Enzymes are well suited to their essential roles in living organisms in three major ways: They have enormous catalytic power, they are highly specific in the reactions they catalyze, and their activity as catalysts can be regulated.

Catalytic Efficiency

As we know, catalysts increase the rate of chemical reactions but are not used up in the process. Although a catalyst actually participates in a chemical reaction, it is not permanently modified and may be used again and again. Enzymes are true catalysts that speed chemical reactions by lowering activation energies (see ■ Figure 10.2) and allowing reactions to achieve equilibrium more rapidly.

Many of the important enzyme-catalyzed reactions are similar to the reactions we studied in the chapters on organic chemistry: ester hydrolysis, alcohol oxidation, and so on. However, laboratory conditions cannot match what happens when these reactions are carried out in the body; enzymes cause such reactions to proceed under mild pH and temperature conditions. In addition, enzyme catalysis within the body can accomplish in seconds reactions that ordinarily take weeks or even months under laboratory conditions.

The influence of enzymes on the rates of reactions essential to life is truly amazing. A good example is the need to move carbon dioxide, a waste product of cellular respiration, out of the body. Before this can be accomplished, the carbon dioxide must be combined with water to form carbonic acid:

➤ **FIGURE 10.1** One miracle of life is that hundreds of chemical reactions take place simultaneously in each living cell; the activity of enzymes in the cells makes it all possible.

$$CO_2 + H_2O \xrightarrow{\text{carbonic anhydrase}} H_2CO_3 \qquad (10.1)$$

In the absence of the appropriate enzyme, carbonic anhydrase, the formation of carbonic acid takes place much too slowly to support the required exchange of carbon dioxide between the blood and the lungs. But in the presence of carbonic anhydrase, this vital reaction proceeds rapidly. Each molecule of the enzyme can catalyze the formation of carbonic acid at the rate of 36 million molecules per minute.

Specificity

Enzyme specificity is a second characteristic that is important in life processes. Unlike other catalysts, enzymes are often quite specific in the type of reaction they catalyze and even the particular substance that will be involved in the reaction. For example, strong acids catalyze the hydrolysis of any amide, the dehydration of any alcohol, and a variety of other processes. However, the

➤ **FIGURE 10.2** An energy diagram for a reaction when uncatalyzed and catalyzed.

The enzyme urease is added to a urea solution containing the indicator phenolphthalein.

As urea is hydrolyzed, forming ammonia, the indicator turns pink. How else might the presence of ammonia be detected?

enzyme urease catalyzes only the hydrolysis of a single amide, urea (Reaction 10.2 and ■ Figure 10.3):

$$H_2N - \overset{\overset{\textstyle O}{\|}}{C} - NH_2 + H_2O \xrightarrow{\text{urease}} CO_2 + 2NH_3 \qquad (10.2)$$

This is an example of **absolute specificity,** the catalyzing of the reaction of one and only one substance. Other enzymes display **relative specificity** by catalyzing the reaction of structurally related substances. For example, the lipases hydrolyze lipids, proteases split up proteins, and phosphatases hydrolyze phosphate esters. **Stereochemical specificity** extends even to enantiomers: D-amino acid oxidase will not catalyze the reactions of L-amino acids.

Regulation

A third significant property of enzymes is that their catalytic behavior can be regulated. Even though each living cell contains thousands of different molecules that could react with each other in an almost unlimited number of ways, only a relatively small number of these possible reactions take place because of the enzymes present. The cell controls the rates of these reactions and the amount of any given product formed by regulating the action of the enzymes.

➤ 10.2 Enzyme Nomenclature and Classification

Some of the earliest discovered enzymes were given names ending with *-in* to indicate their protein composition. For example, three of the digestive enzymes that catalyze protein hydrolysis are named pepsin, trypsin, and chymotrypsin.

The large number of enzymes now known has made it desirable to adopt a systematic nomenclature system known as the International Enzyme Commission (IEC) system. Enzymes are grouped into six major classes on the basis

Absolute specificity
The characteristic of an enzyme that it acts on one and only one substance.

Relative specificity
The characteristic of an enzyme that it acts on several structurally related substances.

Stereochemical specificity
The characteristic of an enzyme that it is able to distinguish between stereoisomers.

TABLE 10.1 The IEC classification of enzymes

Group name	Type of reaction catalyzed
Oxidoreductases	Oxidation–reduction reactions
Transferases	Transfer of functional groups
Hydrolases	Hydrolysis reactions
Lyases	Addition to double bonds or the reverse of that reaction
Isomerases	Isomerization reactions
Ligases	Formation of bonds with ATP cleavage[a]

[a]ATP is discussed in Chapter 12.

Substrate
The substance that undergoes a chemical change catalyzed by an enzyme.

of the reaction catalyzed (see ■ Table 10.1). In the IEC system, each enzyme has an unambiguous (and often long) systematic name that specifies the **substrate** (substance acted on), the functional group acted on, and the type of reaction catalyzed. All IEC names end in -*ase*. The hydrolysis of urea provides a typical example:

$$H_2N - \overset{\overset{\displaystyle O}{\|}}{C} - NH_2 + H_2O \xrightarrow{enzyme} 2NH_3 + CO_2$$

IEC name: urea amidohydrolase
Substrate: urea
Functional group: amide
Type of reaction: hydrolysis

(10.3)

Inborn error of metabolism
A disease in which a genetic change causes a deficiency of a particular protein, often an enzyme.

CHEMISTRY AROUND US • 10.1

ENZYMES AND DISEASE

A number of hereditary diseases appear to result from the absence of enzymes or from the presence of altered enzymes. These diseases are often referred to as **inborn errors of metabolism** (the term *metabolism* refers to all reactions occurring within a living organism).

An example of an inborn error of lipid metabolism is Gaucher's disease, characterized by symptoms that include an enlarged spleen and an enlarged liver. A large amount of glucocerebroside is found in fat deposits in these organs. Normally, these organs contain β-glucosidase, an enzyme that catalyzes the degradation of glucocerebrosides (Section 8.7), so that only small amounts of these lipids are normally present. However, the enzyme activity is impaired in Gaucher's disease, and the glucocerebrosides accumulate. Most people with the adult form of the disease do not require treatment; however, the infantile form produces mental retardation and often death.

The table at right lists some of the diseases caused by enzyme deficiencies:

Disease	Deficient enzyme
Albinism	tyrosinase
Galactosemia	galactose 1-phosphate uridyltransferase
Gaucher's disease	β-glucosidase
Homocystinuria	cystathionine synthetase
Maple syrup urine disease	amino acid decarboxylase
Methemoglobinemia	methemoglobin reductase
Niemann-Pick disease	sphingomyelinase
Phenylketonuria (PKU)	phenylalanine hydroxylase
Tay-Sachs disease	hexosaminidase A

Enzymes are also assigned common names, which are usually shorter and more convenient. Common names are derived by adding *-ase* to the name of the substrate or to a combination of substrate name and type of reaction. For example, urea amidohydrolase is assigned the common name urease:

Substrate: urea
Common name: urea + ase = urease

The enzyme name *alcohol dehydrogenase* is an example of a common name derived from both the name of the substrate and the type of reaction:

Substrate: alcohol (ethyl alcohol)
Reaction type: dehydrogenation (removal of hydrogen)
Common name: alcohol dehydrogenation + ase = alcohol dehydrogenase

Throughout this text, we will refer to most enzymes by their accepted common names.

Predict the substrates for the following enzymes:

a. maltase
b. peptidase
c. glucose 6-phosphate isomerase

Learning Check 10.1

➤ 10.3 Enzyme Cofactors

In Section 9.5, we learned that proteins may be simple (containing only amino acid residues) or conjugated with a prosthetic group present. Many enzymes are simple proteins, whereas many others function only in the presence of specific nonprotein molecules or metal ions. If these nonprotein components are tightly bound to and form an integral part of the enzyme structure, they are true prosthetic groups. Often, however, a nonprotein component is only weakly bound to the enzyme and is easily separated from the protein structure. This type of nonprotein component is referred to as a **cofactor**. When the cofactor is an organic substance, it is called a **coenzyme**. The cofactor may also be an inorganic ion (usually a metal ion). The protein portion of enzymes requiring a cofactor is called the **apoenzyme**. Thus, the combination of an apoenzyme and a cofactor produces an active enzyme:

Apoenzyme + cofactor (coenzyme or inorganic ion) → active enzyme   (10.4)

Typical inorganic ions are metal ions such as Mg^{2+}, Zn^{2+}, and Fe^{2+}. For example, the enzyme carbonic anhydrase functions only when Zn^{2+} is present, and rennin needs Ca^{2+} in order to curdle milk. Numerous other metal ions are essential for proper enzyme function in humans; hence, they are required for good health.

Like metal ions, the small organic molecules that act as coenzymes bind reversibly to an enzyme and are essential for its activity. An interesting feature of coenzymes is that many of them are formed in the body from vitamins (see ■ Table 10.2), which explains why it is necessary to have certain vitamins in the diet for good health. For example, the coenzyme nicotinamide adenine dinucleotide (NAD^+), which is a necessary part of some enzyme-catalyzed oxidation–reduction reactions, is formed from the vitamin precursor nicotinamide. Reaction 10.5 shows the participation of NAD^+ in the oxidation of lactate by the enzyme lactate dehydrogenase (LDH). Like other cofactors,

Cofactor
A nonprotein molecule or ion required by an enzyme for catalytic activity.

Coenzyme
An organic molecule required by an enzyme for catalytic activity.

Apoenzyme
A catalytically inactive protein formed by removal of the cofactor from an active enzyme.

TABLE 10.2	Vitamins and their coenzyme forms	
Vitamin	**Coenzyme form**	**Function**
biotin	biocytin	Carboxyl group removal or transfer
folacin	tetrahydrofolic acid	One-carbon group transfer
lipoic acid	lipoamide	Acyl group transfer
niacin	nicotinamide adenine dinucleotide (NAD^+)	Hydrogen transfer
	nicotinamide adenine dinucleotide phosphate ($NADP^+$)	Hydrogen transfer
pantothenic acid	coenzyme A (CoA)	Acyl group carrier
pyridoxal, pyridoxamine, pyridoxine (B_6 group)	pyridoxal phosphate	Amino group transfer
riboflavin	flavin mononucleotide (FMN) flavin adenine dinucleotide (FAD)	Hydrogen transfer Hydrogen transfer
thiamin (B_1)	thiamin pyrophosphate (TPP)	Aldehyde group transfer
vitamin B_{12}	coenzyme B_{12}	Shift of hydrogen atoms between adjacent carbon atoms; methyl group transfer

NAD^+ is written separately from the enzyme so that the change in its structure may be shown and to emphasize that the enzyme and cofactor are easily separated.

$$CH_3 - \underset{\substack{| \\ \text{lactate}}}{\overset{\substack{OH \\ |}}{CH}} - COO^- + NAD^+ \underset{\text{dehydrogenase}}{\overset{\text{lactate}}{\rightleftharpoons}} CH_3 - \underset{\substack{\| \\ \text{pyruvate}}}{\overset{\substack{O \\ \|}}{C}} - COO^- + NADH + H^+ \quad (10.5)$$

We can see from the reaction that the coenzyme NAD^+ is essential because it is the actual oxidizing agent. NAD^+ is typical of coenzymes that often aid in the transfer of chemical groups from one compound to another. In Reaction 10.5, NAD^+ accepts hydrogen from lactate and will transfer it to other compounds in subsequent reactions. In Chapter 12, we will discuss in greater detail several coenzymes that act as shuttle systems in the exchange of chemical substances among various biochemical pathways.

► 10.4 The Mechanism of Enzyme Action

Although enzymes differ widely in structure and specificity, a general theory has been proposed to account for their catalytic behavior. Because enzyme molecules are large compared with the molecules whose reactions they catalyze (substrate molecules), it has been proposed that substrate and enzyme molecules come in contact and interact over only a small region of the enzyme surface. This region of interaction is called the **active site.**

The binding of a substrate molecule to the active site of an enzyme may occur through hydrophobic attraction, hydrogen bonding, and/or ionic bonding. The complex formed when a substrate and an enzyme bond is called the enzyme–substrate (ES) complex. Once this complex is formed, the conversion of substrate (S) to product (P) may take place:

General reaction:

$$\underset{\text{enzyme}}{E} + \underset{\text{substrate}}{S} \rightleftarrows \underset{\substack{\text{enzyme–}\\\text{substrate}\\\text{complex}}}{ES} \longrightarrow \underset{\text{enzyme}}{E} + \underset{\text{product}}{P} \qquad (10.6)$$

Specific example:

$$\underset{\substack{\text{enzyme} \quad \text{substrate}}}{\text{Sucrase} + \text{sucrose}} \rightleftarrows \left\{\substack{\text{sucrase–}\\\text{sucrose}\\\text{complex}}\right\} \longrightarrow \underset{\substack{\text{enzyme} \qquad \text{products}}}{\text{sucrase} + \text{glucose} + \text{fructose}} \quad (10.7)$$

The chemical transformation of the substrate occurs at the active site, usually aided by enzyme functional groups that participate directly in the making and breaking of chemical bonds. After chemical conversion has occurred, the product is released from the active site, and the enzyme is free for another round of catalysis.

An extension of the theory of ES complex formation is used to explain the high specificity of enzyme activity. According to the **lock-and-key theory,** enzyme surfaces will accommodate only those substrates having specific shapes and sizes. Thus, only specific substrates "fit" a given enzyme and can form complexes with it (see ■ Figure 10.4a).

A limitation of the lock-and-key theory is the implication that enzyme conformations are fixed or rigid. However, research results suggest that the active sites of some enzymes are not rigid. This is taken into account in a modification of the lock-and-key theory known as the **induced-fit theory,** which proposes that enzymes have somewhat flexible conformations that may adapt to incoming substrates. The active site has a shape that becomes complementary to that of the substrate only after the substrate is bound (see Figure 10.4b).

Active site
The location on an enzyme where a substrate is bound and catalysis occurs.

Lock-and-key theory
A theory of enzyme specificity proposing that a substrate has a shape fitting that of the enzyme's active site, as a key fits a lock.

Induced-fit theory
A theory of enzyme action proposing that the conformation of an enzyme changes to accommodate an incoming substrate.

(a) Lock-and-key model

(b) Induced-fit model

➤ **FIGURE 10.4** Models representing enzyme action: (a) In the lock-and-key model, the rigid enzyme and substrate have matching shapes. (b) In the induced-fit model, the flexible enzyme changes shape to match the substrate.

➤ 10.5 Enzyme Activity

Enzyme activity
The rate at which an enzyme catalyzes a reaction.

Turnover number
The number of molecules of substrate acted on by one molecule of enzyme per minute.

Enzyme international unit (IU)
A quantity of enzyme that catalyzes the conversion of 1 μmol of substrate per minute under specified conditions.

Enzyme activity refers in general to the catalytic ability of an enzyme to increase the rate of a reaction. The amazing rate (36 million molecules per minute) at which one molecule of carbonic anhydrase converts carbon dioxide to carbonic acid was mentioned earlier. This rate, called the **turnover number,** is one of the highest known for enzyme systems. More common turnover numbers for enzymes are closer to 10^3/min, or 1000 reactions per minute. Nevertheless, even such low numbers dramatize the speed with which a small number of enzyme molecules can transform a large number of substrate molecules. ■ Table 10.3 gives the turnover numbers of several enzymes.

Experiments that measure enzyme activity are called *enzyme assays.* Assays for blood enzymes are routinely performed in clinical laboratories. Such assays are often done by determining how fast the characteristic color of a product forms or the color of a substrate decreases. Reactions in which protons (H^+) are produced or used up can be followed by measuring how fast the pH of the reacting mixture changes with time.

Because some clinical assays are done many times a day, the procedures for running the assays have been automated and computerized. The determined enzyme activity levels are usually reported in terms of **enzyme international units (IU),** which define enzyme activity as the amount of enzyme that will convert a specified amount of substrate to a product within a certain time.

One standard international unit (1 IU) is the quantity of enzyme that catalyzes the conversion of 1 micromole (μmol) of substrate per minute under specified reaction conditions. Thus, unlike the turnover number, which is a constant characteristic for one molecule of a particular enzyme, international units measure how much enzyme is present. For example, an enzyme preparation with an activity corresponding to 40 IU contains a concentration of enzyme 40 times greater than the standard. This is a useful way to measure enzyme activity because the level of enzyme activity compared with normal activity is significant in the diagnosis of disease (Section 10.9).

Learning Check 10.2

Differentiate between the terms *turnover number* and *enzyme international unit.*

TABLE 10.3	Examples of enzyme turnover numbers	
Enzyme	**Turnover number (per minute)**	**Reaction catalyzed**
carbonic anhydrase	36,000,000	$CO_2 + H_2O \rightleftharpoons H_2CO_3$
catalase	5,600,000	$2H_2O_2 \rightleftharpoons 2H_2O + O_2$
cholinesterase	1,500,000	Hydrolysis of acetylcholine
chymotrypsin	6,000	Hydrolysis of specific peptide bonds
DNA polymerase I	900	Addition of nucleotide monomers to DNA chains
lactate dehydrogenase	60,000	pyruvate + NADH + $H^+ \rightleftharpoons$ lactate + NAD^+
penicillinase	120,000	Hydrolysis of penicillin

➤ 10.6 Factors Affecting Enzyme Activity

Several factors affect the rate of enzyme-catalyzed reactions. The most important factors are enzyme concentration, substrate concentration, temperature, and pH. In this section, we'll look at each of these factors in some detail. In Section 10.7, we'll consider another very important factor, the presence of enzyme inhibitors.

Enzyme Concentration

In an enzyme-catalyzed reaction, the concentration of enzyme is normally very low compared with the concentration of substrate. When the enzyme concentration is increased, the concentration of ES also increases in compliance with reaction rate theory:

$$E + S \rightleftarrows ES$$
increased [E] gives more [ES]

Thus, the availability of more enzyme molecules to catalyze a reaction leads to the formation of more ES and a higher reaction rate. The effect of enzyme concentration on the rate of a reaction is shown in ■ Figure 10.5. As the figure indicates, the rate of a reaction is directly proportional to the concentration of the enzyme—that is, if the enzyme concentration is doubled, the rate of conversion of substrate to product is also doubled.

Substrate Concentration

As shown in ■ Figure 10.6, the concentration of substrate significantly influences the rate of an enzyme-catalyzed reaction. Initially, the rate is responsive to increases in substrate concentration. However, at a certain concentration, the rate levels out and remains constant. This maximum rate (symbolized by V_{max}) occurs because the enzyme is saturated with substrate and cannot work any faster under the conditions imposed.

Temperature

Enzyme-catalyzed reactions, like all chemical reactions, have rates that increase with temperature (see ■ Figure 10.7). However, because enzymes are proteins, there is a temperature limit beyond which the enzyme becomes vulnerable to denaturation. Thus, every enzyme-catalyzed reaction has an **optimum temperature**, usually in the range 25°C–40°C. Above or below

Optimum temperature
The temperature at which enzyme activity is highest.

➤ **FIGURE 10.5** The dependence of reaction-rate (initial velocity) on enzyme concentration.

➤ **FIGURE 10.6** The dependence of reaction-rate (initial velocity) on substrate concentration.

➤ **FIGURE 10.7** The effect of temperature on enzyme-catalyzed reaction rates.

> **FIGURE 10.8** A typical plot of the effect of pH on reaction rate.

Optimum pH
The pH at which enzyme activity is highest.

Learning Check 10.3

Enzyme inhibitor
A substance that decreases the activity of an enzyme.

> **FIGURE 10.9** Pickles are prepared under acidic conditions. Why do they resist spoilage?

that value, the reaction rate will be lower. This effect is illustrated in Figure 10.7.

The Effect of pH

The graph of enzyme activity as a function of pH is somewhat similar to the behavior as a function of temperature (see ■ Figure 10.8). Notice in Figure 10.8 that an enzyme is most effective in a narrow pH range and is less active at pH values lower or higher than this optimum. This variation in enzyme activity with changing pH may be due to the influence of pH on acidic and basic side chains within the active site. In addition, most enzymes are denatured by pH extremes. The resistance of pickled foods (see ■ Figure 10.9) to spoilage is due to the enzymes of microorganisms being less active under acidic conditions.

Many enzymes have an **optimum pH** near 7, the pH of most biological fluids. However, the optimum pH of a few is considerably higher or lower than 7. For example, pepsin, a digestive enzyme of the stomach, shows maximum activity at a pH of about 1.5, the pH of gastric fluids. ■ Table 10.4 lists optimum pH values for several enzymes.

Indicate how each of the following affects the rate of an enzyme-catalyzed reaction:

a. increase in enzyme concentration
b. increase in substrate concentration
c. increase in temperature
d. increase in pH

➤ 10.7 Enzyme Inhibition

An **enzyme inhibitor** is any substance that can decrease the rate of an enzyme-catalyzed reaction. An understanding of enzyme inhibition is important for several reasons. First, the characteristic function of many poisons and some medicines is to inhibit one or more enzymes and to decrease the rates of the reactions they catalyze. Second, some substances normally found in cells inhibit specific enzyme-catalyzed reactions and thereby provide a means for the internal regulation of cellular metabolism.

| TABLE 10.4 | Examples of optimum pH for enzyme activity |

Enzyme	Source	Optimum pH
pepsin	Gastric mucosa	1.5
β-glucosidase	Almond	4.5
sucrase	Intestine	6.2
urease	Soybean	6.8
catalase	Liver	7.3
succinate dehydrogenase	Beef heart	7.6
arginase	Beef liver	9.0
alkaline phosphatase	Bone	9.5

CHEMISTRY AROUND US • 10.2

ENZYME DISCOVERY HEATS UP

Enzymes provide the same advantages to a commercial process as they do to a living organism. They are highly specific and extremely fast-acting catalysts. The use of enzymes in commercial industrial processes is developing rapidly and has a potentially bright future because of the existence of enzymes in microorganisms that thrive in extreme environments. Such enzymes have been called **extremozymes** because of the extreme conditions under which they function. These conditions include temperatures greater than 100°C in hot springs and deep-sea thermal vents, or below 0°C in Antarctic waters. Some enzymes function under immense pressure on the ocean floor, whereas others are unaffected by extremes of pH and salt concentrations in solution.

One goal of current research is to obtain useful enzymes from extremophiles, the microbes that function in these extreme environments. The process used by one company to obtain commercially useful enzymes consists of the following steps:

1. Biomass samples are obtained from the extreme environment.
2. DNA is extracted from the biomass and purified.
3. Fragments of the purified DNA are cloned to produce large libraries of genes.
4. The cloned genes are screened to find any that are coded to produce enzymes.
5. The DNA sequence is determined for any genes that are coded to produce enzymes.
6. The DNA sequence for enzyme-coded genes is cloned to produce large amounts of the gene, which in turn is inserted into organisms and used to produce large amounts of the enzyme.
7. The enzyme is optimized by random DNA mutations.

The last step is particularly interesting. An organism's DNA is randomly mutated over several generations. Offspring from each generation are selected that produce a specific enzyme with activity greater than or different from that of the enzyme from the previous generation. This technique, known as *mutagenesis*, is used to alter enzymes so they will perform new tasks or do old ones better.

Because extremozymes function under conditions that are so far removed from normal, research into their behavior is intense. One of the research goals is to develop ways to use enzyme catalysis in industrial processes that require extreme temperatures or pH. One example is the current use of heat-stable DNA polymerase in the DNA replication technique called polymerase chain reaction (Section 11.3). Another research goal is to develop enzyme-catalyzed industrial processes that don't generate large amounts of undesirable waste material. For example, an esterase that catalyzes the hydrolysis of the antibiotic intermediate *p*-nitrobenzyl ester was developed. That hydrolysis reaction was usually done using catalytic zinc and organic solvents in a process that generated a lot of industrial waste. Some other applications being considered are the use of hydrolases at high temperature in the food-processing industry, and the use of heat-stable enzymes in the production of corn syrup.

http://www.brookscole.com/chemistry/seager_chemistry
At Web site 10.2, more information is provided about unique enzymes that are commercially available.

Extremozymes enable microorganisms to survive in the harsh environment of deep-sea thermal vents.

Extremozyme
A nickname for certain enzymes isolated from microorganisms that thrive in extreme environments.

Enzyme inhibitors are classified into two categories, reversible and irreversible, on the basis of how they behave at the molecular level.

Irreversible Inhibition

An irreversible inhibitor forms a covalent bond with a specific functional group of the enzyme and as a result renders the enzyme inactive. A number of very deadly poisons act as irreversible inhibitors.

The cyanide ion (CN^-) is an example of an irreversible enzyme inhibitor. It is extremely toxic and acts very rapidly. The cyanide ion interferes with the operation of an iron-containing enzyme called cytochrome oxidase. The ability of cells to use oxygen depends on the action of cytochrome oxidase. When the cyanide ion reacts with the iron of this enzyme, it forms a very stable complex (Reaction 10.8), and the enzyme can no longer function properly. As a result, cell respiration stops, causing death in a matter of minutes.

$$\text{Cyt—Fe}^{3+} + \text{CN}^- \rightarrow \text{Cyt—Fe—CN}^{2+}$$

cytochrome   cyanide   stable complex   (10.8)
oxidase   ion

Any antidote for cyanide poisoning must be administered quickly. One antidote is sodium thiosulfate (the "hypo" used in developing photographic film). This substance converts the cyanide ion to a thiocyanate ion that does not bind to the iron of cytochrome oxidase.

$$\text{CN}^- + \text{S}_2\text{O}_3^{2-} \rightarrow \text{SCN}^- + \text{SO}_3^{2-}$$

cyanide   thiosulfate   thiocyanate   sulfite   (10.9)

The toxicity of heavy metals such as mercury and lead is due to their ability to render the protein part of enzymes ineffective. These metals act by combining with the —SH groups found on many enzymes (Reaction 10.10). In addition, as we learned in Chapter 9, metals can cause nonspecific protein denaturation. Both mercury and lead poisoning can cause permanent neurological damage.

$$\begin{array}{c}\text{—SH} \\ \text{—SH}\end{array} + \text{Hg}^{2+} \longrightarrow \begin{array}{c}\text{—S} \\ \hspace{1em} \text{Hg} + 2\text{H}^+ \\ \text{—S}\end{array}$$ (10.10)

active enzyme   inactive enzyme

Heavy-metal poisoning is treated by administering chelating agents—substances that combine with the metal ions and hold them very tightly. One effective antidote is the chelating agent ethylenediaminetetraacetic acid, or EDTA.

The calcium salt of EDTA is administered intravenously. In the body, calcium ions of the salt are displaced by heavy-metal ions, such as lead, that bind to the chelate more tightly. The lead–EDTA complex is soluble in body fluids and is excreted in the urine.

$$\text{CaEDTA}^{2-} + \text{Pb}^{2+} \rightarrow \text{PbEDTA}^{2-} + \text{Ca}^{2+}$$ (10.11)

Not all enzyme inhibitors act as poisons toward the body; some, in fact, are useful therapeutic agents. Sulfa drugs and the group of compounds known as penicillins are two well-known families of **antibiotics** that inhibit specific enzymes essential to the life processes of bacteria. The first penicillin was discovered in 1928 by Alexander Fleming. He noticed that the bacteria on a culture plate did not grow in the vicinity of a mold that had contaminated the culture; the mold was producing penicillin. The general penicillin structure is shown in ■ Table 10.5, along with four widely used forms. Penicillins inhibit transpeptidase, an enzyme that is important in bacterial cell wall construction. The inability to form strong cell walls prevents the bacteria from surviving.

Reversible Inhibition

A reversible inhibitor (in contrast to one that is irreversible) reversibly binds to an enzyme. An equilibrium is established; therefore, the inhibitor can be removed from the enzyme by shifting the equilibrium:

$$\text{E} + \text{I} \rightleftarrows \text{EI}$$

There are two types of reversible inhibitors: competitive and noncompetitive.

A **competitive inhibitor** binds to the active site of an enzyme and thus "competes" with substrate molecules for the active site. Competitive inhibitors often have molecular structures that are similar to the normal substrate of the

http://www.brookscole.com/chemistry/
seager_chemistry
At Web site 10.3 you will find information about lead poisoning from ceramic mugs.

Antibiotic
A substance produced by one microorganism that kills or inhibits the growth of other microorganisms.

http://www.brookscole.com/chemistry/
seager_chemistry
At Web site 10.4, you can learn more about Alexander Fleming.

Competitive inhibitor
An inhibitor that competes with substrate for binding at the active site of the enzyme.

TABLE 10.5 Four widely used penicillins

$$
\begin{array}{c}
\overset{\displaystyle O}{\underset{\displaystyle \parallel}{}} \\
R-C-NH \quad S \quad CH_3 \\
\text{...} \quad CH_3 \\
O \quad N \quad COOH
\end{array}
$$

General penicillin structure

Name	Year marketed	Side chain (R—)	Comments
Penicillin G	1943	⬡—CH₂—	Usually given by injection
Penicillin V	1953	⬡—O—CH₂—	An oral penicillin; resistant to stomach hydrolysis
Methicillin	1960	⬡ (OCH₃, OCH₃)	Given by injection
Ampicillin	1961	⬡—CH— (NH₂)	Given by injection or taken orally; effective against a broad spectrum of organisms

enzyme. The competitive inhibition of succinate dehydrogenase by malonate is a classic example. Succinate dehydrogenase catalyzes the oxidation of the substrate succinate to form fumarate by transferring two hydrogens to the coenzyme FAD:

$$
\begin{array}{ccc}
COO^- & & COO^- \\
| & & | \\
CH_2 & \xrightarrow[\text{dehydrogenase}]{\text{succinate}} & CH \\
| & +\ FAD & \parallel \quad +\ FADH_2 \\
CH_2 & & HC \\
| & & | \\
COO^- & & COO^- \\
\text{succinate} & & \text{fumarate}
\end{array}
\tag{10.12}
$$

Malonate, having a structure similar to succinate, competes for the active site of succinate dehydrogenase and thus inhibits the enzyme:

$$
\begin{array}{cc}
COO^- & COO^- \\
| & | \\
CH_2 & CH_2 \\
| & | \\
CH_2 & COO^- \\
| & \\
COO^- & \\
\text{succinate} & \text{malonate}
\end{array}
$$

> **FIGURE 10.10** Structural relationships of sulfanilamide, *p*-aminobenzoic acid, and folic acid.

The action of sulfa drugs on bacteria is another example of competitive enzyme inhibition. Folic acid, a substance needed for growth by some disease-causing bacteria, is normally synthesized within the bacteria by a chemical process that requires *p*-aminobenzoic acid. Because sulfanilamide, the first sulfa drug, resembles *p*-aminobenzoic acid and competes with it for the active site of the bacterial enzyme involved, it can prevent bacterial growth (see ■ Figure 10.10). This is possible because the enzyme binds readily to either of these molecules. Therefore, the introduction of large quantities of sulfanilamide into a patient's body causes most of the active sites to be bound to the wrong (from the bacterial viewpoint) substrate. Thus, the synthesis of folic acid is stopped or slowed, and the bacteria are prevented from multiplying. Meanwhile, the normal body defenses work to destroy them.

Human beings also require folic acid, but they get it from their diet. Consequently, sulfa drugs exert no toxic effect on humans. However, one danger of sulfa drugs, or of any antibiotic, is that excessive amounts can destroy important intestinal bacteria that perform many symbiotic, life-sustaining functions. Although sulfa drugs have largely been replaced by antibiotics such as the penicillins and tetracyclines, some are still used in treating certain bacterial infections, such as those occurring in the urinary tract.

The nature of competitive inhibition is represented in ■ Figure 10.11. There is competition between the substrate and the inhibitor for the active site. Once the inhibitor combines with the enzyme, the active site is blocked, preventing further catalytic action.

Competitive inhibition can be reversed by increasing the concentration of substrate and letting Le Châtelier's principle operate. This is illustrated by the following equilibria that would exist in a solution containing enzyme (E), substrate (S), and competitive inhibitor (I):

> **FIGURE 10.11** The behavior of competitive inhibitors. The active site is occupied by the inhibitor, and the substrate is prevented from bonding.

OVER THE COUNTER • 10.1

TOPICAL ANTIBIOTICS

Seven compounds have been approved by the FDA for use as active ingredients in nonprescription topical antibiotics. Applied to the skin to help prevent infections resulting from minor cuts, scrapes, and burns, these compounds all inhibit the growth of bacteria, and some actively kill bacteria. The approved compounds are: bacitracin, bacitracin zinc, chlortetracycline hydrochloride, tetracycline hydrochloride, neomycin sulfate, oxytetracycline hydrochloride, and polymyxin B sulfate. The last two must be used in combination with others on the list because of their limited effectiveness against certain microorganisms when used alone. Gramicidin, a compound used in the past, is not allowed because it has the potential to break down red blood cells when absorbed through fresh wounds.

The effectiveness of these topical antimicrobials was supported by the results of a well-controlled study of minor skin injuries or insect bites in 59 children. A streptococcal infection developed in 15 of 32 children who received a placebo, but in only 3 of 27 children who received the topical antibiotic. Of the 15 children who developed an infection after receiving the placebo, 12 needed prescription oral antibiotics to clear up the infection. Only 1 of the 3 children who developed an infection despite using the topical antibiotic required oral antibiotics to fight off the infection.

FDA approval of these products carried with it some labeling requirements concerning their use. First, the amount of ointment or cream applied should be equal in surface area to the tip of a finger. Second, the ointments and creams are only to be used for a short time. These two guidelines are designed to help minimize the development of resistant strains of bacteria. If neomycin, for example, is used repeatedly on large areas of skin, numerous resistant strains of staphylococci can develop. Third, the product label must state that the infected area should be cleaned before applying the ointment or cream. Fourth, the label must contain a warning to discontinue use of the product if the condition persists or worsens. This is an important warning, because about 1 in 20 people are allergic to neomycin, and some have hypersen-

sitive responses to bacitracin. A few people are allergic to tetracycline products, and should discontinue their use immediately if an adverse reaction occurs. Finally, the label must state that these products are not intended for use on existing infections, animal bites, sunburns, puncture wounds, or eye injuries. Also, the products are not appropriate to use on cuts, scrapes, or burns that require medical care, such as deep cuts that require stitches, large wounds, scrapes with embedded particles that can't be washed out, or burns more serious than a small reddened area.

The products currently on the market that contain these antibiotic ingredients include: Baciguent®, Bacitracin Zinc®, Aureomycin®, Neomycin®, Myciguent Cream®, Achromycin®, Lanabiotic®, Medi-Quick Triple Antibiotic®, Clomycin Cream®, Mycitracin Plus Pain Reliever®, Polysporin®, Polysporin Powder®, Niomixin®, Neosporin Original®, and Neosporin Plus Maximum Strength Cream®.

Some of the OTC topical antibiotics.

If the concentration of substrate is increased, then according to Le Châtelier's principle, the first equilibrium should shift to the right. As it does, the concentration of enzyme decreases, which causes equilibrium 2 to shift to the left. The overall effect is that the concentration of ES has increased while the concentration of EI has decreased. Thus, the competition between inhibitor and substrate is won by whichever molecular species is in greater concentration.

A **noncompetitive inhibitor** bears no resemblance to the normal enzyme substrate and binds reversibly to the surface of an enzyme at a site other than the catalytically active site (see ■ Figure 10.12). The interaction between the enzyme and the noncompetitive inhibitor causes the three-dimensional shape of the enzyme and its active site to change. The enzyme no longer binds the normal substrate, or the substrate is improperly bound in a way that prevents the catalytic groups of the active site from participating in catalyzing the reaction.

Unlike competitive inhibition, noncompetitive inhibition cannot be reversed by the addition of more substrate because additional substrate has

Noncompetitive inhibitor
An inhibitor that binds to the enzyme at a location other than the active site.

> **FIGURE 10.12** The behavior of non-competitive inhibitors. The inhibitor bonds to the enzyme at a site other than the active site and changes the shape of the active site.

no effect on the enzyme-bound inhibitor (it can't displace the inhibitor because it can't bond to the site occupied by the inhibitor).

Learning Check 10.4

Compare a noncompetitive and a competitive inhibitor with regard to the following:

a. Resemblance to substrate
b. Binding site on the enzyme
c. The effect of increasing substrate concentration

> 10.8 The Regulation of Enzyme Activity

Enzymes work together in an organized yet complicated way to facilitate all the biochemical reactions needed by a living organism. For an organism to respond to changing conditions and cellular needs, very sensitive controls over enzyme activity are required. We will discuss three mechanisms by which this is accomplished: activation of zymogens, allosteric regulation, and genetic control.

The Activation of Zymogens

Zymogen or proenzyme
The inactive precursor of an enzyme.

A common mechanism for regulating enzyme activity is the synthesis of enzymes in the form of inactive precursors called **zymogens** or **proenzymes.** Some enzymes in their active form would degrade the internal structures of the cells and, ultimately, the cells themselves. Such enzymes are synthesized and stored as zymogens. When the active enzyme is needed, the stored zymo-

---

CHEMISTRY AROUND US • 10.3

ANTIFREEZE POISONING AND COMPETITIVE INHIBITION

About 50 Americans die every year as a result of ingesting ethylene glycol, a main constituent of permanent-type automobile antifreeze. The ethylene glycol itself is not lethally toxic, but an oxidation product, oxalic acid, is. The first metabolic step in the oxidation to oxalic acid is the oxidation of ethylene glycol to an aldehyde in a reaction catalyzed by the enzyme alcohol dehydrogenase. This step is inhibited by administering a nearly intoxicating dose of ethanol to the victim. The ethanol is also a substrate for alcohol dehydrogenase and competitively inhibits the reaction of ethylene glycol. The unoxidized ethylene glycol is then excreted harmlessly. The same approach is used in the treatment of methanol poisoning.

Ethanol competitively inhibits the enzymatic conversion of ethylene glycol to oxalic acid.

$$
\begin{array}{ccc}
\text{CH}_2\text{OH} & & \text{O} \quad\quad \text{H} \\
| & \xrightarrow[\text{this step}]{\substack{\text{ethanol} \\ \text{inhibits}}} & \diagdown \diagup \\
\text{CH}_2\text{OH} & & \text{C} \\
\text{ethylene glycol} & & | \\
 & & \text{CH}_2\text{OH}
\end{array}
\rightarrow \rightarrow \rightarrow
\begin{array}{c}
\text{COOH} \\
| \\
\text{COOH} \\
\text{oxalic acid}
\end{array}
$$

http://www.brookscole.com/chemistry/seager_chemistry
At Web site 10.5 is additional information about the toxicity of ethylene glycol.

TABLE 10.6 Examples of zymogens

Zymogen	Active enzyme	Function
chymotrypsinogen	chymotrypsin	Digestion of proteins
pepsinogen	pepsin	Digestion of proteins
procarboxypeptidase	carboxypeptidase	Digestion of proteins
proelastase	elastase	Digestion of proteins
prothrombin	thrombin	Blood clotting
trypsinogen	trypsin	Digestion of proteins

gen is released from storage and activated at the location of the reaction. Activation usually involves the cleavage of one or more peptide bonds of the zymogen. The digestive enzymes pepsin, trypsin, and chymotrypsin and several of the enzymes involved in blood clotting are produced and activated this way. Trypsinogen, the inactive form of trypsin, is synthesized in the pancreas and activated in the small intestine by the hydrolysis of a single peptide bond. The products are the active enzyme (trypsin) and a hexapeptide:

$$\text{trypsinogen} + H_2O \xrightarrow{\text{enterokinase}} \text{trypsin} + \text{hexapeptide} \qquad (10.13)$$

Several examples of zymogens are listed in ■ Table 10.6.

Allosteric Regulation

A second method of enzyme regulation involves the combination of the enzyme with some other compound such that the three-dimensional conformation of the enzyme is altered and its catalytic activity is changed. Compounds that alter enzymes this way are called **modulators** of the activity; they may increase the activity (activators) or decrease the activity (inhibitors). The noncompetitive inhibitors we studied in Section 10.7 are examples of modulators. Enzymes that have a quaternary protein structure with distinctive binding sites for modulators are referred to as **allosteric enzymes**. These variable-rate enzymes are often located at key control points in cellular processes. In general, **activators** act as signals for increased production, whereas inhibitors are used to stop the formation of excess products.

An excellent example of allosteric regulation—the control of an allosteric enzyme—is the five-step synthesis of the amino acid isoleucine (see ■ Figure 10.13). Threonine deaminase, the enzyme that catalyzes the first step in the

Modulator
A substance that binds to an enzyme at a location other than the active site and alters the catalytic activity.

Allosteric enzyme
An enzyme with quaternary structure whose activity is changed by the binding of modulators.

Activator
A substance that binds to an allosteric enzyme and increases its activity.

➤ **FIGURE 10.13** The allosteric regulation of threonine deaminase by isoleucine, an example of feedback inhibition.

A SUMMARY CHART OF ENZYME INHIBITORS

At this point in your study of enzymes, a summary chart is a useful tool for representing the various types of enzyme inhibitors in a compact, easy-to-review way.

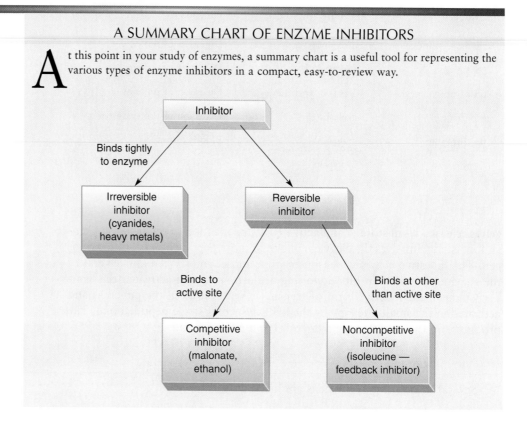

conversion of threonine to isoleucine, is subject to inhibition by the final product, isoleucine. The structures of isoleucine and threonine are quite different, so isoleucine is not a competitive inhibitor. Also, the site to which isoleucine binds to the enzyme is different from the enzyme active site that binds to threonine. This second site, called the allosteric site, specifically recognizes isoleucine, whose presence there induces a change in the conformation of the enzyme such that threonine binds poorly to the active site. Thus, isoleucine exerts an inhibiting effect on the enzyme activity. As a result, the reaction slows as the concentration of isoleucine increases, and no excess isoleucine is produced. When the concentration of isoleucine falls to a low enough level, the enzyme becomes more active, and more isoleucine is synthesized. This type of allosteric regulation in which the enzyme that catalyzes the first step of a series of reactions is inhibited by the final product is called **feedback inhibition.**

The control of enzyme activity by allosteric regulation of key enzymes is of immense benefit to an organism because it allows the concentration of cellular products to be maintained within very narrow limits.

Genetic Control

One way to increase production from an enzyme-catalyzed reaction, given a sufficient supply of substrate, is for a cell to increase the number of enzyme molecules present. The synthesis of all proteins, including enzymes, is under genetic control by nucleic acids (Chapter 11). An example of the genetic control of enzyme activity involves **enzyme induction,** the synthesis of enzymes in response to a temporary need of the cell.

Feedback inhibition
A process in which the end product of a sequence of enzyme-catalyzed reactions inhibits an earlier step in the process.

Enzyme induction
The synthesis of an enzyme in response to a cellular need.

The first demonstrated example of enzyme induction came from studies conducted in the 1950s on the bacterium *Escherichia coli*. β-Galactosidase, an enzyme required for the utilization of lactose, catalyzes the hydrolysis of lactose to D-galactose and D-glucose.

β-lactose   *β-galactoside bond*   β-D-galactose   β-D-glucose   (10.14)

In the absence of lactose in the growth medium, the number of β-galactosidase molecules per cell of *E. coli* is fewer than ten. However, if *E. coli* are added to a lactose-containing medium, within minutes the bacterium begins to produce thousands of molecules of β-galactosidase. This is an example of enzyme induction. If lactose is removed from the growth medium, the production of β-galactosidase again falls back to a low level.

The biological significance of the genetic control of enzyme activity is that it allows an organism to adapt to environmental changes. The coupling of genetic control and allosteric regulation of enzyme activity puts the cellular processes in the body under constant control.

Learning Check 10.5

Explain how each of the following processes is involved in the regulation of enzyme activity:

a. Activation of zymogens
b. Allosteric regulation
c. Genetic control

➤ 10.9 Medical Application of Enzymes

Enzymes in Clinical Diagnosis

Certain enzymes are normally found almost exclusively inside tissue cells and are released into the blood and other biological fluids only when these cells are damaged or destroyed. Because of the normal breakdown and replacement of tissue cells that goes on constantly, the blood serum contains these enzymes but at very low concentrations—often a million times lower than the concentration inside cells. However, blood serum levels of cellular enzymes increase significantly when excessive cell injury or destruction occurs, or when cells grow rapidly as a result of cancer. Changes in blood serum concentrations of specific enzymes can be used clinically to detect cell damage or uncontrolled growth, and even to suggest the site of the damage or cancer. Also, the extent of cell damage can often be estimated by the magnitude of the serum concentration increase above normal levels. For these reasons, the measurement of enzyme concentrations in blood serum and other biological fluids has become a major diagnostic tool, particularly in diagnosing diseases of the heart, liver, pancreas, prostate, and bones. In fact, certain enzyme determinations are performed so often that they have become routine procedures in the clinical chemistry laboratory. ■ Table 10.7 lists some enzymes used in medical diagnosis.

TABLE 10.7 Diagnostically useful assays of blood serum enzymes

Enzyme	Pathological condition
acid phosphatase	Prostate cancer
alkaline phosphatase (ALP)	Liver or bone disease
amylase	Diseases of the pancreas
creatine kinase (CK)	Heart attack
aspartate transaminase (AST)	Heart attack or hepatitis
alanine transaminase (ALT)	Hepatitis
lactate dehydrogenase (LDH)	Heart attack, liver damage
lipase	Acute pancreatitis
lysozyme	Monocytic leukemia

Isoenzyme
A slightly different form of the same enzyme produced by different tissues.

Isoenzymes

Lactate dehydrogenase (LDH) and creatine kinase (CK), two of the enzymes listed in Table 10.7, are particularly useful in clinical diagnosis because each occurs in multiple forms called **isoenzymes.** Although all forms of a particular isoenzyme catalyze the same reaction, their molecular structures are slightly different and their locations within the body tissues may vary.

Probably the most fully studied group of isoenzymes is that of lactate dehydrogenase (LDH). The quaternary structure of the enzyme consists of four subunits that are of two different types. One is labeled H because it is the predominant subunit present in the LDH enzyme found in heart muscle cells. The other, labeled M, predominates in other muscle cells. There are five possible ways to combine these four subunits to form the enzyme (see ■ Figure 10.14). Each combination has slightly different properties, which allows them to be separated and identified by electrophoresis.

■ Table 10.8 shows the distribution of the five LDH isoenzymes in several human tissues. Notice that each type of tissue has a distinct pattern of isoenzyme percentages. Liver and skeletal muscle, for example, contain particularly high percentages of LDH_5. Also notice the differences in LDH_1 and LDH_2 levels in heart and skeletal muscles.

➤ **FIGURE 10.14** Isomeric forms (isoenzymes) of lactate dehydrogenase.

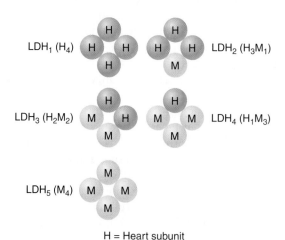

H = Heart subunit
M = Muscle subunit

TABLE 10.8 Tissue distribution of LDH isoenzymes

Tissue	LDH$_1$ (%)	LDH$_2$ (%)	LDH$_3$ (%)	LDH$_4$ (%)	LDH$_5$ (%)
Brain	23	34	30	10	3
Heart	50	36	9	3	2
Kidney	28	34	21	11	6
Liver	4	6	17	16	57
Lung	10	20	30	25	15
Serum	28	41	19	7	5
Skeletal muscle	5	5	10	22	58

Because of the difference in tissue distribution of LDH isoenzymes, serum levels of LDH are used in the diagnosis of a wide range of diseases: anemias involving the rupture of red blood cells, acute liver diseases, congestive heart failure, and muscular diseases such as muscular dystrophy. This wide range of diseases makes an LDH assay a good initial diagnostic test. Elevated levels of LDH$_1$ and LDH$_2$, for example, indicate myocardial infarction, whereas elevated LDH$_5$ indicates possible liver damage. If LDH activity is elevated, then this and other isoenzyme assays can be used to pinpoint the location and type of disease more accurately.

FOR FUTURE REFERENCE   MAGNESIUM CONTENT OF SELECTED FOODS

The body of a 160-pound person contains only about 35 g of magnesium, an amount small enough that magnesium barely qualifies as a major mineral. Despite the small amount, magnesium is involved in hundreds of reactions in the body as a result of being a cofactor in hundreds of enzymes. It plays a major role in the synthesis of high-energy ATP. In this role, it is essential to the following processes in the body: the use of glucose as an energy source; the synthesis of protein, fat, and nucleic acids; and the transport of materials across the membranes of cells. It is also involved in muscle contraction, blood clotting, the prevention of dental cavities, and the normal functioning of the immune system. The typical diet provides only about 75% of the recommended intake for adults in the U.S. However, other sources, such as hard water used for drinking, also provide significant amounts.

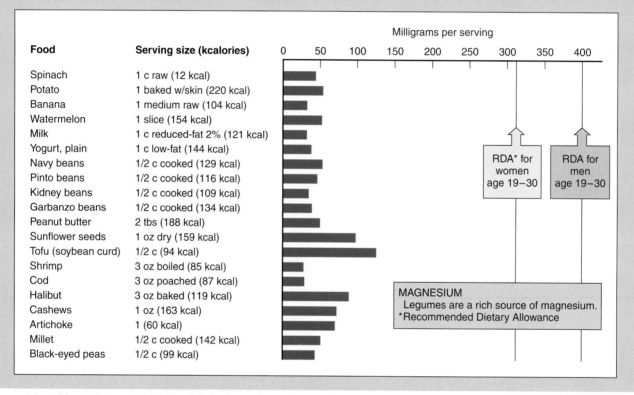

SOURCE: Adapted from Whitney, E. N.; Rolfes, S. R. *Understanding Nutrition*, 8th ed. Belmont, CA: West/Wadsworth, 1999.

CONCEPT SUMMARY

General Characteristics of Enzymes. Enzymes are highly efficient protein catalysts involved in almost every biological reaction. They are often quite specific in terms of the substance acted on and the type of reaction catalyzed.

Enzyme Nomenclature and Classification. Enzymes are grouped into six major classes on the basis of the type of reaction catalyzed. Common names for enzymes often end in *-ase* and are based on the substrate and/or the type of reaction catalyzed.

Enzyme Cofactors. Cofactors are nonprotein molecules required for an enzyme to be active. Cofactors are either organic (coenzymes) or inorganic ions.

The Mechanism of Enzyme Action. The behavior of enzymes is explained by a theory in which the formation of an enzyme–substrate complex is assumed to occur. The specificity of enzymes is explained by the lock-and-key theory and the induced-fit theory.

Enzyme Activity. The catalytic ability of enzymes is described by turnover number and enzyme international units. Experiments that measure enzyme activity are referred to as enzyme assays.

Factors Affecting Enzyme Activity. The catalytic activity of enzymes is influenced by numerous factors. The most important are substrate concentration, enzyme concentration, temperature, and pH.

Enzyme Inhibition. Chemical substances called inhibitors decrease the rates of enzyme-catalyzed reactions. Irreversible inhibitors render enzymes permanently inactive and include several very toxic substances such as the cyanide ion and heavy-metal ions. Reversible inhibitors are of two types: competitive and noncompetitive.

The Regulation of Enzyme Activity. Three mechanisms of cellular control over enzyme activity exist. One method involves the synthesis of enzyme precursors called zymogens, which are activated when needed by the cell. The second mechanism relies on the binding of small molecules (modulators), which increase or decrease enzyme activity. Genetic control of enzyme synthesis, the third method, regulates the amount of enzyme available.

Medical Applications of Enzymes. Numerous enzymes have become useful as aids in diagnostic medicine. The presence of s pecific enzymes in body fluids such as blood has been related to certain pathological conditions.

LEARNING OBJECTIVES

When you have completed your study of this chapter, you should be able to:

1. Describe the general characteristics of enzymes, and explain why enzymes are vital in body chemistry.
2. Determine the function and/or substrate of an enzyme on the basis of its name.
3. Identify the general function of cofactors.
4. Use the lock-and-key theory to explain specificity in enzyme action.
5. List two ways of describing enzyme activity.
6. Identify the factors that affect enzyme activity.
7. Compare the mechanisms of competitive and noncompetitive enzyme inhibition.
8. Describe the three methods of cellular control over enzyme activity.
9. Discuss the importance of measuring enzyme levels in the diagnosis of disease.

KEY TERMS AND CONCEPTS

Absolute specificity (10.1)
Activator (10.8)
Active site (10.4)
Allosteric enzyme (10.8)
Antibiotic (10.7)
Apoenzyme (10.3)
Coenzyme (10.3)
Cofactor (10.3)
Competitive inhibitor (10.7)
Enzyme (10.1)

Enzyme activity (10.5)
Enzyme induction (10.8)
Enzyme inhibitor (10.7)
Enzyme international unit (IU) (10.5)
Extremozyme (10.6)
Feedback inhibition (10.8)
Inborn error of metabolism (10.2)
Induced-fit theory (10.4)
Isoenzymes (10.9)
Lock-and-key theory (10.4)

Modulator (10.8)
Noncompetitive inhibitor (10.7)
Optimum pH (10.6)
Optimum temperature (10.6)
Relative specificity (10.1)
Stereochemical specificity (10.1)
Substrate (10.2)
Turnover number (10.5)
Zymogen or proenzyme (10.8)

KEY REACTIONS

1. Formation of an active enzyme (Section 10.3):

$$\text{apoenzyme} + \text{cofactor (coenzyme or inorganic ion)} \rightarrow \text{active enzyme}$$

Reaction 10.4

2. Mechanism of enzyme action (Section 10.4):

$$E + S \rightleftharpoons ES \rightarrow E + P$$

Reaction 10.6

GENERAL CHARACTERISTICS OF ENZYMES (SECTION 10.1)

10.1 What is the role of enzymes in the body?

10.2 List two ways that enzyme catalysis of a reaction is superior to normal laboratory conditions.

10.3 What is the relationship between an enzyme and the energy of activation for a reaction?

10.4 Define what is meant by the term *absolute specificity*.

10.5 List three ways in which enzymes are particularly well suited for their essential roles in living organisms.

ENZYME NOMENCLATURE AND CLASSIFICATION (SECTION 10.2)

10.6 What name is given to the nomenclature system for enzymes?

10.7 What is the relationship between urea and urease? Between maltose and maltase?

10.8 Match the following enzymes and substrates:

ENZYME	SUBSTRATE
a. urease	fumarate
b. fumarase	arginine
c. fructase	lactose
d. arginase	urea
e. lactase	fructose

10.9 Match the following general enzyme names and reactions catalyzed:

ENZYME	REACTION CATALYZED
a. decarboxylase	formation of ester linkages
b. phosphatase	removal of carboxyl groups from compounds
c. peptidase	hydrolysis of peptide linkages
d. esterase	hydrolysis of phosphate ester linkages

10.10 Because one substrate may undergo a number of reactions, it is often convenient to use an enzyme nomenclature system that includes both the substrate name (or general type) and the type of reaction catalyzed. Identify the substrate and type of reaction for the following enzyme names:
   a. succinate dehydrogenase
   b. L-amino acid reductase
   c. cytochrome oxidase

ENZYME COFACTORS (SECTION 10.3)

10.11 Some enzymes consist of protein plus another component. Which of the terms *cofactor* or *coenzyme* correctly describes each of the following nonprotein components?
   a. an inorganic ion
   b. a nonspecific component

   c. an organic material
   d. nicotinic acid,

10.12 What are the relationships between the terms *cofactor, active enzyme,* and *apoenzyme*?

10.13 State the relationship between vitamins and enzyme activity.

10.14 List some typical inorganic ions that serve as cofactors.

10.15 What vitamins are required for FAD and NAD^+ formation?

THE MECHANISM OF ENZYME ACTION (SECTION 10.4)

10.16 What is an enzyme–substrate complex?

10.17 In what way are the substrate and active site of an enzyme related?

10.18 Compare the lock-and-key theory with the induced-fit theory.

10.19 How is enzyme specificity explained by the lock-and-key theory?

ENZYME ACTIVITY (SECTION 10.5)

10.20 What observations may be used in experiments to determine enzyme activity?

10.21 Define the term *turnover number*.

10.22 What is an enzyme international unit? Why is the international unit a useful method of expressing enzyme activity in medical diagnoses?

FACTORS AFFECTING ENZYME ACTIVITY (SECTION 10.6)

10.23 Use graphs to illustrate enzyme activity as a function of the following:
   a. substrate concentration
   b. enzyme concentration
   c. pH
   d. temperature

10.24 Write a single sentence to summarize the information of each graph in Exercise 10.23.

10.25 What happens to the rate of an enzyme-catalyzed reaction as substrate concentration is raised beyond the saturation point?

10.26 How might V_{max} for an enzyme be determined?

10.27 How would you expect hypothermia to affect enzyme activity in the body?

10.28 Using Table 10.4, list three enzymes and their optimum pH.

ENZYME INHIBITION (SECTION 10.7)

10.29 Distinguish between competitive and noncompetitive enzyme inhibition.

10.30 Distinguish between irreversible and reversible enzyme inhibition.

10.31 How does each of the following irreversible inhibitors interact with enzymes?
a. cyanide
b. heavy-metal ions

10.32 List an antidote for each of the two poisons in Exercise 10.31 and describe how each functions.

THE REGULATION OF ENZYME ACTIVITY (SECTION 10.8)

10.33 Why is the regulation of enzyme activity necessary?

10.34 List three mechanisms for the control of enzyme activity.

10.35 Describe the importance of zymogens in the body. Give an example of an enzyme that has a zymogen.

10.36 What is another name for a zymogen?

10.37 What type of enzyme possesses a binding site for modulators?

10.38 Name and contrast the two types of modulators.

10.39 What is feedback enzyme inhibition?

10.40 What is meant by genetic control of enzyme activity and enzyme induction?

MEDICAL APPLICATION OF ENZYMES (SECTION 10.9)

10.41 Why are enzyme assays of blood serum useful in a clinical diagnosis?

10.42 How is each of the following enzyme assays useful in diagnostic medicine?
a. CK
b. ALP
c. amylase

10.43 What are isoenzymes? List two examples of isoenzymes.

10.44 Why is an LDH assay a good initial diagnostic test?

10.45 What specific enzyme assays are particularly useful in diagnosing heart attacks?

CHEMISTRY AROUND US AND OVER THE COUNTER

10.46 What expression is sometimes used to refer to hereditary diseases caused by the absence of enzymes?

10.47 List two diseases that result from enzyme deficiencies.

10.48 What are extremozymes?

10.49 Why is the use of extremozymes in industrial processes an attractive area for research?

10.50 How much topical antibiotic cream should be applied to a minor cut or insect bite?

10.51 Why should a topical antibiotic cream not be used repeatedly on large areas of skin?

10.52 How does ethanol serve as an antidote for ethylene glycol poisoning?

10.53 In Exercise 10.52, is ethanol acting as a competitive or a noncompetitive inhibitor? Explain.

▷ POINTS TO PONDER AND PURSUE

10.1 Explain how the pasteurization of milk utilizes one of the factors that influence enzyme activity.

10.2 Explain how food preservation by freezing utilizes one of the factors that influence enzyme activity.

10.3 Some pottery glazes contain lead compounds. Why would it be unwise to use such pottery to store or serve food?

10.4 Urease can catalyze the hydrolysis of urea, but not the hydrolysis of methyl urea. Why?

10.5 The active ingredient in meat tenderizer is the enzyme papain, a protease. Explain how treating meat with this material before cooking would make it more tender.

10.6 Why are enzymes that are used for laboratory or clinical work stored in refrigerators?

10.7 Describe the differences between graphs showing temperature versus reaction rate for an enzyme-catalyzed reaction and an uncatalyzed reaction.

10.8 Answer the question associated with Figure 10.3. How else might ammonia be detected?

10.9 In Figure 10.9, it is noted that pickles resist spoilage. Why is that true for this acidic food?

10.10 What is the source of the unique enzymes discussed at Web site 10.2?

10.11 According to Web site 10.3, what types of food cause the greatest leaching of lead from ceramic cups and mugs?

10.12 What enzyme did Alexander Fleming discover in 1921, according to Web site 10.4?

10.13 What conditions are indicative of the early stages of ethylene glycol toxicity? See Web site 10.5.

http://www.brookscole.com/chemistry/seager_chemistry

11

Nucleic Acids and Protein Synthesis

Numerous technologies are available to visualize soft tissues, and organs such as the heart, by using various forms of radiation. However, sonography technologists (sonographers) use sound waves and ultrasound scanning equipment rather than radiation. Sonography is often used to view a fetus as it develops prior to birth. Although ultrasound may reveal a developing fetus, the blueprint for its growth has already been determined through nucleic acids, the topic of this chapter.

http://www.brookscole.com/chemistry/
seager_chemistry
At Web site 11.1, you will find career information about diagnostic medical sonography.

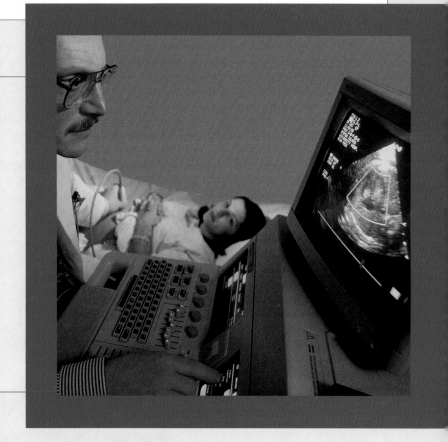

CHAPTER OUTLINE

Nucleic acid
A biomolecule involved in the transfer of genetic information from existing cells to new cells.

O ne of the most remarkable properties of living cells is their ability to produce nearly exact replicas of themselves through hundreds and thousands of generations. Such a process requires that certain types of information be passed unchanged from one generation to the next. The transfer of necessary genetic information to new cells is accomplished by means of biomolecules called **nucleic acids.** The high-molecular-weight compounds represent coded information, much as words represent information in a book. It is the nearly infinite variety of possible structures that enables nucleic acids to represent the huge amount of information that must be transmitted sexually or asexually to reproduce a living organism. This information controls the inherited characteristics of the individuals in the new generation and determines the life processes as well (see ■ Figure 11.1).

➤ 11.1 Components of Nucleic Acids

Ribonucleic acid (RNA)
A nucleic acid found mainly in the cytoplasm of cells.

Deoxyribonucleic acid (DNA)
A nucleic acid found primarily in the nuclei of cells.

Nucleotide
The repeating structural unit or monomer of polymeric nucleic acids.

Nucleic acids are classified into two categories: **ribonucleic acid (RNA)**, found mainly in the cytoplasm of living cells, and **deoxyribonucleic acid (DNA)**, found primarily in the nuclei of cells. Both DNA and RNA are polymers, consisting of long, linear molecules. The repeating structural units, or monomers, of the nucleic acids are called **nucleotides.** Nucleotides, however, are composed of three simpler components: a heterocyclic base, a sugar, and a phosphate (see ■ Figure 11.2). These components will be discussed individually.

Each of the five bases commonly found in nucleic acids are heterocyclic compounds that can be classified as either a pyrimidine or a purine, the parent compounds from which the bases are derived:

pyrimidine   purine

➤ **FIGURE 11.1** Nucleic acids passed from the parents to offspring determine the inherited characteristics.

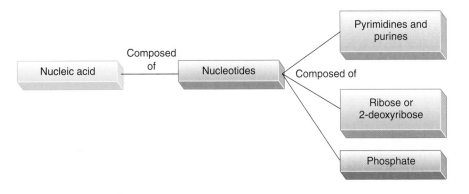

➤ **FIGURE 11.2** The composition of nucleic acids.

The three pyrimidine bases are uracil, thymine, and cytosine, usually abbreviated U, T, and C. Adenine (A) and guanine (G) are the two purine bases. Adenine, guanine, and cytosine are found in both DNA and RNA, but uracil is ordinarily found only in RNA, and thymine only in DNA. Structural formulas of the five bases are given in ■ Figure 11.3.

➤ **FIGURE 11.3** The major bases of nucleic acids.

Pyrimidines

uracil
(only in RNA)

thymine
(only in DNA)

cytosine

Purines

adenine

guanine

The sugar component of RNA is D-ribose, as the name ribonucleic acid implies. In deoxyribonucleic acid (DNA), the sugar is D-deoxyribose. Both sugars occur in the β-configuration.

HO—CH$_2$ ⏤O⏤ OH ⟵ *β-configuration*

OH OH
D-ribose

HO—CH$_2$ ⏤O⏤ OH

OH ⟵ *no — OH group*
D-deoxyribose

Phosphate, the third component of nucleotides, is derived from phosphoric acid (H_3PO_4), which under cellular pH conditions exists in ionic form:

$$^-O—\overset{\overset{\displaystyle O}{\|}}{\underset{\underset{\displaystyle O^-}{|}}{P}}—OH$$

The formation of a nucleotide from these three components is represented in Reaction 11.1:

adenine

$$^{-}O-\overset{\overset{\displaystyle O}{\|}}{\underset{\underset{\displaystyle O^{-}}{|}}{P}}-OH + HO-CH_2 \quad \text{ribose} \quad \longrightarrow \quad ^{-}O-\overset{\overset{\displaystyle O}{\|}}{\underset{\underset{\displaystyle O^{-}}{|}}{P}}-O-CH_2 \quad + 2H_2O \quad (11.1)$$

phosphate

adenosine 5′-monophosphate
(AMP)

In a nucleotide, the base is always attached to the 1′ position of the sugar, and the phosphate is generally located at the 5′ position. The carbon atoms in the sugar are designated with a number followed by a prime to distinguish them from the atoms in the bases. Because the nucleotide of Reaction 11.1 contains ribose as the sugar, the nucleotide must be one found in RNA. The general structure of a nucleotide can be represented as

$$^{-}O-\overset{\overset{\displaystyle O}{\|}}{\underset{\underset{\displaystyle O^{-}}{|}}{P}}-O-CH_2 \quad \text{base}$$

OH present
in RNA

or

| phosphate | | base |
| | | sugar |

➤ 11.2 The Structure of DNA

DNA molecules, among the largest molecules known, contain between 1 and 100 million nucleotide units. The nucleotides are joined together in nucleic acids by phosphate groups that connect the 5′ carbon of one nucleotide to the 3′ carbon of the next. These linkages are referred to as *phosphodiester bonds*. The result is a chain of alternating phosphate and sugar units to which the bases are attached. The sugar–phosphate chain is referred to as the **nucleic acid backbone,** and it is constant throughout the entire DNA molecule (see ■ Figure 11.4). Thus, one DNA molecule differs from another only in the sequence, or order of attachment, of the bases along the backbone. Just as the amino acid sequence of a protein determines the primary structure of the protein, the order of the bases provides the primary structure of DNA.

■ Figure 11.5 shows the structure of a tetranucleotide that represents a partial structural formula of a DNA molecule. The end of the polynucleotide

Nucleic acid backbone
The sugar–phosphate chain that is common to all nucleic acids.

➤ **FIGURE 11.4** The nucleic acid backbone.

> **FIGURE 11.5** The structure of ACGT, a tetranucleotide segment of DNA. The nucleic acid backbone is in color.

segment of DNA in Figure 11.5 that has no nucleotide attached to the 5′ OH is called the 5′ end of the segment. The other end of the chain is the 3′ end. By convention, the sequence of bases along the backbone is read from the 5′ end to the 3′ end. Because the backbone structure along the chain does not vary, a polynucleotide structure may conveniently be abbreviated by giving only the sequence of the bases along the backbone. For example, ACGT represents the tetranucleotide in Figure 11.5.

Draw the structural formula for a trinucleotide portion of DNA with the sequence CTG. Point out the 5′ and 3′ ends of the molecule. Indicate with arrows the phosphodiester linkages. Enclose the nucleotide corresponding to T in a box with dotted lines.

Learning Check 11.1

The Secondary Structure of DNA

The secondary structure of DNA was proposed in 1953 by American molecular biologist James D. Watson and English biologist Francis H. C. Crick (see ■ Figure 11.6). This was perhaps the greatest discovery of modern

➤ **FIGURE 11.6** J. D. Watson (left) and F. H. C. Crick (right), working in the Cavendish Laboratory at Cambridge, England, build scale models of the DNA double helix based on the X-ray crystallography data of M. H. F. Wilkins and Rosalind Franklin.

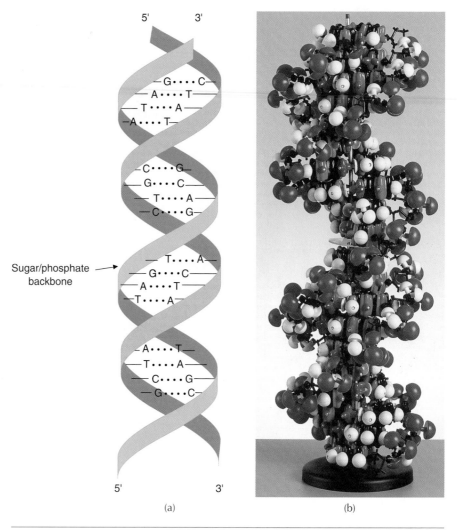

Sugar/phosphate backbone

(a)

(b)

➤ **FIGURE 11.7** The double helix of DNA: (a) a schematic drawing and (b) a three-dimensional molecular model. Both representations show the bases pointing toward the center of the helix away from the sugar–phosphate backbone.

biology, and it earned its discoverers the Nobel Prize in medicine and physiology in 1962.

The model of DNA established by Watson and Crick was based on key contributions of other researchers, including the analysis of the base composition of DNA. The analysis of DNA from many different forms of life revealed an interesting pattern. The relative amounts of each base often varied from one organism to another, but in all DNA the percentages of adenine and thymine were always equal to each other as were the percentages of guanine and cytosine. For example, human DNA contains 20% guanine, 20% cytosine, and 30% adenine, and 30% thymine. Watson and Crick concluded from these and other types of data that DNA is composed of two strands entwined around each other in a double helix, as shown in ■ Figure 11.7.

The two intertwined polynucleotide chains of the DNA double helix run in opposite (antiparallel) directions. Thus, each end of the double helix contains the 5′ end of one chain and the 3′ end of the other. The sugar–phosphate backbone is on the outside of the helix, and the bases point inward. The unique feature of the Watson and Crick structure is the way in which the chains are held together to form the double helix. They theorized that the DNA struc-

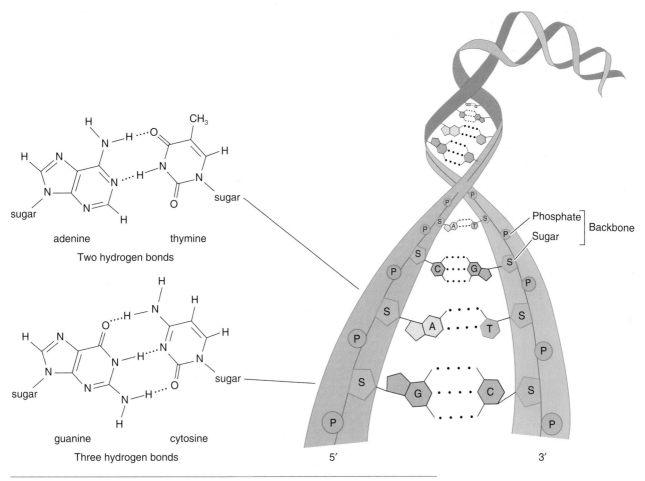

➤ **FIGURE 11.8** Base pairing. Hydrogen bonding between complementary base pairs holds the two strands of a DNA molecule together.

ture is stabilized by hydrogen bonding between the bases that extend inward from the sugar–phosphate backbone (see ■ Figure 11.8). The spacing in the interior of the double helix is such that adenine always hydrogen bonds to thymine, and guanine always hydrogen bonds to cytosine. Thus, in double-helical DNA, wherever there is an adenine on one strand of the helix, there must be a thymine on the other strand, and similarly for guanine and cytosine. The two DNA strands with these matched sequences are said to be **complementary** to each other.

Complementary DNA strands
Two strands of DNA in a double-helical form such that adenine and guanine of one strand are matched and hydrogen-bonded to thymine and cytosine, respectively, of the second strand.

➤ **Example 11.1**

One strand of a DNA molecule has the base sequence CCATTG. What is the base sequence for the complementary strand? GGTAAC

Solution

Three things must be remembered and used to solve this problem: (1) The base sequence of a DNA strand is always written from the 5′ end to the 3′ end; (2) adenine (A) is always paired with its complementary base thymine (T), and guanine (G) is always paired with its complement cytosine (C); and (3) in double-stranded DNA, the two strands run in opposite directions, so that the

OVER THE COUNTER • 11.1

NUCLEIC ACID SUPPLEMENTS

Because of their fundamental roles in life itself, it is not surprising that many claims of health benefits related to these substances have been made. As a result of these claims, DNA, RNA, and their various derivatives are well represented on the dietary supplements shelves of retailers. In addition, numerous shampoos and other cosmetics that contain DNA and RNA are also available. The idea behind the use of such products is that as we age, our DNA/RNA becomes depleted or defective, and must be replaced if good health is to be maintained or restored. It has been suggested that injections or oral supplements could be used to accomplish the replacement.

The health benefits claimed for such therapy include: slowing the aging process, improving memory and mental processes, stimulating the immune system, and fighting cancer. There has been almost universal doubt about such health benefits, especially if the nucleic acids were taken orally. It has been argued that these substances would be destroyed during digestion and therefore could not reach target organs and tissues of the body.

Only very recently have reports been made of scientific evidence that nucleic acid supplementation might provide some benefits in terms of positively influencing the immune response of the body. Evidence to support the claims of slowing the aging process, and improving memory and mental processes is very inconclusive and not promising at this time. There has been very little investigation of the anti-cancer properties of RNA and DNA in their native forms. However, some synthetic forms of the nucleic acid have been studied in this regard and some promising results have been obtained in the treatment of specific types of cancer.

The only negative claim about using nucleic acid supplements is that they may produce excessive uric acid and promote or aggravate gout. Because of this, anyone with gout or a tendency to develop it should not use nucleic acid supplements without first obtaining a physician's approval.

Dietary supplements containing nucleic acids.

5′ end of one strand is associated with the 3′ end of the other strand. These three ideas lead to the following:

Original strand: 5′ end C–C–A–T–T–G 3′ end

Complementary stand: 3′ end G–G–T–A–A–C 5′ end

When we follow convention and write the sequence of the complementary strand in the 5′ to 3′ direction, it becomes

5′ C–A–A–T–G–G 3′

Learning Check 11.2

Write the complementary base sequence for the DNA strand TTACG.

5′ CGTAA 3′

► 11.3 DNA Replication

DNA is responsible for one of the most essential functions of living organisms, the storage and transmission of hereditary information. A human cell normally contains 46 structural units called **chromosomes** (see ■ Figure 11.9). Each chromosome contains one molecule of DNA coiled tightly about a group of small, basic proteins called histones. Individual sections of DNA molecules make up **genes,** the fundamental units of heredity. In Section 11.5, you'll learn

Chromosome
A tightly packed bundle of DNA and protein that is involved in cell division.

Gene
An individual section of a chromosomal DNA molecule that is the fundamental unit of heredity.

Chromosome

(a)

DNA double helix

Histones

(b)

➤ **FIGURE 11.9** Chromosomes. (a) The 46 chromosomes of a human cell. (b) Each chromosome is a protein-coated strand of multicoiled DNA.

that each gene directs the synthesis of a specific protein. The number of genes contained in the structural unit of an organism varies with the type of organism. For example, a virus, the smallest structure known to carry genetic information, is thought to contain from a few to several hundred genes. A bacterial cell, such as *Escherichia coli (E. coli),* contains about 1000 genes, whereas a human cell contains 100,000 or more.

One of the elegant features of the Watson–Crick three-dimensional structure for DNA was that it provided the first satisfactory explanation for the transmission of hereditary information—it explained how DNA is duplicated for the next generation.

The process by which an exact copy of DNA is produced is called **replication.** It occurs when two strands of DNA separate and each serves as a template (pattern) for the construction of its own complement. The process

Replication
The process by which an exact copy of a DNA molecule is produced.

➤ **FIGURE 11.10** A schematic diagram of the replication of DNA.

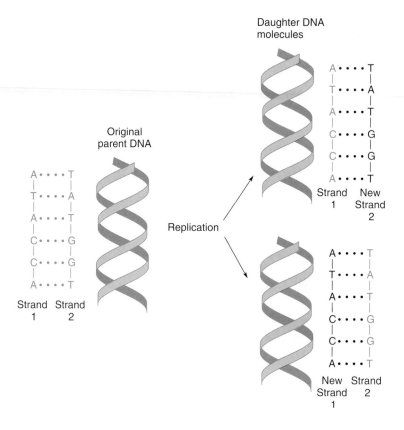

generates new DNA double-stranded molecules that are exact replicas of the original DNA molecule. A schematic diagram of the replication process is presented in ■ Figure 11.10. Notice that the two daughter DNA molecules have exactly the same base sequences as the original parent DNA. Notice, too, that each daughter contains one strand of the parent and one new strand that is complementary to the parent strand. This type of replication is called **semiconservative replication**.

The process of replication can be viewed as occurring in three steps.

Step 1. Unwinding of the double helix: Replication begins when the enzyme helicase catalyzes the separation and unwinding of the nucleic acid strands at a specific point along the DNA helix. In this process, hydrogen bonds between complementary base pairs are broken, and the bases that were formerly in the center of the helix are exposed. The point were this unwinding takes place is called a **replication fork** (see ■ Figure 11.11).

Step 2. Synthesis of DNA segments: DNA replication takes place along both nucleic acid strands separated in Step 1. The process proceeds from the 3′ end toward the 5′ end of the exposed strand (the template). Because the two strands are antiparallel, the synthesis of new nucleic acid strands proceeds toward the replication fork on one exposed strand and away from the replication fork on the other strand (Figure 11.11). New (daughter) DNA strands form as nucleotides, complementary to those on the exposed strands, are linked together under the influence of the enzyme DNA polymerase. Notice that the daughter chain grows from the 5′ end toward the 3′ end as the process moves from the 3′ end of the exposed (template) strand toward the 5′ end. As the synthesis proceeds, a second replication fork is created when a new section of

Semiconservative replication
A replication process that produces DNA molecules containing one strand from the parent and a new strand that is complementary to the strand from the parent.

Replication fork
A point where the double helix of a DNA molecule unwinds during replication.

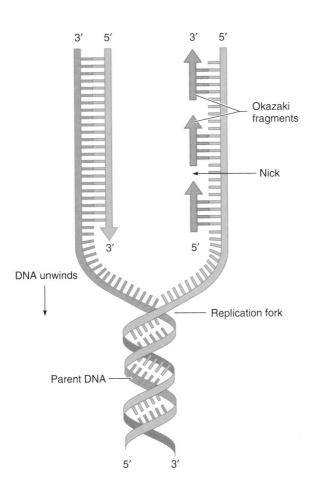

DNA unwinds. The daughter strand that was growing toward the first replication fork continues growing smoothly toward the new fork. However, the other daughter strand was growing away from the first fork. A new segment of this strand begins growing from the new fork but is not initially bound to the segment that grew from the first fork. Thus, as the parent DNA progressively unwinds, this daughter strand is synthesized as a series of fragments that are bound together in Step 3. The gaps or breaks between segments in this daughter strand are called nicks, and the DNA fragments separated by the nicks are called **Okazaki fragments** after their discoverer, Reiji Okazaki (Figure 11.11).

Step 3. Closing the nicks: One daughter DNA strand is synthesized without any nicks, but the Okazaki fragments of the other strand must be joined together. An enzyme called DNA ligase catalyzes this final step of DNA replication. The result is two DNA double-helical molecules that are identical to the original molecule.

Observations with an electron microscope show that the replication of DNA molecules in eukaryotic cells occurs simultaneously at many points along the original DNA molecule. These replication zones blend together as the process of replication continues. This is necessary if long molecules are to be replicated rapidly. For example, it is estimated that the replication of the largest chromosome in *Drosophila* (the fruit fly) would take more than 16 days if there were only one origin for replication. Research results indicate that the actual replication is accomplished in less than 3 minutes because it takes place simultaneously at more than 6000 replication forks per DNA molecule. ■ Figure 11.12 shows a

Okazaki fragment
A DNA fragment produced during replication as a result of strand growth in a direction away from the replication fork.

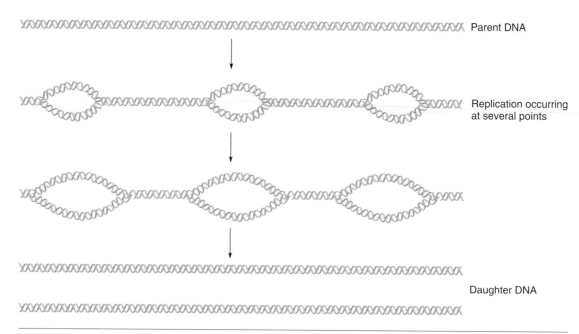

Parent DNA

Replication occurring
at several points

Daughter DNA

➤ **FIGURE 11.12** A schematic representation of eukaryotic chromosome replication. The new DNA is shown in purple.

schematic diagram in which replication is proceeding at several points in the DNA chain.

A knowledge of the DNA replication process led scientist Kary Mullis in 1983 to a revolutionary laboratory technique called the *polymerase chain reaction (PCR)*. The PCR technique mimics the natural process of replication, in which the DNA double helix unwinds. As the two strands separate, DNA polymerase makes a copy using each strand as a template. To perform PCR, a small quantity of the target DNA is added to a test tube along with a buffered solution containing DNA polymerase, the cofactor $MgCl_2$, the four nucleotide building blocks, and primers. The primers are short polynucleotide segments that will bind to the separated DNA strands and serve as starting points for new chain growth. The PCR mixture is taken through three-step replication cycles:

1. Heat (94°C–96°C) is used for one to several minutes to unravel (denature) the DNA into single strands.
2. The tube is cooled to 50°C–65°C for one to several minutes, during which the primers hydrogen-bond to the separated strands of target DNA.
3. The tube is heated to 72°C for one to several minutes, during which time the DNA polymerase synthesizes new strands.

Each cycle doubles the amount of DNA. Following 30 such cycles, a theoretical amplification factor of 1 billion is attained.

Almost overnight, PCR became a standard research technique for detecting all manner of mutations associated with genetic disease (Section 11.9). PCR can also be used to detect the presence of unwanted DNA, as in the case of a bacterial or viral infection. Conventional tests that involve the culture of microorganisms or the use of antibodies can take weeks to perform. PCR offers a fast and simple alternative. The ability of PCR to utilize degraded DNA samples and sometimes the DNA from single cells is of great interest to forensic scientists. PCR has also permitted DNA to be amplified from some unusual sources, such as extinct mammals, Egyptian mummies, and ancient insects trapped in amber.

http://www.brookscole.com/chemistry/
seager_chemistry
At Web site 11.2, you will find more information on the polymerase chain reaction.

➤ **FIGURE 11.13** A portion of RNA that has folded back on itself and formed a double-helical region.

➤ 11.4 Ribonucleic Acid (RNA)

RNA, like DNA, is a long, unbranched polymer consisting of nucleotides joined by $3' \longrightarrow 5'$ phosphodiester bonds. The number of nucleotides in an RNA molecule ranges from as few as 73 to many thousands. The primary structure of RNA differs from that of DNA in two ways. As we learned in Section 11.1, the sugar unit in RNA is ribose rather than deoxyribose. The other difference is that RNA contains the base uracil (U) instead of thymine (T).

The secondary structure of RNA is also different from that of DNA. RNA molecules are single-stranded, except in some viruses. Consequently, an RNA molecule need not contain complementary base ratios of 1:1. However, RNA molecules do contain regions of double-helical structure where they form loops (see ■ Figure 11.13). In these regions, adenine (A) pairs with uracil (U), and guanine (G) pairs with cytosine (C). The proportion of helical regions varies over a wide range depending on the kind of RNA studied, but a value of 50% is typical.

RNA is distributed throughout cells; it is present in the nucleus, the cytoplasm, and in mitochondria (Table 8.3; Section 12.7). Cells contain three kinds of RNA: messenger RNA (mRNA), ribosomal RNA (rRNA), and transfer RNA (tRNA). Each of these kinds of RNA performs an important function in protein synthesis.

Messenger RNA

Messenger RNA (mRNA) functions as a carrier of genetic information from the DNA of the cell nucleus directly to the cytoplasm, where protein synthesis takes place. The bases of mRNA are in a sequence that is complementary to the base sequence of one of the strands of the nuclear DNA. In contrast to DNA, which remains intact and unchanged throughout the life of the cell, mRNA has a short lifetime—usually less than an hour. It is synthesized as needed and then rapidly degraded to the constituent nucleotides.

Messenger RNA (mRNA)
RNA that carries genetic information from the DNA in the cell nucleus to the site of protein synthesis in the cytoplasm.

Ribosomal RNA

Ribosomal RNA (rRNA) constitutes 80–85% of the total RNA of the cell. It is located in the cytoplasm in organelles called **ribosomes**, which are about 65% rRNA and 35% protein. The function of ribosomes as the sites of protein synthesis is discussed in Section 11.8.

Ribosomal RNA (rRNA)
RNA that constitutes about 65% of the material in ribosomes, the sites of protein synthesis.

Ribosome
A subcellular particle that serves as the site of protein synthesis in all organisms.

Transfer RNA

Transfer RNA (tRNA) molecules deliver amino acids, the building blocks of proteins, to the site of protein synthesis. Cells contain at least one specific type of tRNA for each of the 20 common amino acids found in proteins. These tRNA molecules are the smallest of all the nucleic acids, containing 73–93 nucleotides per chain.

Transfer RNA (tRNA)
RNA that delivers individual amino acid molecules to the site of protein synthesis.

CHEMISTRY AROUND US • 11.1

CLONING ANIMALS

It has generally been thought that DNA from mature cells could not start over and guide an egg's development into a complete, multicellular organism. However, this concept may have to be modified. The first breakthrough came in the form of a now-famous sheep named Dolly. In 1997, Ian Wilmut of the Roslin Institute in Scotland and Keith Campbell of PPS Therapeutics shocked the world by announcing that they had used the DNA of an adult sheep (as opposed to an embryo) to create a lamb genetically identical to the adult.

They accomplished this feat by taking nuclei from adult cells (from mammary glands) and injecting them into enucleated eggs (eggs whose own nuclei had been removed). The resulting eggs were then subjected to an electrical pulse that fused the injected nuclei with the eggs and also triggered them to start dividing and develop into embryos. The embryos were later implanted into foster mothers. At first, scientists were skeptical because it was difficult to duplicate the experiment. But recent detailed DNA fingerprinting has proven that Dolly is indeed a clone of an adult sheep.

Sheep are not the only mammals to be cloned. Ryuzo Yanagimachi of the University of Hawaii in Honolulu and Teruhiko Wakayama of the University of Tokyo in Japan have used a variation of the cloning technique of the Scottish researchers to clone mice. Wakayama used a very fine needle to remove the nucleus from the donor cell and inject it into an enucleated cell. Instead of using an electrical pulse to fuse the injected nucleus and activate the egg, the injected cells were allowed to sit undisturbed for up to 6 hours. This allowed the egg cell time to alter the injected

DNA so its developmental genes could be expressed. The division and development of the egg into an embryo was triggered by placing the injected eggs in a culture medium containing a strontium compound. Strontium stimulates the release of calcium from the egg's internal store, which is the same signal that tells fertilized eggs to begin to divide. Also, the adult donor cells used by the Honolulu team were cumulus cells, which surround an egg as it matures.

The resulting cloned mice seemed to be perfectly normal. The researchers have been able to produce healthy offspring by mating the clones with other mice, as well as by making clones from the previous clones. Although the method is not efficient and only produces live mice 2–3% of the time, it is still impressive.

One of the potential uses of cloning is to genetically engineer and then rapidly mass-produce animals that can produce and secrete human proteins in their milk or blood. These *transgenic* animals could produce therapeutic proteins for treating humans and for studying human diseases. Sometimes transgenic animals are infertile, and cloning can be used to continue valuable breeds. Another possible application is *xenotransplantation,* the transplanting of organs from other species into humans. Transplanted foreign organs are promptly rejected by the human immune system, but by using nuclear cell transplant techniques, researchers might modify animal cells to grow organs that would not trigger rejection. Finally, the reprogramming of cells could be studied, which could lead to greater efficiency in generating live embryos, thereby leading to advances in human tissue engineering.

Anticodon

A three-base sequence in tRNA that is complementary to one of the codons in mRNA.

A comparison of RNA characteristics for the three different forms is summarized in ■ Table 11.1.

Because of their small size, a number of tRNAs have been studied extensively. ■ Figure 11.14 shows a representation of the secondary structure of a typical tRNA. This tRNA molecule, like all others, has regions where there is hydrogen bonding between complementary bases, and regions (loops) where there is no hydrogen bonding.

Two regions of tRNA molecules have important functions during protein synthesis. One of these regions, designated the **anticodon**, enables the tRNA

TABLE 11.1 Different forms of RNA molecules in *Escherichia coli (E. coli)*

Class of RNA	% in cells	Number of RNA subtypes	Number of nucleotides
Ribosomal RNA (rRNA)	80	3	120 1700 3700
Transfer RNA (tRNA)	15	60	73–93
Messenger RNA (mRNA)	5	Many	75–3000

➤ **FIGURE 11.14** The typical tRNA cloverleaf structure.

to bind to mRNA during protein synthesis. The second important site is the 3′ end of the molecule, which binds to an amino acid and transports it to the site of protein synthesis. Each amino acid is joined to the 3′ end of its specific tRNA by an ester bond that forms between the carboxyl group of the amino acid and the 3′ hydroxy group of ribose. The reaction is catalyzed by an enzyme that matches tRNA molecules to their proper amino acids. These enzymes, which are very specific for both the structure of the amino acid and the tRNA, rarely cause a bond to form between an amino acid and the wrong tRNA. When a tRNA molecule is attached to its specific amino acid, it is said to be "activated" because it is ready to participate in protein synthesis. ■ Figure 11.15 shows the general structure of an activated tRNA and a schematic representation that will be used in Section 11.8 to describe protein synthesis.

➤ **FIGURE 11.15** An activated tRNA: (a) general structure and (b) a schematic representation.

(a)

(b)

➤ 11.5 The Flow of Genetic Information

Central dogma of molecular biology
The well-established process by which genetic information stored in DNA molecules is expressed in the structure of synthesized proteins.

You learned in Section 11.3 that DNA is the storehouse of genetic information in the cell and that the stored information can be passed on to new cells as DNA undergoes replication. In this section, you will discover how the information stored in DNA is expressed within the cell.

This process of expression is so well established that it is called the **central dogma of molecular biology.** The accepted dogma, or principle, says that genetic information contained in DNA molecules is transferred to RNA molecules. The transferred information of RNA molecules is then expressed in the structure of synthesized proteins. In other words, the genetic information in DNA (genes) directs the synthesis of certain proteins. In fact, there is a specific DNA gene for every protein in the body. DNA does not direct the synthesis of carbohydrates, lipids, or the other nonprotein substances essential for life. However, these other materials are manufactured by the cell through reactions made possible by enzymes (proteins) produced under the direction of DNA. Thus, in this respect, the information stored in DNA really does determine every characteristic of the living organism.

Two steps are involved in the flow of genetic information: transcription and translation. In higher organisms (eukaryotes), the DNA containing the stored information is located in the nucleus of the cell, and protein synthesis occurs in the cytoplasm. Thus, the stored information must first be carried out of the nucleus. This is accomplished by **transcription,** or transferring the necessary information from a DNA molecule onto a molecule of messenger RNA. The appropriately named messenger RNA carries the information (the message) from the nucleus to the site of protein synthesis in the cellular cytoplasm.

In the second step, the mRNA serves as a template on which amino acids are assembled in the proper sequence necessary to produce the specified protein. This takes place when the code or message carried by mRNA is **translated** into an amino acid sequence by tRNA. There is an exact word-to-word translation from mRNA to tRNA. Thus, each word in the mRNA language has a corresponding word in the amino acid language of tRNA. This communicative relationship between mRNA nucleotides and amino acids is called the *genetic code* (Section 11.7). ■ Figure 11.16 summarizes the mechanisms for the flow of genetic information in the cell.

Transcription
The transfer of genetic information from a DNA molecule to a molecule of messenger RNA.

Translation
The conversion of the code carried by messenger RNA into an amino acid sequence of a protein.

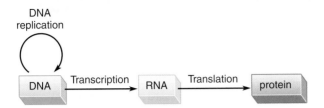

➤ **FIGURE 11.16** The flow of genetic information in the cell.

➤ 11.6 Transcription: RNA Synthesis

An enzyme called RNA polymerase catalyzes the synthesis of RNA. During the first step of the process, the DNA double helix begins to unwind at a point near the gene that is to be transcribed. Only one strand of the DNA molecule is transcribed. Ribonucleotides are linked together along the unwound DNA strand in a sequence determined by complementary base pairing of the DNA strand bases and ribonucleotide bases.

The DNA strand always has one sequence of bases recognized by RNA polymerase as the initiation or starting point. Starting at this point, the enzyme catalyzes the synthesis of mRNA in the 5′ to 3′ direction until it reaches another sequence of bases that designates the termination point. Because the complementary chains of RNA and DNA run in opposite directions, the enzyme must move along the DNA template in the 3′ to 5′ direction (see ■ Figure 11.17). Once the mRNA molecule has been synthesized, it moves away from the DNA template, which then rewinds to form the original double-helical structure. Transfer RNA and ribosomal RNA are synthesized in the same way, with DNA serving as a template.

➤ Example 11.2

Write the sequence for the mRNA that could be synthesized using the following DNA base sequence as a template:

5′ G–C–A–A–C–T–T–G 3′

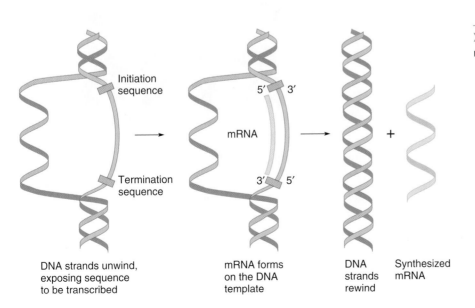

DNA strands unwind, exposing sequence to be transcribed

mRNA forms on the DNA template

DNA strands rewind

Synthesized mRNA

➤ **FIGURE 11.17** The synthesis of mRNA.

IS AIDS FOREVER?

Viruses, often described as bridges between living and non-living material, are infectious particles composed primarily of DNA or RNA, and a protein coat. Because their structures are so simple, viruses are unable to reproduce in the absence of other organisms. However, upon gaining access to the interior of a host cell, viruses can take over the metabolic operations of the cell and force it to synthesize viral molecular components at the expense of normal host cell components. Upon rupture of the cell, the new viruses are released into the intercellular spaces, where they can infect other cells.

AIDS (acquired immunodeficiency syndrome) is a disease that symbolizes the constant threat posed by a new virus. Without a cure or a vaccine, a new virus can cause an epidemic. The human immunodeficiency virus (HIV), which causes AIDS, has an affinity for a specific white blood cell called the helper T cell, which plays a critical role in the body's immune system. When helper T cells are infected with the AIDS virus, they cannot perform their immune function, and the body becomes susceptible to the rare infections and unusual cancers seen in AIDS patients.

The AIDS virus is an example of a **retrovirus,** a virus that uses RNA instead of storing its genetic program in DNA. When the viral RNA enters a helper T cell, the RNA is "reverse-transcribed" into DNA, which is treated by the host cell as its own, and inserted into its chromosomes. The subsequent synthesis of viral nucleic acids and proteins leads to the production of progeny viruses and the death of the cell.

Drugs approved by the FDA to combat HIV fall into one of two categories. One group of compounds are reverse transcriptase inhibitors. Reverse transcriptase is the enzyme that catalyzes the formation of DNA from viral RNA. Azidothymidine (AZT), a compound in this category, is the original anti-HIV drug. It was the first of several drugs developed that contained modified sugars or bases. Unfortunately, when used alone, AZT and its chemical cousins are marginally effective for a period of only several months because the virus quickly develops resistance to their effect. However, the use of AZT to treat pregnant women with AIDS greatly reduces the risk of transmitting the disease to their newborn babies. This group of drugs also includes nevirapine (Viramune) and delavirdine (Rescriptor).

Protease inhibitors, the second group of drugs, block the production of another key viral enzyme, HIV protease. They include the drug saquinavir (Invirase and Fortovase).

Doctors have found that a combination of several drugs is the best therapy. By interfering with many parts of the viral replication cycle, it is more difficult for the virus to have multiple successful mutations that lend resistance. Most doctors prescribe two reverse transcriptase inhibitors and a protease inhibitor because this configuration demonstrated the first positive results. This triple combination has markedly improved the prognosis for HIV-infected individuals. Unfortunately, the treatment is very expensive, costing $10,000–$20,000 a year. It also requires taking 8–32 pills a day at specific times. Because of the expense, only patients in developed countries can benefit from this approach.

To be of more use globally, a vaccine for AIDS is needed. Although some progress has been made toward achieving this goal, a successful vaccine still appears to be some distance in the future. Three different types of vaccine are now being developed. The first type would cause the body to produce anti-HIV antibodies that would circulate in the blood. This is done by exposing the body to viruses or viral material, such as proteins from the surface of a virus or actual killed AIDS viruses. The use of *pseudovirions,* or artificial viruses, has also been considered. A second vaccine type would create an anti-HIV response by the immune system at the cellular level. Live vector viruses (non-HIVs engineered to carry genes encoding HIV proteins) are used to elicit the desired response. The use of naked DNA containing one or more HIV genes is being considered, but this method causes concern about integrating HIV genes into human cells. The third type of vaccine would combine the previous two and attack HIV with both antibodies and responses at the cellular level.

http://www.brookscole.com/chemistry/seager_chemistry
At Web site 11.3, you will find information furnished by the Centers for Disease Control National AIDS Clearing House.

HIV attacking a T cell. The T cell surface is yellow and the core (red) of each virus containing RNA is seen.

Virus
An infectious particle composed of only proteins and DNA or RNA.

Retrovirus
A virus in which RNA directs the synthesis of DNA.

azidothymidine (AZT)

Solution

RNA synthesis begins at the 3′ end of the DNA template and proceeds toward the 5′ end. The complementary RNA strand is formed from the bases C, G, A, and U. Uracil (U) is the complement of adenine (A) on the DNA template.

direction of strand

DNA template: 5′   G–C–A–A–C–T–T–G   3′

New mRNA: 3′   C–G–U–U–G–A–A–C   5′

direction of strand

Writing the sequence of the new mRNA in the 5′ to 3′ direction, it becomes

5′   C–A–A–G–U–U–G–C   3′

Learning Check 11.3

Write the sequence for the mRNA which could be synthesized on the following DNA template:

5′   A–T–T–A–G–C–C–G   3′

Although the general process of transcription operates universally, there are differences in detail between the processes in prokaryotes and eukaryotes. Each gene of prokaryotic cells exists as a continuous segment of a DNA molecule. Transcription of this gene segment produces mRNA that undergoes translation into a protein almost immediately because there is no nuclear membrane in a prokaryote separating DNA from the cytoplasm. In 1977, however, it was discovered that the genes of eukaryotic cells are segments of DNA that are "interrupted" by segments that do not code for amino acids. These DNA segments that carry no amino acid code are called **introns**, and the coded DNA segments are called **exons**.

When transcription occurs in the nuclei of eukaryotic cells, both introns and exons are transcribed. This produces what is called **heterogeneous nuclear RNA (hnRNA)**. This long molecule of hnRNA then undergoes a series of enzyme-catalyzed reactions that cut and splice the hnRNA to produce mRNA (see ■ Figure 11.18). The mRNA resulting from this process contains only the sequence of bases that actually codes for protein synthesis. Although the function of introns in eukaryotic DNA is not yet understood, it is an area of active research.

Intron
A segment of a eukaryotic DNA molecule that carries no codes for amino acids.

Exon
A segment of a eukaryotic DNA molecule that is coded for amino acids.

Heterogeneous nuclear RNA (hnRNA)
RNA produced when both introns and exons of eukaryotic cellular DNA are transcribed.

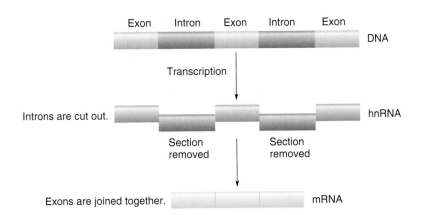

➤ **FIGURE 11.18** Segments of hnRNA formed from introns of eukaryotic cells are removed by special enzymes to produce mRNA.

➤ 11.7 The Genetic Code

Discovery of the Genetic Code

By 1961, it was clear that the sequence of bases in DNA serves to direct the synthesis of mRNA, and that the sequence of bases in mRNA corresponds to the order of amino acids in a particular protein. However, the genetic code, the exact relationship between mRNA sequences and amino acids, was unknown. At least 20 mRNA "words" are needed to represent uniquely each of the 20 amino acids found in proteins. If the mRNA words consisted of a single letter represented by a base (A, C, G, or U), only four amino acids could be uniquely represented. Thus, it was proposed that it is not one mRNA base but a combination of bases that codes for each amino acid. For example, if a code word consists of a sequence of two mRNA bases, there are $4^2 = 16$ possible combinations, so 16 amino acids could be represented uniquely. This is a more extensive code, but it still contains too few words to do the job for 20 amino acids. If the code consists of a sequence of three bases, there are $4^3 = 64$ possible combinations, more than enough to specify uniquely each amino acid in the primary sequence of a protein. Research has confirmed that nature does indeed use three-letter code words (a triplet code) to store and express genetic information. Each sequence of three nucleotide bases that represents code words on mRNA molecules is called a **codon.**

After the discovery of three-letter codons, researchers were anxious to answer the next question: Which triplets of bases (codons) code for which amino acids? In 1961, Marshall Nirenberg and his coworkers attempted to break the code in a very ingenious way. They made a synthetic molecule of mRNA consisting of uracil bases only. Thus, this mRNA contained only one codon, the triplet UUU. They incubated this synthetic mRNA with ribosomes, amino acids, tRNAs, and the appropriate enzymes for protein synthesis. The exciting result of this experiment was that a polypeptide was synthesized that contained only phenylalanine. Thus, the first word of the genetic code had been deciphered; UUU = phenylalanine.

A series of similar experiments by Nirenberg and other researchers followed, and by 1967 the entire genetic code had been broken. The complete code is shown in ■ Table 11.2.

Characteristics of the Genetic Code

The first important characteristic of the genetic code is that it applies almost universally. With very minor exceptions, the same amino acid is represented by the same three-base codon (or codons) in every organism.

Another interesting feature of the genetic code is apparent from Table 11.2. Most of the amino acids are represented by more than one codon, a condition known as *degeneracy*. Only methionine and tryptophan are represented by single codons. Leucine, serine, and arginine are the most degenerate, with each one represented by six codons. The remaining 15 amino acids are each coded for by at least two codons. Even though most amino acids are represented by more than one codon, it is significant that the reverse is not true; no single codon represents more than one amino acid. Each three-base codon represents one and only one amino acid.

Only 61 of the possible 64 base triplets represent amino acids. The remaining three (UAA, UAG, and UGA) are signals for chain terminations. They tell the protein-synthesizing process when the primary structure of the synthesized protein is complete and it is time to stop adding amino acids to the chain. These three codons are indicated in Table 11.2 by stop signals.

Codon

A sequence of three nucleotide bases that represents a code word on mRNA molecules.

TABLE 11.2 The genetic code: mRNA codons for each of the 20 amino acids

Amino acid	Codons	Number of codons
alanine	GCA, GCC, GCG, GCU	4
arginine	AGA, AGG, CGA, CGC, CGG, CGU	6
asparagine	AAC, AAU	2
aspartic acid	GAC, GAU	2
cysteine	UGC, UGU	2
glutamic acid	GAA, GAG	2
glutamine	CAA, CAG	2
glycine	GGA, GGC, GGG, GGU	4
histidine	CAC, CAU	2
isoleucine	AUA, AUC, AUU	3
leucine	CUA, CUC, CUG, CUU, UUA, UUG	6
lysine	AAA, AAG	2
methionine, initiation	AUG	1
phenylalanine	UUC, UUU	2
proline	CCA, CCC, CCG, CCU	4
serine	UCA, UCC, UCG, UCU, AGC, AGU	6
threonine	ACA, ACC, ACG, ACU	4
tryptophan	UGG	1
tyrosine	UAC, UAU	2
valine	GUA, GUC, GUG, GUU	4
Stop signals	UAG, UAA, UGA	3
Total number of codons		64

> **STUDY SKILLS 11.1**

REMEMBERING KEY WORDS

Three very important processes discussed in this chapter are replication, transcription, and translation. New terminology is always easier to learn if we can make associations with something we already know. See if these ideas help. To replicate means to copy or perhaps to repeat. Associate the word *replicate* with *repeat*.

For the processes of transcription and translation, think of an analogy with the English language. If you were to transcribe your lecture notes, you would rewrite them in the same language. In the same way, transcription of DNA to mRNA is a rewriting of the information using the same nucleic acid language. If you translated your lecture notes, you would convert them to another language, such as German. Translation of mRNA to an amino acid sequence is converting information from a nucleic acid language to a language of proteins.

TABLE 11.3 General characteristics of the genetic code

Characteristic	Example
Codons are three-letter words	GCA = alanine
The code is degenerate	GCA, GCC, GCG, GCU all represent alanine
The code is precise	GCC represents only alanine
Chain initiation is coded	AUG
Chain termination is coded	UAA, UAG, and UGA
The code is almost universal	GCA = alanine, perhaps, in all organisms

The presence of stop signals in the code implies that start signals must also exist. There is only one initiation (start) codon, and it is AUG, the codon for the amino acid methionine. AUG functions as an initiation codon only when it occurs as the first codon of a sequence. When this happens, protein synthesis begins at that point (Section 11.8).

These general features of the genetic code are summarized in ■ Table 11.3.

➤ 11.8 Translation and Protein Synthesis

To this point, you have become familiar with the molecules that participate in protein synthesis and the genetic code, the language that directs the synthesis. We now investigate the actual process by which polypeptide chains are assembled. There are three major stages in protein synthesis: initiation of the polypeptide chain, elongation of the chain, and termination of the completed polypeptide chain.

Initiation of the Polypeptide Chain

The first amino acid to be involved in protein synthesis in prokaryotic (bacterial) cells is a derivative of methionine. This compound, N-formylmethionine, initiates the growing polypeptide chain as the N-terminal amino acid. The fact that most proteins do not have N-formylmethionine as the N-terminal amino acid indicates that when protein synthesis is completed, the N-formylmethionine is cleaved from the finished protein.

$$
\begin{array}{ccc}
& \text{O} & & & \text{O} \\
& \parallel & & & \parallel \\
\text{H}-\text{C}-\text{NH}-\text{CH}-\text{C}-\text{O}^- \\
& & & | \\
\underbrace{}_{formyl\ group} & & & \text{CH}_2 \\
& & & | \\
& & & \text{CH}_2-\text{S}-\text{CH}_3
\end{array}
$$

N-formylmethionine
(fMet)

A ribosome comprises two subunits, a large subunit and a small subunit. The initiation process begins when mRNA is aligned on the surface of a small ribosomal subunit in such a way that the initiating codon, AUG, occupies a specific site on the ribosome called the P site (peptidyl site). When the AUG

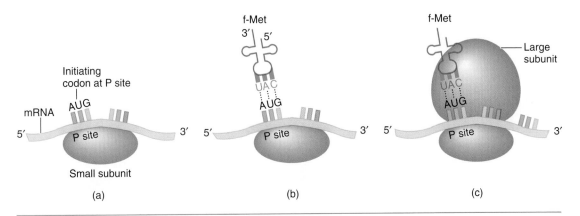

> **FIGURE 11.19** Initiation complex formation. (a) mRNA aligns on a small ribosomal subunit so that AUG, the initiating codon, is at the ribosomal P site. (b) A tRNA with an attached *N*-formylmethionine forms hydrogen bonds with the codon. (c) A large ribosomal subunit completes the initiation complex.

codon is used this way to initiate synthesis of a polypeptide chain in prokaryotic cells, it represents *N*-formylmethionine (f-Met) instead of methionine. When AUG is located anywhere else on the mRNA, it simply represents methionine. For the eukaryotic cells of humans, AUG always specifies methionine even when it is the initiating codon. Next, a tRNA molecule with its attached f-Met binds to the codon through hydrogen bonds. The resulting complex binds to the large ribosomal subunit to form a unit called an *initiation complex* (see ■ Figure 11.19).

Elongation of the Chain

A second site, called the A site (aminoacyl site), is located on the mRNA–ribosome complex next to the P site. The A site is where an incoming tRNA carrying the next amino acid will bond. Each of the tRNA molecules representing the 20 amino acids can try to fit the A site, but only the one with the correct anticodon that is complementary to the next codon on the mRNA will fit properly.

Once at the A site, the second amino acid (in this case, phenylalaline) is linked to *N*-formylmethionine by a peptide bond whose formation is catalyzed by the enzyme peptidyl transferase (see ■ Figure 11.20). After the peptide bond forms, the tRNA bound to the P site is "empty," and the growing polypeptide chain is now attached to the tRNA bound to the A site.

> **FIGURE 11.20** The amino acid at the P site bonds through a peptide bond to the amino acid at the A site.

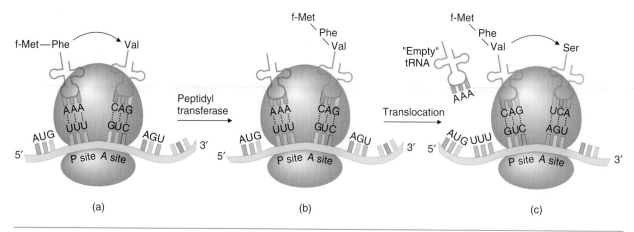

(a)　　　　　　　　　　(b)　　　　　　　　　　(c)

➤ **FIGURE 11.21** Polypeptide chain elongation. (a) The P site is occupied by the tRNA with the growing peptide chain, and Val—tRNA is located at the A site. (b) The formation of a peptide bond between Val and the dipeptide f-Met—Phe takes place under the influence of peptidyl transferase. (c) During translocation when the ribosome shifts to the right, the empty tRNA leaves, the polypeptide—tRNA moves to the P site, and the next tRNA carrying Ser arrives at the A site.

In the next phase of elongation, the whole ribosome moves one codon along the mRNA toward the 3′ end. As the ribosome moves, the empty tRNA is released from the P site, and tRNA attached to peptide chain moves from the A site and takes its place on the P site. This movement of the ribosome along the mRNA is called *translocation* and makes the A site available to receive the next tRNA with the proper anticodon. The amino acid carried by this tRNA bonds to the peptide chain, and the elongation process is repeated. This occurs over and over until the entire polypeptide chain is synthesized. The elongation process is represented in ■ Figure 11.21 for the synthesis of the tripeptide f-Met—Phe—Val.

The Termination of Polypeptide Synthesis

The chain elongation process continues and polypeptide synthesis continues until the ribosome complex reaches a stop codon (UAA, UAG, or UGA) on mRNA. At that point, a specific protein known as a termination factor binds to the stop codon and catalyzes the hydrolysis of the completed polypeptide chain from the final tRNA. The "empty" ribosome dissociates and can then bind to another strand of mRNA to once again begin the process of protein synthesis.

Several ribosomes can move along a single strand of mRNA one after another (see ■ Figure 11.22). Thus, several identical polypeptide chains can be synthesized almost simultaneously from a single mRNA molecule. This markedly increases the efficiency of utilization of the mRNA. Such complexes of several ribosomes and mRNA are called **polyribosomes** or **polysomes**. Growing polypeptide chains extend from the ribosomes into the cellular cytoplasm and spontaneously fold to assume characteristic three-dimensional secondary and tertiary configurations.

Polyribosome or polysome
A complex of mRNA and several ribosomes.

Learning Check 11.4

Write the primary structure of the polypeptide produced during translation of the following mRNA sequence:

5′　　　AUG–CAC–CAU–GUA–UUG–UGU–UAG　　　3′

Large subunit of ribosome

mRNA

5′

Small subunit of ribosome

Growing protein

Secondary structure begins to form

Complete protein

Ribosome subunits released

3′

Ribosomes move along mRNA

➤ **FIGURE 11.22** A polyribosome, several ribosomes proceeding simultaneously along mRNA.

➤ 11.9 Mutations

In Section 11.3, we pointed out that the base-pairing mechanism proves a nearly perfect way to copy a DNA molecule during replication. However, not even the copying mechanism involved in DNA replication is totally error-free. It is estimated that, on average, one in every 10^{10} bases of DNA (i.e., 1 in 10 billion) is incorrect. Any change resulting in an incorrect sequence of bases on DNA is called a **mutation**. The faithful transcription of mutated DNA leads to an incorrect base sequence in mRNA. This can lead to an incorrect amino acid sequence for a protein, or possibly the failure of a protein to be synthesized at all.

Some mutations occur naturally during DNA replication; others can be induced by environmental factors such as ionizing radiation (X rays, ultraviolet light, gamma rays, etc.). Repeated exposure to X rays greatly increases the rate of mutation. Thus, patients are given X rays only when necessary, and technicians who administer X rays remain behind protective barriers. A large number of chemicals (e.g., nitrous acid and dimethyl sulfate) can also induce mutations by reacting with DNA. Such chemicals are called **mutagens**.

Some mutations might be beneficial to an organism by making it more capable of surviving in its environment. Mutations may also occur without affecting an organism. Conversely, mutations may be lethal or produce genetic diseases whenever an important protein (or enzyme) is not correctly synthesized. Sickle-cell disease, phenylketonuria (PKU), hemophilia, muscular dystrophy, and many other disorders are results of such mutations that have become permanently incorporated into the genetic makeup of certain individuals.

Mutation
Any change resulting in an incorrect base sequence on DNA.

Mutagen
A chemical that induces mutations by reacting with DNA.

➤ 11.10 Recombinant DNA

Remarkable technology is available that allows segments of DNA from one organism to be introduced into the genetic material of another organism. The resulting new DNA (containing the foreign segment) is referred to as **recombinant DNA**. The application of this technology, commonly called *genetic engineering,* has produced major advances in human health care and holds the promise for a future of exciting advances in biology, agriculture (see ■ Figure 11.23), and many other areas of study.

An early success of genetic engineering was the introduction of the gene for human insulin into the DNA of the common bacterium *E. coli*. The bacterium then transcribed and translated the information carried by the gene and

Recombinant DNA
DNA of an organism that contains genetic material from another organism.

➤ **FIGURE 11.23** This tomato, a beneficiary of genetic engineering, has an increased shelf life.

produced the protein hormone. By culturing such *E. coli*, it has become possible to produce and market large quantities of human insulin. The availability of human insulin is very important for diabetics allergic to the insulin traditionally used, which is isolated from the pancreatic tissue of slaughtered pigs or cattle. ■ Table 11.4 lists several other medically important materials that have been produced through genetic engineering.

Enzymes Used in Genetic Engineering

The discovery of restriction enzymes in the 1960s and 1970s made genetic engineering possible. Restriction enzymes, which are found in a wide variety of bacterial cells, catalyze the cleaving of DNA molecules. These enzymes are normally part of a mechanism that protects certain bacteria from invasion by foreign DNA such as that of a virus. In these bacteria, some of the bases in their DNA have methyl groups attached. For example, the following compounds are found:

1-methylguanine   5-methylcytosine

The methylated DNA of these bacteria is left untouched by the restriction enzymes, but foreign DNA that lacks the methylated bases undergoes rapid cleavage of both strands and thus becomes nonfunctional. Because there is a "restriction" on the type of DNA allowed in the bacterial cell, the protective enzymes are called **restriction enzymes.**

Restriction enzymes act at sites on DNA called palindromes. In language, a palindrome is any word or phrase that reads the same in either direction, such as "radar" or "Madam, I'm Adam." For double-stranded DNA, a palindrome is a section in which the two strands have the same sequence but running in opposite directions. Examples of DNA palindromes follow; the arrows indicate the points of attack by restriction enzymes:

$$5' \quad C-C-G-C-G-G \quad 3' \qquad 5' \quad G-G-A-T-C-C \quad 3'$$
$$3' \quad G-G-C-G-C-C \quad 5' \qquad 3' \quad C-C-T-A-G-G \quad 5'$$

Restriction enzyme
A protective enzyme found in some bacteria that catalyzes the cleaving of all but a few specific types of DNA.

TABLE 11.4	Some substances produced by genetic engineering
Substance	**Use**
Human insulin	Treatment of diabetes mellitus
Human growth hormone	Treatment of dwarfism
Interferon	Fights viral infections
Hepatitis B vaccine	Protection against hepatitis
Malaria vaccine	Protection against malaria
Tissue plasminogen activator	Promotes the dissolving of blood clots

At least 100 restriction enzymes are known, and each catalyzes DNA cleavage in a specific and predictable way. These enzymes are the tools used to take DNA apart and reduce it to fragments of known size and nucleotide sequence.

Another set of enzymes important in genetic engineering, called DNA ligases, has been known since 1967. These enzymes normally function to connect DNA fragments during replication, and they are used in genetic engineering to put together pieces of DNA produced by restriction enzymes.

Plasmids

The introduction of a new DNA segment (gene) into a bacterial cell requires the assistance of a DNA carrier called a **vector**. The vector is often a circular piece of double-stranded DNA called a **plasmid.** Plasmids range in size

Vector
A carrier of foreign DNA into a cell.

Plasmid
Circular, double-stranded DNA found in the cytoplasm of bacterial cells.

CHEMISTRY AROUND US • 11.3

DNA AND THE LAW

DNA is obviously a key chemical in all living organisms because of its functions of passing along hereditary information to offspring and directing all life processes. DNA is also playing key roles in some advances made in forensic science and the understanding of the nature of numerous hereditary diseases.

DNA profiling—also called DNA typing or *DNA fingerprinting*—is based on the fact that although most of the DNA molecule is common to all humans, parts of it differ from person to person. In fact, 99.9% of our DNA is invariable, but the other 0.1% differs from one person to the next. This might not seem like much of a difference, but it is equivalent to 3 million bases. Therefore, with the exception of identical twins, the DNA of every individual is unique. Unlike a conventional fingerprint, which occurs only on the fingertips, a DNA fingerprint is the same for every cell, tissue, and organ of a person. Consequently, DNA fingerprinting is rapidly becoming the method of choice for identifying and distinguishing among individual human beings.

FBI and police labs around the United States have begun to use DNA fingerprints to link suspects to biological evidence—blood, semen, or hair—found at the scene of a crime. Another important use of DNA fingerprints in the justice system is to establish paternity in custody and child-support litigation. In these applications, DNA fingerprints bring nearly perfect accuracy to parental identification. The U.S. armed services have begun a program to record DNA fingerprints for all personnel, in case they are needed later to identify casualties. The DNA method will be far superior to the dog tags and dental records currently in use.

The technique of DNA fingerprinting involves cleaving DNA molecules at specific points using restriction enzymes. The mixture of DNA fragments is then placed on a gel, where it separates into specific patterns when an electric current is applied. This process is called *electrophoresis*. The patterns are made visible by radioactive reagents attached to the fragments that are exposed to X-ray film. The film shows a specific pattern of dark bands when it is developed.

The ability to obtain DNA fingerprints from exceedingly small samples has become possible by using the polymerase chain reaction (PCR) technique (Section 11.3). With PCR, billions of copies of a DNA sample can be prepared in a few hours. PCR has enabled researchers to study extremely faint and tiny traces of DNA found in biological samples, such as specks of dried blood, strands of hair, chips of bone, or even from fingerprints. PCR amplification of the DNA in a single sperm cell has been used to link suspects to rape victims. A single cell from saliva on the stamp of a letter bomb would provide enough DNA to help identify the person who licked the stamp.

http://www.brookscole.com/chemistry/seager_chemistry
At Web site 11.4, you will find an article on the use of DNA fingerprinting in human health and in society.

DNA fingerprinting: An autoradiograph of DNA.

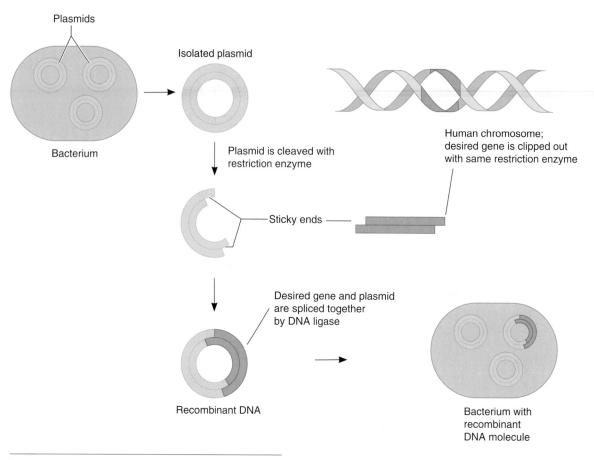

➤ **FIGURE 11.24** The formation of recombinant DNA.

from 2000 to several hundred thousand nucleotides and are found in the cytoplasm of bacterial cells. Plasmids function as accessories to chromosomes by carrying genes for the inactivation of antibiotics and the production of toxins. They have the unusual ability to replicate independently of chromosomal DNA. A typical bacterial cell contains about 20 plasmids and one or two chromosomes.

The Formation of Recombinant DNA

The recombinant DNA technique begins with the isolation of a plasmid from a bacterium. A restriction enzyme is added to the plasmid, which is cleaved at a specific site:

$$\overset{\text{restriction}}{\underset{\text{enzyme}}{\longrightarrow}}$$

∾G – G – A – T – C – C∾   ∾G   G – A – T – C – C∾
∾C – C – T – A – G – G∾   ∾C – C – T – A – G   G∾

sticky ends

Because a plasmid is circular, cleaving it this way produces a double-stranded chain with two ends (see ■ Figure 11.24). These are called "sticky ends" because each has one strand in which several bases are unpaired and ready to pair up with a complementary section if available.

The next step is to provide the sticky ends with complementary sections for pairing. A human chromosome is cleaved into several fragments using the same restriction enzyme. Because the same enzyme is used, the human DNA is cleaved such that the same sticky ends result:

$$G - A - T - C - C \sim\sim\sim\sim\sim G$$
$$\text{human DNA segment}$$
$$G \sim\sim\sim\sim\sim C - C - T - A - G$$

To splice the human gene into the plasmid, the two are brought together under conditions suitable for the formation of hydrogen bonds between the complementary bases of the sticky ends. The breaks in the strands are then joined by using DNA ligase, and the plasmid once again becomes a circular piece of double-stranded DNA (Figure 11.24); recombinant DNA has been formed.

Bacterial cells are bathed in a solution containing recombinant DNA plasmids, which diffuse into the cells. When the bacteria reproduce, they replicate all the genes, including the recombinant DNA plasmid. Bacteria multiply quickly, and soon a large number of bacteria, all containing the modified plasmid, are manufacturing the new protein as directed by the recombinant DNA.

The use of this technology makes possible (in principle) the large-scale production of virtually any polypeptide or protein. By remembering the many essential roles these substances play in the body, it is easy to predict that genetic engineering will make tremendous contributions to improving health care in the future.

Genome
The sum of all the genetic material (chromosomes) of a cell; a person's genetic blueprint.

CHEMISTRY AROUND US • 11.4

DECODING THE HUMAN GENOME

The human **genome** is the sum total of the approximately 80,000 different genes that determine the genetic characteristics of every individual. It is estimated that the human genome consists of about 3 billion base pairs, assembled in 23 pairs of chromosomes. The federally funded Human Genome Project is a research effort to develop a map of the correct location of each of the 80,000 human genes on the chromosomes. The motivation for this enormous task is to provide a key to understanding the nearly 4000 known genetic disorders and countless diseases that may originate in part with genetic alteration, as well as the astonishing variety of normal human traits.

The potential benefits of mapping the human genome are so great that researchers around the world are engaged in the task. New strategies for identifying genes using chemical reagents or probes that cause regions of DNA to fluoresce have greatly accelerated gene identification. At the same time, instrumental techniques have advanced to the point that more than 9000 bases of DNA can be sequenced every hour.

However, the Human Genome Project may have a competitor. Armed with advanced sequencing machines that are ten times more efficient than current sequencing machines, J. Craig Venter of the Institute for Genomic Research (TIGR) and Mike Hunkapiller of Perkin-Elmer Applied Biosystems claim that they can sequence the entire human genome in 3 years instead of the 15 years that the Human Genome Project is projected to take. Instead of assigning various segments of the genome to hundreds of scientists, as the Human Genome Project does, Venter plans to use what he calls whole-genome shotgunning.

His method involves the use of sound waves to chop the entire genome into millions of pieces. The mix containing these pieces will then be fed into 230 newly developed sequencing machines. New software developed by TIGR scientists will then be used to interpret the data from the sequencing machines and reassemble the millions of pieces of chopped-up genome into a coherent whole. Many scientists doubt that this method will work.

They worry that even if a whole genome is assembled, it will be incomplete. However, Venter points to past successes of this approach in sequencing the DNA of seven microorganisms, including the bacterium that causes ulcers.

In any event, one of the first tangible benefits of the ongoing Human Genome Project was obtained in 1992. An international team of researchers identified the genetic origin of the most common form of muscular dystrophy. More recently, scientists have identified genetic links to a number of other serious conditions, including amyotrophic lateral sclerosis (Lou Gehrig's disease), Huntington's disease, melanoma (skin cancer), Alzheimer's disease, breast cancer, and colon cancer.

The discovery of altered genes could pave the way for improved methods of diagnosing the diseases they cause. Blood tests are available to detect many hereditary disorders, including cystic fibrosis, Huntington's disease, and hemophilia. The genetic test for Huntington's disease took only a few months to develop once the genetic link was identified. Genetic testing for hereditary breast and colon cancers may soon be available. However, genetic testing is not without controversy. Some experts have warned that genetic data indicating susceptibility to certain diseases could be used to deny people health insurance or employment.

A promising approach to the treatment of genetic disorders is gene therapy. In this technique, the absent or altered gene is replaced by a normal working gene, so that the body can make the correct enzyme or protein and thereby eliminate the cause of the disease. Gene therapy trials have already begun in the treatment of cystic fibrosis, several types of cancer, AIDS, rheumatoid arthritis, hemophilia, and numerous other diseases. Gene therapy could potentially revolutionize modern medicine.

http://www.brookscole.com/chemistry/seager_chemistry
At Web site 11.5, you will find an article on the expectations of the Human Genome Project.

> **FOR FUTURE REFERENCE**   DRUGS THAT MAY ADVERSELY AFFECT THE FETUS AND NEWBORN INFANT

Numerous factors in addition to inherited genetic characteristics may affect the proper development and health of a human fetus or a newborn infant, including drugs used by the mother during pregnancy. As the understanding of drug interactions with the unborn and newborn has grown, so has the list of drugs that can adversely affect them. A current list is given below. Based on current knowledge, it is strongly recommended that only drugs that provide essential benefits can be used during a pregnancy.

Drugs that *probably* cause adverse effects on the unborn when taken during the *first trimester*

Aminopterin
Anticonvulsants
Antithyroid drugs
Cytarabine
Danazol
Diethylstilbestrol
Ethanol (large amounts and for long periods)
Etretinate
Finasteride
Fluorouracil
HMG-CoA reductase inhibitors
Iodides
Isotretinoin
Kanamycin
Mercaptopurine
Methotrexate
Misoprostol
Opioid analgesics
Progestins
Quinine
Streptomycin
Testosterone
Warfarin

Drugs that *possibly* cause adverse effects on the unborn when taken during the *first trimester*

Angiotensin-converting enzyme inhibitors
Busulfan
Chlorambucil
Estrogens
Lithium
Mebendazole
Monoamine oxidase inhibitors
Oral contraceptives
Piperazine
Rifampin
Tetracyclines

Drugs that *probably* cause adverse effects on the unborn when taken during the *second* and *third trimesters*

Amiodarone
Androgens
Angiotensin-converting enzyme inhibitors
Antithyroid drugs
Aspirin
Benzodiazepines
Chloramphenicol
Estrogens
Ethanol (large amounts and for long periods)
Finasteride
HMG-CoA reductase inhibitors
Iodides
Kanamycin
Lithium
Nonsteroidal antiinflammatory drugs (NSAIDs)
Opioid analgesics
Phenothiazines
Progestins
Rifampin
Streptomycin
Sulfonamides
Sulfonylureas
Tetracyclines
Thiazide diuretics
Tricycli antidepressants
Warfarin

Drugs that *possibly* cause adverse effects on the unborn when taken during the *second* and *third trimesters*

Acetozolamide
Clemastine
Diphenhydramine
Ethacrynic acid
Fluoroquinolones
Haloperidol
Hydroxyzine
Promethazine

SOURCE: Ryback, J.J.; Long, J.W. *The Essential Guide to Prescription Drugs.* New York: Harper and Perennial Publishing, 1999.

CONCEPT SUMMARY

Components of Nucleic Acids. Nucleic acids are classified into two categories: ribonucleic acids (RNA) and deoxyribonucleic acids (DNA). Both types are polymers made up of monomers called nucleotides. All nucleotides are composed of a pyrimidine or purine base, a sugar, and phosphate. The sugar component of RNA is ribose, and that of DNA is deoxyribose. The bases adenine, guanine, and cytosine are found in all nucleic acids. Uracil is found only in RNA and thymine only in DNA.

The Structure of DNA. The nucleotides of DNA are joined by linkages between phosphate groups and sugars. The resulting sugar–phosphate backbone is the same for all DNA molecules, but the order of attached bases along the backbone varies. This order of nucleotides with attached bases is the primary structure of nucleic acids. The secondary structure is a double-stranded helix held together by hydrogen bonds between complementary base pairs on the strands.

DNA Replication. The replication of DNA occurs when a double strand of DNA unwinds at specific points. The exposed bases match up with complementary bases of nucleotides. The nucleotides bind together to form two new strands that are complementary to the strands that separated. Thus, the two new DNA molecules each contain one old strand and one new strand.

Ribonucleic Acid (RNA). Three forms of ribonucleic acid are found in cells: messenger RNA, ribosomal RNA, and transfer RNA. Each form serves an important function during protein synthesis. All RNA molecules are single-stranded, but some contain loops or folds.

The Flow of Genetic Information. The flow of genetic information occurs in two steps called transcription and translation. In transcription, information stored in DNA molecules is passed to molecules of messenger RNA. In translation, the messenger RNA serves as a template that directs the assembly of amino acids into proteins.

Transcription: RNA Synthesis. The various RNAs are synthesized in much the same way as DNA is replicated. Nucleotides with complementary bases align themselves against one strand of a partially unwound DNA segment that contains the genetic information that is to be transcribed. The aligned nucleotides bond together to form the RNA. In eukaryotic cells, the produced RNA is heterogeneous and is cut and spliced after being synthesized to produce the functional RNA.

The Genetic Code. The genetic code is a series of three-letter words that represent the amino acids of proteins as well as start and stop signals for protein synthesis. The letters of the words are the bases found on mRNA, and the words on mRNA are called codons. The genetic code is the same for all organisms and is degenerate for most amino acids.

Translation and Protein Synthesis. The translation step in the flow of genetic information results in the synthesis of proteins. The synthesis takes place when properly coded mRNA forms a complex with the component of a ribosome. Transfer RNA molecules carrying amino acids align themselves along the mRNA in an order representing the correct primary structure of the protein. The order is determined by the matching of complementary codons on the mRNA to anticodons on the tRNA. The amino acids sequentially bond together to form the protein, which then spontaneously forms characteristic secondary and tertiary structures.

Mutations. Any change that results in an incorrect sequence of bases on DNA is called a mutation. Some mutations occur naturally during DNA replication, whereas others are induced by environmental factors. Some mutations are beneficial to organisms; others may be lethal or result in genetic diseases.

Recombinant DNA. The discovery and application of restriction enzymes and DNA ligases have resulted in a technology called genetic engineering. The primary activities of genetic engineers are the isolation of genes (DNA) that code for specific useful proteins and the introduction of these genes into the DNA of bacteria. The new (recombinant) DNA in the rapidly reproducing bacteria mediates production of the useful protein, which is then isolated for use.

LEARNING OBJECTIVES

When you have completed your study of this chapter, you should be able to

1. Identify the components of nucleotides and correctly classify the sugars and bases.
2. Describe the structure of DNA.
3. Outline the process of DNA replication.
4. Contrast the structure of DNA and RNA and list the function of the three types of cellular RNA.
5. Describe the process by which RNA is synthesized in cells.
6. Explain how the genetic code functions in the flow of genetic information.
7. Outline the process by which proteins are synthesized in cells.
8. Describe how genetic mutations occur and how they influence organisms.
9. Describe the technology used to produce recombinant DNA.

KEY TERMS AND CONCEPTS

Anticodon (11.4)
Central dogma of molecular biology (11.5)
Chromosome (11.3)
Codon (11.7)
Complementary DNA strands (11.2)
Deoxyribonucleic acid (DNA) (11.1)
Exon (11.6)
Gene (11.3)
Genome (11.10)
Heterogeneous nuclear RNA (hnRNA) (11.6)

Intron (11.6)
Messenger RNA (mRNA) (11.4)
Mutagen (11.9)
Mutation (11.9)
Nucleic acid (Introduction)
Nucleic acid backbone (11.2)
Nucleotide (11.1)
Okazaki fragment (11.3)
Plasmid (11.10)
Polyribosome (polysome) (11.8)
Recombinant DNA (11.10)
Replication (11.3)

Replication fork (11.3)
Restriction enzyme (11.10)
Retrovirus (11.6)
Ribonucleic acid (RNA) (11.1)
Ribosomal RNA (rRNA) (11.4)
Ribosome (11.4)
Semiconservative replication (11.3)
Transcription (11.5)
Transfer RNA (tRNA) (11.4)
Translation (11.5)
Vector (11.10)
Virus (11.6)

COMPONENTS OF NUCLEIC ACIDS (SECTION 11.1)

11.1 What is the principal location of DNA within the eukaryotic cell?

11.2 Which pentose sugar is present in DNA? In RNA?

11.3 Name the three components of nucleotides.

11.4 Indicate whether each of the following is a pyrimidine or a purine:
a. guanine
b. thymine
c. uracil
d. cytosine
e. adenine

11.5 Which bases are found in DNA? In RNA?

11.6 Write the structural formula for the nucleotide thymidine 5′-monophosphate. The base component is thymine.

THE STRUCTURE OF DNA (SECTION 11.2)

11.7 In what way might two DNA molecules that contain the same number of nucleotides differ?

11.8 Identify the 3′ and 5′ ends of the DNA segment AGTCAT.

11.9 Describe the secondary structure of DNA as proposed by Watson and Crick.

11.10 What data obtained from the chemical analysis of DNA supported the idea of complementary base pairing in DNA proposed by Watson and Crick?

11.11 Describe the role of hydrogen bonding in the secondary structure of DNA.

11.12 How many total hydrogen bonds would exist between the following strands of DNA and their complementary strands?
a. GCATGC
b. TATGGC

11.13 A strand of DNA has the base sequence GCTTAG. Write the base sequence for the complementary strand.

DNA REPLICATION (SECTION 11.3)

11.14 What is a chromosome? How many chromosomes are in a human cell?

11.15 List the steps involved in DNA replication.

11.16 Describe the function of the enzyme helicase in the replication of DNA.

11.17 What enzymes are involved in DNA replication?

11.18 In what direction is a new DNA strand formed?

11.19 Explain how the synthesis of a DNA daughter strand growing toward a replication fork differs from the synthesis of a daughter strand growing away from a replication fork.

11.20 The base sequence CTTAAG represents a portion of a single strand of DNA. Represent the complete double-stranded molecule for this portion of the strand and use the representation to illustrate the replication of the DNA strand. Be sure to clearly identify the nucleotide bases involved, new strands formed, and the daughter DNA molecules.

11.21 Explain the origin of Okazaki fragments.

RIBONUCLEIC ACID (RNA) (SECTION 11.4)

11.22 List the three cellular regions where RNA is found.

11.23 Compare the secondary structures of RNA and DNA.

11.24 Must the ratio of guanine to cytosine be 1:1 in RNA? Explain.

11.25 Briefly describe the characteristics and functions of the three types of cellular RNA.

11.26 Which of the kinds of RNA delivers amino acids to the site of protein synthesis?

THE FLOW OF GENETIC INFORMATION (SECTION 11.5)

11.27 What is the central dogma of molecular biology?

11.28 In the flow of genetic information, what is meant by the terms *transcription* and *translation?*

TRANSCRIPTION: RNA SYNTHESIS (SECTION 11.6)

11.29 Briefly describe the synthesis of mRNA in the transcription process.

11.30 Write the base sequence for the mRNA that would be formed during transcription from the DNA strand with the base sequence GCCATATTG.

11.31 Describe the differences in the transcription and translation processes as they occur in prokaryotic cells and eukaryotic cells.

11.32 What is the relationship between exons, introns, and hnRNA?

THE GENETIC CODE (SECTION 11.7)

11.33 What is a codon?

11.34 Using the information in Table 11.2, determine the amino acid that is coded for by each of the following codons:
a. CAU
b. UCA
c. UCU

11.35 Describe the experiment that allowed researchers to first identify the codon for a specific amino acid.

11.36 Which of the following statements about the genetic code are true and which are false? Correct each false statement.
a. Each codon is composed of four bases.
b. Some amino acids are represented by more than one codon.
c. All codons represent an amino acid.
d. Each living species is thought to have its own unique genetic code.
e. The codon AUG at the beginning of a sequence is a signal for protein synthesis to begin at that codon.

f. It is not known whether or not the code contains stop signals for protein synthesis.

TRANSLATION AND PROTEIN SYNTHESIS (SECTION 11.8)

11.37 The β-chain of hemoglobin is a protein that contains 146 amino acid residues. What minimum number of nucleotides must be present in a strand of mRNA that is coded for this protein?

11.38 List the three major stages in protein synthesis.

11.39 Does protein synthesis begin with the N-terminal or with the C-terminal amino acid?

11.40 What is meant by the terms A site and P site?

11.41 Beginning with DNA, describe specifically the coding and synthesis of the following tetrapeptide that represents the first four amino acid residues of the hormone oxytocin: Gly–Leu–Pro–Cys. Be sure to include processes such as formation of mRNA (use correct codons, etc.), attachment of mRNA to a ribosome, attachment of tRNA to mRNA-ribosome complex, and so on.

MUTATIONS (SECTION 11.9)

11.42 What is a genetic mutation?

11.43 Briefly explain how genetic mutations can do the following:
a. Harm an organism
b. Help an organism

11.44 What is the result of a genetic mutation that causes the mRNA sequence GCC to be replaced by CCC?

RECOMBINANT DNA (SECTION 11.10)

11.45 Explain the function and importance of restriction enzymes and DNA ligase in the formation of recombinant DNA.

11.46 Explain how recombinant DNA differs from normal DNA.

11.47 Explain what plasmids are and how they are used to get bacteria to synthesize a new protein that they normally do not synthesize.

11.48 List three substances likely to be produced on a large scale by genetic engineering, and give an important use for each.

CHEMISTRY AROUND US AND OVER THE COUNTER

11.49 Describe how a virus destroys a host cell.

11.50 What two categories of drugs have been approved by the FDA to combat HIV?

11.51 What is meant by a retrovirus?

11.52 Why is the AIDS virus so lethal?

11.53 Explain why DNA fingerprinting is a powerful forensic tool.

11.54 Describe the advantages of coupling PCR techniques with DNA fingerprinting.

11.55 Approximately how many genes are believed to constitute the human genome?

11.56 A map of the human genome is viewed as being of tremendous importance. Explain.

11.57 Why might a cream containing hydrocortisone be effective in treating an insect sting?

11.58 What cellular source was used to provide the DNA that led to the cloned sheep Dolly?

11.59 What might be a possible use for transgenic animals?

POINTS TO PONDER AND PURSUE

11.1 Genetic engineering shows great promise for the future but has been controversial at times. Discuss with some classmates the pros and cons of genetic engineering. List two benefits and two concerns that come from your discussion.

11.2 Two samples of DNA are compared, and one has a greater percentage of guanine-cytosine base pairs. How should that greater percentage affect the attractive forces holding the double helix together?

11.3 If DNA replication were not semiconservative, what might be another possible structure for the daughter DNA molecules?

11.4 The genetic code contains three stop signals and 61 codons that code for 20 amino acids. From the standpoint of mutations, why are we fortunate that the genetic code does not have only the required 20 amino acid codons and 44 stop signals?

11.5 Azidothymidine (AZT), a drug used to fight the HIV virus,

is believed to act as an enzyme inhibitor. What type of enzyme inhibition is most likely caused by this drug?

11.6 If DNA specifies only the primary structure of a protein, how does the correct three-dimensional protein structure develop?

11.7 Identify the bacterium that produces the Taq polymerase and the source from which the bacterium was isolated, according to Web site 11.2.

11.8 According to Web site 11.4, how many steps are required in the laboratory procedure for DNA fingerprinting?

11.9 Using Web site 11.5, list three general areas where the Human Genome Project will likely make contributions.

http://www.brookscole.com/chemistry/seager_chemistry

Nutrition and Energy for Life

In addition to working in health care facilities, dietitians are employed in other capacities, providing nutritional counseling to individuals and groups. Large institutions such as schools and businesses often rely on dietitians to set up and supervise food services. This chapter will familiarize you with the nutritional requirements for carbohydrates, lipids, and proteins and provide an overview of energy production from food.

http://www.brookscole.com/chemistry/
seager_chemistry
At Web site 12.1, you will find extensive career information for dietitians.

CHAPTER OUTLINE

Nutrition
An applied science that studies food, water, and other nutrients and the ways living organisms use them.

The science of **nutrition** is an applied field that focuses on the study of food, water, and other nutrients and the ways in which living organisms utilize them. Nutrition scientists use the principles of nutrition to obtain answers to questions that have a great deal of practical significance. For example, they might try to determine the proper components of a sound diet, the best way to maintain proper body weight, or the foods and other nutrients needed by people with specific illnesses or injuries.

The fuels of the human body are the sugars, lipids, and proteins derived from food (see ■ Figure 12.1). The reactions that release energy from these substances are among the body's most important biochemical processes. In this chapter, we will study an introduction to nutrition and how cells extract energy from food.

➤ 12.1 Nutritional Requirements

A human body must be supplied with appropriate nutrients if it is to function properly and remain healthy. Some nutrients are required in relatively large amounts (gram quantities daily) because they provide energy and materials required to repair damaged tissue or build new tissue. These **macronutrients** are the carbohydrates, lipids, and proteins contained in food. **Micronutrients** are required by the body in only small amounts (milligrams or micrograms daily). The small quantities needed suggest correctly that at least some micronutrients are utilized in enzymes. Micronutrients are classified as either vitamins or minerals.

Macronutrient
A substance needed by the body in relatively large amounts.

Micronutrient
A substance needed by the body only in small amounts.

In addition to macro- and micronutrients, the human body must also receive appropriate amounts of water and fiber. The importance of water becomes obvious when we learn that 45–75% of the human body mass is water. Fiber is an indigestible plant material composed primarily of cellulose. Fiber makes no contribution to the body in the form of macro- or micronutrients, but it prevents or relieves constipation by absorbing water and softening the stool for easier elimination.

Reference Daily Intakes (RDI)
A set of standards for protein, vitamins, and minerals used on food labels as part of the Daily Values.

A number of countries of the world have established nutritional guidelines for their citizens in an attempt to improve and maintain good national health. In the United States, the Nutrition Labeling and Education Act of 1990 brought sweeping changes to the regulations that define what is required on a food label. The official guidelines are called **Reference Daily Intakes (RDI)** for proteins and

➤ **FIGURE 12.1** Carbohydrates, lipids, and proteins from food supply the energy for all of our activities.

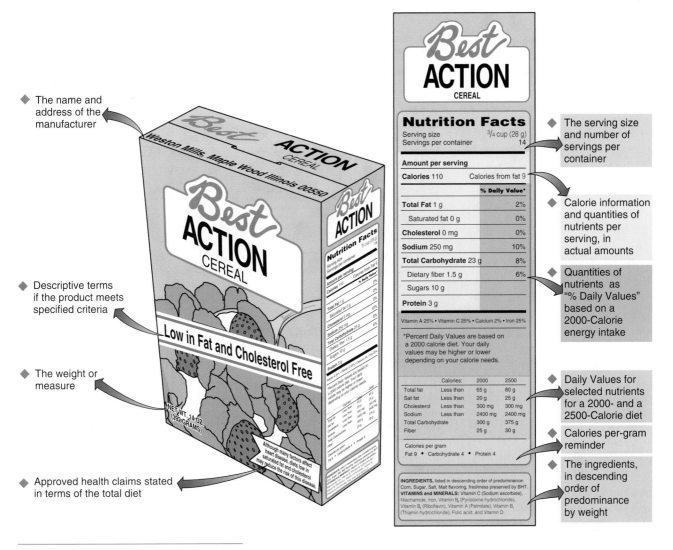

The name and address of the manufacturer

Descriptive terms if the product meets specified criteria

The weight or measure

Approved health claims stated in terms of the total diet

The serving size and number of servings per container

Calorie information and quantities of nutrients per serving, in actual amounts

Quantities of nutrients as "% Daily Values" based on a 2000-Calorie energy intake

Daily Values for selected nutrients for a 2000- and a 2500-Calorie diet

Calories per-gram reminder

The ingredients, in descending order of predominance by weight

Nutrition Facts

Serving size 3/4 cup (28 g)
Servings per container 14

Amount per serving

Calories 110 Calories from fat 9

% Daily Value*

Total Fat 1 g	2%
Saturated fat 0 g	0%
Cholesterol 0 mg	0%
Sodium 250 mg	10%
Total Carbohydrate 23 g	8%
Dietary fiber 1.5 g	6%
Sugars 10 g	
Protein 3 g	

Vitamin A 25% • Vitamin C 25% • Calcium 2% • Iron 25%

*Percent Daily Values are based on a 2000 calorie diet. Your daily values may be higher or lower depending on your calorie needs.

		Calories:	2000	2500
Total fat	Less than		65 g	80 g
Sat fat	Less than		20 g	25 g
Cholesterol	Less than		300 mg	300 mg
Sodium	Less than		2400 mg	2400 mg
Total Carbohydrate			300 g	375 g
Fiber			25 g	30 g

Calories per gram
Fat 9 • Carbohydrate 4 • Protein 4

INGREDIENTS, listed in descending order of predominance: Corn, Sugar, Salt, Malt flavoring, freshness preserved by BHT. VITAMINS and MINERALS: Vitamin C (Sodium ascorbate), Niacinamide, Iron, Vitamin B₆ (Pyridoxine hydrochloride), Vitamin B₂ (Riboflavin), Vitamin A (Palmitate), Vitamin B₁ (Thiamin hydrochloride), Folic acid, and Vitamin D.

➤ **FIGURE 12.2** An example of a food label.

19 vitamins and minerals, and **Daily Reference Values (DRV)** for other nutrients of public health importance. For simplicity, all reference values on food labels are referred to as **Daily Values (DV)** (see ■ Figure 12.2). The Food and Drug Administration (FDA) decided to use 2000 Calories as a standard for energy intake in calculating the DRV values. A 2000-Calorie diet is considered about right for many adults. Recall from Chapter 1 that the nutritional Calorie (written with a capital C) is equal to 1 kilocalorie (kcal) of energy. The guidelines are reviewed and revised about every 5 years. The U.S. Department of Agriculture (USDA) has issued the Food Guide Pyramid, designed to replace the old Four Basic Food Groups posters with new recommendations (see ■ Figure 12.3).

➤ 12.2 The Macronutrients

Carbohydrates

Carbohydrates are ideal energy sources for most body functions and also provide useful materials for the synthesis of cell and tissue components. These facts, plus the relatively low cost and ready availability of carbohydrates, have led to their worldwide use as the main dietary source of energy.

Daily Reference Values (DRV)
A set of standards for nutrients and food components (such as fat and fiber) that have important relationships with health; used on food labels as part of the Daily Values.

Daily Values (DV)
Reference values developed by the FDA specifically for use on food labels. The Daily Values represent two sets of standards: Reference Daily Intakes (RDI) and Daily Reference Values (DRV).

http://www.brookscole.com/chemistry/seager_chemistry
At Web site 12.2, you will find useful information on food labeling.

http://www.brookscole.com/chemistry/seager_chemistry
At Web site 12.3, you will find additional information on the Food Guide Pyramid.

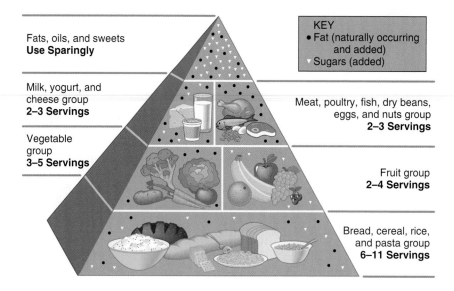

FIGURE 12.3 The Food Guide Pyramid developed by the U.S. Department of Agriculture. The circles and triangles show fats, oils, and added sugars in foods.

Despite their importance as a dominant source of energy, many people consider foods rich in carbohydrates to be inferior, at least in part because of their reputation of being fattening. It is now recognized that this reputation is generally not deserved. Most of the excess calories associated with eating carbohydrates are actually due to the high-calorie foods eaten *with* the carbohydrates—for example, potatoes and bread are often eaten with butter, a high-energy lipid.

Dietary carbohydrates are often classified as simple or complex. **Simple carbohydrates** are the sugars we classified earlier as monosaccharides and disaccharides (Chapter 7). **Complex carbohydrates** consist essentially of the polysaccharides amylose and amylopectin, which are collectively called starch. Cellulose, another polysaccharide, is also a complex carbohydrate; however, because it cannot be digested by humans, it serves a nonnutritive role as fiber.

Numerous nutritional studies include recommendations about dietary carbohydrates in their reports. One conclusion common to most of these studies is that a typical American diet does not include enough complex carbohydrates. These studies recommend that about 58% of daily calories should come from carbohydrate food. Currently, only about 46% of the typical diet comes from carbohydrates, and too much of that total comes from simple carbohydrates (see ■ Figure 12.4).

Lipids

Lipids, or fats, like carbohydrates, have a somewhat negative dietary image. In part, this is because the word *fat* has several meanings, and its relationship to health is sometimes misunderstood. In our modern society, some people think of fat as something to be avoided at all costs, regardless of body size.

As we saw in Chapter 8, a number of different substances are classified as lipids. However, about 95% of the lipids in foods and in our bodies are triglycerides, in which three fatty acid molecules are bonded to a molecule of glycerol by ester linkages. The fatty acids in triglycerides can be either saturated or unsaturated. Generally, triglycerides containing a high percentage of unsaturated fatty acids are liquids at room temperature and are called oils (see ■ Figure 12.5). A higher concentration of saturated fatty acids causes triglyceride melting points to increase, and they are solids (fats) at room temperature.

Lipids are important dietary constituents for a number of reasons: They are a concentrated source of energy and provide more than twice the energy

Simple carbohydrates
Monosaccharides and disaccharides, commonly called sugars.

Complex carbohydrates
The polysaccharides amylose and amylopectin, collectively called starch.

➤ **FIGURE 12.5** Oil-fried foods are rich in triglycerides.

➤ **FIGURE 12.4** (a) The composition of a typical American diet; (b) the composition of a healthy diet with lower fat intake and reduced levels of fat.

of an equal mass of carbohydrate; they contain some fat-soluble vitamins and help carry them through the body; and some fatty acids needed by the body cannot be synthesized and must come from the diet—they are the essential fatty acids.

In addition to metabolic needs, dietary lipids are desirable for other reasons. Lipids improve the texture of foods, and absorb and retain flavors, thus making foods more palatable. Lipids are digested more slowly than other foods and therefore prolong **satiety,** a feeling of satisfaction and fullness after a meal.

The essential fatty acids are the polyunsaturated linoleic and linolenic acids. Generally, vegetable oils are good sources of unsaturated fatty acids (see ■ Figure 12.6). Some oils that are especially rich in linoleic acid come from corn, cottonseed, soybean, and wheat germ. Infants especially need linoleic acid for good health and growth; human breast milk contains a higher percentage of it than does cow's milk.

We still do not have a complete understanding of the relationship between dietary lipids and health. A moderate amount of fat is needed in everyone's diet, but many people consume much more than required. Research results indicate a correlation between the consumption of too much fat, and fat of the wrong type (saturated fatty acids), and two of the greatest health problems: obesity and cardiovascular disease. Because of these results, there is a recommended reduction in the percentage of Calories obtained from fats— from the national average of almost 42% to no more than 30%.

Satiety
A state of satisfaction or fullness.

> **FIGURE 12.6** The fatty acid composition of common food fats.

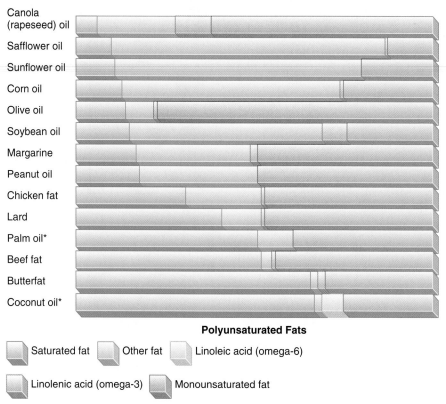

Dietary Fat

Canola (rapeseed) oil
Safflower oil
Sunflower oil
Corn oil
Olive oil
Soybean oil
Margarine
Peanut oil
Chicken fat
Lard
Palm oil*
Beef fat
Butterfat
Coconut oil*

Polyunsaturated Fats

Saturated fat   Other fat   Linoleic acid (omega-6)

Linolenic acid (omega-3)   Monounsaturated fat

*Data on linoleic and linolenic acids for palm and coconut oils not available.

TABLE 12.1	The RDI for protein
Group	**RDI (g)**
Pregnant women	60
Nursing mothers	65
Infants under age 1	14
Children age 1 to 4	16
Adults	50

Essential amino acid
An amino acid that cannot be synthesized within the body at a rate adequate to meet metabolic needs.

Proteins

Proteins are the only macronutrients for which an RDI has been established. The RDI varies for different groups of people, as shown in ■ Table 12.1, which shows that children have a greater need for protein than do adults. This reflects one of the important uses of protein in the body, the production of new tissue as the body grows. In addition, proteins are needed for the maintenance and repair of cells and for the production of enzymes, hormones, and other important nitrogen-containing compounds of the body. Proteins can also be used to supply energy; they provide about 4 Calories/gram, the same as carbohydrates. It is recommended that 12% of the Calories obtained from food be obtained from dietary proteins (Figure 12.4).

As we learned in Chapter 9, proteins are natural polymers of amino acids joined by amide (or peptide) linkages. On digestion, proteins are broken down to individual amino acids that are absorbed into the body's amino acid pool and used for the purposes mentioned earlier. The **essential amino acids** listed in ■ Table 12.2 must be obtained from the diet because they cannot be synthesized in the body in sufficient amounts to satisfy the body's needs.

The minimum quantity of each essential amino acid needed per day by an individual will vary, depending on the use in the body. For example, results of one study showed that young men need a daily average of only 7 g (0.034 mol) of tryptophan but 31 g (0.21 mol) of methionine. According to this study, the body of an average young man requires about seven times as many moles of methionine as it does tryptophan to carry out the processes involving amino acids.

CHEMISTRY AROUND US • 12.1

VEGETARIAN DIETS

People whose diets classify them as vegetarians usually fall into one of three main categories. Lacto-ovo-vegetarians abstain from all meat, but do eat dairy products and eggs. This is the most common type of vegetarian diet followed in the United States. Lacto-vegetarians eat no meat or eggs, but include dairy products in their diet. Vegans eat nothing that is derived from animals, including meat, fish, poultry, dairy products, eggs, and honey. Smaller numbers of people follow other types of vegetarian diets. Pescovegetarians eat fish, and semivegetarians avoid only red meat products. Fruitarians follow a type of vegan diet in which very few processed or cooked foods are eaten. The diet consists mainly of raw vegetables and fruits, grains, and nuts. Only foods that can be harvested without killing the plant are used. Thus, apples and squash are acceptable, but carrots are not.

The more restrictive a person's diet is, the more carefully it must be monitored to make certain it is nutritionally adequate. For example, it is especially important to make sure any vegetar-

ian diet followed by growing children or pregnant women contains all the necessary vitamins and nutrients. Vitamin supplements are often recommended in such cases, especially for vegetarians who avoid dairy products. Some typical supplements used are vitamin B_{12}, vitamin D, calcium, iron, and zinc.

A proper diet can have a positive influence on health. For example, hypertension can be treated by using a combination diet that is high in fruits and vegetables, includes low-fat dairy foods, and is low in sodium as well as saturated and total fat.

The reasons people choose to become vegetarians are widely varied. Some do so in order to avoid certain foods that cause health problems, including allergic reactions, hypertension, or lactose intolerance. Others choose to be vegetarians for economic, ecological, religious, or ethical reasons. For example, some people feel strongly that animals should not be killed to provide for human needs. For whatever reasons they are followed, well-designed vegetarian diets can provide the nutrition needed for a healthy life.

Proteins in foods are classified as **complete proteins** if they contain all the essential amino acids in the proper proportions needed by the body. Protein foods that don't meet this requirement are classified as incomplete. Several protein-containing foods are shown in ■ Figure 12.7.

Complete protein
Protein in food that contains all essential amino acids in the proportions needed by the body.

➤ 12.3 Micronutrients I: Vitamins

Vitamins are organic micronutrients that the body cannot produce in amounts needed for good health. The highly polar nature of some vitamins renders them water-soluble. Nine water-soluble vitamins have been identified. Some are designated by names, some by letters, and some by letters and numbers. ■ Table

TABLE 12.2	The essential amino acids
Histidine	Phenylalanine
Isoleucine	Threonine
Leucine	Tryptophan
Lysine	Valine
Methionine	

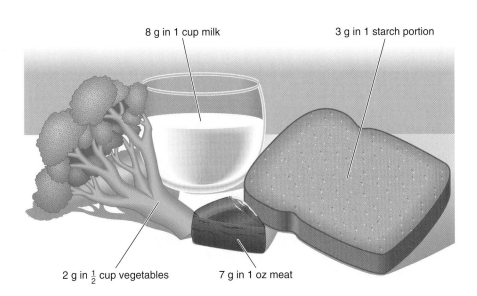

8 g in 1 cup milk

3 g in 1 starch portion

2 g in $\frac{1}{2}$ cup vegetables

7 g in 1 oz meat

➤ **FIGURE 12.7** The protein content of several foods.

TABLE 12.3 Vitamin sources, functions, and deficiency conditions

Vitamin	Dietary sources	Functions	Deficiency conditions
Water-soluble			
B_1 (thiamin)	Bread, beans, nuts, milk, peas, pork	Coenzyme in decarboxylation reactions	Beriberi: nausea, severe exhaustion, paralysis
B_2 (riboflavin)	Milk, meat, eggs, dark green vegetables, bread, beans, peas	Forms the coenzymes FMN and FAD, which are hydrogen transporters	Dermatitis (skin problems)
Niacin	Meat, whole grains, poultry, fish	Forms the coenzyme NAD^+, which is a hydride transporter	Pellagra: weak muscles, no appetite, diarrhea, dermatitis
B_6 (pyridoxine)	Meats, whole grains, poultry, fish	Coenzyme form carries amino and carboxyl groups	Dermatitis, nervous disorders
B_{12} (cobalamin)	Meat, fish, eggs, milk	Coenzyme in amino acid metabolism	Rare except in vegetarians; pernicious anemia
Folic acid	Leafy green vegetables, peas, beans	Coenzyme in methyl group transfers	Anemia
Pantothenic acid	All plants and animals, nuts; whole-grain cereals	Part of coenzyme A, acyl group carrier	Anemia
Biotin	Found widely; egg yolk, liver, yeast, nuts	Coenzyme form used in fatty acid synthesis	Dermatitis, muscle weakness
C (ascorbic acid)	Citrus fruits, tomatoes, green pepper, strawberries, leafy green vegetables	Synthesis of collagen for connective tissue	Scurvy: tender tissues; weak, bleeding gums; swollen joints
Fat-soluble			
A (retinol)	Eggs, butter, cheese, dark green and deep orange vegetables	Synthesis of visual pigments	Inflamed eye membranes, night blindness, scaliness of skin
D (calciferol)	Fish-liver oils, fortified milk	Regulation of calcium and phosphorus metabolism	Rickets (malformation of the bones)
E (tocopherol)	Whole-grain cereals, margarine, vegetable oil	Prevention of oxidation of vitamin A and fatty acids	Breakage of red blood cells
K	Cabbage, potatoes, peas, leafy green vegetables	Synthesis of blood-clotting substances	Blood-clotting disorders

12.3 presents the names nutritionists use (the "correct" names), along with other common names you will likely encounter in books and on product labels. With the exception of vitamin C, all water-soluble vitamins have been shown to function as coenzymes.

Some vitamins (A, D, E, and K) have very nonpolar molecular structures and therefore dissolve only in nonpolar solvents. In the body, the nonpolar solvents are the lipids we have classified as fats, so these vitamins are called fat-soluble. The fat-soluble vitamins have diverse functions in the body (Table 11.3). Care must be taken to avoid overdoses of the fat-soluble vitamins. Toxic effects are known to occur, especially with vitamin A, when excess amounts of these vitamins accumulate in body tissue. Excesses of water-soluble vitamins are excreted readily through the kidneys and are not normally a problem.

CHEMISTRY AROUND US • 12.2

EATING DISORDERS

The eating disorders *anorexia nervosa* and *bulimia nervosa* have increased dramatically in the United States in a way that parallels an increased national obsession with being thin. People with anorexia follow eating habits that cause them to lose dangerous amounts of body weight. This disorder occurs most often among individuals in the 14–19 age group, and about 90% are female. People with anorexia generally have the following characteristics:

1. The fear of becoming fat and losing control of body size or shape; an obsession with being thin.
2. A preoccupation with food, possibly as a result of the body's response to starvation.
3. Poor body image, a low opinion of body shape, and an over-estimation of their body's size.
4. One or more psychological problems, such as depression, low self-esteem, anxiety, and especially obsessive-compulsive behavior.
5. In women, medical problems such as amenorrhea (the absence of menstruation); in men and women, low body temperature, low blood pressure, reduced heart rate, metabolic and electrolyte imbalances, and nutritional deficiencies.

Binge eating followed by purging is a characteristic behavior of people suffering from bulimia nervosa. Bulimia is most common among those in the 15–21 age group, and, again, over 90% are female. The purging takes the form of vomiting or excessively using laxatives, and is done in an attempt to avoid gaining weight. Most people with bulimia accomplish this goal and maintain a normal body weight. After binge eating, they suffer feelings of self-reproach and guilt, which are relieved to some extent by a decrease in the feeling of fullness that accompanies the purging. People with bulimia typically have mood swings, uncontrollable impulses, and depression. Some also experience problems with their teeth, which are damaged by the stomach acid contained in vomit, and females may also experience amenorrhea.

Several key factors put an individual at risk of developing an eating disorder. Our society emphasizes thinness and discriminates against obesity. People who seem to be most vulnerable to eating disorders are those whose gender, ethnicity, occupation, or other interests especially value and reward thinness. Another factor may be the family. Members of families that overemphasize thinness or that become overly involved in each other's lives may be at risk of developing an eating disorder. Other contributing factors include mood disorders, the inability to differentiate internal needs, and finally, the hypothalamus in the brain. The hypothalamus secretes certain chemicals that help control eating patterns. According to one theory, every person has a certain set-point for body weight that is determined by heredity and early eating habits. When body weight deviates from the set-point, the body compensates by adjusting eating patterns to help the body return to the set-point weight.

Fortunately, there are treatments for eating disorders. After attempting to help patients establish normal eating habits and regain their health, therapists then try to correct the problems that led to the disorder. For anorexia, a strategy using supportive nursing care is sometimes used. The first goal of this care is to gradually increase the caloric content of the food in the diet, and gradually increase the body weight back to normal. Therapy to address the underlying psychological and/or family problems that led to the disorder follows. For bulimia patients, several strategies are often combined, including individual insight therapy, group therapy, behavioral therapy, and sometimes the use of antidepressant medications.

http://www.brookscole.com/chemistry/seager_chemistry
More information on eating disorders is given at Web site 12.4.

➤ 12.4 Micronutrients II: Minerals

The word **mineral** is generally used to describe inorganic substances, often as if they were in the elemental form. Thus, we might hear that a food is a good source of phosphorus, when in fact the food does not contain elemental phosphorus, a caustic nonmetal, but probably contains phosphate ions such as HPO_4^{2-} and $H_2PO_4^{-}$ or, more likely, organic substances that contain phosphorus, such as nucleic acids, and phospholipids.

In the body, the elements classified as minerals are never used in elemental form, but rather in the form of ions or compounds. They are classified as **major minerals** and **trace minerals** on the basis of the amount present in the body, as shown in ■ Figure 12.8. In general, the major minerals are required in larger daily amounts than are the trace minerals. For example, the RDIs for calcium and phosphorus, the major minerals found in the body in largest amounts, are 800 mg. The RDIs for iron and iodine, two of the trace minerals, are 10–18 mg and 0.12 mg, respectively. By way of comparison, the RDIs for vitamin C and niacin are, respectively, 55–60 mg and 18 mg.

Mineral
A metal or nonmetal used in the body in the form of ions or compounds.

Major mineral
A mineral found in the body in quantities greater than 5 g.

Trace mineral
A mineral found in the body in quantities less than 5 g.

➤ **FIGURE 12.8** Minerals in a 60-kg person. The major minerals are those present in amounts larger than 5 g (a teaspoon); only four of the numerous trace minerals are shown. Only calcium and phosphorus appear in amounts larger than a pound (454 g).

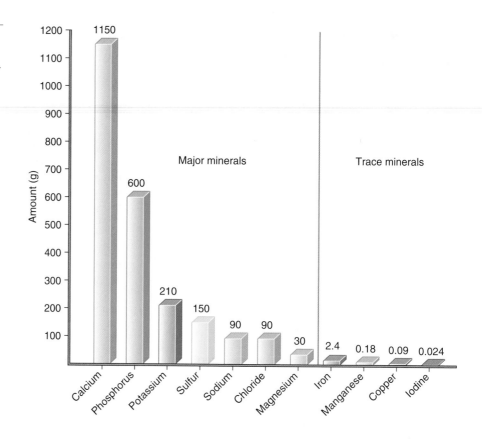

As shown in ■ Table 12.4, the functions of minerals are consistent with their classification as major or trace and with the amount required daily in the diet. For example, compounds of some major minerals (Ca and P) are the primary inorganic structural components of bones and teeth. Other major minerals (Na, K, Cl, and Mg) form principal ions that are distributed throughout the body's various fluids. Some trace minerals are components of vitamins (Co), enzymes (Zn and Se), hormones (I), or specialized proteins (Fe and Cu). Thus, we see that even though trace minerals are required in small quantities, their involvement in critical enzymes, hormones, and the like makes them equally as important for good health as the major minerals.

➤ 12.5 The Flow of Energy in the Biosphere

The sun is the ultimate source of energy used in all biological processes. The sun's enormous energy output is the result of the fusion of hydrogen atoms into helium atoms. The net reaction for this process is

$$4H \xrightarrow{\text{nuclear fusion}} He + energy \tag{12.1}$$

A portion of the liberated energy reaches the earth as sunlight and is absorbed by chlorophyll pigments in plants. In the plants, the energy drives the reactions of photosynthesis that convert carbon dioxide and water first into glucose, and then into starch, triglycerides, and other storage forms of energy. The net reaction for glucose production is

$$6CO_2 + 6H_2O + energy \xrightarrow{\text{photosynthesis}} C_6H_{12}O_6 + 6O_2 \tag{12.2}$$
$$\text{glucose}$$

TABLE 12.4 Major and trace mineral sources, functions, and deficiency conditions

Mineral	Dietary sources	Functions	Deficiency conditions
Major minerals			
Calcium	Dairy foods, dark green vegetables	Bone and teeth formation, blood clotting, nerve impulse transmission	Stunted growth, rickets, weak and brittle bones
Chlorine	Table salt, seafood, meat	HCl in gastric juice, acid–base balance	Muscle cramps, apathy, reduced appetite
Magnesium	Whole-grain cereals, meat, nuts, milk, legumes	Activation of enzymes, protein synthesis	Inhibited growth, weakness, spasms
Phosphorus	Milk, cheese, meat, fish, grains, legumes, nuts	Enzyme component, acid–base balance, bone and tooth formation	Weakness, calcium loss, weak bones
Potassium	Meat, milk, many fruits, cereals, legumes	Acid–base and water balance, nerve function	Muscle weakness, paralysis
Sodium	Most foods except fruit	Acid–base and water balance, nerve function	Muscle cramps, apathy, reduced appetite
Sulfur	Protein foods	Component of proteins	Deficiencies are very rare
Trace minerals			
Arsenic[a]	Many foods	Growth and reproduction	Poor growth and reproduction
Cobalt	Meat, liver, dairy foods	Component of vitamin B_{12}	Pernicious anemia (vitamin-deficiency symptom)
Copper	Drinking water, liver, grains, legumes, nuts	Component of numerous enzymes, hemoglobin formation	Anemia, fragility of arteries
Chromium	Fats, vegetable oils, grains, meat	Enhances insulin action	Reduced ability to metabolize glucose
Fluorine	Drinking water, seafood, onions, spinach	Maintenance of bones and teeth	Higher frequency of tooth decay
Iodine	Iodized salt, fish, dairy products	Component of thyroid hormones	Hypothyroidism, goiter
Iron	Liver, lean meat, whole grains, dark green vegetables	Component of enzymes and hemoglobin	Anemia
Manganese	Grains, beet greens, legumes, fruit	Component of enzymes	Deficiencies are rare
Molybdenum	Legumes, cereals, organ meats, dark green vegetables	Component of enzymes	Deficiencies are rare
Nickel	Many foods	Needed for health of numerous tissues	Organ damage, deficiencies are rare
Selenium	Grains, meats, poultry, milk	Component of enzymes	Deficiencies are rare
Silicon[a]	Many foods	Bone calcification	Poor bone development, deficiencies are rare
Tin[a]	Many foods	Needed for growth	Poor growth
Vanadium[a]	Many foods	Growth, bone development, reproduction	Poor growth, bone development and reproduction
Zinc	Milk, liver, shellfish, wheat bran	Component of numerous enzymes	Poor growth, lack of sexual maturation, loss of appetite, abnormal glucose tolerance

[a]Need and deficiency symptoms determined for animals, probable in humans, but not yet recognized.

➤ **FIGURE 12.9** Energy flow in the biosphere.

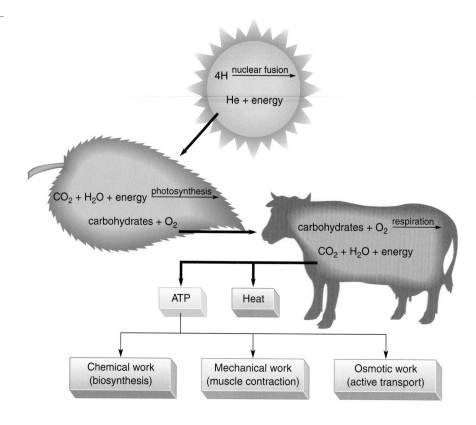

Cellular respiration

The entire process involved in the use of oxygen by cells.

All animals obtain energy either directly or indirectly from these energy stores in plants.

During **cellular respiration**, both plants and animals combine these energy-rich compounds with oxygen from the air, producing carbon dioxide and water, and releasing energy. Cellular respiration, which should not be confused with pulmonary respiration (breathing), is represented as

$$\text{glucose and other storage forms of energy} + O_2 \xrightarrow{\text{respiration}} CO_2 + H_2O + \text{energy captured} \quad (12.3)$$

A portion of the energy released during respiration is captured within the cells in the form of adenosine triphosphate (ATP), a chemical carrier of energy, from which energy can be obtained and used directly for the performance of biological work. The remainder of the energy from respiration is liberated as heat. Steps in the flow of energy in the biosphere are summarized in ■ Figure 12.9.

Notice in Figure 12.9 that atmospheric carbon dioxide is the initial source of carbon for carbohydrates. Another important feature of Figure 12.9 is that overall the respiration process is the reverse of the photosynthetic process. Thus, some of the earth's carbon compounds are repeatedly recycled by living organisms (see ■ Figure 12.10).

➤ **FIGURE 12.10** A simplified carbon cycle.

➤ 12.6 Metabolism and an Overview of Energy Production

Metabolism

Living cells are very active chemically, with thousands of different reactions occurring at the same time. The sum total of all the chemical reactions involved in maintaining a living cell is called **metabolism**. The reactions of metabolism are divided into two categories, catabolism and anabolism. **Catabolism** consists of all reactions that lead to the breakdown of biomolecules. **Anabolism** includes all reactions that lead to the synthesis of biomolecules. In general, energy is released during catabolism and used up during anabolism.

A characteristic of all living organisms is a high degree of order in their structures and chemical processes. Even the simplest organisms exhibit an enormous array of chemical reactions that are organized into orderly, well-regulated sequences known as **metabolic pathways.** Each metabolic pathway consists of a series of consecutive chemical reactions or steps that convert a starting material into an end product. For example, two pathways that will be covered in Chapter 13 are the citric acid cycle and the electron transport chain.

Fortunately for those who study the biochemistry of living systems, a great many similarities exist among the major metabolic pathways found in all life forms. These similarities enable scientists to study the metabolism of simple organisms and use the results to help explain the corresponding metabolic pathways in more complex organisms, including humans. This text concentrates on human biochemistry, but you should realize that much of what is said about human metabolism can be applied equally well to the metabolism of most other organisms.

The Catabolism of Food

The three stages involved in the catabolic extraction of energy from food are illustrated in ■ Figure 12.11.

Stage I is digestion, in which large, complex molecules are chemically broken into relatively small, simple ones (see ■ Figure 12.12). The most common reaction in digestion is hydrolysis, in which proteins are converted to amino acids, carbohydrates are converted to monosaccharides (primarily glucose, galactose, and fructose), and fats are converted to fatty acids and glycerol. The smaller molecules are then absorbed into the body through the lining of the small intestine.

In Stage II, the small molecules from digestion are degraded to even simpler units, primarily the two-carbon acetyl portion of acetyl coenzyme A (acetyl CoA):

$$CH_3-\underset{\substack{\| \\ }}{\overset{\substack{O \\ \|}}{C}}-S-CoA$$

acetyl   coenzyme A

Some energy is released in the second stage, but much more is produced during the oxidation of the acetyl units of acetyl CoA in the third stage. Stage III consists of the citric acid cycle followed by electron transport and oxidative phosphorylation. Because the reactions of Stage III are the same regardless of the type of food being degraded, it is sometimes referred to as the **common catabolic pathway.** Energy released during Stage III appears in the form of energy-rich molecules of ATP. The whole purpose of the catabolic pathway is to convert the chemical energy in foods to molecules of ATP. The common

Metabolism
The sum of all reactions occurring in an organism.

Catabolism
All reactions involved in the breakdown of biomolecules.

Anabolism
All reactions involved in the synthesis of biomolecules.

Metabolic pathway
A sequence of reactions used to produce one product or accomplish one process.

Common catabolic pathway
The reactions of the citric acid cycle plus those of the electron transport chain and oxidative phosphorylation.

➤ **FIGURE 12.11** The three stages in the extraction of energy from food.

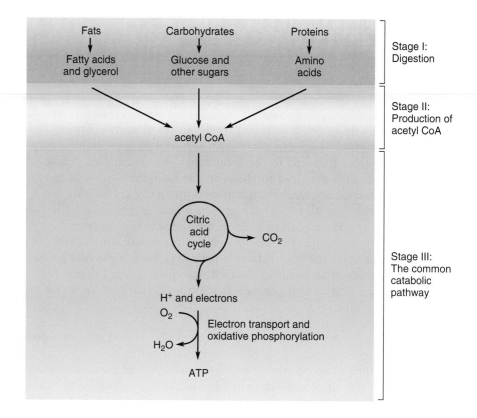

catabolic pathway will be covered in Chapter 13. Chapters 13–15 discuss the ways in which carbohydrates, lipids, and proteins provide molecules that are degraded by reactions of the common catabolic pathway.

➤ 12.7 ATP: The Primary Energy Carrier

Certain bonds in ATP store the energy released during the oxidation of carbohydrates, lipids, and proteins. ATP molecules act as energy carriers and deliver the energy to the parts of the cell where energy is needed to power muscle contraction, biosynthesis, and other cellular work.

The Structure and Hydrolysis of ATP

The structure of the ATP molecules is relatively simple (see ■ Figure 12.13). It consists of a heterocyclic base, adenine, bonded to the sugar ribose. Taken together, this portion of the molecule is called adenosine. The triphosphate portion is bonded to the ribose to give adenosine triphosphate, or ATP. At the cell pH of 7.4, all the protons of the triphosphate group are ionized, giving the ATP molecule a charge of -4. In the cell, ATP is complexed with Mg^{2+} in a 1:1 ratio. Thus, the 1:1 complex still exhibits a net charge of -2. Two

➤ **FIGURE 12.12** The digestion of food involves enzyme-catalyzed hydrolysis reactions.

$$\text{proteins} \xrightarrow{\text{hydrolysis}} \text{amino acids}$$

$$\text{disaccharides and polysaccharides} \xrightarrow{\text{hydrolysis}} \text{monosaccharides}$$

$$\text{fats and oils} \xrightarrow{\text{hydrolysis}} \text{fatty acids + glycerol}$$

> **FIGURE 12.13** The structure of ATP.

other triphosphates—GTP, containing the base guanine, and UTP, containing the base uracil—are important in carbohydrate metabolism (Chapter 13).

The triphosphate end of ATP is the part of the molecule that is important in the transfer of biochemical energy. The key reaction in this energy delivery system is the transfer of a phosphoryl group, $—PO_3^{2-}$, from ATP to another molecule. For example, during the hydrolysis of ATP in water, a phosphoryl group is transferred from ATP to water. The products of this hydrolysis are adenosine diphosphate (ADP) and a phosphate ion, often referred to as an inorganic phosphate, P_i, or simply as *phosphate*.

$$\text{adenosine} - O - \overset{\overset{\displaystyle O}{\|}}{\underset{\underset{\displaystyle O^-}{|}}{P}} - O - \overset{\overset{\displaystyle O}{\|}}{\underset{\underset{\displaystyle O^-}{|}}{P}} - O - \overset{\overset{\displaystyle O}{\|}}{\underset{\underset{\displaystyle O^-}{|}}{P}} - O^- + H - OH \longrightarrow$$

ATP

$$\text{adenosine} - O - \overset{\overset{\displaystyle O}{\|}}{\underset{\underset{\displaystyle O^-}{|}}{P}} - O - \overset{\overset{\displaystyle O}{\|}}{\underset{\underset{\displaystyle O^-}{|}}{P}} - O^- + HO - \overset{\overset{\displaystyle O}{\|}}{\underset{\underset{\displaystyle O^-}{|}}{P}} - O^- + H^+ \qquad (12.4)$$

ADP   inorganic phosphate (P_i)

The transfer of a phosphoryl group from ATP to water is accompanied by a release of energy. This energy is called free energy and is represented by the symbol ΔG. When energy is released by a reaction or process, ΔG will

> **STUDY SKILLS 12.1**

BIOPROCESSES

The last four chapters of the text complete our study of the molecular basis of life. This chapter and the next three build on your knowledge of the biomolecules (carbohydrates, lipids, proteins, and nucleic acids) and examine the processes that support living organisms. Your attention should now be directed toward gaining an understanding of these processes, where they occur in the cell, their purpose, and, in some cases, key summary reactions. A number of detailed pathways such as glycolysis, the fatty acid spiral, and the urea cycle will be presented. Don't let the numerous reactions in these pathways intimidate you. Although each reaction could be looked at individually, we have tried to help you focus your attention by emphasizing the overall importance of each pathway by identifying its purpose, cellular location, starting material, products, energy requirements, and regulation.

TABLE 12.5 Free energy changes for the hydrolysis of some phosphate compounds

Reaction	$\Delta G^{\circ\prime}$ (kcal/mol)
Phosphoenolpyruvate + H_2O → pyruvate + phosphate	−14.8
Carbamoyl phosphate + H_2O → ammonia + CO_2 + phosphate	−12.3
1,3-bisphosphoglycerate + H_2O → 3-phosphoglycerate + phosphate	−11.8
Phosphocreatine + H_2O → creatine + phosphate	−10.3
ATP + H_2O → ADP + phosphate	−7.3
ADP + H_2O → AMP + phosphate	−7.3
Glucose 1-phosphate + H_2O → glucose + phosphate	−5.0
AMP + H_2O → adenosine + phosphate	−3.4
Glucose 6-phosphate + H_2O → glucose + phosphate	−3.3
Glycerol 3-phosphate + H_2O → glycerol + phosphate	−2.2

High-energy compound

A substance that on hydrolysis liberates a great amount of free energy.

have a negative value. When energy is absorbed, ΔG will have a positive value. When ΔG is measured under standard conditions (temperature, concentration, etc.), it is represented by ΔG°, and a superscript prime ($\Delta G^{\circ\prime}$) is added if ΔG° is measured at body conditions. The liberated free energy is available for use by the cell to carry out processes requiring an input of energy. This energy-releasing hydrolysis reaction is represented by the following equation:

$$ATP + H_2O \rightarrow ADP + P_i + H^+ \qquad \Delta G^{\circ\prime} = -7.3 \text{ kcal/mol} \qquad (12.5)$$

Many other phosphate-containing compounds found in biological systems also liberate energy on hydrolysis. Several are listed in ■ Table 12.5, along with values for their free standard energy of hydrolysis.

Compounds that liberate a great amount of free energy on hydrolysis are called **high-energy compounds.** Remember, a negative sign for $\Delta G^{\circ\prime}$ means the energy is liberated when the reaction takes place. Thus, ATP ($\Delta G^{\circ\prime} = -7.3$ kcal/mol) is considered a high-energy compound, but not the only one found in cells, as you can see from Table 12.5.

The hydrolysis of ATP to ADP is the principal energy-releasing reaction for ATP. However, some other hydrolysis reactions also play important roles in energy transfer. An example is the hydrolysis of ATP to adenosine monophosphate (AMP) and pyrophosphate (PP_i):

$$(12.6)$$

The reaction may also be written as

$$ATP + H_2O \rightarrow AMP + PP_i + 2H^+ \qquad \Delta G^{\circ\prime} = -8.0 \text{ kcal/mol} \qquad (12.7)$$

TABLE 12.6	Hydrolyses of some ATP-related compounds	
Reaction		$\Delta G^{\circ\prime}$ (kcal/mol)
$ATP + H_2O$	$\rightarrow \quad ADP + P_i + H^+$	-7.3
$ATP + H_2O$	$\rightarrow \quad AMP + PP_i + 2H^+$	-8.0
$PP_i + H_2O$	$\rightarrow \quad 2P_i$	-6.5
$ADP + H_2O$	$\rightarrow \quad AMP + P_i + H^+$	-7.3

When the hydrolysis of ATP to AMP occurs during metabolism, it is usually followed by the immediate hydrolysis of the pyrophosphate, which releases even more free energy:

$$^-O - \overset{\overset{O}{\|}}{\underset{\underset{O^-}{|}}{P}} - O - \overset{\overset{O}{\|}}{\underset{\underset{O^-}{|}}{P}} - O^- + H_2O \rightarrow 2HO - \overset{\overset{O}{\|}}{\underset{\underset{O^-}{|}}{P}} - O^- \quad \Delta G^{\circ\prime} = -6.5 \text{ kcal/mol} \quad (12.8)$$

$$\underset{PP_i}{\qquad\qquad} \underset{P_i}{\qquad\qquad\qquad\qquad\qquad\qquad}$$

The hydrolyses of ATP and related compounds are summarized in ■ Table 12.6.

The ATP–ADP Cycle

In biological systems, ATP functions as an immediate donor of free energy rather than as a storage form of free energy. In a typical cell, an ATP molecule is hydrolyzed within 1 minute following its formation. Thus, the turnover rate of ATP is very high. For example, a resting human body hydrolyzes ATP at the rate of about 40 kg every 24 hours. During strenuous exertion, this rate of ATP utilization may be as high as 0.5 kg per minute. Motion, biosynthesis, and other forms of cellular work can occur only if ATP is continuously regenerated from ADP. ■ Figure 12.14 depicts the central role served by the ATP–ADP cycle in linking energy production with energy utilization.

Mitochondria

Enzymes that catalyze the formation of ATP, as well as the other processes in the common catabolic pathway, are all located within the cell in organelles called **mitochondria** (see ■ Figure 12.15). Mitochondria are often called

Mitochondrion
A cellular organelle where reactions of the common catabolic pathway occur.

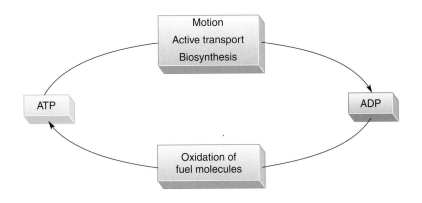

➤ **FIGURE 12.14** The ATP–ADP cycle. ATP, which supplies the energy for cellular work, is continuously regenerated from ADP during the oxidation of fuel molecules.

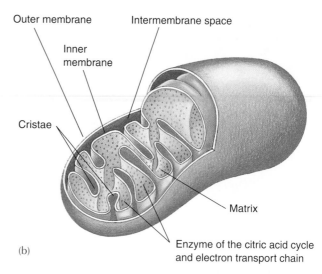

(a)

(b)

➤ **FIGURE 12.15** The mitochondrion: (a) a photomicrograph; (b) a schematic drawing, cut away to reveal the internal organization.

cellular "power stations" because they are the sites for most ATP synthesis in the cells.

A mitochondrion contains both inner and outer membranes. As a result of extensive folding, the inner membrane has a surface area that is many times that of the outer membrane. The folds of the inner membrane are called *cristae*, and the space that surrounds them is called the *matrix*. The enzymes for ATP synthesis (electron transport and oxidative phosphorylation) are located within the inner membrane. The enzymes for the citric acid cycle are found within the matrix, attached or near to the surface of the inner membrane. Thus, all the enzymes involved in the common catabolic pathway are in close proximity.

➤ 12.8 Important Coenzymes in the Common Catabolic Pathway

As discussed in Section 10.3, coenzymes often participate in enzyme-catalyzed reactions by acting as shuttle systems in the transfer of chemical groups from one compound to another. For example, a coenzyme may accept hydrogen from one compound and later donate hydrogen to another compound. We also pointed out that many important coenzymes are formed from vitamins. The three substances discussed in this section exhibit both of these coenzyme characteristics.

Coenzyme A

Coenzyme A is a central compound in metabolism. It is a part of acetyl coenzyme A (acetyl CoA), the substance formed from all foods as they pass through Stage II of catabolism (see ■ Figure 12.16).

➤ **FIGURE 12.16** Stage II of the oxidation of foods. Acetyl CoA is formed as the pathways of catabolism converge.

$$
\left.\begin{array}{l}\text{Fatty acids and glycerol}\\\text{Glucose and other sugars}\\\text{Amino acids}\end{array}\right\} \longrightarrow \overset{\displaystyle O}{\underset{\substack{\text{acetyl coenzyme A}\\\text{(acetyl CoA)}}}{CH_3 - \overset{\|}{C} - S - CoA}}
$$

OVER THE COUNTER • 12.1

CREATINE SUPPLEMENTS: THE JURY IS STILL OUT

Creatine has become a popular dietary supplement among professional athletes and other individuals interested in building muscles. Users find that supplemental creatine in their diet helps them retain strength and size, and it increases their energy level. It also seems to help some athletes recover from injury more quickly. Preliminary scientific research results support the claim that muscle size and strength increase significantly faster in creatine users than in nonusers when both groups follow the same training and nutritional programs.

However, the advantages are accompanied by negative side effects. Supplemental creatine in the diet may influence some electrolyte levels in body tissues. In muscle tissue, the ratio of calcium ions to phosphate ions may be reversed, causing a disruption of the muscle's normal contraction-relaxation processes. As a result, a muscle might contract and shorten when it should be relaxing and lengthening. This can cause muscle cramping and in some cases serious muscle strains or even muscle tears.

Electrolyte imbalance can also contribute to dehydration and other heat-related illnesses. Electrolyte levels influence the fluid balance in the body, and when muscle cells retain increased levels of fluids, the blood plasma volume of the body is reduced. This has a negative effect on the body's ability to dissipate heat, a potentially serious condition in hot, humid environments.

Research results concerning the use of dietary creatine supplements are preliminary and are generally based on short-term studies. Some factors have not been studied, such as the effects of combining creatine supplementation with strength training and a conditioning program. Some research seems to indicate that creatine may be more useful in developing "ornamental" muscle, important to bodybuilding, rather than "functional" muscle, which adds strength and is important in athletics.

http://www.brookscole.com/chemistry/seager_chemistry
More information on creatine supplementation is given at Web site 12.5.

Three of the creatine supplements being marketed.

Coenzyme A is derived from the B vitamin pantothenic acid (see ■ Figure 12.17). It also contains two other components, a phosphate derivative of ADP and β-mercaptoethylamine. A key feature in the structure of coenzyme A is the presence of the sulfhydryl group (—SH), which is the reactive portion of the molecule. For this reason, coenzyme A is often abbreviated as CoA—SH.

The letter *A* was included in the coenzyme name to signify its participation in the transfer of acetyl groups. It is now known that coenzyme A transfers acyl groups in general rather than just the acetyl group. This is demonstrated in fatty acid oxidations (Section 14.4), fatty acid synthesis (Section 14.7), and other biological reactions. Acyl groups are linked to coenzyme A through the sulfur atom in a **thioester bond.**

$$
\begin{array}{cc}
\overset{\displaystyle O}{\underset{\displaystyle \|}{}} & \overset{\displaystyle O}{\underset{\displaystyle \|}{}} \\
R-C-S-CoA & CH_3-C-S-CoA \\
\text{acyl CoA} & \text{acetyl CoA}
\end{array}
$$

Thioester bond
The carbon-sulfur bond of the
$$-\overset{O}{\overset{\|}{C}}-S-$$ functional group.

The Coenzymes NAD⁺ and FAD

The majority of energy for ATP synthesis is released when the oxygen we breathe is reduced. During that process, oxygen accepts electrons and hydro-

FIGURE 12.17 The structure of coenzyme A. The sulfhydryl group (—SH) is the reactive site.

gen ions, and water is produced. The electrons come from the oxidation of fuel molecules, such as glucose, but the electrons are not transferred directly to the oxygen from the fuel molecules or their breakdown products. Instead, these substrates first transfer the electrons to special coenzyme carriers. The reduced forms of these coenzymes then transfer the electrons to oxygen through the reactions of the electron transport chain. ATP is formed from ADP and P_i as a result of this flow of electrons.

The first of these coenzymes, nicotinamide adenine dinucleotide (NAD^+), is a derivative of ADP and the vitamin nicotinamide (see ■ Figure 12.18). The reactive site of NAD^+ is located in the nicotinamide portion of the molecule. During the oxidation of a substrate, the nicotinamide ring of NAD^+ accepts two electrons and one proton, which forms the reduced coenzyme NADH.

$$+ 2e^- + H^+ \rightleftharpoons \tag{12.9}$$

NAD$^+$ (oxidized form)   NADH (reduced form)

A typical cellular reaction in which NAD^+ serves as an electron acceptor is the oxidation of an alcohol:

General reaction:

$$R-\underset{\underset{H}{|}}{\overset{\overset{OH}{|}}{C}}-H + NAD^+ \rightarrow R-\overset{\overset{O}{\|}}{C}-H + NADH + H^+ \tag{12.10}$$

alcohol   aldehyde

➤ **FIGURE 12.18** The structure of nicotinamide adenine dinucleotide (NAD^+). The reactive site is in the nicotinamide part of the molecule.

Specific example:

$$CH_3CH_2-OH + NAD^+ \xrightarrow[\text{dehydrogenase}]{\text{alcohol}} CH_3-\overset{\overset{\displaystyle O}{\|}}{C}-H + NADH + H^+ \quad (12.11)$$

ethanol   acetaldehyde

In this reaction, one hydrogen atom of the alcohol substrate is directly transferred to NAD^+, whereas the other appears in solution as H^+. Both electrons lost by the alcohol have been transferred to the nicotinamide ring in NADH.

Biochemical reactions in which coenzymes participate are often written in a more concise way. For example, Reaction 12.10 may be written

$$R-\overset{\overset{\displaystyle OH}{|}}{\underset{\underset{\displaystyle H}{|}}{C}}-H \xrightarrow{\hspace{2cm}} R-\overset{\overset{\displaystyle O}{\|}}{C}-H \quad (12.12)$$

$$NAD^+ \quad NADH + H^+$$

This format allows for the separate emphasis of the oxidation reaction and the involvement of a coenzyme.

Write the following reaction in the normal form of a chemical equation:

Learning Check 12.1

$$\underset{\displaystyle CH_2-COO^-}{\overset{\displaystyle OH}{\underset{|}{\overset{|}{CH}-COO^-}}} \xrightarrow{\hspace{2cm}} \underset{\displaystyle CH_2-COO^-}{\overset{\displaystyle O}{\overset{\|}{C}-COO^-}}$$

$$NAD^+ \quad NADH + H^+$$

Flavin adenine dinucleotide (FAD) is the other major electron carrier in the oxidation of fuel molecules. This coenzyme is a derivative of ADP and the

➤ **FIGURE 12.19** The structure of flavin adenine dinucleotide (FAD). The reactive portion (shown in blue) is in the riboflavin ring system.

Riboflavin

Adenine

Ribose

vitamin riboflavin. The reactive site is located within the riboflavin ring system (see ■ Figure 12.19).

As in the reduction of NAD^+, substrates of enzymes that use FAD as the coenzyme give up two electrons to the coenzyme. However, unlike NAD^+, the FAD coenzyme accepts both of the hydrogen atoms lost by the substrate. Thus, the abbreviation for the reduced form of FAD is $FADH_2$:

$$+ 2H^+ + 2e^- \rightleftharpoons \qquad (12.13)$$

FAD
(oxidized form)

$FADH_2$
(reduced form)

The substrates for oxidation reactions in which FAD is the coenzyme differ from those involving NAD^+. FAD is often involved in oxidation reactions in which a $-CH_2-CH_2-$ portion of the substrate is oxidized to a double bond:

General reaction:

$$R-\underset{\underset{H}{|}}{\overset{\overset{H}{|}}{C}}-\underset{\underset{H}{|}}{\overset{\overset{H}{|}}{C}}-R + FAD \longrightarrow R-CH=CH-R + FADH_2 \qquad (12.14)$$

saturated
(reduced)

unsaturated
(oxidized)

Specific example:

$$CH_3CH_2CH_2 - \overset{\overset{\displaystyle O}{\|}}{C} - S - CoA \longrightarrow CH_3CH = CH - \overset{\overset{\displaystyle O}{\|}}{C} - SCoA \quad (12.15)$$

butyryl CoA

FAD   FADH$_2$

Complete the following reactions:

a. $^-OOC - CH_2 - CH_2 - COO^- + FAD \longrightarrow$

b. $CH_3 - \overset{\overset{\displaystyle OH}{|}}{CH} - COO^- + NAD^+ \longrightarrow$

> ## FOR FUTURE REFERENCE   FOOD LABEL TERMS

Many consumers use food labels to obtain a variety of information, such as the ingredients, the serving size, and nutritional information. The FDA defines the words that may be used on a label to describe the nutrient content of a product. At times, the use of related terms such as *fat free, low fat,* and *less fat* can be confusing, even though the terms have specific meanings according to FDA rules. Below is a list of commonly used label terms and their meanings.

GENERAL TERMS	MEANING
Free	"Nutritionally trivial" and unlikely to have a physiological consequence; synonyms include "without," "no," and "zero." A food that does not contain a nutrient naturally may make such a claim, but only as it applies to all similar foods (for example, "applesauce, a fat-free food").
Healthy	A food that is low in fat, saturated fat, cholesterol, and sodium and that contains at least 10% of the RDA for vitamin A, vitamin C, iron, calcium, protein, or fiber.
High	Must contain 20% or more of the RDA for a given nutrient per serving; synonyms include "rich in" or "excellent source."
Less	Must contain at least 25% less of a given nutrient or kilocalories than the comparison food (see individual nutrients below); synonyms include "fewer" and "reduced."
Light or lite	Any use of the term other than as defined specifically later in this table must specify what it is referring to (for example, "light in color" or "light in texture").
Low	An amount that allows frequent consumption of a food without exceeding the RDA for the nutrient. A food that is naturally low in a nutrient may make such a claim, but only as it applies to all similar foods (for example, "fresh cauliflower, a low-sodium food"); synonyms include "little," "few," and "low source of."
More	Must contain a least 10% more of the RDA value for a given nutrient than the comparison food; synonyms include "added" and "extra."
Good source of	Product provides between 10 and 19% of the RDA for a given nutrient per serving.

ENERGY	
kcalorie-free	Contains fewer than 5 kcal per serving.
Light	Contains one-third fewer kcal per serving than the comparison food.
Low kcalorie	Contains 40 kcal or less per serving.
Reduced kcalorie	Contains at least 25% fewer kcal per serving than the comparison food.

(continued)

(continued)

FAT AND CHOLESTEROL

Percent fat-free	May be used only if the product meets the definition of *low fat* or *fat-free* and must reflect the amount of fat in 100 g (for example, a food that contains 2.5 g of fat per 50 g can claim to be "95 percent fat-free").
Fat-free	Contains less than 0.5 g of fat per serving (and no added fat or oil); synonyms include "zero fat," "no fat," and "nonfat."
Low fat	Contains 3 g or less of fat per serving.
Less fat	Contains 25% or less fat per serving than the comparison food.
Saturated fat-free	Contains less than 0.5 g of saturated fat and less than 0.5 g of *trans* fatty acids per serving.
Low saturated fat	Contains 1 g or less of saturated fat per serving.
Less saturated fat	Contains 25% or less saturated fat per serving than the comparison food.
Cholesterol-free	Contains less than 2 mg of cholesterol and 2 g or less of saturated fat per serving.
Low cholesterol	Contains 20 mg or less of cholesterol and 2 g or less of saturated fat per serving.
Less cholesterol	Contains 25% or less cholesterol per serving than the comparison food (reflecting a reduction of at least 20 mg per serving), and 2 g or less of saturated fat per serving.
Extra lean	Contains less than 5 g of fat, 2 g of saturated fat, and 95 mg of cholesterol per serving and per 100 g of meat, poultry, and seafood.
Lean	Contains less than 10 g of fat, 4.5 g of saturated fat, and 95 mg of cholesterol per serving and per 100 g of meat, poultry, and seafood.
Light	Contains 50% or less of the fat than in the comparison food (for example, 50% less fat than our regular cookies).

CARBOHYDRATES: FIBER AND SUGAR

High fiber	Must contain 5 g or more of fiber per serving. A high-fiber claim made on a food than contains more than 3 g of fat per serving and per 100 g of food must also declare the amount of total fat.
Sugar-free	Contains less than 0.5 g of sugar per serving.

SODIUM

Sodium-free and **salt-free**	Contains less than 5 mg of sodium per serving.
Low sodium	Contains 140 mg or less of sodium per serving.
Light	A low-kcalorie, low-fat food with a 50% reduction in sodium.
Light in sodium	Contains less than 50% of the sodium of the corresponding food.
Very low sodium	Contains 35 mg or less of sodium per serving.

NOTE: Foods containing more than 13 g of total fat per serving or per 50 g of food must indicate those contents immediately after a cholesterol claim. Also, as can be seen from the table, all cholesterol claims are prohibited when the food contains more than 2 g of saturated fat per serving.

SOURCE: Adapted from Whitney, E. N.; Rolfes, S. R. *Understanding Nutrition,* 8th ed. Belmont, CA: West/Wadsworth, 1999.

Nutritional Requirements. Every human body requires certain amounts of various nutrients, water, and fiber to function properly. The amounts needed vary with a number of factors such as body size, age, and sex. Various countries have established nutritional guidelines in attempts to maintain good health for their citizens. In the United States, the Reference Daily Intake (RDI) and Daily Reference Values (DRV) are designed to represent appropriate nutrient intake for 95% of the population.

The Macronutrients. Macronutrients are substances required by the body in relatively large amounts. They are used by the body to provide energy and the materials necessary to form new or replacement tissue. The macronutrients are carbohydrates, lipids, and proteins. An RDI has been established only for proteins, but various groups have recommended the amounts of carbohydrates, lipids, and proteins that should be included in the diet.

Micronutrients I: Vitamins. Vitamins are organic micronutrients that the body cannot produce in the amounts needed for good health. A number of vitamins have a high water solubility resulting from the highly polar nature of their molecules. All but one of the water-soluble vitamins are known to function as coenzymes in the body and are involved in many important metabolic processes. Fat-soluble vitamins have nonpolar molecular structures. As a result, they are insoluble in water but soluble in fat or other nonpolar solvents. Fat-soluble vitamins act somewhat like hormones in the body. Fat-soluble vitamins do not dissolve in water-based body fluids, so they are not excreted through the kidneys. Amounts in excess of bodily requirements are stored in body fat. Thus, it is much easier to produce toxic effects by overdosing with fat-soluble vitamins than with water-soluble vitamins.

Micronutrients II: Minerals. Minerals are inorganic substances needed by the body. Those present in the body in amounts equal to or greater than 5 g are called major minerals. Those present in smaller amounts are called trace minerals. Minerals perform many useful functions in the body.

The Flow of Energy in the Biosphere. Energy for all living processes comes from the sun. A portion of the sun's energy is trapped in the process of photosynthesis, which produces carbohydrates. We derive energy from the foods we eat: carbohydrates, lipids, and proteins. As these substances are oxidized, molecules of ATP are formed that supply the energy needed for cellular processes.

Metabolism and an Overview of Energy Production. Metabolism, the sum of all cellular reactions, involves the breakdown (catabolism) and synthesis (anabolism) of molecules. There are three stages in the catabolism of foods to provide energy. In Stage I, digestion converts foods into smaller molecules. In Stage II, these smaller molecules are converted into two-carbon acetyl units that combine with coenzyme A, forming acetyl CoA. Stage III, which is called the common catabolic pathway, consists of the citric acid cycle followed by the electron transport chain and oxidative phosphorylation. The main function of catabolism is to produce ATP molecules.

ATP: The Primary Energy Carrier. Molecular ATP is the link between energy production and energy use in cells. ATP is called a high-energy compound because of the large amount of free energy liberated on hydrolysis. Cells are able to harness the free energy liberated by ATP to carry out cellular work. ATP is consumed immediately following its formation and is regenerated from ADP as fuel molecules are oxidized. Mitochondria play a key role in energy production because they house the enzymes for both the citric acid cycle and the electron transport chain.

Important Coenzymes in the Common Catabolic Pathway. Three very important coenzymes are involved in catabolism: coenzyme A, NAD^+, and FAD. Coenzyme A binds to the two-carbon fragments produced in Stage II to form acetyl CoA, the direct fuel of the citric acid cycle. NAD^+ and FAD are the oxidizing agents that participate in the oxidation reactions of the citric acid cycle. They transport hydrogen atoms and electrons from the citric acid cycle to the electron transport chain.

When you have completed your study of this chapter, you should be able to:

1. Describe the differences between macronutrients and micronutrients in terms of amounts required and their functions in the body.
2. Show that you understand the concept of a Daily Value.
3. Describe the primary functions in the body of each macronutrient.
4. Distinguish between and classify vitamins as water-soluble or fat-soluble on the basis of name and behavior in the body.
5. List at least one good food source for each of the water-soluble and fat-soluble vitamins.
6. List at least one food source and a primary function in the body for each major mineral.
7. Describe the major steps in the flow of energy in the biosphere.
8. Differentiate between metabolism, anabolism, and catabolism.
9. Outline the three stages in the extraction of energy from food.
10. Explain how ATP plays a central role in the production and use of cellular energy.
11. Explain the role of coenzymes in the common catabolic pathway.

Anabolism (12.6)
Catabolism (12.6)
Cellular respiration (12.5)
Common catabolic pathway (12.6)
Complete protein (12.2)
Complex carbohydrate (12.2)
Daily Reference Values (DRV) (12.1)
Daily Values (DV) (12.1)

Essential amino acid (12.2)
High-energy compound (12.7)
Macronutrient (12.1)
Major mineral (12.4)
Metabolic pathway (12.6)
Metabolism (12.6)
Micronutrient (12.1)
Mineral (12.4)

Mitochondrion (12.7)
Nutrition (Introduction)
Reference Daily Intakes (RDI) (12.1)
Satiety (12.2)
Simple carbohydrates (12.2)
Thioester bond (12.8)
Trace mineral (12.4)

KEY REACTIONS

1. Photosynthesis (Section 12.5):

$$6CO_2 + 6H_2O + energy \rightarrow C_6H_{12}O_6 + 6O_2$$

Reaction 12.2

2. Hydrolysis of ATP (Section 12.7):

$$ATP + H_2O \rightarrow ADP + P_i + H^+$$

Reaction 12.5

3. Oxidation by NAD^+ (Section 12.8):

$$R - \overset{\overset{\displaystyle OH}{|}}{\underset{\underset{\displaystyle H}{|}}{C}} - H + NAD^+ \longrightarrow R - \overset{\overset{\displaystyle O}{\|}}{C} - H + NADH + H^+$$

Reaction 12.10

4. Oxidation by FAD (Section 12.8):

$$R - \overset{\overset{\displaystyle H}{|}}{\underset{\underset{\displaystyle H}{|}}{C}} - \overset{\overset{\displaystyle H}{|}}{\underset{\underset{\displaystyle H}{|}}{C}} - R + FAD \longrightarrow R - CH = CH - R + FADH_2$$

Reaction 12.14

EXERCISES

NUTRITIONAL REQUIREMENTS (SECTION 12.1)

12.1 What is the primary difference between a macronutrient and a micronutrient?

12.2 What is the principal component of dietary fiber?

12.3 Explain the importance of sufficient fiber in the diet.

THE MACRONUTRIENTS (SECTION 12.2)

12.4 List two general functions in the body for each of the macronutrients.

12.5 For each of the following macronutrients, list a single food item that would be a good source:
a. carbohydrate (simple)
b. carbohydrate (complex)
c. lipid
d. protein

12.6 List the types of macronutrients found in each of the following food items. List the nutrients in approximate decreasing order of abundance (most abundant first, etc.).
a. Potato chips
b. Buttered toast
c. Plain toast with jam
d. Cheese sandwich
e. A lean steak
f. A fried egg

12.7 Why is linoleic acid called an essential fatty acid for humans?

12.8 Use Figure 12.2 and list the Daily Values (DV) for fat, saturated fat, total carbohydrate, and fiber based upon a 2000-Calorie diet.

MICRONUTRIENTS I: VITAMINS (SECTION 12.3)

12.9 What general function is served by eight of the water-soluble vitamins?

12.10 Why is there more concern about large doses of fat-soluble vitamins than of water-soluble vitamins?

12.11 What fat-soluble vitamin or deficiency is associated with the following?
   a. Night blindness
   b. Blood clotting
   c. Calcium and phosphorus use in forming bones and teeth
   d. Preventing oxidation of fatty acids

12.12 What water-soluble vitamin deficiency is associated with the following?
   a. Scurvy
   b. Beriberi
   c. Pernicious anemia
   d. Pellagra

MICRONUTRIENTS II: MINERALS (SECTION 12.4)

12.13 What is the difference between a vitamin and a mineral?

12.14 What determines whether a mineral is classified as a major or trace mineral?

12.15 List at least one specific function for each major mineral of the body.

12.16 What are the general functions of trace minerals in the body?

THE FLOW OF ENERGY IN THE BIOSPHERE (SECTION 12.5)

12.17 What is the source of energy for life on earth?

12.18 Write a net equation for the photosynthetic process.

12.19 Explain how the earth's supply of carbon compounds may be recycled by the processes of photosynthesis and respiration.

12.20 Describe in general terms the steps in the flow of energy from the sun to molecules of ATP.

METABOLISM AND AN OVERVIEW OF ENERGY PRODUCTION (SECTION 12.6)

12.21 Classify catabolism and anabolism as synthetic or breakdown processes.

12.22 What is a metabolic pathway?

12.23 Outline the three stages in the extraction of energy from food.

12.24 Which stage of energy production is concerned primarily with the following:
   a. Formation of ATP
   b. Digestion of fuel molecules
   c. Consumption of O_2
   d. Generation of acetyl CoA

12.25 Explain why Stage III of energy production is referred to as the common catabolic pathway.

12.26 In terms of energy production, what is the main purpose of the catabolic pathway?

ATP: THE PRIMARY ENERGY CARRIER (SECTION 12.7)

12.27 What do the symbols ATP, ADP, and AMP represent?

12.28 Explain why ATP is referred to as the primary energy carrier.

12.29 Using the partial structure shown below, write the structure of ATP, ADP, and AMP:

$$-\overset{\overset{\displaystyle O}{\|}}{\underset{\underset{\displaystyle O^-}{|}}{P}} - O - \text{adenosine}$$

12.30 What do the symbols P_i and PP_i represent?

12.31 Using symbolic formulas such as ADP and PP_i, write equations for the hydrolysis of ATP to ADP and the hydrolysis of ATP to AMP.

12.32 Using Table 12.5, explain why phosphoenolpyruvate is called a high-energy compound but glycerol 3-phosphate is not.

12.33 What does the symbol $\Delta G°'$ represent? How does the sign for $\Delta G°'$ indicate whether a reaction is exergonic or endergonic?

12.34 Which portion of the ATP molecule is particularly responsible for its being described as a high-energy compound?

12.35 Why is it more accurate to refer to ATP as a carrier or donor of free energy rather than as a storage form of free energy?

12.36 What is the role of mitochondria in the use of energy by living organisms?

12.37 Describe the importance and function of the ATP–ADP cycle.

12.38 Describe the structure of a mitochondrion and identify the location of enzymes important in energy production.

IMPORTANT COENZYMES IN THE COMMON CATABOLIC PATHWAY (SECTION 12.8)

12.39 What roles do the following coenzymes serve in the catabolic oxidation of foods?
   a. coenzyme A   **b.** FAD   **c.** NAD^+

12.40 What coenzymes are produced by the body from the following vitamins?
   a. riboflavin
   b. pantothenic acid
   c. nicotinamide

12.41 Write the abbreviations for the oxidized and reduced forms of
   a. flavin adenine dinucleotide
   b. nicotinamide adenine dinucleotide

12.42 Which compound is oxidized in the following reaction? Which compound is reduced?

$$^-OOC - CH_2 - CH_2 - COO^- + FAD \longrightarrow$$

$$^-OOC - CH = CH - COO^- + FADH_2$$

12.43 Write balanced reactions for the following oxidations using NAD$^+$:

a. $CH_3 - \underset{\underset{\displaystyle OH}{|}}{CH} - COO^- \rightarrow CH_3 - \underset{\overset{\displaystyle O}{\|}}{C} - COO^-$

b. $CH_3 - OH \rightarrow H - \underset{\overset{\displaystyle O}{\|}}{C} - H$

12.44 Draw abbreviated structural formulas to show the reactive portion of FAD and the reduced compound FADH$_2$.

CHEMISTRY AROUND US AND OVER THE COUNTER

12.45 What foods are eaten by lacto-ovo-vegetarians?

12.46 What vitamin supplements are often recommended for vegetarians?

12.47 In what age group is bulimia most common?

12.48 What behavior is characteristic of individuals suffering from anorexia nervosa?

12.49 Why are creatine supplements of interest to some athletes?

▶ POINTS TO PONDER AND PURSUE

12.1 At one time a system of MDR (minimum daily requirement) values for nutrients was used. Why do you think Daily Values (DV) might be more informative and useful than MDRs?

12.2 Briefly explain why you think it would be undesirable or beneficial (decide which position to defend in each case) to follow a diet that is (a) extremely low in carbohydrates, (b) extremely low in lipids, or (c) extremely low in proteins.

12.3 Donuts, even those without frosting, are high in triglycerides. Explain.

12.4 One of the oils listed in Figure 12.6 is palm oil, which for several years was used in fast-food restaurants to fry potatoes. Why was this use considered an unhealthy practice?

12.5 In Figure 12.7, it is observed that vegetables are a source of protein. We most often think of meats, grains, and dairy products as sources of protein, but not vegetables. Use Table 9.4 to list the types of protein that might be found in vegetables.

12.6 According to Web site 12.2, what is the purpose of the new food label?

12.7 What volume of milk constitutes one serving on the Food Guide Pyramid? See Web site 12.3.

12.8 Using Web site 12.4, how many general eating disorders are discussed?

12.9 According to Web site 12.5, what substances in the body are used to synthesize creatine?

http://www.brookscole.com/chemistry/seager_chemistry

Carbohydrate Metabolism

Dental hygienists provide a variety of health care services. A well-known example is the cleaning of teeth (prophylaxis). Dental hygienists also teach patients how to properly use toothbrushes, dental floss, and other dental aids that contribute to the maintenance of healthy teeth and gums. In addition, they help patients understand how nutrition affects dental health. In this chapter, you will be introduced to several nutritionally important topics, including the production and regulation of energy from carbohydrates.

http://www.brookscole.com/chemistry/seager_chemistry
At Web site 13.1, you will find more information about the career of a dental hygienist.

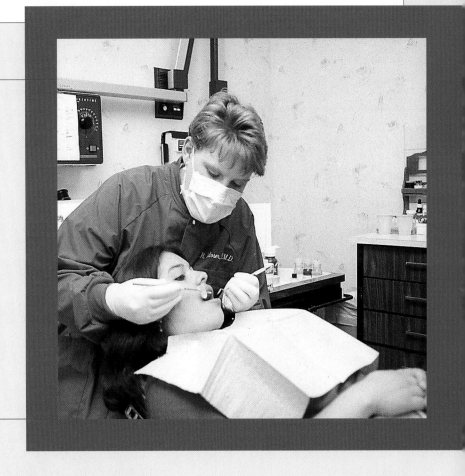

CHAPTER OUTLINE

- The Digestion of Carbohydrates
- Blood Glucose
- Glycolysis
- The Fates of Pyruvate
- The Citric Acid Cycle
- The Electron Transport Chain
- Oxidative Phosphorylation
- The Complete Oxidation of Glucose
- Glycogen Metabolism
- Gluconeogenesis
- The Hormonal Control of Carbohydrate Metabolism

Carbohydrates play major roles in cell metabolism. The central substance in carbohydrate metabolism is glucose; it is the key food molecule for most organisms. As the major carbohydrate metabolic pathways are discussed, you will see that each one involves either the use or the synthesis of glucose. You will also see how the chemical degradation of glucose provides energy in the cell, how glucose is stored, how the blood glucose level is regulated, and how diabetes results from changes in blood glucose regulation mechanisms.

➤ 13.1 The Digestion of Carbohydrates

The major function of dietary carbohydrates is to serve as an energy source. In a typical American diet, carbohydrates provide about 45–55% of the daily energy needs. Fats and proteins furnish the remaining energy. During carbohydrate digestion, di- and polysaccharides are hydrolyzed to monosaccharides, primarily glucose, fructose, and galactose:

$$\text{polysaccharides} + \text{H}_2\text{O} \xrightarrow{\text{digestion}} \text{glucose} \tag{13.1}$$

$$\text{sucrose} + \text{H}_2\text{O} \xrightarrow{\text{digestion}} \text{glucose} + \text{fructose} \tag{13.2}$$

$$\text{lactose} + \text{H}_2\text{O} \xrightarrow{\text{digestion}} \text{glucose} + \text{galactose} \tag{13.3}$$

$$\text{maltose} + \text{H}_2\text{O} \xrightarrow{\text{digestion}} \text{glucose} \tag{13.4}$$

After digestion is completed, glucose, fructose, and galactose are absorbed into the bloodstream through the lining of the small intestine and transported to the liver. In the liver, fructose and galactose are rapidly converted to glucose or to compounds that are metabolized by the same pathway as glucose.

➤ 13.2 Blood Glucose

Glucose is by far the most plentiful monosaccharide found in blood, and the term *blood sugar* usually refers to glucose. In adults, the normal **blood sugar level** measured after a fast of 8–12 hours is in the range of 70–110 mg/100 mL. (In clinical reports, such levels are often expressed in units of milligrams per deciliter, mg/dL.) The blood sugar level reaches a maximum of approximately 140–160 mg/100 mL about 1 hour after a carbohydrate-containing meal. It returns to normal after 2–2½ hours.

If a blood sugar level is below the normal fasting level, a condition called **hypoglycemia** exists. Because glucose is the only nutrient normally used by the brain for energy, mild hypoglycemia leads to dizziness and fainting as brain cells are deprived of energy. Severe hypoglycemia can cause convulsions and shock. When the blood glucose concentration is above normal, the condition is referred to as **hyperglycemia**. If blood glucose levels exceed approximately 108 mg/100 mL, the sugar is not completely reabsorbed by the kidneys, and glucose is excreted in the urine (see ■ Figure 13.1). The blood glucose level at which this occurs is called the **renal threshold,** and the condition in which glucose appears in the urine is called **glucosuria**. Prolonged hyperglycemia at a glucosuric level must be considered serious because it indicates that something is wrong with the body's normal ability to control the blood sugar level.

The liver is the key organ involved in regulating the blood glucose level. The liver responds to the increase in blood glucose that follows a meal by removing glucose from the bloodstream. The removed glucose is primarily converted to glycogen or triglycerides for storage. Similarly, when blood glucose levels are low, the liver responds by converting stored glycogen to glucose and by synthesizing new glucose from noncarbohydrate substances.

➤ **FIGURE 13.1** Blood glucose levels.

Blood sugar level
The amount of glucose present in blood, normally expressed as milligrams per 100 mL of blood.

http://www.brookscole.com/chemistry/seager_chemistry
Web site 13.2 offers more information on hypoglycemia.

Hypoglycemia
A lower-than-normal blood sugar level.

Hyperglycemia
A higher-than-normal blood sugar level.

Renal threshold
The blood glucose level at which glucose begins to be excreted in the urine.

Glucosuria
A condition in which elevated blood sugar levels result in the excretion of glucose in the urine.

➤ 13.3 Glycolysis

Glycolysis is an important part of the catabolic process by which the body gets energy from carbohydrates. It consists of a series of ten reactions, with the net result being the conversion of a glucose molecule into two molecules of pyruvate. All enzymes for glycolysis are present in cellular cytoplasm. The ten reactions of glycolysis catalyzed by these enzymes are represented in ■ Figure 13.2. The addition of all these reactions gives a net reaction for glycolysis:

Glycolysis
A series of reactions by which glucose is oxidized to pyruvate.

$$\text{glucose} + 2P_i + 2ADP + 2NAD^+ \rightarrow$$
$$2 \text{ pyruvate} + 2ATP + 2NADH + 2H^+ + H_2O \quad (13.5)$$

➤ **FIGURE 13.2** A summary of glycolysis. The Ⓟ stands for $\begin{smallmatrix} O^- \\ | \\ -P=O \\ | \\ O^- \end{smallmatrix}$, and the prefix "bis-" is used to denote two of these groups.

This equation indicates a net gain of 2 mol of ATP for every mole of glucose converted to pyruvate. In addition, 2 mol of the coenzyme NADH is formed when NAD^+ serves as an oxidizing agent in glycolysis. Fructose and galactose from digested food are both converted into intermediates that enter into the glycolysis pathway. Fructose enters glycolysis as dihydroxyacetone phosphate and glyceraldehyde 3-phosphate, whereas galactose enters in the form of glucose 6-phosphate.

The Regulation of Glycolysis

The glycolytic pathway, like all metabolic pathways, is controlled constantly. The precise regulation of the concentrations of pathway intermediates is efficient and makes possible the rapid adjustment of the output level of pathway products as the need arises.

The glycolytic pathway is regulated by three enzymes: hexokinase, phosphofructokinase, and pyruvate kinase. The regulation of the pathway is shown in ■ Figure 13.3.

Hexokinase catalyzes the conversion of glucose to glucose 6-phosphate and thus initiates the glycolytic pathway. The enzyme is inhibited by a high concentration of glucose 6-phosphate, the product of the reaction it catalyzes.

Lactose intolerance
The inability to digest milk and other products containing lactose.

OVER THE COUNTER • 13.1

LACTOSE INTOLERANCE

Lactose makes up about 40% of the nutrients in an infant's diet during the first year of life. The small intestines of infants and small children contain a particular form of lactase, the enzyme that catalyzes the digestion of lactose. This form of the enzyme is very active and helps them digest the lactose in their diet easily. The digestive system of adults usually contains a less active form of the enzyme, and in cultures where milk is not used much as an adult food, lactase deficiency is prevalent. Even in the United States, about 20% of the adult population suffers from **lactose intolerance** to some degree.

When lactose is not digested in the small intestine, it passes into the colon (large intestine), where it causes water to be drawn by osmosis from the interstitial fluid into the colon itself. Bacteria in the colon act on the lactose and produce gases and organic acids. The presence in the colon of excess water, gases, and organic acids leads to the symptoms of lactose intolerance—abdominal distention (a bloated feeling), cramps, and diarrhea.

The main treatment for lactose intolerance is dietary management; patients must be aware of the amount of lactose in all products they consume. If lactose intolerance is suspected, the first step is total abstention from lactose, to determine whether the symptoms really are caused by the enzyme. Dietary restrictions can be modified once the cause has been identified, and a person's tolerance level can be determined by slowly adding lactose-containing foods to the diet. If total abstention from lactose is required, it may be necessary to add calcium supplements to the diet to replace the calcium normally contained in dairy foods.

Hard cheeses like cheddar and Swiss are quite low in lactose and can often be tolerated, except in extreme cases. Even though yogurt has a high lactose content, it is better tolerated than milk. This is probably because the live bacteria in yogurt release lactase when the bacteria are digested by proteolytic enzymes and stomach acid. The released lactase then helps digest the lactose. In addition to monitoring their diet, some people with lactose intolerance use commercially available products that contain lactase. Available in liquid or tablet form, these products are taken just before eating or along with food that contains lactose. Some of the lactase products are Lactaid®, DairyEase®, and Lactrase®.

Many prescription and OTC drugs contain lactose as a carrier or filler. The FDA provides a list identifying such drugs, and people with lactose intolerance should consult the list to avoid inadvertently using a medication that contains lactose.

http://www.brookscole.com/chemistry/seager_chemistry
At Web site 13.3, you will find additional information on lactose intolerance.

Two commercial products containing lactase.

Glucose

Feedback inhibition → hexokinase (inhibited by glucose 6-phosphate)

glucose 6-phosphate

fructose 6-phosphate

phosphofructokinase (inhibited by ATP and citrate; activated by ADP and AMP)

fructose 1,6-bisphosphate

phosphoenolpyruvate

pyruvate kinase (inhibited by ATP)

Pyruvate

> **FIGURE 13.3** The regulation of the glycolysis pathway. Glucose 6-phosphate is a noncompetitive inhibitor of hexokinase. Phosphofructokinase and pyruvate kinase are also allosteric enzymes subject to control.

Thus, the phosphorylation of glucose is controlled by feedback inhibition (Section 10.8).

A second and very important control point for glycolysis shown in Figure 13.3 is the step catalyzed by phosphofructokinase. Once glucose is converted to fructose 1,6-bisphosphate, its carbon skeleton must continue through the glycolytic pathway. The phosphorylation of fructose 6-phosphate is a committed step that will occur as long as the enzyme is active. As an allosteric enzyme (Section 10.8), phosphofructokinase is inhibited by high concentrations of ATP and citrate, and it is activated by high concentrations of ADP and AMP.

The third control point is the last step of glycolysis, the conversion of phosphoenolpyruvate to pyruvate catalyzed by pyruvate kinase. This enzyme, too, is an allosteric enzyme that is inhibited by higher concentrations of ATP.

To understand how these controls interact, remember that when the glycolytic pathway is operating at a maximum rate, so also are the citric acid cycle and the electron transport chain. As a result of their activity, large quantities of ATP are produced. The entire pathway continues to operate at a high rate, as long as the produced ATP is used up by processes within the cell. However, if ATP use decreases, the concentration of ATP increases, and more diffuses to the binding sites of the allosteric enzymes sensitive to ATP. Thus, the activity of these enzymes decreases. When phosphofructokinase is inhibited in this way, the entire glycolysis pathway slows or stops. As a result, the concentration of glucose 6-phosphate increases, thus inhibiting hexokinase and the initial step of the glycolysis pathway.

Conversely, when ATP concentrations are low, the concentrations of AMP and ADP are high. The AMP and ADP function as activators (Section 10.8) that influence phosphofructokinase and thereby speed up the entire glycolysis pathway.

Explain how each of the following enzymes helps regulate the glycolysis pathway:

a. hexokinase
b. phosphofructokinase
c. pyruvate kinase

Learning Check 13.1

➤ 13.4 The Fates of Pyruvate

The sequence of reactions that convert glucose to pyruvate is very similar in all organisms and in all kinds of cells. However, the fate of the pyruvate as it is used to generate energy is variable. Three different fates of pyruvate are considered here.

As glucose is oxidized to pyruvate in glycolysis, NAD^+ is reduced to NADH. The need for a continuous supply of NAD^+ for glycolysis is a key to understanding the fates of pyruvate.

$$\text{glucose} + 2NAD^+ \longrightarrow 2 \text{ pyruvate} + 2NADH + 2H^+ \quad (13.6)$$

$$2ADP + 2P_i \quad 2ATP$$

In each case, pyruvate is metabolized in ways that regenerate NAD^+ so that glycolysis can continue.

Oxidation to Acetyl CoA

Aerobic
In the presence of oxygen.

The mitochondrial membrane is permeable to pyruvate formed by glycolysis in the cytoplasm, and under **aerobic** conditions (a plentiful oxygen supply), pyruvate is converted inside the mitochondria to acetyl CoA.

$$\underset{\text{pyruvate}}{CH_3 - \overset{\overset{\displaystyle O}{\|}}{C} - COO^-} + CoA - S - H + NAD^+ \xrightarrow[\text{complex}]{\overset{\text{pyruvate}}{\text{dehydrogenase}}} \underset{\text{acetyl CoA}}{CH_3 - \overset{\overset{\displaystyle O}{\|}}{C} - S - CoA} + NADH + CO_2 \quad (13.7)$$

Acetyl CoA formed in this reaction can enter the citric acid cycle on its way to complete oxidation to CO_2. The NAD^+ needed for the conversion of pyruvate to acetyl CoA (Reaction 13.7) and for the operation of the glycolysis pathway (Step 6) is regenerated when NADH transfers its electrons to O_2 in the electron transport chain.

Not all acetyl CoA generated metabolically enters the citric acid cycle. Some also serves as a starting material for biosynthesis (Section 14.7). The fatty acid components of triglycerides, for example, are synthesized using acetyl CoA. Thus, as most people are aware, the intake of excess dietary carbohydrates can lead to an increase in body fat storage.

Reduction to Lactate

Anaerobic
In the absence of oxygen.

Under **anaerobic** conditions, such as those that accompany strenuous or long-term muscle activity, the cellular supply of oxygen is not adequate for the reoxidation of NADH to NAD^+. Under such conditions, the cells begin reducing pyruvate to lactate as a means of regenerating NAD^+:

$$\underset{\text{pyruvate}}{CH_3 - \overset{\overset{\displaystyle O}{\|}}{C} - COO^-} + NADH + H^+ \underset{\text{dehydrogenase}}{\overset{\text{lactate}}{\rightleftharpoons}} \underset{\text{lactate}}{CH_3 - \overset{\overset{\displaystyle OH}{|}}{CH} - COO^-} + NAD^+ \quad (13.8)$$

When Reaction 13.8 is added to the net reaction of glycolysis, an overall reaction for **lactate fermentation** results:

Lactate fermentation
The production of lactate from glucose.

$$C_6H_{12}O_6 + 2ADP + 2P_i \xrightarrow[\text{fermentation}]{\text{lactate}} 2CH_3 - \underset{\text{lactate}}{\overset{\overset{\displaystyle OH}{|}}{CH} - COO^-} + 2ATP \quad (13.9)$$

Notice that the NAD^+ produced when pyruvate is reduced (Reaction 13.8) is used up and does not appear in the lactate fermentation (Reaction 13.9). Although much more energy is available from the complete oxidation of pyruvate under aerobic conditions, the two ATPs produced from lactate fermentation are sufficient to sustain the life of anaerobic microorganisms. In human metabolism, those two molecules of ATP also play a critical role by furnishing energy when cellular supplies of oxygen are insufficient for ATP production through the complete oxidation of pyruvate.

Reduction to Ethanol

Several organisms, including yeast, have developed an alternative pathway for the regeneration of NAD^+ under anaerobic conditions. The process, called **alcoholic fermentation**, involves the decarboxylation of pyruvate to form acetaldehyde:

$$CH_3-\overset{\overset{\displaystyle O}{\|}}{C}-COO^- + H^+ \xrightarrow[\text{decarboxylase}]{\text{pyruvate}} CH_3-\overset{\overset{\displaystyle O}{\|}}{C}-H + CO_2 \qquad (13.10)$$
$$\text{pyruvate} \qquad\qquad\qquad\qquad \text{acetaldehyde}$$

The carbon dioxide produced in this reaction causes beer to foam and provides carbonation for naturally fermented wines and champagnes (see ■ Figure 13.4).

The second step in alcoholic fermentation is the reduction of acetaldehyde to ethanol by NADH:

$$CH_3-\overset{\overset{\displaystyle O}{\|}}{C}-H + NADH + H^+ \underset{\text{dehydrogenase}}{\overset{\text{alcohol}}{\rightleftharpoons}} CH_3-\overset{\overset{\displaystyle OH}{|}}{\underset{\underset{\displaystyle H}{|}}{C}}-H + NAD^+ \qquad (13.11)$$
$$\text{acetaldehyde} \qquad\qquad\qquad\qquad \text{ethanol}$$

The NAD^+ regenerated by this reaction may then be used in the glycolysis pathway.

An overall reaction for alcoholic fermentation is obtained by combining the reaction for the conversion of pyruvate to ethanol and the reactions of glycolysis. Notice that because NAD^+ is used up in glycolysis but regenerated in ethanol formation, it does not appear in the overall equation:

$$C_6H_{12}O_6 + 2ADP + 2P_i \xrightarrow[\text{fermentation}]{\text{alcoholic}} 2CH_3CH_2-OH + 2CO_2 + 2ATP \qquad (13.12)$$
$$\text{glucose} \qquad\qquad\qquad\qquad \text{ethanol}$$

■ Figure 13.5 summarizes the various pathways pyruvate can follow as it is used to regenerate NAD^+.

Alcoholic fermentation
The conversion of glucose to ethanol.

➤ **FIGURE 13.4** Bread, wine, and cheese are all produced through fermentation processes.

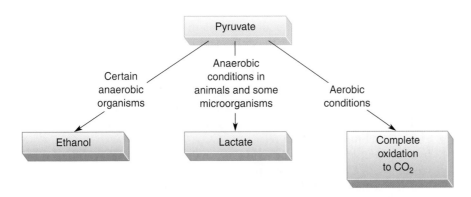

➤ **FIGURE 13.5** The fates of pyruvate.

Learning Check 13.2

Why is it important that each biochemical fate of pyruvate involve the production of NAD^+?

Citric acid cycle
A series of reactions in which acetyl CoA is oxidized to carbon dioxide and reduced forms of coenzymes FAD and NAD^+ are produced.

> **FIGURE 13.6** Sir Hans Adolf Krebs (1900–1981), German-born British biochemist, received the 1953 Nobel Prize in physiology or medicine.

➤ 13.5 The Citric Acid Cycle

Stage III in the oxidation of fuel molecules (Section 12.6) begins when the two-carbon acetyl units (of acetyl CoA) enter the **citric acid cycle**. This process is called the citric acid cycle because one of the key intermediates is citric acid. However, it is also called the tricarboxylic acid cycle in reference to the three carboxylic acid groups in citric acid, and the Krebs cycle in honor of Sir Hans A. Krebs, who deduced its reaction sequence in 1937 (see ■ Figure 13.6).

The citric acid cycle is the principal process for generating the reduced coenzymes NADH and $FADH_2$, which are necessary for the reduction of oxygen and ATP synthesis in the electron transport chain. Another important function of the citric acid cycle as a source of intermediates for biosynthesis is discussed in Section 14.11.

Reactions of the Citric Acid Cycle

The reactions of the citric acid cycle occur within the matrix of the mitochondrion. ■ Figure 13.7 summarizes the details of the eight reactions that make up the citric acid cycle. Of particular importance are those reactions in which molecules of NADH and $FADH_2$ are produced. Notice that a two-carbon unit (the acetyl group) enters the cycle and that two carbon atoms are liberated in the form of two CO_2 molecules, one each in Reactions 3 and 4. Thus, this series of reactions both begins and ends with a C_4 compound, oxaloacetate. This is why the pathway is called a cycle. In each trip around the cycle, the starting material is regenerated and the reactions can proceed again as long as more acetyl CoA is available as fuel.

A Summary of the Citric Acid Cycle

The individual reactions of the cycle can be added to give an overall net equation:

$$\text{acetyl CoA} + 3NAD^+ + FAD + GDP + P_i + 2H_2O \rightarrow$$
$$2CO_2 + \text{CoA—S—H} + 3NADH + 2H^+ + FADH_2 + GTP \quad (13.13)$$

Some important features of the citric acid cycle are the following:

1. Acetyl CoA, available from the breakdown of carbohydrates, lipids, and amino acids, is the fuel of the citric acid cycle.
2. The operation of the cycle requires a supply of the oxidizing agents NAD^+ and FAD. The cycle is dependent on reactions of the electron transport chain to supply the necessary NAD^+ and FAD (Section 13.6). Because oxygen is the final acceptor of electrons in the electron transport chain, the continued operation of the citric acid cycle depends ultimately on an adequate supply of oxygen.
3. Two carbon atoms enter the cycle as an acetyl unit, and two carbon atoms leave the cycle as two molecules of CO_2. However, the carbon atoms leaving the cycle correspond to carbon atoms that entered in the previous cycle; there is a one-cycle delay between the entry of two carbon atoms as an acetyl unit and their release as CO_2.
4. In each complete cycle, four oxidation–reduction reactions produce three molecules of NADH and one molecule of $FADH_2$.
5. One molecule of the high-energy phosphate compound guanosine triphosphate (GTP, Section 12.7) is generated.

➤ **FIGURE 13.7** The citric acid cycle.

Regulation of the Citric Acid Cycle

The rate at which the citric acid cycle operates is precisely adjusted to meet cellular needs for ATP. One important control point is the very first step in the cycle, the synthesis of citrate. The catalyst for this reaction, citrate synthetase, is an allosteric enzyme that is inhibited by ATP and NADH and activated by ADP. When cellular needs for ATP are met, ATP interacts with citrate synthetase and reduces its affinity for acetyl CoA. Thus, ATP acts as a modulator (inhibitor) of citrate synthetase and inhibits the entry of acetyl CoA into the citric acid cycle. Similarly, NADH acts as an inhibitor and signals the

CHEMISTRY AROUND US • 13.1

LACTATE ACCUMULATION

During vigorous exercise, the oxygen supply to muscles can become too low to support increased aerobic conversion of pyruvate to acetyl CoA. Under such conditions, a shift occurs from aerobic pyruvate oxidation to anaerobic production of lactate as the means for producing ATP. The resulting accumulation of lactate causes the muscle pain and cramps that accompany physical exhaustion. The buildup of lactate also causes a slight decrease in blood pH, which triggers an increase in the rate and depth of breathing, a response that provides more oxygen to the cells. The combination of pain from lactate accumulation and the maximum rate of breathing represents a limit to physical activities. Regular training and exercise increase this limit by improving the efficiency with which oxygen is delivered to the cells. The muscle cells of a well-conditioned athlete can maintain the aerobic oxidation of pyruvate for a much longer period of activity than can those of a nonathlete.

The shift to anaerobic lactate formation also occurs in the heart muscle during an episode of cardiac arrest, in which the oxygen-carrying blood supply to the heart is cut off by the blockage of an artery leading to the heart muscles. Aerobic oxidation of pyruvate and its ATP production are stopped. Glycolysis proceeds at an accelerated rate to meet the energy needs, and lactate accumulates. The heart muscle contracts, producing a cramp, and stops beating. Massaging the heart muscle relieves such cramps, just as it does for skeletal muscles. Thus, it is sometimes possible to start the heart beating by massaging it.

Premature infants, whose lungs are not fully developed, are given pure oxygen to breathe in order to prevent a shift to anaerobic lactate production. In addition, they are given bicarbonate solutions intravenously to combat the increasing acidity of the blood due to lactate buildup.

http://www.brookscole.com/chemistry/seager_chemistry
Web site 13.4 provides useful information on the human body's lactate threshold.

Vigorous exercise causes a buildup of lactate in tissues.

citric acid cycle to slow the production of NADH and $FADH_2$. On the other hand, when ATP levels are low, ADP levels are usually high. ADP, an activator of citrate synthetase, stimulates the entry of acetyl CoA into the citric acid cycle and the production of NADH and $FADH_2$.

Isocitrate dehydrogenase (Reaction 3), a second controlling enzyme, is an allosteric enzyme that is activated by ADP and inhibited by NADH in very much the same way as is citrate synthetase. The third control point (Step 4) is catalyzed by the α-ketoglutarate dehydrogenase complex. This group of enzymes is inhibited by succinyl CoA and NADH, products of the reaction it catalyzes, and also by ATP. These control points are indicated in ■ Figure 13.8. In short, the entry of acetyl CoA into the citric acid cycle and the rate at which the cycle operates are reduced when cellular ATP levels are high. When ATP supplies are low (and ADP levels are high), the cycle is stimulated.

➤ 13.6 The Electron Transport Chain

The reduced coenzymes NADH and $FADH_2$ are end products of the citric acid cycle. In the final stage of food oxidation, the hydrogen ions and electrons carried by these coenzymes combine with oxygen and form water.

$$4H^+ + 4e^- + O_2 \rightarrow 2H_2O$$

(13.14)

acetyl CoA

Inhibited by ATP and NADH
Activated by ADP

oxaloacetate

(1) citrate

malate

fumarate

isocitrate

(3)

Inhibited by NADH
Activated by ADP

succinate

α-ketoglutarate

succinyl CoA (4)

Inhibited by succinyl CoA,
NADH, and ATP

> **FIGURE 13.8** Control points in the citric acid cycle.

This simple energy-releasing process does not take place in a single step. Instead, it occurs as a series of steps that involve a number of enzymes and cofactors located within the inner membrane of the mitochondria. Electrons from the reduced coenzymes are passed from one electron carrier to another within the membrane in assembly-line fashion and are finally combined with molecular oxygen, the final electron acceptor. The series of reactions involved in the process is called the **electron transport chain.** As illustrated in ■ Figure 13.9, there are several intermediate electron carriers between NADH and molecular oxygen. The electron carriers are lined up in order of increasing affinity for electrons.

The first electron carrier in the electron transport chain is an enzyme that contains a tightly bound coenzyme. The coenzyme has a structure similar to FAD. The enzyme formed by the combination of this coenzyme with a protein is called flavin mononucleotide (FMD). Two electrons and one H^+ ion from NADH plus another H^+ ion from a mitochondrion pass to FMN, then to an iron–sulfur (Fe—S) protein, and then to coenzyme Q (CoQ). CoQ is also the entry point into the electron transport chain for the two electrons and two H^+ ions from $FADH_2$. As NADH and $FADH_2$ release their hydrogen atoms and electrons, NAD^+ and FAD are regenerated for reuse in the citric acid cycle.

Four of the five remaining electron carriers are **cytochromes** (abbreviated cyt), which are structurally related proteins that contain an iron group. CoQ

Electron transport chain
A series of reactions in which protons and electrons from the oxidation of foods are used to reduce molecular oxygen to water.

Cytochrome
An iron-containing enzyme located in the electron transport chain.

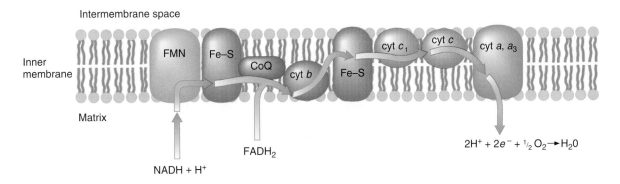

> **FIGURE 13.9** The electron transport chain. The heavy arrows show the path of the electrons as they pass from one carrier to the next.

passes along the two electrons to two molecules of cytochrome *b*, and the two H^+ ions are given up to the mitochondrion. The electrons are passed along the chain, and in the final step, an oxygen atom accepts the electrons and combines with two H^+ ions to form water.

➤ 13.7 Oxidative Phosphorylation

As electrons are transported along the electron transport chain, a significant amount of free energy is released (52.6 kcal/mol). Some of this energy is conserved by the synthesis of ATP from ADP and P_i. Because this synthesis of ATP (a phosphorylation reaction) is linked to the oxidation of NADH and $FADH_2$, it is termed **oxidative phosphorylation**. It takes place at three different locations of the electron transport chain (see ■ Figure 13.10).

For each mole of NADH oxidized in the electron transport chain, 2.5 mol of ATP are formed. The free energy required for the phosphorylation of ADP is 7.3 kcal/mol, or 18 kcal for 2.5 mol, of ATP. Because the oxidation of NADH liberates 52.6 kcal/mol, coupling the oxidation and phosphorylation reactions conserves 18 kcal/52.6 kcal, or approximately 34% of the energy released.

The electron donor $FADH_2$ enters the electron transport chain later than does NADH. Thus, for every $FADH_2$ oxidized to FAD, only 1.5 ATP molecules are formed during oxidative phosphorylation, and approximately 25% of the released free energy is conserved as ATP.

Oxidative phosphorylation
A process coupled with the electron transport chain whereby ADP is converted to ATP.

➤ **FIGURE 13.10** ATP is synthesized at three sites within the electron transport chain. Energy decreases at other locations result in the release of heat energy.

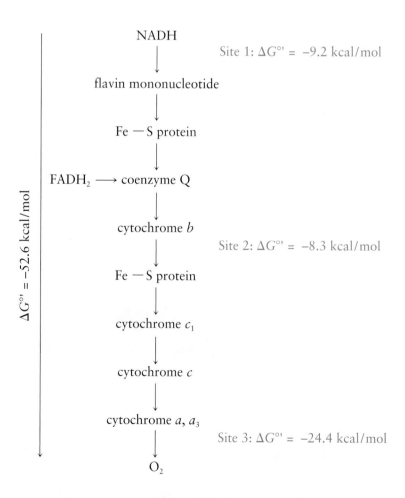

Now we can calculate the energy yield for the entire catabolic pathway (citric acid cycle, electron transport chain, and oxidative phosphorylation combined). According to Section 13.5, every acetyl CoA entering the citric acid cycle produces 3 NADH and 1 FADH₂ plus 1 GTP, which is equivalent in energy to 1 ATP. Thus, 10 ATP molecules are formed per molecule of acetyl CoA catabolized.

3 NADH ultimately produce	7.5 ATP
1 FADH₂ ultimately produces	1.5 ATP
1 GTP is equivalent to	1 ATP
	10 ATP total produced

The complex mechanism by which the cell couples the oxidations of the electron transport chain and the synthesis of ATP is believed to involve a flow of protons. This explanation, called the **chemiosmotic hypothesis,** proposes that the flow of electrons through the electron transport chain causes protons to be "pumped" from the matrix across the inner membrane and into the space between the inner and outer mitochondrial membranes (see ■ Figure 13.11). This creates a difference in proton concentration across the inner mitochondrial membrane, along with an electrical potential difference. As a result of the concentration and potential differences, protons flow back through the membrane through a channel formed by the enzyme F_1-ATPase. The flow of protons through this enzyme is believed to drive the phosphorylation reaction.

Chemiosmotic hypothesis
The postulate that a proton flow across the inner mitochondrial membrane during operation of the electron transport chain provides energy for ATP synthesis.

$$ADP + P_i \xrightarrow[\;F_1\text{-ATPase}\;]{H^+ \text{ flow through}} ATP \qquad (13.15)$$

How many molecules of ATP are produced from the complete catabolism of the following?

a. Two molecules of FADH₂
b. Two molecules of NADH
c. Two molecules of acetyl CoA

Learning Check 13.3

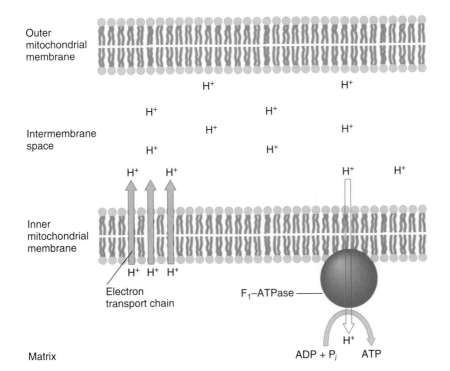

➤ **FIGURE 13.11** The chemiosmotic hypothesis. Electron transport pumps protons (H^+) out of the matrix. The formation of ATP accompanies the flow of protons through F_1-ATPase back into the matrix.

Outer mitochondrial membrane

H^+ H^+ H^+ H^+ H^+ H^+

Intermembrane space H^+ H^+ H^+ H^+

H^+ H^+ H^+ H^+

Inner mitochondrial membrane

H^+ H^+ H^+

Electron transport chain

F_1–ATPase

H^+

Matrix

ADP + P_i ATP

➤ 13.8 The Complete Oxidation of Glucose

The information given to this point about glycolysis, the citric acid cycle, and the electron transport chain almost makes it possible to calculate the ATP yield from the total oxidation of 1 mol of glucose. However, one more factor must be considered. NADH produced in the cytoplasm during glycolysis does not pass through the mitochondrial membrane to the site of the electron transport chain. Brain and muscle cells employ a transport mechanism that passes electrons from the cytoplasmic NADH through the membrane to FAD molecules inside the mitochondria, forming $FADH_2$. In Section 13.7, we pointed out that one molecule of $FADH_2$ gives 1.5 molecules of ATP. Thus, with this transport mechanism, one molecule of cytoplasmic NADH generates only 1.5 molecules of ATP, whereas 2.5 molecules of ATP come from each molecule of mitochondrial NADH. A more efficient shuttle mechanism is found in liver, heart, and kidney cells, where one cytoplasmic NADH results in one mitochondrial NADH. Thus, 2.5 molecules of ATP are formed in this case.

This production of ATP from the complete aerobic catabolism of 1 mol of glucose in the liver is summarized in ■ Table 13.1. Notice from the table that of the 32 mol of ATP generated, the great majority (28 mol) is formed as a result of oxidative phosphorylation. In Section 13.4, we learned that only 2 mol of ATP is produced per mole of glucose by lactate fermentation and by alcoholic fermentation. Thus, in terms of ATP production from fuels, the complete aerobic oxidation of glucose is 16 times more efficient than either lactate or alcoholic fermentation.

TABLE 13.1 The ATP yield from the complete oxidation of one mol of glucose

Reaction	Comments	Yield of ATP (moles)
Cytoplasmic reactions		
Glycolysis		
glucose → glucose 6-phosphate	1 mol consumes 1 ATP	−1
glucose 6-phosphate → fructose 1,6-bisphosphate	1 mol consumes 1 ATP	−1
glyceraldehyde 3-phosphate → 1,3-bisphosphoglycerate	2 mol produces 2 mol of cytoplasmic NADH	
1,3-bisphosphoglycerate → 3-phosphoglycerate	2 mol produces 2 ATP	+2
phosphoenolpyruvate → pyruvate	2 mol produces 2 ATP	+2
Mitochondrial reactions		
Oxidation of pyruvate		
pyruvate → acetyl CoA + CO_2	2 mol produces 2 NADH	
Citric acid cycle		
isocitrate → α-ketoglutarate + CO_2	2 mol produces 2 NADH	
α-ketoglutarate → succinyl CoA + CO_2	2 mol produces 2 NADH	
succinyl CoA → succinate	2 mol produces 2 GTP	+2
succinate → fumarate	2 mol produces 2 $FADH_2$	
malate → oxaloacetate	2 mol produces 2 NADH	
Electron transport chain and oxidative phosphorylation		
2 cytoplasmic NADH formed in glycolysis	each yields 2.5 ATP	+5[a]
2 NADH formed in the oxidation of pyruvate	each yields 2.5 ATP	+5
2 $FADH_2$ formed in the citric acid cycle	each yields 1.5 ATP	+3
6 NADH formed in the citric acid cycle	each yields 2.5 ATP	+15
	Net yield of ATP	+32

[a]In muscle and brain cells, 3 ATP are produced in this step and only 30 ATP overall.

Another useful way to look at the efficiency of the complete oxidation of glucose is to compare the amount of free energy available in glucose and the amount of free energy stored in synthesized ATP.

Glucose oxidation:

$$C_6H_{12}O_6 + 6O_2 \rightarrow 6CO_2 + 6H_2O \qquad \Delta G^{\circ\prime} = -686 \text{ kcal/mol} \qquad (13.16)$$

ATP synthesis:

$$32ADP + 32P_i \rightarrow 32ATP + 32H_2O \qquad \Delta G^{\circ\prime} = +234 \text{ kcal/mol} \qquad (13.17)$$

Overall reaction:

$$C_6H_{12}O_6 + 6O_2 + 32ADP + 32P_i \rightarrow 6CO_2 + 32ATP + 38H_2O \qquad (13.18)$$

$$\text{Total: } \Delta G^{\circ\prime} = -452 \text{ kcal/mol}$$

Thus, glucose oxidation liberates 686 kcal/mol, whereas the synthesis of 32 mol of ATP stores 234 kcal/mol. The efficiency of energy stored is

$$\frac{\text{energy stored}}{\text{energy available}} \times 100 = \frac{234 \text{ kcal/mol}}{686 \text{ kcal/mol}} \times 100 = 34.1\%$$

It is remarkable that living cells can capture 34% of the released free energy and make it available to do biochemical work. By contrast, an automobile engine makes available only about 20–30% of the energy actually released by burning gasoline.

► 13.9 Glycogen Metabolism

Glycogen Synthesis

Glucose consumed in excess of immediate body requirements is converted to glycogen (Section 7.9), which is stored primarily in the liver and muscle tissue. The liver of an average adult can store about 110 g of glycogen, and the muscles can store about 245 g. The process by which glucose is converted to glycogen is called **glycogenesis**. Glycogenesis can occur in all cells, but it is an especially important function of liver and muscle cells. The net result of this anabolic process is the bonding of glucose units to a growing glycogen chain, with the hydrolysis of a high-energy nucleotide, UTP (uridine triphosphate, Section 12.7), providing the energy. Each glucose unit lengthening the chain forms a new $\alpha(1 \rightarrow 4)$ linkage. The $\alpha(1 \rightarrow 6)$ branch points are inserted by a "branching enzyme."

The overall process of glycogenesis is summarized by Reaction 13.19:

Glycogenesis
The synthesis of glycogen from glucose.

$$n \text{ glucose} \xrightarrow{\hspace{3cm}} (\text{glucose})_n$$
$$\text{glycogen} \qquad (13.19)$$
$$UTP \qquad UDP + P_i$$

Glycogen Breakdown

The breakdown of glycogen back to glucose is called **glycogenolysis.** Although major amounts of glycogen are stored in both muscle tissue and the liver, glycogenolysis can occur in the liver (and kidney and intestinal cells) but not in muscle tissue because one essential enzyme (glucose 6-phosphatase) is lacking. The first step in glycogen breakdown is the cleavage of $\alpha(1 \rightarrow 4)$ linkages,

Glycogenolysis
The breakdown of glycogen to glucose.

KEY NUMBERS FOR ATP CALCULATIONS

It is important to be able to determine the energy yield in terms of the number of ATP molecules produced by the stepwise catabolism of glucose. Some of the relationships used in this determination will be used in the next chapter to calculate the energy yield from fatty acid catabolism. The three key substances to focus on are acetyl CoA, NADH, and FADH$_2$. Every mitochondrial NADH produces 2.5 ATP molecules, and every FADH$_2$ produces 1.5 ATP molecules. Every acetyl CoA that passes through the citric acid cycle gives 3 NADH, 1 FADH$_2$, and 1 GTP, which in turn produce a total of 10 ATP molecules. These reactions are outlined here.

Catabolic step		Energy production
Glycolysis (1 glucose $\longrightarrow$ 2 pyruvate)	produces	2NADH (cytoplasmic) + 2 ATP
		$\xrightarrow{\text{liver, heart, kidneys}}$ 5 ATP
		$\xrightarrow{\text{brain, muscle}}$ (3 ATP)
Oxidation of 2 pyruvates to 2 acetyl CoA	produces	2 NADH (mitochondrial) $\rightarrow$ 5 ATP
Citric acid cycle, one turn	produces	3 NADH $\rightarrow$ 7.5 ATP
		1 FADH$_2$ $\rightarrow$ 1.5 ATP
		1 GTP $\rightarrow$ 1 ATP
		10 ATP
Citric acid cycle, two turns	produce	$2 \times 10 = 20$ ATP
		Total/glucose molecule = 32 ATP (30 ATP)

catalyzed by glycogen phosphorylase. In this reaction, glucose units are released from the glycogen chain as glucose 1-phosphate:

$$(glucose)_n + P_i \rightarrow (glucose)_{n-1} + \text{glucose 1-phosphate} \qquad (13.20)$$
$$\text{glycogen} \qquad \text{glycogen with one}$$
$$\text{fewer glucose unit}$$

Also associated with the phosphorylase enzyme is a second enzyme, called the debranching enzyme. This debranching enzyme catalyzes the hydrolysis of the $\alpha(1 \rightarrow 6)$ linkages, eliminating the branch points. The phosphorylase can then continue to act on the rest of the chain.

In the second reaction, phosphoglucomutase catalyzes the isomerization of glucose 1-phosphate to glucose 6-phosphate:

$$\text{glucose 1-phosphate} \rightleftarrows \text{glucose 6-phosphate} \qquad (13.21)$$

The final step in glycogen breakdown is the hydrolysis of glucose 6-phosphate to produce free glucose. The enzyme for this reaction, glucose 6-phosphatase, is the one found only in liver, kidney, and intestinal cells:

$$\text{glucose 6-phosphate} + H_2O \rightarrow \text{glucose} + P_i \qquad (13.22)$$

Without glucose 6-phosphatase, muscle cells cannot form free glucose from glycogen. However, they can carry out the first two steps of glycogenolysis to produce glucose 6-phosphate. This form of glucose is the first intermediate in the glycolysis pathway and can be used to produce energy. In this way, muscle cells utilize glycogen only for energy production. On the other hand, an important function of the liver is to maintain a relatively constant level of blood glucose. Thus, it has the capacity to degrade glycogen all the way to glucose, which is released into the blood during muscular activity and between meals.

CHEMISTRY AROUND US • 13.2

CARBOHYDRATE LOADING AND ATHLETIC ENDURANCE

Observations made in the early 1960s indicated that muscle endurance was limited by the amount of glycogen stored in the muscles. This observation led to the practice of *carbohydrate loading* by athletes seeking to increase their endurance. Why was endurance limited by glycogen stores when muscles can also use free fatty acids from circulating blood as a source of energy? The answer is that fatty acids can only be utilized aerobically. Glycogen, on the other hand, can be used equally well as an energy source both aerobically and anaerobically. Thus, when glycogen stores are depleted by muscle exertion and only fatty acids are available, the rate at which energy is supplied to the muscles drops to a value determined by the supply of oxygen to the muscles. This rate of energy release is less than the rate from glycogen when it was operating both aerobically and anaerobically, and the muscles tire quickly.

Carbohydrate loading became very popular with long-distance runners, cyclists, cross-country skiers, and other athletes who depended on endurance to compete competitively. Initially, it was thought that glycogen stores in the body could be increased only by tricking the body by abrupt dietary changes. A typical carbohydrate-loading approach was to participate in very heavy exercise or training for 3–4 days, while eating a low-carbohydrate diet. This was followed by 1–2 days of light exercise or training on a high-carbohydrate diet. The initial low-carbohydrate diet coupled with vigorous muscle activity caused the muscle glycogen to become depleted. A rebound effect took place when a high-carbohydrate diet and light exercise were substituted. The rebound effect resulted in the production of higher levels of growth hormone and insulin in the body, and a corresponding increase in glycogen stores. The technique caused glycogen stores to be about double the normal levels, and to increase muscle endurance significantly. In one study, test subjects on a high-fat, high-protein diet were found to have less than 1.6 g of glycogen per 100 g of muscle and could perform a standardized workload for 1 hour. When the same subjects were put on a high-carbohydrate diet for 3 days, their glycogen stores jumped to 4 g per 100 g of muscle, and they could perform the same standardized workload for up to 4 hours.

The technique worked, but the athletes often felt irritable and lethargic during the low-carbohydrate phase, and the high-fat diet contradicted current recommendations for good health. Recent studies show that regular training, coupled with a diet high in complex carbohydrates (starches) and low in fat, increases glycogen stores without the need for sudden dietary changes. Currently, it is recommended that endurance athletes follow a diet high in carbohydrates during training. Then, about 2–3 days before competing, they increase their carbohydrate intake to about 70% of total Calories, and reduce their training intensity. Studies indicate that this approach increases muscle glycogen storage to levels comparable to those achieved using carbohydrate loading.

Learning Check 13.4

What enzyme enables the liver to form glucose from glycogen? Why can't muscles carry out the same conversion?

➤ 13.10 Gluconeogenesis

The supply of glucose in the form of liver and muscle glycogen can be depleted by about 12–18 hours of fasting. These stores of glucose are also depleted in a short time as a result of heavy work or strenuous exercise (see ■ Figure 13.12). Nerve tissue, including the brain, would be deprived of glucose under such conditions if the only source was glycogen. Fortunately, a metabolic pathway exists that overcomes this problem.

Glucose can be synthesized from noncarbohydrate materials in a process called **gluconeogenesis**. When carbohydrate intake is low and when glycogen stores are depleted, the carbon skeletons of lactate, glycerol (derived from the hydrolysis of fats), and certain amino acids are used to synthesize pyruvate, which is then converted to glucose:

Gluconeogenesis
The synthesis of glucose from noncarbohydrate molecules.

$$(\text{lactate, certain amino acids, glycerol}) \rightarrow \text{pyruvate} \rightarrow \text{glucose} \qquad (13.23)$$

About 90% of gluconeogenesis takes place in the liver. Very little occurs in the brain, skeletal muscle, or heart, even though these organs have a high demand for glucose. Thus, gluconeogenesis taking place in the liver helps maintain the blood glucose level so that tissues needing glucose can extract it from the blood.

Gluconeogenesis involving lactate is especially important. As might be expected, this process follows a separate and different pathway from glycolysis. During active exercise, lactate levels increase in muscle tissue, and the compound diffuses out of the tissue into the blood. It is transported to the liver, where lactate dehydrogenase, the enzyme that catalyzes lactate formation in muscle, converts it back to pyruvate:

$$\underset{\text{lactate}}{CH_3-\overset{\overset{\displaystyle OH}{|}}{CH}-COO^-} + NAD^+ \rightleftharpoons \underset{\text{pyruvate}}{CH_3-\overset{\overset{\displaystyle O}{\|}}{C}-COO^-} + NADH + H^+ \quad (13.24)$$

The pyruvate is then converted to glucose by the gluconeogenesis pathway, and the glucose enters the blood. In this way, the liver increases a low blood glucose level and makes glucose available to muscle. This cyclic process, involving the transport of lactate from muscle to the liver, the resynthesis of glucose by gluconeogenesis, and the return of glucose to muscle tissue, is called the **Cori cycle** (see ■ Figure 13.13). ■ Figure 13.14 shows the relationship of gluconeogenesis to the other major pathways of glucose metabolism.

Cori cycle

The process in which glucose is converted to lactate in muscle tissue, lactate is reconverted to glucose in the liver, and glucose is returned to the muscle.

➤ 13.11 The Hormonal Control of Carbohydrate Metabolism

It is important that metabolic pathways be responsive to cellular conditions so that energy is not wasted in producing unneeded materials. Besides the regulation of enzymes at key control points (Section 13.3), the body also uses three important regulatory hormones: epinephrine, glucagon, and insulin.

In Section 13.2, we pointed out that when carbohydrates are consumed, the blood glucose level increases. This increase signals the β-cells of the pancreas to release a small amount of the hormone insulin into the bloodstream.

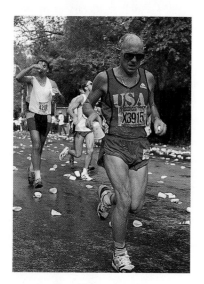

➤ **FIGURE 13.12** Gluconeogenesis provides a source of glucose when glycogen supplies are exhausted after vigorous exercise.

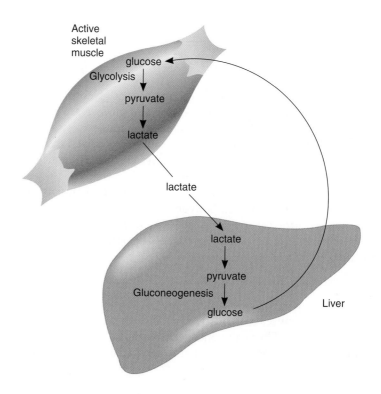

➤ **FIGURE 13.13** The Cori cycle, showing the relationship between glycolysis (in the muscle) and gluconeogenesis (in the liver).

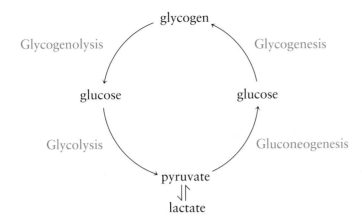

> ➤ **FIGURE 13.14** A summary of the major pathways in glucose metabolism. Note that the breakdown processes end in *-lysis,* whereas the synthesis pathways end in *-genesis.*

Diabetes mellitus
A family of diseases in which the uptake of blood glucose by tissues is impaired.

Insulin shock
A severe hypoglycemia produced by the injection of too much insulin.

CHEMISTRY AROUND US • 13.3

DIABETES MELLITUS

The best-known and one of the most prevalent metabolic disorders in humans is **diabetes mellitus.** Diabetes is the fourth-leading cause of death by disease in the United States. This condition is actually a family of diseases in which the uptake of blood glucose by tissues is impaired. The two main forms are Type I (insulin-dependent or juvenile-onset) diabetes and Type II (non-insulin-dependent or adult-onset) diabetes. In Type I diabetes, which usually appears in children before the age of 10, practically no insulin is produced by the β-cells of the pancreas. The disease can be controlled through daily injections of insulin and a strict dietary program. The insulin dose must be carefully regulated, though; too much insulin produces severe hypoglycemia, called **insulin shock.** Diabetics experiencing insulin shock may black out or lapse into a coma. People with diabetes often carry candy bars for a quick source of glucose if they feel the onset of insulin shock.

Type II diabetes is usually diagnosed in adulthood. In this form of the disease, a normal amount of insulin is secreted, but the body does not respond to it properly. This may result from a decrease in the number of insulin receptors on cell surfaces or from a modified structure in the receptor. More than 13 million Americans have Type II diabetes. About half the people with diabetes don't realize they have it because they don't recognize the meaning of symptoms such as unexplained thirst and frequent urination. Obesity is often associated with Type II diabetes. The disease can be treated by a carefully regulated dietary and exercise regimen.

In both types of diabetes, the consequences are the same. The body's failure to make or respond to insulin causes glucose levels to rise rapidly to hyperglycemic levels, followed by glucosuria (Section 13.2). The inability of cells to satisfy their energy needs by normal glucose metabolism sets off a chain reaction of increased fatty acid metabolism (Section 14.6) and lowered blood pH that, if left unchecked, can eventually lead to blindness, kidney failure, and death.

For many Type II diabetics, drugs such as tolbutamide are used to stimulate the pancreas to increase the release of insulin. Insulin from the pancreases of slaughtered sheep, pigs, or cows has been used to treat diabetics who must receive insulin daily. Pig insulin differs from human insulin by only one amino acid, but even this slight difference can sometimes cause complications. One of the fruits of recombinant DNA technology has been the production of human insulin from bacterial cultures (Section 11.10). Other areas of diabetes research where advances are being made include the implantation in the patient of insulin-releasing systems and the transplantation of living pancreatic cells.

http://www.brookscole.com/chemistry/seager_chemistry
At Web site 13.5, you will find patient information documents on diabetes.

Type I and Type II diabetes mellitus can be controlled with insulin injections and an appropriate diet.

Glucagon   Insulin

Stored   Blood
glycogen   sugar
level   level

➤ **FIGURE 13.15** The biochemical balance between two opposing hormones, insulin and glucagon.

➤ **FIGURE 13.16** Epinephrine helps the body mobilize glucose for energy in emergency situations. In this situation, both the lion and the photographer are probably utilizing epinephrine.

The protein enhances the absorption of glucose from the blood into cells of active tissue such as skeletal and heart muscles (Section 9.6). In addition, insulin increases the rate of synthesis of glycogen, fatty acids, and proteins and stimulates glycolysis. Each of these activities uses up glucose. As a result of insulin released into the bloodstream, the blood glucose level begins to decrease within 1 hour after carbohydrates have been ingested and returns to normal within about 3 hours.

Glucagon, a second polypeptide hormone, is synthesized and secreted by the α-cells of the pancreas. Glucagon activates the breakdown of glycogen (glycogenolysis) in the liver and thus counteracts the effect of insulin by raising blood glucose levels. ■ Figure 13.15 emphasizes the fact that insulin and glucagon work in opposition to each other and that the blood sugar level at any given time depends, in part, on the biochemical balance between these two hormones. Blood levels of glucose are also influenced to some degree by growth hormone and adrenal cortex steroids.

The third hormone affecting glucose metabolism is epinephrine, a catecholamine synthesized by the adrenal gland (Section 6.5). Epinephrine stimulates glycogen breakdown in muscles and, to a smaller extent, in the liver. This glycogenolysis reaction—usually a response to pain, anger, or fear—provides energy for a sudden burst of muscular activity (see ■ Figure 13.16). ■ Table 13.2 summarizes the hormonal control of glycogen.

TABLE 13.2	The hormonal control of glycogen		
Hormone	Source	Effect on glycogen	Impact on blood glucose
Insulin	β-cells of pancreas	Increases formation	Lowers blood glucose levels
Glucagon	α-cells of pancreas	Activates breakdown of liver glycogen	Raises blood glucose levels
Epinephrine	Adrenal medulla	Stimulates breakdown	Raises blood glucose levels

FOR FUTURE REFERENCE   NIACIN-RICH FOODS

Niacin is an important part of the coenzymes NAD and NADP. In this form, niacin plays an essential role in energy-transfer reactions in the body, especially in the metabolism of glucose, fat, and alcohol. Niacin occurs in two forms, nicotinic acid and nicotinamide (also called niacinamide). In the body, nicotinic acid is easily converted to nicotinamide, which is the primary form of niacin in the blood. The body is able to synthesize niacin from the amino acid tryptophan. For this reason, the niacin content of food is given in niacin equivalents (NE). Thus, a food that contains 1 mg of niacin per serving and enough tryptophan for the body to produce 1 mg of niacin would contain the equivalent of 2 mg of niacin, and would be reported as having 2 mg NE. The chart below lists selected foods that have a minimum NE of 1 mg per serving.

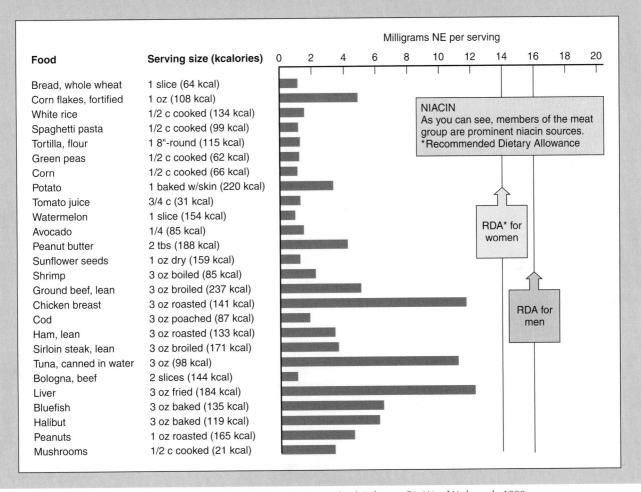

Food	Serving size (kcalories)
Bread, whole wheat	1 slice (64 kcal)
Corn flakes, fortified	1 oz (108 kcal)
White rice	1/2 c cooked (134 kcal)
Spaghetti pasta	1/2 c cooked (99 kcal)
Tortilla, flour	1 8"-round (115 kcal)
Green peas	1/2 c cooked (62 kcal)
Corn	1/2 c cooked (66 kcal)
Potato	1 baked w/skin (220 kcal)
Tomato juice	3/4 c (31 kcal)
Watermelon	1 slice (154 kcal)
Avocado	1/4 (85 kcal)
Peanut butter	2 tbs (188 kcal)
Sunflower seeds	1 oz dry (159 kcal)
Shrimp	3 oz boiled (85 kcal)
Ground beef, lean	3 oz broiled (237 kcal)
Chicken breast	3 oz roasted (141 kcal)
Cod	3 oz poached (87 kcal)
Ham, lean	3 oz roasted (133 kcal)
Sirloin steak, lean	3 oz broiled (171 kcal)
Tuna, canned in water	3 oz (98 kcal)
Bologna, beef	2 slices (144 kcal)
Liver	3 oz fried (184 kcal)
Bluefish	3 oz baked (135 kcal)
Halibut	3 oz baked (119 kcal)
Peanuts	1 oz roasted (165 kcal)
Mushrooms	1/2 c cooked (21 kcal)

Milligrams NE per serving

NIACIN
As you can see, members of the meat group are prominent niacin sources.
*Recommended Dietary Allowance

RDA* for women

RDA for men

SOURCE: Adapted from Whitney, E. N.; Rolfes, S. R. *Understanding Nutrition,* 8th ed. Belmont, CA: West/Wadsworth, 1999.

CONCEPT SUMMARY

The Digestion of Carbohydrates. Glucose is the key food molecule for most organisms, and it is the central substance in carbohydrate metabolism. During digestion, carbohydrates are hydrolyzed to the monosaccharides glucose, fructose, and galactose, which are absorbed into the bloodstream through the lining of the small intestine.

Blood Glucose. Glucose is the most abundant carbohydrate in the blood. The liver regulates blood glucose levels so that a sufficient concentration is always available to meet the body's energy needs. A lower-than-normal blood glucose level is referred to as hypoglycemia; a higher-than-normal blood glucose level is termed hyperglycemia.

Glycolysis. Glycolysis, a series of ten reactions that occur in the cytoplasm, is a process in which one glucose molecule is converted into two molecules of pyruvate. Two molecules of ATP and two molecules of NADH are produced in the process. Both galactose and fructose are converted into intermediates that enter the glycolysis pathway. The glycolysis pathway is controlled by the regulation of three key enzymes in the reaction sequence.

The Fates of Pyruvate. Pyruvate, the product of glycolysis, has three different fates. Under aerobic conditions, pyruvate enters mitochondria and is converted to acetyl CoA. The acetyl CoA enters the citric acid cycle. Products of the citric acid cycle enter the electron transport chain, where oxidation is completed and ATP is synthesized. A second fate of pyruvate is conversion to lactate. This takes place in anaerobic microorganisms and in humans when glycolysis occurs faster than the oxygen-dependent citric acid cycle and electron transport chain can operate. Some microorganisms convert pyruvate to ethanol, the third fate of pyruvate. In each of these three processes, NAD^+ is regenerated so that glycolysis can continue.

The Citric Acid Cycle. This pathway consists of eight reactions that process incoming molecules of acetyl CoA. The carbon atoms leave the cycle in the form of molecules of CO_2. The hydrogen atoms and electrons leave the cycle in the form of reduced coenzymes NADH and $FADH_2$. The cycle is regulated by three allosteric enzymes in response to cellular levels of ATP. One acetyl CoA molecule entering the citric acid cycle produces three molecules of NADH, one of $FADH_2$, and one of GTP.

The Electron Transport Chain. This pathway involves a series of reactions that pass electrons from NADH and $FADH_2$ to molecular oxygen. Each carrier in the series has an increasing affinity for electrons. Four of the carriers, which are referred to as cytochromes, contain iron, which accepts and then transfers the electrons. As NADH and $FADH_2$ release their hydrogen atoms and electrons, NAD^+ and FAD are regenerated for return to the citric acid cycle.

Oxidative Phosphorylation. At three sites within the electron transport chain, the decrease in free energy is sufficient to convert ADP to ATP. One molecule of NADH produces 2.5 molecules of ATP. $FADH_2$, which enters the chain one step later, produces 1.5 molecules of ATP. According to the chemiosmotic hypothesis, the synthesis of ATP takes place because of a flow of protons across the inner mitochondrial membrane.

The Complete Oxidation of Glucose. The complete oxidation of 1 mol of glucose produces 32 (or 30) mol of ATP. In contrast, lactate and alcoholic fermentation produce only 2 mol of ATP.

Glycogen Metabolism. When blood glucose levels are high, excess glucose is converted into glycogen by the process of glycogenesis. The glycogen is stored within the liver and muscle tissue. The breakdown of glycogen occurs when muscles need energy and when the liver is restoring a low blood sugar level to normal.

Gluconeogenesis. Lactate, certain amino acids, and glycerol can be converted into glucose. This process, called gluconeogenesis, takes place in the liver when glycogen supplies are being depleted and when carbohydrate intake is low. Gluconeogenesis from lactate is especially important during periods of high muscle activity; the liver converts excess lactate from the muscles into glucose, which is then cycled back to the muscles. This process is called the Cori cycle.

The Hormonal Control of Carbohydrate Metabolism. Three hormones exert major interactive effects on carbohydrate metabolism: insulin, glucagon, and epinephrine. Insulin promotes the uptake and utilization of glucose by the cells. Thus, insulin lowers blood sugar levels. Glucagon stimulates the conversion of glycogen to glucose and thus raises blood glucose levels. Epinephrine is released in response to anger, fear, and excitement. It also stimulates the release of glucose from glycogen and raises blood glucose levels.

LEARNING OBJECTIVES

When you have completed your study of this chapter, you should be able to:

1. Identify the products of carbohydrate digestion.
2. Explain the importance to the body of maintaining proper blood sugar levels.
3. List the net reactants and products of the glycolysis pathway.
4. Describe how the glycolysis pathway is regulated in response to cellular needs.
5. Name the three fates of pyruvate.
6. Identify the two major functions of the citric acid cycle.
7. Describe how the citric acid cycle is regulated in response to cellular energy needs.
8. Explain the function of the electron transport chain and describe how electrons move down the chain.
9. List the major feature of the chemiosmotic hypothesis.
10. Calculate the amount of ATP produced by the complete oxidation of a mole of glucose.
11. Explain the importance of the processes of glycogenesis and glycogenolysis.
12. Describe gluconeogenesis and the operation of the Cori cycle.
13. Describe how hormones regulate carbohydrate metabolism.

KEY TERMS AND CONCEPTS

Aerobic (13.4)
Alcoholic fermentation (13.4)
Anaerobic (13.4)
Blood sugar level (13.2)
Chemiosmotic hypothesis (13.7)
Citric acid cycle (13.5)
Cori cycle (13.10)
Cytochrome (13.6)

Diabetes mellitus (13.11)
Electron transport chain (13.6)
Gluconeogenesis (13.10)
Glucosuria (13.2)
Glycogenesis (13.9)
Glycogenolysis (13.9)
Glycolysis (13.3)

Hyperglycemia (13.2)
Hypoglycemia (13.2)
Insulin shock (13.11)
Lactate fermentation (13.4)
Lactose intolerance (13.3)
Oxidative phosphorylation (13.7)
Renal threshold (13.2)

KEY REACTIONS

1. Glycolysis (Section 13.3):

$$\text{glucose} + 2P_i + 2ADP + 2NAD^+ \rightarrow 2 \text{ pyruvate} + 2ATP + 2NADH + 2H^+ + H_2O$$ Reaction 13.5

2. Oxidation of pyruvate to acetyl CoA (Section 13.4):

$$CH_3-\overset{O}{\overset{\|}{C}}-COO^- + CoA-S-H + NAD^+ \longrightarrow CH_3-\overset{O}{\overset{\|}{C}}-S-CoA + NADH + CO_2$$ Reaction 13.7

3. Reduction of pyruvate to lactate (Section 13.4):

$$CH_3-\overset{O}{\overset{\|}{C}}-COO^- + NADH + H^+ \rightleftharpoons CH_3-\overset{OH}{\overset{|}{CH}}-COO^- + NAD^+$$ Reaction 13.8

4. Net reaction of lactate fermentation (Section 13.4):

$$C_6H_{12}O_6 + 2ADP + 2P_i \longrightarrow 2CH_3-\overset{OH}{\overset{|}{CH}}-COO^- + 2ATP$$ Reaction 13.9

5. Reduction of pyruvate to ethanol (Section 13.4):

$$CH_3-\overset{O}{\overset{\|}{C}}-COO^- + 2H^+ + NADH \longrightarrow CH_3CH_2-OH + NAD^+ + CO_2$$ Reactions 13.10 and 13.11

6. Net reaction of alcoholic fermentation (Section 13.4):

$$C_6H_{12}O_6 + 2ADP + 2P_i \rightarrow 2CH_3CH_2-OH + 2CO_2 + 2ATP$$ Reaction 13.12

7. The citric acid cycle (Section 13.5):

$$\text{acetyl CoA} + 3NAD^+ + FAD + GDP + P_i + 2H_2O \rightarrow 2CO_2 + CoA-S-H + 3NADH + 2H^+ + FADH_2 + GTP$$ Reaction 13.13

8. The electron transport chain (Section 13.6):

$$4H^+ + 4e^- + O_2 \rightarrow 2H_2O$$ Reaction 13.14

9. Oxidative phosphorylation (Section 13.7):

$$ADP + P_i \rightarrow ATP$$ Reaction 13.15

10. Complete oxidation of glucose (Section 13.8):

$$C_6H_{12}O_6 + 6O_2 + 32ADP + 32P_i \rightarrow 6CO_2 + 32ATP + 38H_2O$$ Reaction 13.18

11. Glycogenesis (Section 13.9):

$$\text{glucose} \rightarrow \text{glycogen}$$

12. Glycogenolysis (Section 13.9):

$$\text{glycogen} \rightarrow \text{glucose}$$

13. Gluconeogenesis (Section 13.10):

Reaction 13.23

EXERCISES

THE DIGESTION OF CARBOHYDRATES (SECTION 13.1)

13.1 Why is glucose considered the pivotal compound in carbohydrate metabolism?

13.2 Name the products produced by the digestion of
a. starch
b. lactose
c. sucrose

13.3 What type of reaction is involved as carbohydrates undergo digestion?

BLOOD GLUCOSE (SECTION 13.2)

13.4 Describe what is meant by the terms *blood sugar level* and *normal fasting level*.

13.5 What range of concentrations for glucose in blood is considered a normal fasting level?

13.6 How do each of the following terms relate to blood sugar level?
a. Hypoglycemia
b. Hyperglycemia
c. Renal threshold
d. Glucosuria

13.7 Describe how the liver helps regulate blood glucose levels.

13.8 Explain why severe hypoglycemia can be very serious.

GLYCOLYSIS (SECTION 13.3)

13.9 What is the main purpose of glycolysis?

13.10 What is the starting material for glycolysis? The product?

13.11 In what part of the cell does glycolysis take place?

13.12 How many steps in glycolysis require ATP? How many steps produce ATP?

13.13 Which coenzyme serves as an oxidizing agent in glycolysis?

13.14 Number the carbon atoms of glucose 1 through 6, and show the location of each carbon in the two molecules of pyruvate produced by glycolysis.

13.15 What is the third control point in glycolysis? How is it responsive to cellular needs?

13.16 What effect does a high concentration of glucose 6-phosphate have on glycolysis?

13.17 Why is the conversion of fructose 6-phosphate to fructose 1,6-bisphosphate sometimes called the "committed step" of glycolysis?

THE FATES OF PYRUVATE (SECTION 13.4)

13.18 Distinguish between the terms *aerobic* and *anaerobic*.

13.19 Compare the fate of pyruvate in the body under (a) aerobic conditions and (b) anaerobic conditions.

13.20 Explain how lactate formation allows glycolysis to continue under anaerobic conditions.

13.21 Write a summary equation for the conversion of glucose to
a. lactate
b. pyruvate
c. ethanol

13.22 Explain how alcoholic fermentation allows yeast to survive with limited oxygen supplies.

THE CITRIC ACID CYCLE (SECTION 13.5)

13.23 What are two other names for the citric acid cycle?

13.24 What is the primary function of the citric acid cycle in ATP production? What other vital role is served by the citric acid cycle?

13.25 Why is it appropriate to refer to the sequence of reactions in the citric acid cycle as a cycle?

13.26 Describe the citric acid cycle by identifying the following:
a. The fuel needed by the cycle
b. The form in which carbon atoms leave the cycle
c. The form in which hydrogen atoms and electrons leave the cycle

13.27 What way are the processes in the citric acid cycle and the electron transport chain interrelated?

13.28 How many molecules of each of the following are produced by one run through the citric acid cycle?
a. NADH   b. $FADH_2$ c. GTP   d. CO_2

13.29 Write reactions for the steps in the citric acid cycle that involve oxidation by NAD^+.

13.30 Write a reaction for the step in the citric acid cycle that involves oxidation by FAD.

13.31 What is the structural difference between ATP and GTP? Explain why GTP can function as a high-energy compound.

13.32 The reactions in the citric acid cycle involve enzymes.
a. Identify the enzymes at the three control points.
b. Each control-point enzyme is inhibited by NADH and ATP. Explain how this enables the cell to be responsive to energy needs.
c. Two of the control-point enzymes are activated by ADP. Explain the benefit of this regulation to the cell.

THE ELECTRON TRANSPORT CHAIN (SECTION 13.6)

13.33 What is the primary function of the electron transport chain?

13.34 Which coenzymes bring electrons to the electron transport chain?

13.35 List the participants of the electron transport chain in the order in which they accept electrons.

13.36 What is the role of the cytochromes in the electron transport chain?

13.37 Write the reaction for the last step of the electron transport chain.

OXIDATIVE PHOSPHORYLATION (SECTION 13.7)

13.38 The third stage in the oxidation of foods to provide energy involves oxidative phosphorylation. What is oxidized? What is phosphorylated?

13.39 How many ATP molecules are synthesized when NADH passes electrons to molecular oxygen?

13.40 How many ATP molecules are synthesized when $FADH_2$ passes electrons to molecular oxygen?

13.41 Explain what is meant by the statement: "Oxidation of NADH in the electron transport chain results in 34% energy conservation."

13.42 How many ATP molecules can be formed by the oxidation of six acetyl CoA molecules in the common catabolic pathway?

13.43 Identify the steps in the electron transport chain where ATP is synthesized.

13.44 The change in free energy for the complete oxidation of glucose to CO_2 and H_2O is -686 kcal per mole of glucose. If all this free energy could be used to drive the synthesis of ATP, how many moles of ATP could be formed from 1 mol of glucose?

13.45 According to the chemiosmotic theory, ATP synthesis results from the flow of what particles? Across what membrane does this flow occur?

13.46 What enzyme plays a key role in ATP synthesis?

THE COMPLETE OXIDATION OF GLUCOSE (SECTION 13.8)

13.47 Explain why NADH produced in muscles during glycolysis ultimately gives rise to just 1.5 molecules of ATP.

13.48 A total of 32 mol of ATP can be produced by the complete oxidation of 1 mol of glucose in the liver. Indicate the number of these moles of ATP produced by
a. glycolysis
b. the citric acid cycle
c. the electron transport chain and oxidative phosphorylation

13.49 The total amount of energy available from 1 mol of glucose is 686 kcal. What percentage of this is conserved in ATP production by the following?
a. Glycolysis
b. Lactate formation
c. Total oxidation of glucose to CO_2

GLYCOGEN METABOLISM (SECTION 13.9)

13.50 Where is glycogen stored in the body?

13.51 Compare the terms *glycogenesis* and *glycogenolysis*.

13.52 What high-energy compound is involved in the conversion of glucose to glycogen?

13.53 Explain how liver glycogen is important in controlling the blood sugar level.

13.54 Why can liver glycogen, but not muscle glycogen, be used to raise blood sugar levels?

13.55 How does glycogen enter the glycolysis pathway?

13.56 What organ serves as the principal site for gluconeogenesis?

GLUCONEOGENESIS (SECTION 13.10)

13.57 Describe the importance of gluconeogenesis.

13.58 What are the sources of compounds that undergo gluconeogenesis?

13.59 Explain the operation of the Cori cycle and its importance to muscles.

THE HORMONAL CONTROL OF CARBOHYDRATE METABOLISM (SECTION 13.11)

13.60 Describe the influence of glucagon and insulin on the following:
a. Blood sugar level
b. Glycogen formation

13.61 What effect does epinephrine have on carbohydrate metabolism?

OVER THE COUNTER AND CHEMISTRY AROUND US

13.62 Describe the cause and symptoms of lactose intolerance.

13.63 What is the main treatment for lactose intolerance?

13.64 Explain why you breathe heavily when you engage in strenuous exercise.

13.65 Why do premature infants sometimes exhibit lactate accumulation?

13.66 What is carbohydrate loading?

13.67 Why does the amount of glycogen storage have an impact on muscle endurance?

13.68 Contrast the characteristics of Type I and Type II diabetes.

13.69 What is meant by insulin shock?

13.70 What are the consequences of unchecked diabetes?

POINTS TO PONDER AND PURSUE

13.1 The citric acid cycle does not utilize oxygen, yet it cannot function without oxygen. Why?

13.2 Malonate (^-OOC—CH_2—COO^-) is an extremely effective competitive inhibitor of the citric acid cycle. Which enzyme do you think malonate inhibits? Explain.

13.3 A friend started to make wine by adding yeast to fresh grape juice and placing the mixture in a sealed bottle. Several days later the bottle exploded. Explain.

13.4 Two students are debating whether 30 or 32 molecules of ATP can be formed by the catabolism of one molecule of glucose. Who is correct?

13.5 Explain why monitoring blood lactate levels might be a useful technique in following the conditioning level of an Olympic runner.

13.6 You may have experienced a rush of epinephrine during a recent exam or athletic event. Are there other physiological responses to epinephrine in addition to higher blood sugar level?

13.7 In Figure 13.4, it is noted that making bread involves a fermentation process. Why, then, is the alcohol content of bread zero?

13.8 Based on information at Web site 13.2, what plasma glucose concentration is almost always associated with a serious abnormality?

13.9 According to Web site 13.3, what percentage of Asian Americans are lactose intolerant?

13.10 According to Web site 13.4, what would be the implication on exercise if the body could not produce lactate?

13.11 At Web site 13.5, using the link "Diabetes Statistics," determine how many undiagnosed cases of diabetes are estimated to exist in the United States.

http://www.brookscole.com/chemistry/seager_chemistry

Lipid and Amino Acid Metabolism

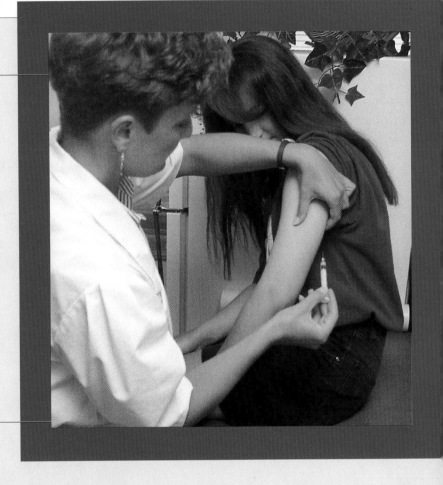

Besides the familiar work nurses do in hospitals, they also practice as community or public health nurses, school nurses, and occupational nurses. Nurse practitioners are trained to perform a number of procedures that are generally done only by physicians. Nurse practitioners who set up an independent practice in rural areas sometimes provide the only source of health information and immediate medical treatment for the people living there. In this chapter, you will encounter a number of special topics that deal with health issues such as blood lipid levels and the disease PKU.

http://www.brookscole.com/chemistry/seager_chemistry
Web site 14.1 provides career information for nurse practitioners.

Triglycerides (fats and oils) are important dietary sources of energy. Fat also functions as the major form of energy storage in humans. Fat, with a caloric value of 9 Calorie/g, contains more than twice the energy per gram of glycogen or starch. Thus, fat is a more concentrated and efficient storage form of energy than carbohydrates. Because it is water insoluble, fat can be stored in much larger quantities in our bodies. For example, the limited carbohydrate reserves in the body are depleted after about 1 day without food, but stored fat can provide needed calories (but not other nutrients) in an individual for 30–40 days.

Amino acids perform three vital functions in the human body: They are the building blocks for the synthesis of proteins, they provide carbon and nitrogen atoms for the synthesis of other biomolecules, and they can be sources of energy.

In this chapter, we discuss the storage, degradation, and synthesis of lipids. We also study the degradation and synthesis of amino acids and the relationships between amino acid metabolism and the metabolism of carbohydrates and lipids.

➤ 14.1 Blood Lipids

During digestion, triglycerides are hydrolyzed to glycerol, fatty acids, and monoglycerides (one fatty acid attached to glycerol). Phosphoglycerides are also hydrolyzed to their component substances. These smaller molecules, along with cholesterol, are absorbed into cells of the intestinal mucosa, where resyn-

➤ **FIGURE 14.1** The digestion and absorption of triglycerides.

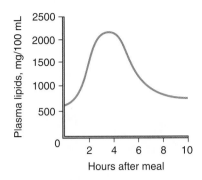

> **FIGURE 14.2** The rise in the blood lipids following a meal containing fat.

thesis of the triglycerides and phosphoglycerides occurs. For transport within the aqueous environment of the lymph and blood, these insoluble lipids are complexed with proteins to form lipoprotein aggregates called **chylomicrons**. The chylomicrons (which contain triglycerides, phosphoglycerides, and cholesterol) pass into the lymph system and then into the bloodstream. Chylomicrons are modified by the liver into smaller lipoprotein particles, the form in which most lipids are transported to various parts of the body by the bloodstream (see ■ Figure 14.1).

To some extent, the behavior of blood lipids parallels that of blood sugar. The concentration of both substances increases following a meal and returns to normal as a result of storage in fat depots and of oxidation to provide energy. As shown in ■ Figure 14.2, the concentration of plasma lipids usually begins to rise within 2 hours after a meal, reaches a peak in 4–6 hours, and then drops rather rapidly to a normal level.

One general method of classifying lipoproteins is by density. Because lipids are less dense than proteins, lipoprotein particles with a higher proportion of lipids are less dense than those with a lower proportion. Chylomicrons are the least dense because they have the highest ratio of lipid to protein. ■ Figure 14.3 is a schematic model of a low-density lipoprotein (LDL). The compositions of the four classes of lipoprotein are compared in ■ Figure 14.4. We see from Figure 14.4 that the protein content of chylomicrons is significantly lower than that of other plasma lipoproteins, and that this low percentage results in a density that is less than 0.95 g/mL. Notice that the so-called high-density lipoproteins (HDL) have a protein mass that is 45–50% of the total mass of the particles. The densities of HDL particles have correspondingly high values, ranging 1.06–1.21 g/mL.

Chylomicron
A lipoprotein aggregate found in the lymph and the bloodstream.

➤ 14.2 Fat Mobilization

Carbohydrates from dietary sources and glycogen catabolism are the compounds utilized preferentially for energy production by some tissues. The brain and active skeletal muscles are examples of such tissues. But as we have seen, the body's stores of glycogen are depleted after only a few hours of fasting. When this happens, fatty acids stored in triglycerides are called on as energy sources. Even when glycogen supplies are adequate, many body cells, such as those of resting muscle and liver, utilize energy derived from the breakdown of fatty acids. This helps conserve the body's glycogen stores and glucose for use by brain cells and red blood cells. Red blood cells cannot oxidize fatty acids because red blood cells have no mitochondria, the site of fatty acid oxidation.

> **FIGURE 14.3** A schematic model of low-density lipoprotein (LDL).

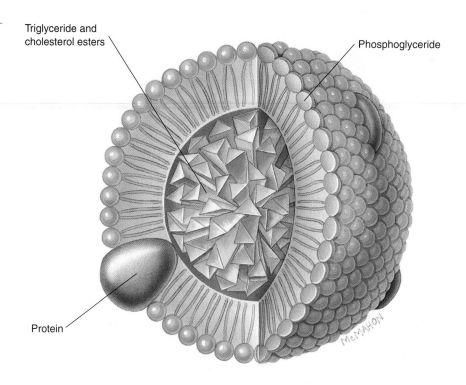

Brain cells are bathed in cerebrospinal fluid and do not obtain nutrients directly from the blood (the blood-brain barrier). Glucose and many other substances can cross the barrier into the cerebrospinal fluid, but fatty acids cannot.

When body cells need fatty acids for energy, the endocrine system is stimulated to produce several hormones, including epinephrine, which interact

> **FIGURE 14.4** Relative amounts of cholesterol, protein, triglyceride, and phosphoglyceride in the four classes of lipoproteins.

OVER THE COUNTER • 14.1

CHOLESTEROL-LOWERING DRUGS

In Chemistry Around Us 8.2, we explored cholesterol and its relationship to healthy hearts. One of many kinds of lipids found in the blood, cholesterol is essential for several of the body's processes. Cholesterol is involved in the formation of cell membranes, the insulation of nerves, the synthesis of several hormones, and the digestion of food.

Cholesterol and other lipids are insoluble in water. In order for them to be transported in the water-based blood, they are combined with proteins to form lipoproteins, which have a higher water solubility than the lipids. Lipoproteins are classified according to their densities, and two that involve cholesterol are called high-density lipoprotein (HDL) and low-density lipoprotein (LDL). Low-density lipoproteins carry more cholesterol in the bloodstream than any other type of lipoprotein. It has been determined that a high level of LDL in the blood increases the risk of coronary heart disease. The level of LDL in the blood can often be reduced by changes in lifestyle, such as increasing the amount of exercise, reducing the intake of dietary fat, stopping smoking, and losing weight.

When lifestyle changes fail to reduce the level of blood LDL enough, several types of cholesterol-lowering drugs are available. Niacin is the oldest medication and one of the few OTC treatments for high cholesterol levels. It is a useful lipid-modifying drug because it (1) decreases LDL cholesterol, total cholesterol, and triglycerides, and (2) raises HDL cholesterol. The chief constraints against niacin use have been flushing, gastrointestinal discomfort, and metabolic effects including liver toxicity.

The resin drugs have been in use for about 20 years and include cholestyramine (Questran®) and colestipol (Colestid®). These drugs bind with bile acids in the digestive tract. Bile acids are digestive aids that are synthesized from cholesterol in the liver. When they are tied up and removed from the digestive processes, the liver synthesizes more. Because cholesterol is used in the synthesis, less is available to be released into the blood to form LDL.

A second type of cholesterol-lowering drug acts by reducing the production of triglycerides, which are involved in the formation of LDL. With reduced triglyceride levels, less LDL is formed and less cholesterol circulates in the blood. The drug gemfibrozil (Lopid®), or large doses of niacin, a vitamin, function this way to lower blood cholesterol levels.

A third type of cholesterol-lowering drug is called statins. Statins function in the liver to block the synthesis of cholesterol. As a result, the cholesterol level in liver cells is reduced, thereby causing the cells to remove cholesterol from the circulating

blood, thus lowering the blood cholesterol level. These drugs can also help the body reabsorb cholesterol from plaque that has formed in blood vessels. This helps to slowly break down the plaque and unplug the vessels.

Statin drugs can reduce LDL cholesterol by up to 40% depending on the dosage, and they are the only type of lipid-lowering drug that has proven to reduce the risk of death from cardiovascular disease. They have also been shown to reduce the risk of a second heart attack. Examples of the statin drugs are fluvastatin (Lescol®), lovastatin (Mevacor®), simvastatin (Zocor®), pravastatin (Pravachol®), and atorvastatin (Lipitor®).

lovastatin

http://www.brookscole.com/chemistry/seager_chemistry
Web site 14.2 provides information on HDL and LDL and on how to lower your cholesterol level.

A few of the cholesterol-lowering drugs available by prescription.

---

with **adipose tissue.** Earlier we saw that epinephrine stimulates the conversion of glycogen to glucose. In the fat cells of adipose tissue, epinephrine stimulates the hydrolysis of triglycerides to fatty acids and glycerol, which enter the bloodstream. This process is called **fat mobilization.** In the blood, mobilized fatty acids form a lipoprotein with the plasma protein called serum albumin. In this form, the fatty acids are transported to the tissue cells that need them. The glycerol produced by the hydrolysis of the triglyceride is water soluble, so it dissolves in the blood and is also transported to cells that need it.

Adipose tissue

A kind of connective tissue where triglycerides are stored.

Fat mobilization

The hydrolysis of stored triglycerides, followed by the entry of fatty acids and glycerol into the bloodstream.

➤ 14.3 Glycerol Metabolism

The glycerol hydrolyzed from fats can provide energy to cells. Glycerol is converted by cells to dihydroxyacetone phosphate in two steps:

$$
\begin{array}{ccccc}
\text{H}_2\text{C}-\text{OH} & & \text{H}_2\text{C}-\text{OH} & & \text{H}_2\text{C}-\text{OH} \\
| & & | & & | \\
\text{HC}-\text{OH} & \xrightarrow{} & \text{HC}-\text{OH} & \xrightarrow{} & \text{C}=\text{O} \longrightarrow \text{Glycolysis} \\
| & & | & & | \\
\text{H}_2\text{C}-\text{OH} & & \text{H}_2\text{C}-\text{OPO}_3^{2-} & & \text{H}_2\text{C}-\text{OPO}_3^{2-} \\
& \text{ATP}\quad\text{ADP} & & \text{NAD}^+\quad\text{NADH}+\text{H}^+ & \\
\text{glycerol} & & \text{glycerol} & & \text{dihydroxyacetone} \\
& & \text{3-phosphate} & & \text{phosphate}
\end{array}
$$

(14.1)

Dihydroxyacetone phosphate is one of the chemical intermediates of glycolysis (Section 13.3). Thus, glycerol, by entering glycolysis, can be converted to pyruvate and contribute to cellular energy production. The pyruvate formed from glycerol can also be converted to glucose through gluconeogenesis (Section 13.10).

➤ 14.4 The Oxidation of Fatty Acids

Fatty acids that enter tissue cells cannot be oxidized to produce energy until they pass through the mitochondrial membrane. This cannot occur until the fatty acid is converted into fatty acyl CoA by reaction with coenzyme A. Fatty acyl

refers to the $R-\overset{\overset{\text{O}}{\|}}{C}-$ portion of the molecule.

$$
\underset{\text{fatty acid}}{R-\overset{\overset{\text{O}}{\|}}{C}-\text{OH}} + \text{HS}-\text{CoA} \xrightarrow{} \underset{\text{fatty acyl CoA}}{R-\overset{\overset{\text{O}}{\|}}{C}-\text{S}-\text{CoA}} + \text{H}_2\text{O}
$$

(14.2)

$$\text{ATP}\quad\text{AMP}+\text{PP}_i$$

This reaction, which is catalyzed by acyl CoA synthetase, is referred to as activation of the fatty acid because the fatty acyl CoA is a high-energy compound. The energy needed for its synthesis is provided by the hydrolysis of ATP to AMP and PP_i and the subsequent hydrolysis of PP_i to 2P_i. The process of forming fatty acyl CoA molecules demonstrates how the energy stored in the phosphate bonds of ATP can be used to drive chemical processes that normally could not occur. The cost to the cell of activating one molecule of fatty acid is the loss of two high-energy bonds in an ATP molecule.

The fatty acyl CoA molecules that enter mitochondria are then degraded in a catabolic process called **β-oxidation**. During β-oxidation, the second (or beta) carbon down the chain from the carbonyl of the fatty acyl CoA molecule is oxidized to a ketone:

β-oxidation process
A pathway in which fatty acids are broken down into molecules of acetyl CoA.

$$
\underset{\text{fatty acyl CoA}}{\overset{\beta}{R}-\overset{\overset{\text{H}}{|}}{\underset{\overset{|}{\text{H}}}{C}}-\overset{\alpha}{\underset{1}{C}}\text{H}_2-\overset{\overset{\text{O}}{\|}}{C}-\text{S}-\text{CoA}} \xrightarrow{\text{Oxidation}} R-\overset{\overset{\text{O}}{\|}}{\underset{\beta}{C}}-\text{CH}_2-\overset{\overset{\text{O}}{\|}}{C}-\text{S}-\text{CoA}
$$

(14.3)

Four reactions are involved in oxidizing the β-carbon to a ketone. We will use the fatty acyl CoA formed from one 18-carbon acid, stearic acid, as an example of the reactions. We will also keep track of any NADH or FADH$_2$ produced in the

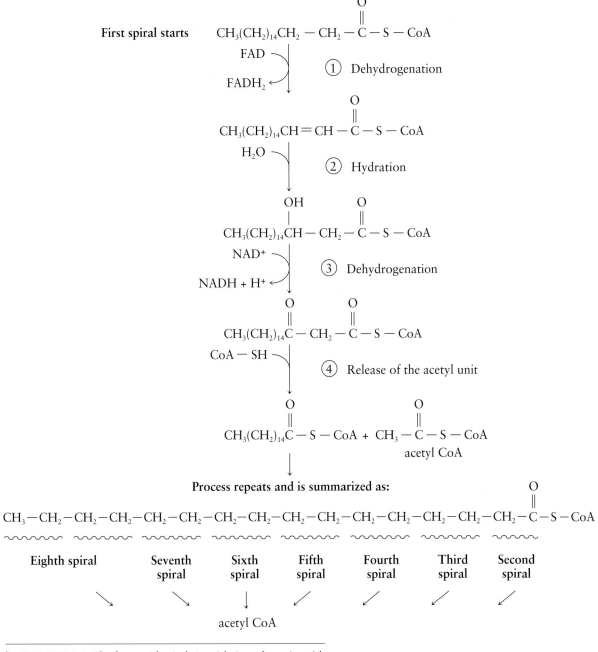

➤ **FIGURE 14.5** The fatty acid spiral: β-oxidation of stearic acid.

reactions because these molecules can enter the respiration process (electron transport chain). The reactions involved in β-oxidation are shown in ■ Figure 14.5.

In the final step of β-oxidation (Step 4), the chain is broken between the α- and β-carbons by reaction with coenzyme A. As a result, a new fatty acyl CoA is formed with a chain shortened by two carbon atoms, and a unit of acetyl CoA is released:

The new fatty acyl compound enters the β-oxidation process at Step 1, and the sequence is repeated until the fatty acyl CoA is completely degraded to acetyl CoA (Figure 14.5). The fatty acyl CoA molecule that starts each run through the β-oxidation process is two carbons shorter than the one going through the previous run. For this reason, the β-oxidation pathway for fatty acid degradation to acetyl CoA is often called the fatty acid spiral (rather than cycle).

Every run through the spiral produces one molecule each of acetyl CoA, NADH, and FADH$_2$ until the fatty acyl CoA is only four carbons long:

$$
\begin{array}{c}
\underset{\substack{\text{fatty acyl CoA}\\\text{entering the spiral}}}{R-CH_2CH_2-\overset{\displaystyle O}{\overset{\|}{C}}-S-CoA} + NAD^+ + FAD + H_2O + CoA-SH \longrightarrow
\end{array}
$$

(14.5)

$$
\underset{\substack{\text{fatty acyl CoA}\\\text{2 carbons shorter}}}{R-\overset{\displaystyle O}{\overset{\|}{C}}-S-CoA} + \underset{\text{acetyl CoA}}{CH_3\overset{\displaystyle O}{\overset{\|}{C}}-S-CoA} + NADH + H^+ + FADH_2
$$

In the last spiral, the four-carbon chain of butyryl CoA passes through the β-oxidation sequence, and it produces one molecule of FADH$_2$, one molecule of NADH, and two molecules of acetyl CoA:

$$
\underset{\text{butyryl CoA}}{CH_3-CH_2-CH_2-\overset{\displaystyle O}{\overset{\|}{C}}-S-CoA} + NAD^+ + H_2O + FAD + CoA-SH \longrightarrow
$$

(14.6)

$$
\underset{\text{2 acetyl CoA molecules}}{2CH_3-\overset{\displaystyle O}{\overset{\|}{C}}-S-CoA} + NADH + H^+ + FADH_2
$$

In other words, the complete conversion of a fatty acyl CoA to two-carbon fragments of acetyl CoA always produces one more molecule of acetyl CoA than of FADH$_2$ or NADH. Thus, the breakdown of stearic acid requires eight passes through the β-oxidation sequence and produces nine molecules of acetyl CoA, buy only eight molecules of FADH$_2$ and eight molecules of NADH.

Net reaction:

$$
CH_3(CH_2)_{16}\overset{\displaystyle O}{\overset{\|}{C}}-S-CoA + 8FAD + 8NAD^+ + 8H_2O + 8CoA-SH \longrightarrow
$$

(14.7)

$$
9CH_3\overset{\displaystyle O}{\overset{\|}{C}}-S-CoA + 8FADH_2 + 8NADH + 8H^+
$$

Learning Check 14.1

Capric acid (ten carbon atoms) is converted to acetyl CoA through β-oxidation. What are the yields of the following?

a. acetyl CoA   **b.** FADH$_2$ **c.** NADH

➤ 14.5 The Energy from Fatty Acids

To compare the energy yield from fatty acids with that of glucose, let's continue our discussion with stearic acid, a typical and abundant fatty acid.

We start with the activation of stearic acid by reacting it with coenzyme A to form stearoyl CoA. The energy needed to cause this reaction to occur comes

TABLE 14.1 Energy produced by the oxidation of stearic acid

Oxidation product or step	ATP/unit	Total ATP
Activiation step		−2
9 acetyl CoA	10	90
8 FADH$_2$	1.5	12
8 (NADH + H$^+$)	2.5	20
Total ATP from the 18-carbon fatty acid		120

from the hydrolysis of ATP to AMP and PP$_i$ and the subsequent hydrolysis of PP$_i$ to 2P$_i$. This is equivalent to hydrolyzing two molecules of ATP to ADP.

As a stearoyl CoA molecule (18 carbons) passes through the β-oxidation spiral, 9 acetyl CoA, 8 FADH$_2$, and 8 NADH molecules are produced. Acetyl CoA produced in the fatty acid spiral can enter the citric acid cycle (followed by the electron transport chain), where each molecule of acetyl CoA results in the production of 10 ATP molecules. In addition, when the FADH$_2$ and NADH molecules enter the electron transport chain, each FADH$_2$ yields 1.5 ATP molecules, and each NADH yields 2.5 molecules. The calculations are summarized in ■ Table 14.1, which shows a total of 120 molecules of ATP formed from the 18-carbon fatty acid.

It is instructive to compare the energy yield from the complete oxidation of fatty acids with that obtained from an equivalent amount of glucose because both are important constituents of the diet. In Section 13.8, we saw that the oxidation of a single glucose molecule produces 32 ATP molecules. The complete oxidation of 3 six-carbon glucose molecules (18 carbons) yields 3 × 32, or 96, ATP molecules. Thus, on the basis of equal numbers of carbons, lipids are nearly 25% more efficient than carbohydrates as energy-storage systems.

On a mass basis, the difference is even more striking. One mole of stearic acid weighs 284 g, and, as we have seen, it yields 120 mol of ATP on complete degradation. Three moles of glucose weighs 540 g and yields 96 mol of ATP. On this basis, 284 g of glucose would generate 50 mol of ATP. On an equal mass basis, lipids contain more than twice the energy of carbohydrates. The reason for this difference is that fatty acids are a more reduced form of fuel, with a number of —CH$_2$— groups as plentiful sources of H atoms for the energy-yielding oxidation process. Glucose, in contrast, is already partially oxidized, with an oxygen atom located on each carbon atom.

Learning Check 14.2

Calculate the number of molecules of ATP produced by the complete oxidation of palmitic acid (16 carbon atoms) to carbon dioxide and water.

➤ 14.6 Ketone Bodies

It is generally agreed by nutrition scientists that a diet balanced in carbohydrates and fats as energy sources is best for good health. When such a balanced diet is followed, most of the acetyl CoA produced by fatty acid metabolism is processed through the citric acid cycle.

KEY NUMBERS FOR ATP CALCULATIONS

The calculation of the energy yield in terms of ATP molecules from the metabolism of a fatty acid is a useful exercise because it may help you better understand this metabolic process. The three key substances to account for in such calculations are acetyl CoA, NADH, and FADH$_2$. Because acetyl CoA consists of two carbon atoms transported by coenzyme A, the number of acetyl CoA molecules formed is determined by the following equation:

$$\text{acetyl CoA} = \frac{\text{number of fatty acid carbons}}{2}$$

For example, a ten-carbon fatty acid would produce 10/2, or 5, acetyl CoA molecules.

Every trip through the fatty acid spiral produces one NADH + H$^+$ molecule and one FADH$_2$ molecule. Thus,

Number of trips = molecules NADH = molecules FADH$_2$

Furthermore, the number of trips equals one less than the number of acetyl CoA molecules produced (because the last trip of the spiral produces two molecules of acetyl CoA). Thus, for the ten-carbon fatty acid above, five acetyl CoA molecules are produced during four trips through the spiral. Also, the four trips produce four NADH + H$^+$ molecules and four FADH$_2$ molecules.

Finally, remember the multipliers: Every acetyl CoA gives rise to 10 ATPs, every NADH produces 2.5 ATPs, and every FADH$_2$ gives 1.5 ATPs. The summary for the ten-carbon fatty acid is

(5 acetyl CoA) $\times$ 10	= 50 ATP
(4 NADH + H$^+$) $\times$ 2.5	= 10 ATP
(4 FADH$_2$) $\times$ 1.5	= 6 ATP
Activation step	= −2 ATP
Total ATP	= 64 ATP

Ketone bodies

Three compounds—acetoacetate, β-hydroxybutyrate, and acetone—formed from acetyl CoA.

Under certain conditions, the balance between carbohydrate and fatty acid metabolism is lost. During a fast, for example, fatty acids from stored fats become the body's primary energy source. With a minimum amount of cellular glucose available, the level of glycolysis decreases, and a reduced amount of oxaloacetate is synthesized. In addition, oxaloacetate is used for gluconeogenesis to a greater-than-normal extent as the cells react to make their own glucose. The lack of oxaloacetate reduces the activity of the citric acid cycle. As a result of these events, more acetyl CoA is produced by fatty acid oxidation than can be processed through the citric acid cycle. As the concentration of acetyl CoA builds up, the excess is converted within the liver to three substances called **ketone bodies**—acetoacetate, β-hydroxybutyrate, and acetone.

$$\underset{\text{acetoacetate}}{CH_3-\overset{\overset{O}{\|}}{C}-CH_2-\overset{\overset{O}{\|}}{C}-O^-} \qquad \underset{\substack{\text{β-hydroxybutyrate} \\ \text{(not a ketone)}}}{CH_3-\overset{\overset{OH}{|}}{CH}-CH_2-\overset{\overset{O}{\|}}{C}-O^-} \qquad \underset{\text{acetone}}{CH_3-\overset{\overset{O}{\|}}{C}-CH_3}$$

The ketone bodies are carried by the blood to body tissues, mainly the brain, heart, and skeletal muscle, where they may be oxidized to meet energy needs (see ■ Figure 14.6). You can see from their structural formulas that β-hydroxybutyrate does not actually contain a ketone group, but it is nevertheless included in the designation because it is always produced along with the other two. Under normal conditions, the concentration of ketone bodies in the blood averages a very low 0.5 mg/100 mL.

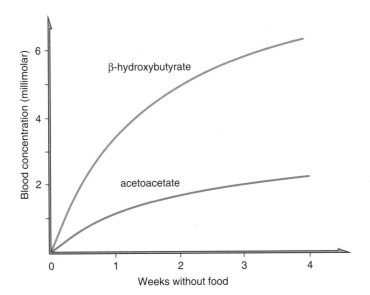

➤ **FIGURE 14.6** The effect of starvation on blood levels of β-hydroxybutyrate and acetoacetate.

Diabetes mellitus, like a long fast or starvation, also produces an imbalance in carbohydrate and lipid metabolism. Even though blood glucose reaches hyperglycemic levels, a deficiency of insulin prevents the glucose from entering tissue cells in sufficient amounts to meet cellular energy needs. The resulting increase in fatty acid metabolism leads to excessive production of acetyl CoA and a substantial increase in the level of ketone bodies in the blood of untreated diabetics. A concentration higher than about 20 mg/100 mL of blood is called **ketonemia** ("ketones in the blood"). At a level of about 70 mg/100 mL of blood, the renal threshold for ketone bodies is exceeded, and ketone bodies are excreted in the urine. This condition is called **ketonuria** ("ketones in the urine"). A routine check of urine for ketone bodies is useful in the diagnosis of diabetes. Occasionally, the concentration of acetone in the blood reaches levels that cause it to be expelled through the lungs, and the odor of acetone can be detected on the breath. When ketonemia, ketonuria, and **acetone breath** exist simultaneously, the condition is called **ketosis**.

Two of the ketone bodies are acids, and their accumulation in the blood as ketosis worsens results in a condition of **acidosis** called **ketoacidosis** because it is caused by ketone bodies in the blood. If ketoacidosis is not controlled, the person becomes severely dehydrated because the kidneys excrete excessive amounts of water in response to low blood pH. Prolonged ketoacidosis leads to general debilitation, coma, and even death.

Patients suffering from diabetes-related ketosis are usually given insulin as a first step in their treatment. The insulin restores normal glucose metabolism and reduces the rate of formation of ketone bodies. If the patient is also suffering severe dehydration, the return to fluid and acid–base balance is speeded up by the intravenous administration of solutions containing sodium bicarbonate.

Ketonemia
An elevated level of ketone bodies in the blood.

Ketonuria
The presence of ketone bodies in the urine.

Acetone breath
A condition in which acetone can be detected in the breath.

Ketosis
A condition in which ketonemia, ketonuria, and acetone breath exist together.

Acidosis
A low blood pH.

Ketoacidosis
A low blood pH due to elevated levels of ketone bodies.

http://www.brookscole.com/chemistry/seager_chemistry
At Web site 14.3, you will find further information on diabetic ketoacidosis.

➤ 14.7 Fatty Acid Synthesis

When organisms (including humans) take in more nutrients than are needed for energy requirements, the excesses are not excreted; they are converted first to fatty acids and then to body fat. Most of the conversion reactions take place in the liver, adipose tissue, and mammary glands. The mammary glands become especially active in the process during lactation.

As is true in the opposing processes of glycolysis and gluconeogenesis, the pathways for the opposing processes of fatty acid degradation and fatty acid synthesis are not simply the reverse of each other. The processes are separate from each other, utilize different enzyme systems, and are even located in different cellular compartments. Degradation by the β-oxidation pathway takes place in cellular mitochondria, whereas biosynthesis of fatty acids occurs in the cytoplasm.

However, one aspect of fatty acid synthesis and fatty acid degradation is the same. Both processes take place in units of two carbon atoms as a result of involving acetyl CoA. Thus, fatty acid chains are built up two carbons at a time during synthesis and broken down two carbons at a time during β-oxidation.

Acetyl CoA is generated inside mitochondria and so must be transported into the cytoplasm if it is to be used to synthesize fatty acids. This transport is done by first reacting acetyl CoA with oxaloacetate (the first step of the citric acid cycle). The product is citrate:

$$\text{acetyl CoA} + \text{oxaloacetate} + H_2O \rightarrow \text{citrate} + \text{CoA}—\text{SH} \qquad (14.8)$$

Mitochondrial membranes contain a citrate transport system that enables excess citrate to move out of the mitochondria into the cytoplasm. Once in the cytoplasm, the citrate reacts to regenerate acetyl CoA and oxaloacetate.

Fatty acid synthesis occurs by way of a rather complex series of reactions catalyzed by a multienzyme complex called the *fatty acid synthetase system*. This system is made up of six enzymes and an additional protein, acyl carrier protein (ACP), to which all intermediates are attached.

The summary equation for the synthesis of palmitic acid from acetyl CoA demonstrates that a great deal of energy, 7 ATP + 14 NADPH (a phosphate derivative of NADH), is required:

$$8CH_3—\overset{\overset{\displaystyle O}{\|}}{C}—S—CoA + 14NADPH + 13H^+ + 7ATP \longrightarrow$$

acetyl CoA

$$(14.9)$$

$$CH_3(CH_2)_{14}COO^- + 8CoA—SH + 6H_2O + 14NADP^+ + 7ADP + 7P_i$$

palmitate

This large input of energy is stored in the synthesized fatty acids and is one of the reasons it is so difficult to lose excess weight due to fat. After synthesis, the fatty acids are incorporated into triglycerides and stored in the form of fat in adipose tissues.

The liver is the most important organ involved in fatty acid and triglyceride synthesis. It is able to modify body fats by lengthening or shortening and saturating or unsaturating the fatty acid chains. The only fatty acids that cannot be synthesized by the body are those that are polyunsaturated. However, linoleic acid (two double bonds) and linolenic acid (three double bonds) from the diet can be converted to other polyunsaturated fatty acids. This utilization of linoleic and linolenic acids as sources of other polyunsaturated fatty acids is the basis for classifying them as essential fatty acids for humans (Section 8.2).

The human body can convert glucose to fatty acids (via acetyl CoA), but it cannot convert fatty acids to glucose. Our cells contain no enzyme that can catalyze the conversion of acetyl CoA to pyruvate, a compound required for gluconeogenesis (Figure 13.13). However, plants and some bacteria do possess such enzymes and convert fats to carbohydrates as part of their normal metabolism.

➤ 14.8 Amino Acid Metabolism

In terms of amount used, the most important function of amino acids is to provide building blocks for the synthesis of proteins in the body. It is estimated that about 75% of amino acid utilization in a normal, healthy adult is for this function. The maintenance of body proteins must occur constantly in the body because tissue proteins break down regularly from normal wear and tear, as well as from diseases and injuries. The amino acids used in this maintenance come from proteins that are eaten and hydrolyzed during digestion, from the body's own degraded tissue, and from the synthesis in the liver of certain amino acids. The amino acids from these three sources constitute what is called the **amino acid pool** of the body. This cellular supply of amino acids is constantly being restocked to allow the synthesis of new proteins and other necessary metabolic processes to take place as needed.

The dynamic process in which body proteins are continuously hydrolyzed and resynthesized is called **protein turnover**. The turnover rate, or life expectancy, of body proteins is a measure of how fast the proteins are broken down and resynthesized. The turnover rate is usually expressed as a half-life. The use of radioactive amino acids has enabled researchers to estimate half-lives by measuring the exchange rate between body proteins and the amino acid pool. For example, the half-life of liver proteins is about 10 days. This means that over a 10-day period, half the proteins in the liver are hydrolyzed to amino acids and replaced by equivalent proteins. Plasma proteins also have a half-life of about 10 days, hemoglobin about 120 days, and muscle protein about 180 days. The half-life of collagen, a protein of connective tissue, is considerably longer—some estimates are as high as 1000 days. Other proteins, particularly enzyme and polypeptide hormones, have much shorter half-lives of only a few minutes. Once it is released from the pancreas, insulin has a half-life estimated to be only 7–10 minutes.

Thus, the stability of body protein is more apparent than real because, as we have seen, synthesized proteins do not become a permanent part of the body structure or function. The frequent turnover of proteins is advantageous, for it allows the body to continually renew important molecules and to respond quickly to its own changing needs.

In addition to tissue protein synthesis, there is a constant draw on the amino acid pool for the synthesis of other nitrogen-containing biomolecules. These molecules include the purine and pyrimidine bases of nucleic acids; heme structures for hemoglobin and myoglobin; choline and ethanolamine, which are the building blocks of phospholipids; and neurotransmitters such as acetylcholine and dopamine. Like proteins, these compounds are also constantly being degraded and replaced. ■ Table 14.2 lists some of

Amino acid pool
The total supply of amino acids in the body.

Protein turnover
The continuing process in which body proteins are hydrolyzed and resynthesized.

TABLE 14.2	Compounds derived from amino acids	
Amino acid	**Product**	**Function**
Tyrosine	dopamine	Neurotransmitter
	norepinephrine	Neurotransmitter, hormone
	epinephrine	Hormone
	melanin	Skin pigmentation
Tryptophan	serotonin	Neurotransmitter
Histidine	histamine	Involved in allergic reactions
Serine	ethanolamine	Required in cephalin synthesis
Cysteine	taurine	A compound of bile salts

➤ **FIGURE 14.7** The major metabolic pathways of amino acids.

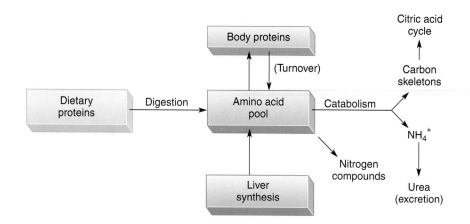

the important nitrogen-containing compounds that are synthesized from amino acids.

Unlike carbohydrates and fatty acids, amino acids in excess of immediate body requirements cannot be stored for later use. They are degraded and, depending on the organism, their nitrogen atoms are either converted to ammonium ions, urea, or uric acid and excreted. Their carbon skeletons are converted to pyruvate, acetyl CoA, or one of the intermediates of the citric acid cycle and used for energy production, the synthesis of glucose through gluconeogenesis (Section 13.10), or conversion to triglycerides. The various pathways involved in amino acid metabolism are summarized in ■ Figure 14.7.

➤ **FIGURE 14.7** The major metabolic pathways of amino acids.

➤ 14.9 Amino Acid Catabolism: The Fate of the Nitrogen Atoms

As previously noted, excess amino acids cannot be stored in the body but are catabolized for energy production, and the nitrogen is either excreted or used to synthesize other nitrogen-containing compounds. In this section, we will study the chemical pathways involved with the nitrogen. The chemical fate of the carbon skeleton of the amino acids is discussed in Section 14.10.

In essence, there are three stages in nitrogen catabolism (see ■ Figure 14.8). These processes (which occur in the liver) are transamination, deamination, and urea formation.

Stage 1: Transamination

In the tissues, amino groups freely move from one amino acid to another under the catalytic influence of enzymes called aminotransferases, or more commonly, **transaminases**. A key reaction for amino acids undergoing catabolism is a **transamination** involving the transfer of amino groups to α-ketoglutarate.

Transaminase
An enzyme that catalyzes the transfer of an amino group.

Transamination
The enzyme-catalyzed transfer of an amino group to a keto acid.

➤ **FIGURE 14.8** An overview of the catabolic fate of nitrogen atoms in amino acids.

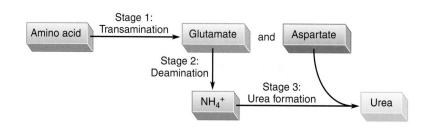

The carbon skeleton of the amino acid remains behind as an α-keto acid:

General reaction:

$$\overset{\overset{\displaystyle NH_3^+}{|}}{R-CH-COO^-} + {}^-OOC-CH_2CH_2-\overset{\overset{\displaystyle O}{||}}{C}-COO^- \xrightarrow{\text{transaminase}} \rightleftharpoons$$

a donor
α-amino acid

α-ketoglutarate
an α-keto acid

(14.10)

$$R-\overset{\overset{\displaystyle O}{||}}{C}-COO^- + {}^-OOC-CH_2CH_2-\overset{\overset{\displaystyle NH_3^+}{|}}{CH}-COO^-$$

a new
α-keto acid

glutamate
the new amino acid

Specific example:

$$CH_3-\overset{\overset{\displaystyle NH_3^+}{|}}{CH}-COO^- + {}^-OOC-CH_2CH_2-\overset{\overset{\displaystyle O}{||}}{C}-COO^- \xrightarrow[\rightleftharpoons]{\text{alanine transaminase}}$$

alanine

α-ketoglutarate

(14.11)

$$CH_3-\overset{\overset{\displaystyle O}{||}}{C}-COO^- + {}^-OOC-CH_2CH_2-\overset{\overset{\displaystyle NH_3^+}{|}}{CH}-COO^-$$

pyruvate

glutamate

Notice that in the reaction, the $-NH_3^+$ group trades places with a carbonyl oxygen.

A second important example of transamination is the production of aspartate, which is used in Stage 3, urea formation. In this case, the amino group is transferred to oxaloacetate:

Specific example:

$$CH_3-\overset{\overset{\displaystyle NH_3^+}{|}}{\underset{\underset{\displaystyle CH_3}{|}}{CH}}-CH-COO^- + {}^-OOC-CH_2-\overset{\overset{\displaystyle O}{||}}{C}-COO^- \rightleftharpoons$$

valine

oxaloacetate

(14.12)

$$CH_3-\underset{\underset{\displaystyle CH_3}{|}}{CH}-\overset{\overset{\displaystyle O}{||}}{C}-COO^- + {}^-OOC-CH_2-\overset{\overset{\displaystyle NH_3^+}{|}}{CH}-COO^-$$

aspartate

The net effect of the transamination reactions in the catabolism of amino acids is to use nitrogen from a variety of amino acids to form glutamate and aspartate. In Section 14.11, we'll see that transamination is also an important method in the biosynthesis of amino acids.

Complete the following transamination reaction:

Learning Check 14.3

$$HO-\underset{\text{tyrosine}}{\boxed{\bigcirc}}-CH_2-\overset{\overset{\displaystyle NH_3^+}{|}}{CH}-COO^- + {}^-OOC-CH_2CH_2-\overset{\overset{\displaystyle O}{||}}{C}-COO^- \rightleftharpoons$$

Stage 2: Deamination

This phase of amino acid catabolism uses the glutamate produced in Stage 1. The enzyme glutamate dehydrogenase catalyzes the removal of the amino

Oxidative deamination

An oxidation process resulting in the removal of an amino group.

group as an ammonium ion and regenerates α-ketoglutarate, which can again participate in the first stage (transamination). This reaction is the principal source of NH_4^+ in humans. Because the deamination results in the oxidation of glutamate, it is called **oxidative deamination**:

$$^-OOC-CH_2CH_2-\overset{\overset{\displaystyle NH_3^+}{|}}{CH}-COO^- + NAD^+ + H_2O \rightleftharpoons$$

glutamate

$$(14.13)$$

$$NH_4^+ + {}^-OOC-CH_2CH_2-\overset{\overset{\displaystyle O}{\|}}{C}-COO^- + NADH + H^+$$

α-ketoglutarate

The NADH produced in this second stage enters the electron transport chain and eventually produces 2.5 ATP molecules. Other amino acids in addition to glutamate may also be catabolized by oxidative deamination. The reactions are catalyzed by enzymes in the liver called amino acid oxidases. In each case, an α-keto acid and NH_4^+ are the products.

Stage 3: Urea Formation

Urea cycle

A metabolic pathway in which ammonia is converted to urea.

The ammonium ions released by the glutamate dehydrogenase reaction are toxic to the body and must be prevented from accumulating. In the **urea cycle**, which occurs only in the liver, ammonium ions are converted to urea (see ■ Figure 14.9). This compound is less toxic than ammonium ions and can be allowed to concentrate until it is convenient to excrete it in urine.

CHEMISTRY AROUND US • 14.1

VEGETARIAN DIETS AND PROTEIN REQUIREMENTS

In Chemistry Around Us 12.1, we suggested that well-designed vegetarian diets can supply the nutritional requirements needed for good health. However, different vegetarian diets present different challenges in this regard. Those following a vegan diet are confronted with the problem of obtaining enough protein and Calories to maintain a healthy body. The Caloric content of vegetables (30–50 Calories per 100 g) is much lower than that of meats (150–300 Cal per 100 g). Thus, on a weight basis, five to six times more vegetables than meat must be consumed to provide equal numbers of Calories. The protein challenge originates from three characteristics of plant protein:

1. Most plant protein is of low biological value.
2. Most plant products contain much less protein than meat (1–2 g per 100 g versus 15–20 g per 100 g).
3. Some types of plant protein are incompletely digested.

Despite these problems, well-designed vegetarian diets usually provide enough Calories and protein for an average adult. Also, the reduced Calories of a vegan diet may be a benefit to some people.

However, everyone is not an average adult, and some special needs should be considered when designing a vegetarian diet. For example, adult males require about 0.8 g of protein and 40 Cal per kilogram of body weight. But a young child may require two to three times that amount for proper growth and development. Pregnant women typically need an additional 10 g of protein and 300 Cal per day, while lactating women need 15 g of protein and 300 Cal more per day than an average adult female. Unless these special requirements are considered when planning a diet, individuals in these categories run the risk of protein-energy malnutrition. The following information indicates that these special needs are not always addressed completely. It has been found that children of vegetarian mothers have a lower birth weight than children of mothers who consume a mixed diet. Also, children on vegetarian diets generally have a lower growth rate for the first 5 years, but generally catch up by the age of 10.

When a vegetarian diet is being planned for small children, several things should be kept in mind:

1. Eggs and milk should be included in the diet, if possible.
2. The diet should include liberal amounts of high-energy vegetable foods, such as nuts, grains, dried beans, and dried fruits.
3. The diet should be rich in high-protein vegetables that have complementary amino acid patterns, and any missing amino acids should be provided by supplements.

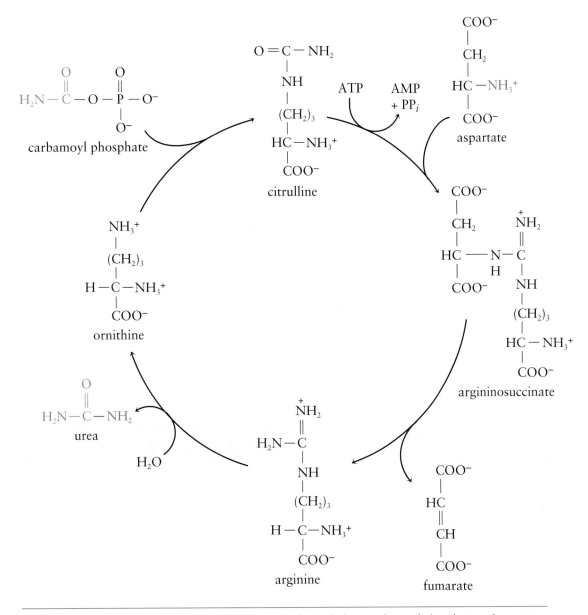

➤ **FIGURE 14.9** The urea cycle. Nitrogen atoms enter the cycle from carbamoyl phosphate and aspartate.

The urea cycle processes the ammonium ions in the form of carbamoyl phosphate, the fuel for the urea cycle. This compound is synthesized in the mitochondria from ammonium ions and carbon dioxide.

$$NH_4^+ + CO_2 + 2ATP + H_2O \xrightarrow[\text{synthetase}]{\substack{\text{carbamoyl} \\ \text{phosphate}}} H_2N-\overset{O}{\overset{\|}{C}}-O-\overset{O}{\overset{\|}{\underset{O^-}{P}}}-O^- + 2ADP + P_i + 3H^+ \quad (14.14)$$

carbamoyl phosphate

As we see, the energy for the synthesis of one molecule of carbamoyl phosphate is derived from two molecules of ATP. Figure 14.9 summarizes the reactions of the urea cycle. The net reaction for carbamoyl phosphate formation and the urea cycle is

$$NH_4^+ + CO_2 + 3ATP + 2H_2O +$$

$$\underset{\text{aspartate}}{\text{COO}^- \mid CH_2 \mid CH-NH_3^+ \mid COO^-} \longrightarrow$$

(14.15)

$$H_2N-\overset{O}{\overset{\|}{C}}-NH_2 + \underset{\text{fumarate}}{\overset{H}{\underset{^-OOC}{\diagup}}C=C\overset{COO^-}{\underset{H}{\diagup}}} + 5H^+ + 2ADP + 2P_i + AMP + PP_i$$

urea   fumarate

After urea is formed, it diffuses out of liver cells into the blood, the kidneys filter it out, and it is excreted in the urine. Normal urine from an adult usually contains about 25–30 g of urea daily, although the exact amount varies with the protein content of the diet. The direct excretion of NH_4^+ accounts for a small but important amount of total urinary nitrogen. Ammonium ions can be excreted along with acidic ions, a mechanism that helps the kidneys control the acid–base balance of body fluids.

CHEMISTRY AROUND US • 14.2

PHENYLKETONURIA (PKU)

At least 35 different hereditary changes in amino acid metabolism have been identified. One of the more well known is an inherited disease associated with abnormal aromatic amino acid metabolism. In *phenylketonuria (PKU)*, there is a lack of the enzyme phenylalanine hydroxylase. As a result, phenylalanine cannot be converted to tyrosine, leading to the accumulation of phenylalanine and its metabolites (phenylpyruvate and phenylacetate) in the tissues and blood:

$$\bigcirc\!\!-CH_2-\overset{O}{\overset{\|}{C}}-COO^-$$

phenylpyruvate

$$\bigcirc\!\!-CH_2-COO^-$$

phenylacetate

Abnormally high levels of phenylpyruvate and phenylacetate damage developing brain cells, causing severe mental retardation. It is estimated that 1 out of every 20,000 newborns is afflicted with PKU. A federal regulation requires that all infants be tested for this disease, which can be detected very easily by analyzing blood or urine samples for phenylalanine or phenylpyruvic acid.

The disease is treated by restricting the amount of phenylalanine in the diet of the infant starting 2–3 weeks after birth. The diet is also supplemented with tyrosine because it is not being formed in the PKU patient. Strict dietary management is not as critical after the age of 6 because the brain has matured by this time and is not as susceptible to the toxic effects of phenylalanine metabolites.

http://www.brookscole.com/chemistry/seager_chemistry
Web site 14.4 provides information and news about PKU.

Infants with PKU have an excellent chance for a normal healthy life, if the disease is detected and treated early. Here a nurse is taking blood samples from an infant's foot to be used in a simple chemical test for PKU.

a. What is the primary function of the urea cycle?
b. What is the fuel for the urea cycle?

➤ 14.10 Amino Acid Catabolism: The Fate of the Carbon Skeleton

After removal of the amino group of an amino acid is accomplished by transamination or oxidative deamination (Section 14.4), the remaining carbon skeleton undergoes catabolism and is converted into one of several products. We will not study the various pathways followed, but we will focus on the final form the skeleton assumes and the metabolic fate of these forms.

After the amino group is gone, the skeletons of all 20 amino acids are degraded into either pyruvate, acetyl CoA, acetoacetyl CoA (which is degraded to acetyl CoA), or various substances that are intermediates in the citric acid cycle. As shown in ■ Figure 14.10, all these degraded forms of the carbon skeletons are a part of or can enter the citric acid cycle and thus may be very important in the production of energy.

Even though the degraded carbon skeletons can be used to produce energy via the citric acid cycle, not all are used this way because cellular needs might dictate other uses. For example, once a carbon skeleton is catabolized to pyruvate, there are two possible uses for it: the production of energy or the synthesis of glucose through gluconeogenesis (Section 13.10). Amino acids that yield a skeleton that is degraded into pyruvate or intermediates of the citric acid cycle can be used to make glucose and are classified as **glucogenic amino acids**. Amino acids with skeletons that are degraded into acetyl CoA or acetoacetyl CoA cannot be converted into glucose but can, in addition to energy production, be used to make ketone bodies and fatty acids. Such amino acids are classified as **ketogenic**. Figure 14.10 shows this catabolic classification of the 20 amino acids. Notice that some amino acids appear in more than one location in the diagram.

Glucogenic amino acid
An amino acid whose carbon skeleton can be converted metabolically to an intermediate used in the synthesis of glucose.

Ketogenic amino acid
An amino acid whose carbon skeleton can be converted metabolically to acetyl CoA or acetoacetyl CoA.

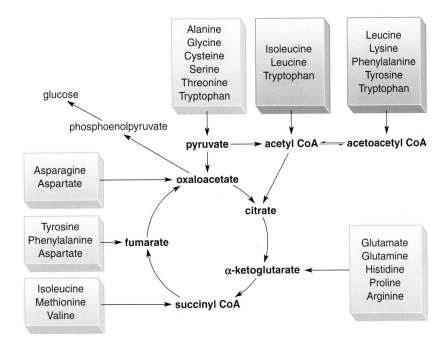

➤ **FIGURE 14.10** Fates of the carbon atoms of amino acids. Glucogenic amino acids are shaded yellow, and ketogenic amino acids are shaded red.

CHEMISTRY AROUND US • 14.3

PROTEIN AND AMINO ACID SUPPLEMENTS

Driven in part by advertisements in newspapers, magazines, and other mass media, the market for protein and amino acid supplements has grown significantly in recent years. People purchase and use these products in the (usually unfounded) belief that numerous benefits will be derived by the practice.

Athletes and body builders take protein supplements to enhance muscle building. Dieters trying to lose weight by reducing their Caloric intake use protein supplements in an attempt to prevent their Calorie-deprived bodies from using vital body proteins as a source of energy. Some people believe that protein supplements, often taken in the form of gelatin, will add strength to their hair and fingernails. No scientific evidence supports any of these claims.

Beliefs in the ability of individual amino acid supplements to cure a number of ailments are also widely held, and in a few instances, these beliefs are supported to some extent by research. Reports in the media of the effectiveness of lysine to prevent and treat herpes infections that cause mouth and genital cold sores has led to the widespread public use of lysine supplements. Similar reports that touted the use and effectiveness of tryptophan to treat pain, depression, and insomnia had a similar effect on the sales of tryptophan supplements. In both of these instances, the reports were based on preliminary research results. A careful analysis of the research indicates that lysine may suppress herpes infections in some individuals and seems to be safe when no more than 3 total per day are taken in divided doses with meals. Tryptophan seems to be effective in the treatment of pain and insomnia. However, these uses are still experimental, and the FDA issued a recall of all tryptophan-containing products following an outbreak of a rare blood disorder among more than 1500 people who took tryptophan supplements. There are also claims that phenylalanine and leucine are effective pain relievers, and that tyrosine cures depression and insomnia. No scientific studies are available to support these claims.

Protein and amino acid supplements are quite expensive, and in some cases their use can adversely affect your health. The FDA has stated that the ingestion of large doses of amino acids is dangerous, especially if only one type of amino acid is involved. Animal studies indicate that amino acid imbalance causes a variety of ailments. Food protein is the best source of a balanced amino acid supply for the body. It is not surprising that the advice most often given by nutrition scientists is to save the money you might be tempted to spend on protein or amino acid supplements, and concentrate instead on eating a well-balanced diet that provides a balance of proteins and other nutrients needed by the body.

http://www.brookscole.com/chemistry/seager_chemistry
For additional information on amino acid supplements, see Web site 14.5.

Protein and amino acid supplements are common items in drug stores and health food stores.

Glucogenic amino acids play a special role in the metabolic changes that occur during starvation. To maintain the blood glucose level, the body first uses the glycogen stored in the liver. After 12–18 hours without food, the glycogen supplies are exhausted. Glucose is then synthesized from glucogenic amino acids through the process of gluconeogenesis. Hydrolysis of proteins from body tissues, particularly from muscle, supplies the amino acids for gluconeogenesis. Because each protein in the body has an important function, the body can meet its energy needs for only a limited time by sacrificing proteins.

▶ 14.11 Amino Acid Biosynthesis

As we have already seen, the liver is highly active biochemically. In biosynthesis, it is responsible for producing most of the amino acids the body can synthesize. Amino acids that can be synthesized in the amounts needed by

TABLE 14.3 Essential and nonessential amino acids	
Essential	**Nonessential**
Histidine	Alanine
Isoleucine	Arginine
Leucine	Asparagine
Lysine	Aspartate
Methionine	Cysteine
Phenylalanine	Glutamate
Threonine	Glutamine
Tryptophan	Glycine
Valine	Proline
	Serine
	Tyrosine

the body are called **nonessential amino acids** (see ■ Table 14.3). As explained in Chapter 12, essential amino acids either cannot be made or cannot be made in large enough amounts to meet bodily needs and must be included in our diet.

Nonessential amino acid
An amino acid that can be synthesized within the body in adequate amounts.

The key starting materials for the synthesis of ten nonessential amino acids are intermediates of the glycolysis pathway and the citric acid cycle (see ■ Figure 14.11). Tyrosine, the only nonessential amino acid with an aromatic side chain, is produced from the essential amino acid phenylalanine in a conversion that requires a single oxidation step catalyzed by the enzyme phenylalanine hydroxylase:

$$
\text{phenylalanine} + O_2 + \text{NADPH} + H^+ \xrightarrow[\text{hydroxylase}]{\text{phenylalanine}} \text{tyrosine} + \text{NADP}^+ + H_2O \tag{14.16}
$$

Three nonessential amino acids (glutamate, alanine, and aspartate) are synthesized from α-keto acids via reactions catalyzed by transaminases. For example, alanine is produced from pyruvate and glutamate (which furnishes the amino group):

$$
\underset{\text{pyruvate}}{CH_3-\overset{O}{\overset{\|}{C}}-COO^-} + \text{glutamate} \underset{\text{transaminase}}{\overset{\text{glutamic pyruvic}}{\rightleftharpoons}} \underset{\text{alanine}}{CH_3-\overset{NH_3^+}{\overset{|}{C}H}-COO^-} + \text{α-ketoglutarate} \tag{14.17}
$$

Thus, we see that transaminases perform two vital functions in amino acid metabolism. By taking part in the biosynthesis of nonessential amino acids, they provide a means to help readjust the relative proportions of amino acids to meet the particular needs of the body. This is a vital function because most of our diets do not contain amino acids in the exact proportions the body requires. Also, as we noted in Section 14.9, transamination reactions allow the nitrogen atoms of all amino acids to be transferred to

➤ **FIGURE 14.11** A summary of amino acid biosynthesis.

α-keto acids to form glutamate and aspartate when disposal of nitrogen is necessary.

Two other nonessential amino acids, asparagine and glutamine, are formed from aspartate and glutamate by reaction of the side-chain carboxylate groups with ammonium ions:

$$^-O - \overset{\overset{\text{O}}{\|}}{C} - CH_2 - \overset{\overset{NH_3^+}{|}}{CH} - COO^- + NH_4^+ \xrightarrow{\quad ATP \quad ADP + P_i \quad} H_2N - \overset{\overset{\text{O}}{\|}}{C} - CH_2 - \overset{\overset{NH_3^+}{|}}{CH} - COO^- \quad (14.18)$$

aspartate   asparagine

$$^-O - \overset{\overset{\text{O}}{\|}}{C} - CH_2 - CH_2 - \overset{\overset{NH_3^+}{|}}{CH} - COO^- + NH_4^+ \xrightarrow{\quad ATP \quad ADP + P_i \quad} H_2N - \overset{\overset{\text{O}}{\|}}{C} - CH_2CH_2 - \overset{\overset{NH_3^+}{|}}{CH} - COO^- \quad (14.19)$$

glutamate   glutamine

So far, we have shown how six of the eleven nonessential amino acids are synthesized. The syntheses of the remaining five are more complex, and we will not consider them here.

FOR FUTURE REFERENCE   DIETARY SOURCES OF RIBOFLAVIN

Riboflavin, like niacin, functions in the body in the form of two coenzymes, FMN and FAD. These enzymes are involved in processes that release energy from nutrients. The vitamin is also important in maintaining normal vision and healthy skin. The following chart lists selected foods that provide a minimum of 0.1 mg of riboflavin per serving. It is obvious from the chart that the best sources are dairy foods and meat.

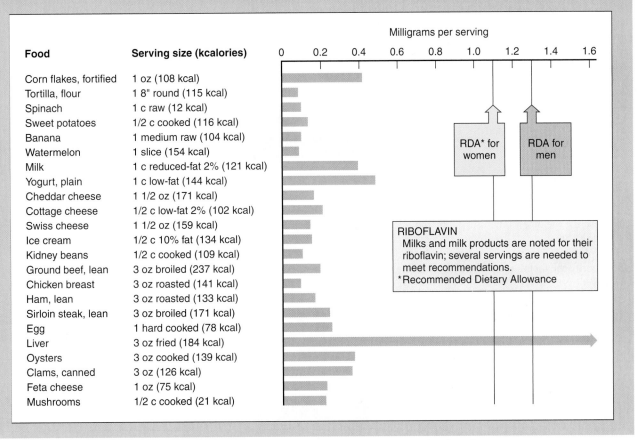

Food	Serving size (kcalories)
Corn flakes, fortified	1 oz (108 kcal)
Tortilla, flour	1 8" round (115 kcal)
Spinach	1 c raw (12 kcal)
Sweet potatoes	1/2 c cooked (116 kcal)
Banana	1 medium raw (104 kcal)
Watermelon	1 slice (154 kcal)
Milk	1 c reduced-fat 2% (121 kcal)
Yogurt, plain	1 c low-fat (144 kcal)
Cheddar cheese	1 1/2 oz (171 kcal)
Cottage cheese	1/2 c low-fat 2% (102 kcal)
Swiss cheese	1 1/2 oz (159 kcal)
Ice cream	1/2 c 10% fat (134 kcal)
Kidney beans	1/2 c cooked (109 kcal)
Ground beef, lean	3 oz broiled (237 kcal)
Chicken breast	3 oz roasted (141 kcal)
Ham, lean	3 oz roasted (133 kcal)
Sirloin steak, lean	3 oz broiled (171 kcal)
Egg	1 hard cooked (78 kcal)
Liver	3 oz fried (184 kcal)
Oysters	3 oz cooked (139 kcal)
Clams, canned	3 oz (126 kcal)
Feta cheese	1 oz (75 kcal)
Mushrooms	1/2 c cooked (21 kcal)

RDA* for women

RDA for men

RIBOFLAVIN
Milks and milk products are noted for their riboflavin; several servings are needed to meet recommendations.
*Recommended Dietary Allowance

SOURCE: Adapted from Whitney, E.N.; Rolfes, S.R. *Understanding Nutrition,* 8th ed. Belmont, CA: West/Wadsworth, 1999.

CONCEPT SUMMARY

Blood Lipids. Fatty acids, glycerol, and monoglycerides that result from lipid digestion are absorbed into the lymph, recombined into triglycerides, and deposited into the bloodstream as lipoprotein aggregates.

Fat Mobilization. When stored fats are needed for energy, hydrolysis reactions liberate the fatty acids and glycerol. These component substances of fat then enter the bloodstream and travel to tissues, where they are utilized.

Glycerol Metabolism. The glycerol available from fat mobilization is first phosphorylated and then oxidized to dihydroxyacetone phosphate, an intermediate of glycosis. By having an entry point into the glycolysis pathway, glycerol can ultimately be converted into glucose or oxidized to CO_2 and H_2O.

The Oxidation of Fatty Acids. The catabolism of fatty acids begins in the cytoplasm, where they are activated by combining with CoA—SH. After being transported into mitochondria, the degradation of fatty acids occurs by the β-oxidation pathway. The β-oxidation pathway produces reduced coenzymes (FADH$_2$ and NADH) and cleaves the fatty acid chain into two-carbon fragments bound to coenzyme A (acetyl CoA).

The Energy from Fatty Acids. The energy available from a fatty acid for ATP synthesis can be calculated by determining the amount of acetyl CoA, FADH$_2$, and NADH produced. On an equal mass basis, fatty acids contain more than twice the energy of glucose.

Ketone Bodies. Acetoacetate, β-hydroxybutyrate, and acetone are known as ketone bodies. They are synthesized in the liver from acetyl CoA. During starvation and in unchecked diabetes, the level of ketone bodies becomes very high, leading to several conditions: ketonemia, ketonuria, acetone breath, and ketosis.

Fatty Acid Synthesis. This process occurs in the cytoplasm and uses acetyl CoA as a starting material. Two-carbon fragments are added to growing fatty acid molecules as energy is supplied by ATP and NADPH.

Amino Acid Metabolism. Amino acids and proteins are not stored within the body. Amino acids absorbed from digested food or synthesized in the body become a part of a temporary supply or pool that may be used in metabolic processes. The turnover, or life expectancy, of proteins is usually expressed in half-lives. Protein half-lives range from several minutes to hundreds of days. Some amino acids in the body are converted into relatively simple but vital nitrogen-containing compounds. The catabolism of amino acids produces intermediates used for energy production, synthesis of glucose, or formation of triglycerides.

Amino Acid Catabolism: The Fate of the Nitrogen Atoms. As an amino acid undergoes catabolism, the amino group may be transferred to an α-keto acid through a process called transamination. In the process of deamination, an amino acid is converted into a keto acid and ammonia. The urea cycle converts toxic ammonia molecules into urea for excretion.

Amino Acid Catabolism: The Fate of the Carbon Skeleton. Amino acids are classified as glucogenic or ketogenic based on their catabolic products. Glucogenic amino acids are degraded into pyruvate or intermediates of the citric acid cycle and can be used for glucose synthesis. Ketogenic amino acids are degraded into acetoacetyl CoA and acetyl CoA.

Amino Acid Biosynthesis. The body can synthesize amino acids, known as nonessential amino acids. The key starting materials are intermediates of the glycolysis pathway and the citric acid cycle, and phenylalanine in the case of tyrosine.

LEARNING OBJECTIVES

When you have completed your study of this chapter, you should be able to:

1. Describe the digestion, absorption, and distribution of lipids in the body.
2. Explain what happens during fat mobilization.
3. Identify the metabolic pathway by which glycerol is catabolized.
4. Outline the steps of β-oxidation for fatty acids.
5. Determine the amount of ATP produced by the complete catabolism of a fatty acid.
6. Name the three ketone bodies and list the conditions that cause their overproduction.
7. Describe the pathway for fatty acid synthesis.
8. Describe the source and function of the body's amino acid pool.
9. Write equations for transamination and deamination reactions.
10. Explain the overall results of the urea cycle.
11. Describe how amino acids can be used for energy production, the synthesis of triglycerides, and gluconeogenesis.
12. State the relationship between intermediates of carbohydrate metabolism and the synthesis of nonessential amino acids.

KEY TERMS AND CONCEPTS

Acetone breath (14.6)
Acidosis (14.6)
Adipose tissue (14.2)
Amino acid pool (14.8)
β-oxidation process (14.4)
Chylomicron (14.1)
Fat mobilization (14.2)

Glucogenic amino acid (14.10)
Ketoacidosis (14.6)
Ketogenic amino acid (14.10)
Ketone bodies (14.6)
Ketonemia (14.6)
Ketonuria (14.6)
Ketosis (14.6)

Nonessential amino acid (14.11)
Oxidative deamination (14.9)
Protein turnover (14.8)
Transaminase (14.9)
Transamination (14.9)
Urea cycle (14.9)

KEY REACTIONS

1. Digestion of triglycerides (Section 14.1):

$$\text{triglyceride} + H_2O \xrightarrow{\text{lipase}} \text{fatty acids} + \text{glycerol} + \text{monoglycerides}$$

2. Fat mobilization (Section 14.2):

$$\text{triglyceride} + 3H_2O \xrightarrow{\text{lipase}} 3 \text{ fatty acids} + \text{glycerol}$$

3. Activation of fatty acids (Section 14.4):

$$R-\overset{\overset{\displaystyle O}{\|}}{C}-OH + HS-CoA \longrightarrow R-\overset{\overset{\displaystyle O}{\|}}{C}-S-CoA + H_2O$$

fatty acid — fatty acyl CoA

$$ATP \qquad AMP + PP_i$$

Reaction 14.2

4. Oxidation of fatty acids (Section 14.4):

$$R-CH_2CH_2-\overset{\overset{\displaystyle O}{\|}}{C}-S-CoA + NAD^+ + FAD + H_2O + CoA-SH \longrightarrow$$

$$R-\overset{\overset{\displaystyle O}{\|}}{C}-S-CoA + CH_3\overset{\overset{\displaystyle O}{\|}}{C}-S-CoA + NADH + H^+ + FADH_2$$

Reaction 14.5

5. Fatty acid synthesis (Section 14.7):

$$8CH_3-\overset{\overset{\displaystyle O}{\|}}{C}-S-CoA + 14NADPH + 13H^+ + 7ATP \longrightarrow$$

$$CH_3(CH_2)_{14}COO^- + 8CoA-SH + 6H_2O + 14NADP^+ + 7ADP + 7P_i$$

Reaction 14.9

6. Transamination (Section 14.9):

$$R-\overset{\overset{\displaystyle NH_3^+}{|}}{CH}-COO^- + {}^-OOC-CH_2CH_2-\overset{\overset{\displaystyle O}{\|}}{C}-COO^- \rightleftharpoons R-\overset{\overset{\displaystyle O}{\|}}{C}-COO^- + {}^-OOC-CH_2CH_2-\overset{\overset{\displaystyle NH_3^+}{|}}{CH}-COO^-$$

Reaction 14.10

7. Oxidative deamination (Section 14.9):

$$^-OOC-CH_2CH_2-\overset{\overset{\displaystyle NH_3^+}{|}}{CH}-COO^- + NAD^+ + H_2O \rightleftharpoons NH_4^+ + {}^-OOC-CH_2CH_2-\overset{\overset{\displaystyle O}{\|}}{C}-COO^- + NADH + H^+$$

Reaction 14.11

8. Urea formation (Section 14.9):

$$NH_4^+ + CO_2 + 3ATP + 2H_2O + \underset{\begin{matrix}|\\CH_2\\|\\CH-NH_3^+\\|\\COO^-\\ \text{aspartate}\end{matrix}}{COO^-} \longrightarrow \underset{\text{urea}}{H_2N-\overset{\overset{\displaystyle O}{\|}}{C}-NH_2} + \underset{\text{fumarate}}{\overset{H}{\underset{{}^-OOC}{}}C=C\overset{COO^-}{\underset{H}{}}} + 2H^+ + 2ADP + 2P_i + AMP + PP_i$$

Reaction 14.15

EXERCISES

BLOOD LIPIDS (SECTION 14.1)

14.1 How does the Caloric value of fat compare with the Caloric value of glycogen and starch?

14.2 What are the products of triglyceride digestion?

14.3 What happens to the products of triglyceride digestion as they pass into the cells of the intestinal mucosa?

14.4 What are chylomicrons?

14.5 Describe what happens to the level of blood lipids after a meal rich in fat.

14.6 List the major types of lipoproteins.

FAT MOBILIZATION (SECTION 14.2)

14.7 Describe what is meant by the term *fat mobilization*.

14.8 Describe the relationship between epinephrine and fat mobilization.

14.9 Explain why fatty acids cannot serve as fuel for the brain.

14.10 Which cells utilize fatty acids over glucose to satisfy their energy needs?

GLYCEROL METABOLISM (SECTION 14.3)

14.11 At what point does glycerol enter the glycolysis pathway?

14.12 Describe the two fates of glycerol after it has been converted to an intermediate of glycolysis.

THE OXIDATION OF FATTY ACIDS (SECTION 14.4)

14.13 Where in the cell does fatty acid catabolism take place?

14.14 How is the fatty acid prepared for catabolism? Where in the cell does fatty acid activation take place?

14.15 Why is the oxidation of fatty acids referred to as β-oxidation?

14.16 Describe the role of FAD and NAD^+ in the oxidation of fatty acids.

14.17 Label the α- and β-positions in the following acids:

a. $CH_3CH_2CH_2CH_2CH_2 - \overset{\overset{O}{\|}}{C} - OH$

b. $CH_3(CH_2)_6\overset{\overset{OH}{|}}{C}H - CH_2 - \overset{\overset{O}{\|}}{C} - OH$

14.18 Write an equation for the activation of palmitic acid (16 carbons).

14.19 Write formulas for the products that result when activated palmitic acid is taken through the following:
a. One trip of the fatty acid spiral
b. Two trips of the fatty acid spiral

14.20 Why is the oxidation of fatty acids referred to as the fatty acid spiral rather than the fatty acid cycle?

14.21 How many molecules of acetyl CoA are produced in the catabolism of a molecule of myristic acid, a 14-carbon carboxylic acid? How many times does the spiral reaction sequence occur?

14.22 Complete the following equations for one turn of the fatty acid spiral:

a. $CH_3CH_2CH_2CH_2CH_2 - \overset{\overset{O}{\|}}{C} - S - CoA + FAD \longrightarrow$
___ + ___

b. ____ + H_2O → ___
c. ____ + NAD^+ → ___ + ___ + ___
d. ____ + CoA—SH → ___ + ___

14.23 Outline the four reactions that occur in the fatty acid spiral.

THE ENERGY FROM FATTY ACIDS (SECTION 14.5)

14.24 Determine the ATP yield per molecule of acetyl CoA that enters the citric acid cycle.

14.25 What is the energy cost in ATP molecules to activate a fatty acid?

14.26 Calculate the number of ATP molecules that can be produced from the complete oxidation of a ten-carbon fatty acid to carbon dioxide and water.

14.27 Compare the energy available (in ATP molecules) from a six-carbon fatty acid to the energy from a molecule of glucose.

KETONE BODIES (SECTION 14.6)

14.28 List the three substances known as ketone bodies.

14.29 Explain how excessive ketone bodies may form in the following:
a. During starvation
b. In patients with diabetes mellitus

14.30 Where are ketone bodies formed? What tissues utilize ketone bodies to meet energy needs?

14.31 Differentiate between (a) ketonemia, (b) ketonuria, (c) acetone breath, and (d) ketosis.

14.32 Why is ketosis frequently accompanied by acidosis?

FATTY ACID SYNTHESIS (SECTION 14.7)

14.33 Where within a liver cell does fatty acid synthesis occur?

14.34 What substance supplies the carbons for fatty acid synthesis?

14.35 What substances furnish the energy for fatty acid synthesis?

14.36 Why can the liver convert glucose to fatty acids but cannot convert fatty acids to glucose?

AMINO ACID METABOLISM (SECTION 14.8)

14.37 List three vital functions served by amino acids in the body.

14.38 What name is given to the cellular supply of amino acids?

14.39 What are the sources of amino acids in the pool?

14.40 What term is used to denote the breakdown and resynthesis of body proteins?

14.41 What percentage of amino acids is used to synthesize proteins?

14.42 Name four important biomolecules other than proteins that are synthesized from amino acids.

14.43 Compare the half-life of plasma proteins with the half-life of enzymes.

AMINO ACID CATABOLISM: THE FATE OF THE NITROGEN ATOMS (SECTION 14.9)

14.44 Match the terms *transamination* and *deamination* to the following:
a. An amino group is removed from an amino acid and donated to an α-keto acid.
b. Ammonium ion is produced.
c. New amino acids are synthesized from other amino acids.
d. A keto acid is produced from an amino acid.

14.45 Complete the following transamination reaction and name the amino acid that is produced:

$$\text{⬡—CH}_2\text{—C—COO}^- + \text{H}_3\overset{+}{\text{N}}\text{—CH—COO}^- \xrightarrow{\text{transaminase}}$$

with O double-bonded to the C, and the right-hand structure having side chain:
CH₂ — CH₂ — COO⁻

14.46 Write an equation to illustrate transamination between leucine and pyruvate.

14.47 Complete the following oxidative deamination reaction:

$$\text{⬡—CH}_2\text{—CH—COO}^- + \text{NAD}^+ + \text{H}_2\text{O} \xrightarrow{\text{amino acid oxidase}}$$

with NH₃⁺ on the CH.

14.48 Write an equation to illustrate the oxidative deamination of glycine.

14.49 By writing two reactions, the first a transamination and the second an oxidative deamination, show how the amino group of alanine can be removed as NH_4^+.

14.50 In what organ of the body is urea synthesized?

14.51 What are the sources of the carbon atom and two nitrogen atoms in urea?

14.52 Name the compound that enters the urea cycle by combining with ornithine.

14.53 How much energy (in number of ATP molecules) is expended to synthesize a molecule of urea?

14.54 Write a summary equation of the formation of urea.

AMINO ACID CATABOLISM: THE FATE OF THE CARBON SKELETON (SECTION 14.10)

14.55 Differentiate between glucogenic and ketogenic amino acids.

14.56 Is any amino acid both glucogenic and ketogenic?

14.57 Identify the five locations at which the carbon skeletons of amino acids enter the citric acid cycle.

14.58 Based on the following conversions, classify each amino acid as glucogenic or ketogenic.
a. aspartate → oxaloacetate
b. leucine → acetyle CoA
c. tyrosine → acetoacetyl CoA

14.59 What special role is played by glucogenic amino acids during fasting or starvation?

AMINO ACID BIOSYNTHESIS (SECTION 14.11)

14.60 Differentiate between essential and nonessential amino acids.

14.61 What amino acids are synthesized from glutamate?

14.62 List two general sources of intermediates for the biosynthesis of amino acids.

CHEMISTRY AROUND US AND OVER THE COUNTER

14.63 What is the relationship between cholesterol, LDL, and heart attacks?

14.64 List three essential roles served by cholesterol.

14.65 What lifestyle changes can reduce the level of LDL in the blood?

14.66 Why may vegetarians have the problem of obtaining enough Calories to maintain a healthy body?

14.67 List three categories of individuals that may run the risk of protein-energy malnutrition while on vegetarian diets.

14.68 What is the best source of a balanced amino acid supply for the body?

14.69 What enzyme is lacking in persons with PKU?

14.70 Describe the treatment used for PKU.

POINTS TO PONDER AND PURSUE

14.1 Cellular reactions for the biosynthesis and catabolism of amino acids are not simply the reversals of the same metabolic pathways. Why is this an advantage for the cell?

14.2 Why do you think the half-lives of enzymes are comparatively short?

14.3 Why do you think the half-life of collagen is comparatively long?

14.4 How might HDL levels suggest a partial explanation for the observation that women as a group have fewer heart attacks than do men as a group?

14.5 A student calculates the net yield of ATP molecules from a 12-carbon fatty acid to be 80. What did the student probably overlook?

14.6 Using Web site 14.3, what condition may cause an excessive loss of free water and electrolytes?

14.7 At Web site 14.4, under the heading "What is PKU?", what is the advice on how long a special diet should be maintained?

http://www.brookscole.com/chemistry/seager_chemistry

Body Fluids

Blood cells of the body are the focus of work done by blood bank technologists and hematology laboratory technicians. Because they perform blood cell counts and identify the kinds of blood cells found in blood, bone marrow, and other body fluids, the technicians are often involved in the diagnosis of leukemias, anemias, infections, and other diseases that affect blood cells. Blood chemistry is one of the major topics of this chapter.

http://www.brookscole.com/chemistry/seager_chemistry
Web site 15.1 offers career information on blood bank technologists.

CHAPTER OUTLINE

o carry out its complex function, a living cell requires a steady supply of reactants, such as nutrients and oxygen. A cell also requires a reliable system for removing the resulting waste products—carbon dioxide and water, for example. Both of these conditions are met by the simple process of diffusion. Reactants can diffuse into the cell through the thin cell membrane, and the products of cell function can diffuse out of the cell and be absorbed by the surrounding liquid.

However, a complex organism, which consists of many closely packed cells, cannot depend solely on diffusion. Some circulating system must bring necessary materials to the cells and carry away the wastes. Otherwise, the liquid surrounding the cells would soon be depleted of reactants and saturated with wastes.

➤ 15.1 A Comparison of Body Fluids

The average adult body contains about 42 L of fluids, which accounts for two-thirds of the total body weight. Most of these fluids are compartmentalized in three regions of the body: the interior of cells, tissue spaces between cells (including the lymph vessels), and the blood vessels. The majority, about 28 L, is located inside the cells and is called **intracellular fluid.** All body fluids not located inside the cells are collectively known as **extracellular fluids.** Thus, fluids in two of the fluid compartments are extracellular: the **interstitial fluid,** which fills the space between tissue cells and moves in lymph vessels, and the fluid of the bloodstream—**plasma.** Interstitial fluid constitutes about 20% (10.5 L) of the total body fluid, and the plasma (nearly 3.5 L) makes up about 7% of the total. Other body fluids that occur in lesser amounts are urine, digestive juices, and cerebrospinal fluid. The distribution of the various body fluids is illustrated in ■ Figure 15.1. Extracellular fluid provides a relatively constant environment for the cells and transports substances to and from the cells. Intracellular fluid, on the other hand, is the medium in which the vital life-maintaining reactions take place.

Chemically, the two extracellular fluids, plasma and interstitial fluid, are nearly identical. The only difference is that plasma contains an appreciable amount of protein not found in interstitial fluid. Intracellular fluid is chemically different from extracellular fluid. Intracellular fluid contains larger amounts of protein and a much different ionic population (see ■ Figure 15.2). From Figure 15.2, we can see that:

1. The principal cation of plasma and interstitial fluid is Na^+. The principal cation of intracellular fluid is K^+ (actually, about 98% of the body's potassium is found inside cells).
2. The principal anion of extracellular fluid is chloride, Cl^-, whereas phosphate (mainly HPO_4^{2-}) is the main anion inside cells.
3. Intracellular fluid contains more than four times as much protein as plasma, and there is remarkably little protein in interstitial fluid.

Intracellular fluid
Body fluid located inside cells.

Extracellular fluid
Body fluid located outside cells.

Interstitial fluid
The fluid surrounding individual tissue cells.

Plasma
The liquid portion of whole blood.

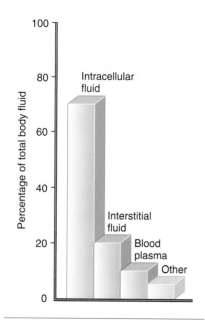

➤ **FIGURE 15.1** The distribution of body fluids.

➤ 15.2 Oxygen and Carbon Dioxide Transport

One of the most important functions of blood is to carry oxygen from the lungs to the tissues. An average adult at rest requires about 350 mL of oxygen per minute. The limited solubility of oxygen in plasma allows only about

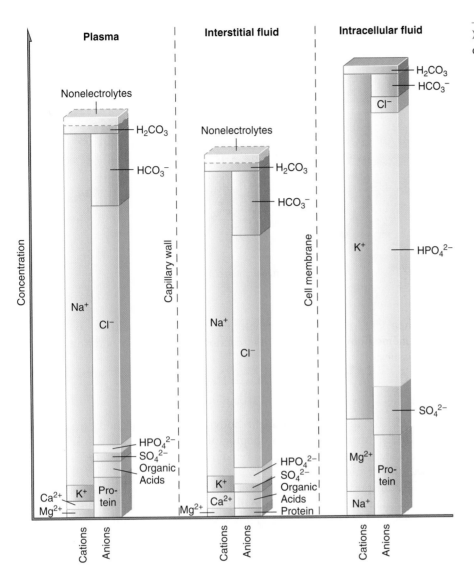

Plasma

Nonelectrolytes

H_2CO_3

HCO_3^-

Na^+

Cl^-

Concentration

HPO_4^{2-}
SO_4^{2-}
Organic Acids

Ca^{2+} | K^+ | Protein
Mg^{2+}

Capillary wall

Cations | Anions

Interstitial fluid

Nonelectrolytes

H_2CO_3

HCO_3^-

Na^+

Cl^-

HPO_4^{2-}
SO_4^{2-}
Organic Acids
Protein

K^+
Mg^{2+} | Ca^{2+}

Cations | Anions

Intracellular fluid

H_2CO_3
HCO_3^-
Cl^-

K^+

HPO_4^{2-}

Cell membrane

SO_4^{2-}

Mg^{2+} | Protein

Na^+

Cations | Anions

> **FIGURE 15.2** The relative chemical composition of the major body fluids.

2% of this total to be dissolved and transported in solution. About 98% of the needed oxygen is carried by red blood cells (see ■ Figure 15.3) in the form of an oxygen-hemoglobin combination called **oxyhemoglobin.** The nonoxygenated hemoglobin is called **deoxyhemoglobin,** or simply **hemoglobin.** Normal human blood contains about 15 g of hemoglobin per 100 mL.

Because of its physiological importance, hemoglobin has been studied extensively. It is known to be a conjugated protein containing four chains (the globin portion of the molecule) and four heme groups. Each heme group consists of an Fe^{2+} ion in the center of a nitrogenous heterocyclic structure (see ■ Figure 15.4). The iron in oxyhemoglobin is bonded at six positions, as shown in Figure 15.4. The four nitrogens are part of the heme group, and the histidine is a part of the globin molecule. The sixth bond is between iron and oxygen. Because each hemoglobin molecule contains four heme groups, it can bind a total of four O_2 molecules. When one of the four heme groups binds an oxygen molecule, the affinity of the other heme groups for oxygen is increased. The reversible binding that takes place between hemoglobin (HHb) and molecular oxygen is represented like this:

$$HHb + O_2 \rightleftharpoons HbO_2^- + H^+$$
$$\text{hemoglobin} \quad \text{oxyhemoglobin}$$

(15.1)

Oxyhemoglobin
Hemoglobin combined with oxygen.

Deoxyhemoglobin (hemoglobin)
Nonoxygenated hemoglobin.

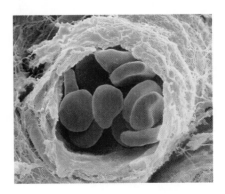

> **FIGURE 15.3** Red blood cells passing through a narrow capillary.

➤ **FIGURE 15.4** The iron in oxyhemoglobin.

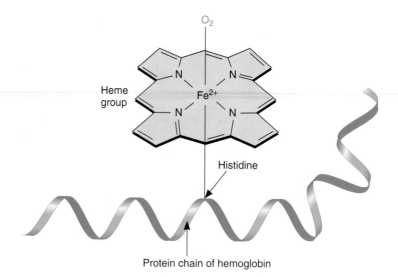

Carbaminohemoglobin
Hemoglobin combined with carbon dioxide.

When CO_2 is present, hemoglobin can combine with it to form **carbaminohemoglobin.** This reversible reaction is favored by conditions found in body tissues:

$$HHb + CO_2 \rightleftharpoons HHbCO_2$$
$$\text{carbaminohemoglobin}$$

(15.2)

About 25% of the total CO_2 in the body is carried from body tissues to the lungs in the form of carbaminohemoglobin. Another 5% dissolves in the plasma and travels in solution; the remaining 70% is transported in the form of bicarbonate ions (HCO_3^-).

The transport of oxygen to tissues begins when inhaled air fills the alveoli of the lungs. These clusters of small sacs are surrounded by many tiny blood capillaries, and it is here that oxygen is absorbed into the blood. ■ Figure 15.5

➤ **FIGURE 15.5** At the lungs, carbon dioxide of red blood cells is exchanged for oxygen.

OVER THE COUNTER • 15.1

EPO: THE ENDURANCE BOOSTER

Erythropoietin (EPO) is a prescription drug used by some athletes who participate in endurance events, such as long-distance running and cross-country skiing. Although using the substance in athletic events is prohibited, EPO is virtually undetectable in drug tests performed on the athletes. EPO is a synthetic hormone that stimulates the body to overproduce red blood cells. The red blood cells contain hemoglobin, which transports oxygen to the cells. An increased number of red blood cells translates into more oxygen being transported to muscle cells, and an increased ability of the muscles to continue doing what would otherwise be anaerobic work.

The drug is usually taken 2–4 weeks before a competition to allow the body to build up the level of red blood cells. The use of the drug accomplishes through chemical means the same thing that was accomplished by the former practice of blood packing. When this practice was allegedly used, an athlete would, over time, have blood removed and stored. The athlete's body would replace the blood that was removed. Then, a few days before the athletic competition, the red blood cells would be removed from the stored blood and injected back into the athlete's bloodstream. The increased number of red blood cells would in turn increase the oxygen-carrying capacity of the blood, thereby giv-

ing the athlete an endurance edge. Suspicions and allegations of blood packing by athletes reached their high point just before and during the 1968 summer Olympic Games, held in Mexico City. The high altitude and thin air of the host city was thought to give an advantage to athletes who normally lived and trained at higher altitudes. It was suggested that blood packing was used by some athletes to compensate. Because any athletes employing blood packing used only their own blood, no blood or urine tests could prove any of the allegations to be true.

Natural EPO is secreted from the kidneys and is transported to the marrow of the bones where it stimulates red blood cell production. Artificial EPO is injected into a vein or just under the skin. The excessive stimulation of red blood cell production causes the blood to become thicker than normal, a condition that can lead to some serious side effects. There is an increased chance for blood clots to form, which can lead to strokes or heart attacks. Any athlete considering the use of EPO or the old technique of blood packing would be well advised to carefully weigh the worth of winning a medal against having a debilitating stroke or heart attack.

http://www.brookscole.com/chemistry/seager_chemistry
At Web site 15.2, more information is provided on the drug EPO.

---

represents the processes that take place as oxygen is absorbed and carbon dioxide is released. The main steps (circled numbers in the figure) are:

1. Oxygen partial pressure is higher in the alveoli than in the red blood cells. As a result, oxygen diffuses from the alveoli through the capillary wall and into the red blood cells.

2. Inside the red blood cells, oxygen reacts with hemoglobin to form oxyhemoglobin and H^+ ions. Oxyhemoglobin formation is favored (according to Le Châtelier's principle) by a high partial pressure of oxygen, a slight cooling of the blood and an increase in pH (decrease in H^+ ions)—conditions encountered by the blood as it passes through the capillaries of the lungs.

3. Bicarbonate ions (HCO_3^-) diffuse from the plasma into the red blood cells.

4. The bicarbonate ions are replaced in the plasma by chloride ions that diffuse out of the blood cells. This step, called the **chloride shift**, maintains charge balance and osmotic pressure relationships between the plasma and red blood cells.

5. Protons released by hemoglobin as it is oxygenated (Step 2) react with bicarbonate ions to form carbonic acid.

6. An enzyme in the red blood cells, carbonic anhydrase, promotes the breakdown of carbonic acid to water and carbon dioxide. The reaction is reversible, but the low pressure of carbon dioxide in the lungs favors the formation of carbon dioxide and water (Le Châtelier's principle, again).

7. Carbaminohemoglobin ($HHbCO_2$) breaks apart to yield hemoglobin and carbon dioxide.

8. Because the carbon dioxide pressure is high in the red blood cells but low in the alveoli, carbon dioxide molecules diffuse out of the red blood cells and into the lung. The carbon dioxide and some of the water formed in Step 6 are expelled in the exhaled air.

Chloride shift
The movement of chloride ions from one fluid to another to maintain charge neutrality.

➤ **FIGURE 15.6** At the tissue cells, oxygen of red blood cells is exchanged for carbon dioxide.

Oxygenated red blood cells are carried by the bloodstream into the capillaries surrounding the cells of active tissues. ■ Figure 15.6 shows the steps involved in delivering oxygen to the tissues and accepting carbon dioxide for transport to the lungs. These steps are essentially the reverse of those at the lungs:

1. Tissue cells use oxygen to produce energy; carbon dioxide and water are formed as by-products. The concentration of carbon dioxide is higher in the tissue cells than in the red blood cells, so carbon dioxide diffuses from the tissue cells into the interstitial fluid and then into the red blood cells.
2. Inside the red blood cells, carbon dioxide rapidly reacts with water in the presence of carbonic anhydrase to produce carbonic acid.
3. Carbonic acid dissociates to give hydrogen ions and bicarbonate ions.
4. The bicarbonate ions diffuse into the plasma. The bicarbonate ions in the plasma are the form in which most carbon dioxide moves from tissue cells to the lungs.
5. The chloride shift again takes place (this time in the opposite direction) to maintain electrolyte balance.
6. A smaller amount (25%) of carbon dioxide from Step 1 reacts with hemoglobin to form carbaminohemoglobin ($HHbCO_2$) for transport to the lungs.
7. The increase in concentration of hydrogen ions inside the red blood cells promotes a reaction with oxyhemoglobin, which releases oxygen.
8. The free oxygen diffuses out of the red blood cells through the plasma, capillary membrane, interstitial fluid, and into the tissue cells.

Learning Check 15.1

Decide whether each of the following red blood cell reactions occurs at the lungs or at the site of the tissue cells:

a. $HHbCO_2 \rightleftarrows HHb + CO_2$
b. $H_2O + CO_2 \rightleftarrows H_2CO_3$
c. $H^+ + HCO_3^- \rightleftarrows H_2CO_3$
d. $HbO_2^- + H^+ \rightleftarrows HHb + O_2$

CHEMISTRY AROUND US • 15.1

BIOLOGICAL BANDAGES

When a blood vessel is opened by an injury, the blood comes in contact with tissue that is different from the tissue that lines the normal vessel. This contact initiates a series of reactions that lead to the formation of a blood clot which seals the opening in the vessel. A critical step in the clot-formation process occurs when fibrinogen and thrombin come in contact and interact. Thrombin converts fibrinogen into fibrin, which then binds together to form a strong fibrous mass. While this takes place, plasminogen is also formed and works with other compounds to remove clots from the blood. This process prevents clots from growing too large and spreading throughout the body.

These processes are the basis for a new bandage that shows promise in the treatment of serious injuries that cause rapid loss of blood. The bandage contains layers of freeze-dried fibrinogen and thrombin packed onto a gauzelike structure made of absorbable surgical fiber. When the bandage is put on an injury inside the body, the thrombin and fibrinogen interact to form a clot which helps stop the injury from bleeding. The gauze is eventually absorbed by the body, and the clot naturally dissolves as the injury heals.

Military medical teams have tested the bandage on torn livers, an injury that often results in massive and fatal bleeding. When a blood-filled liver is subjected to a large impact, it bursts, much like a water balloon. The typical result is a star-shaped rip extending from the front to the back of the liver that bleeds profusely. In tests of the bandage on such injuries, the massive bleeding was stopped in seconds.

The same idea has been applied to the development of a foam that mixes fibrinogen and thrombin as it is injected into a stab or gunshot wound. The foam expands into the injury and forms a clot that stops the bleeding. This type of foam might make it possible for victims of serious trauma to survive longer as they are being transported to an operating room. Sometimes a few additional minutes can make the difference between bleeding to death and receiving the necessary surgical care to survive.

The FDA is currently reviewing the clotting mixture, which is known as fibrin sealant. Before it is approved for use, a significant amount of study will be done to ensure that it does not cause problems for patients. However, the outlook for approval looks promising because the sealant does not cause clots to form until it comes in contact with blood, and the clots that do form are natural clots created by the body. These biological bandages may save many lives in the future.

http://www.brookscole.com/chemistry/seager_chemistry
Web site 15.3 provides additional information on the use of biological bandages.

➤ 15.3 Chemical Transport to the Cells

For substances to be chemically transported in the body, they must become part of the moving bloodstream. They may dissolve in the water-based plasma (as do sugars, amino acids, ions, and gases to some extent), they may become chemically bonded to cellular components (as do oxygen and carbon dioxide with the hemoglobin of blood cells), or they may form a suspension in the plasma of the blood (as do lipids).

Materials transported by the bloodstream to the cells must move through the capillary wall into the interstitial fluid and through the cell membrane into the cell. The reverse must take place when waste products move from inside tissue cells into the bloodstream. During these processes, capillary walls behave as selectively permeable membranes or filters, allowing water containing dissolved nutrients—including oxygen—to pass in one direction and water containing dissolved wastes to pass in the other direction.

The movement of water and dissolved materials through the capillary walls is governed by two factors—by the pressure of blood against the capillary walls and by the differences in protein concentration on each side of the capillary walls.

The concentration of protein in plasma is much higher than the protein concentration of the interstitial fluid outside the blood vessels. This concentration difference results in an osmotic pressure of about 18 torr, which tends to move fluid into the bloodstream. However, the pumping action of the heart puts the blood under pressure (blood pressure) that is greater at the arterial end of a capillary (about 32 torr) than at the venous end (about 12 torr).

> **FIGURE 15.7** The flow of fluid between capillaries and interstitial spaces.

Because the blood pressure at the arterial end is higher than the osmotic pressure, the tendency is for a net flow to occur from the capillary into the interstitial fluid (see ■ Figure 15.7). The fluid that leaves the capillary contains the dissolved nutrients, oxygen, hormones, and vitamins needed by the tissue cells.

As the blood moves through the capillary, its pressure decreases, until at the venous end, it becomes lower than the osmotic pressure, and a net flow of fluid takes place from the interstitial fluid into the capillary. The incoming fluid contains the metabolic waste products, such as carbon dioxide and excess water. This mechanism allows the exchange of nutrients and wastes and maintains fluid balance between the blood and interstitial space.

TABLE 15.1	Major constituents of normal urine

Constituent[a]	Grams per 24 hours[b]
Amino acids	0.80
Urea	25.0
Creatinine	1.5
Uric acid	0.7
H^+	pH 4.5–8.0
Na^+	3.0
K^+	1.7
NH_4^+	0.8
Ca^{2+}	0.2
Mg^{2+}	0.15
Cl^-	6.3
SO_4^{2-}	1.4
$H_2PO_4^-$	1.2
HCO_3^-	0–3
Other compounds	2–3

[a]The 24-hour volume and composition of urine vary widely, depending on the fluid intake and diet.
[b]These data are for an average 24-hour specimen with a total volume of 1400 mL.

➤ 15.4 The Constituents of Urine

Urine contains about 96% water and 4% dissolved organic and inorganic waste products. Approximately 40–50 g of dissolved solids are contained in the daily urine output of an adult (see ■ Table 15.1). The urine of a healthy person has a pH of 4.5–8.0, with 6.6 being a reasonable average for a person on an ordinary diet. Fruits and vegetables tend to make the urine alkaline, whereas high-protein foods tend to make the urine acidic.

The composition of urine is an excellent indicator of the state of health. This accounts for the importance of urinalysis as part of a physical checkup or a diagnosis. Normal constituents are considered abnormal when they are eliminated in excess amounts. ■ Table 15.2 lists some conditions that lead to abnormally high urinary levels of normal constituents; some other substances (such as pus or blood), which are considered abnormal in *any* measurable amount, are also listed.

A urine specimen is often collected from a patient entering a hospital and is checked with a paper test strip that contains bands of different reagents that will react with the different abnormal constituents (see ■ Figure 15.8). This one simple test quickly checks for indications of pathological conditions.

➤ 15.5 Fluid and Electrolyte Balance

The body is 45–75% water (by weight), the percentage depending on a number of factors, including age, sex, and especially the amount of body fat. In general, the percentage of water increases as fat content decreases.

Fluid balance within the body is maintained not only by a balance in the total amount but by a normal and stable distribution of fluid in the body's three fluid-containing areas. Thus, the amounts of fluid inside the cells, in the interstitial spaces, and in the blood vessels must remain relatively constant if a fluid balance is to exist.

TABLE 15.2   Some abnormal urine constituents

Abnormal constituent	Condition	Causes
Glucose (in large amounts)	Glucosuria	Diabetes mellitus, renal diabetes, alimentary glycosuria
Protein	Proteinuria or albuminuria	Kidney damage, nephritis, bladder infection
Ketone bodies	Ketonuria	Diabetes mellitus, starvation, high-fat diets
Pus (leukocytes)		Kidney or bladder infection
Hemoglobin	Hemoglobinuria	Excessive hemolysis of red blood cells
Red blood cells	Hematuria	Hemorrhage in the urinary tract
Bile pigments (in large amounts)	Jaundice	Blockage of bile duct, hepatitis, cirrhosis

➤ **FIGURE 15.8** Urinalysis test strips. A urine sample is tested for several components simultaneously by dipping a test strip into a sample and comparing the reagent areas to a color chart.

Fluid and electrolyte balance are interdependent. If one deviates from normal, so does the other. In modern medical practice, a great deal of attention is given to these matters. Many of today's hospital patients receive some form of fluid and electrolyte balance therapy.

An important and obvious principle of fluid balance is that fluid output and intake must be equal. Water from food and drink is absorbed into the body from the digestive tract. Besides this well-known source, water is also derived within the cells from food metabolism. This by-product water enters the bloodstream along with the water that is absorbed from the digestive tract.

Water intake is regulated in part by the thirst mechanism (see ■ Figure 15.9). When the body loses large amounts of water, salivary secretions decrease, and a dry feeling develops in the mouth. This and other sensations are recognized as thirst, and water is drunk to relieve the condition. The fluid intake compensates for the fluid lost, and balance is reestablished.

Water normally leaves the body through the kidneys (urine), lungs (water vapor in expired air), skin (by diffusion and perspiration), and intestines (feces) (see ■ Table 15.3). Abnormally high fluid losses, and possibly dehydration, can be caused by hyperventilation, excessive sweating, vomiting, or diarrhea.

Fluid balance in the body is maintained or restored primarily by variations in urine output. Normally, this amounts to about 1400 mL/day, but the

http://www.brookscole.com/chemistry/seager_chemistry
At Web site 15.4, more information is provided on the control of fluid and electrolyte balance.

TABLE 15.3 Typical normal values for water entry and exit

Intake (mL)	
Ingested liquids	1500
Water in foods	700
Water formed by catabolism	200
Total	2400

Output (mL)	
Kidneys (urine)	1400
Lungs (water in expired air)	350
Skin By diffusion	350
By perspiration	100
Intestines (in feces)	200
Total	2400

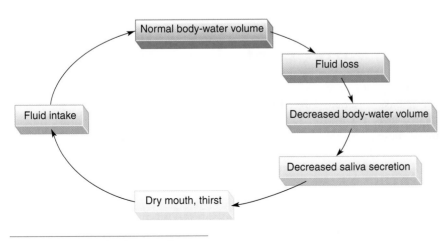

➤ **FIGURE 15.9** The thirst mechanism.

total excretion adjusts according to water intake. The primary factor that controls urine production is the rate of water reabsorption from the renal tubules in the kidneys. This rate is regulated chiefly by the pituitary hormone vasopressin and by the adrenal cortex hormone aldosterone.

Vasopressin, also called antidiuretic hormone (ADH), exerts its action by affecting the permeability of the renal tubules to water. In the absence of vasopressin, less water is reabsorbed into the bloodstream, and a large volume of light yellow, dilute urine results. Patients with the rare disease diabetes insipidus do not produce vasopressin; their urine is extremely voluminous and dilute. In the presence of vasopressin, the tubules become more permeable to water that is drawn into the capillaries, thus increasing blood volume. The reabsorption of water from the tubules produces a low volume of dark yellow, concentrated urine. (Diabetes insipidus is very different from diabetes mellitus, mentioned elsewhere in the text.)

With the help of vasopressin, a healthy person can vary the intake of water widely and yet preserve a stable overall concentration of substances in the blood. But sometimes the water level of the body gets dangerously low, from not drinking enough water or excessive water loss caused by diarrhea or excessive perspiration. Under such circumstances, the adrenal gland secretes the steroid hormone aldosterone. At the kidneys, aldosterone stimulates the reabsorption of sodium ions (electrolytes) from the renal tubes into the blood. Chloride ions follow the sodium ions to maintain electrical neutrality, and water follows the sodium chloride. Thus, aldosterone secretion conserves both salt and water in the body. When the fluid level of the body returns to normal, the amount of aldosterone secreted decreases.

➤ 15.6 Acid–Base Balance

Blood pH normally remains within the rather narrow range of 7.35–7.45. A decrease in pH, called acidosis, or an increase, called **alkalosis,** is serious and requires prompt attention. Death can result when the pH value falls below 6.8 or rises above 7.8 (see ■ Figure 15.10). Acidosis, an abnormally low blood pH, was introduced in Section 14.6.

Large amounts of acids and smaller amounts of bases normally enter the blood. Some mechanisms must neutralize or eliminate these substances if the blood pH is to remain constant. In practice, a constant pH is maintained by the interactive operation of three systems: buffer, respiratory, and urinary. Only when all three parts of this complex mechanism function properly can the acid–base balance be maintained.

➤ 15.7 Buffer Control of Blood pH

A buffer system prevents marked changes in the pH of a solution when an acid or base is added. Three major buffer systems of the blood are the bicarbonate buffer, the phosphate buffer, and the plasma proteins. The most important of these is the bicarbonate buffer system, consisting of a mixture of bicarbonate ions (HCO_3^-) and carbonic acid (H_2CO_3).

Buffer systems work by neutralizing H^+ or OH^- ions that come from the dissociation of acids or bases added to a system. As an example of buffers in action, consider the fate of lactic acid (HLac) entering the bloodstream. When lactic acid dissolves in the blood, it dissociates to produce H^+ and lactate ions (Lac^-). The liberated H^+ ions drive the pH of the blood down. The bicar-

Alkalosis
An abnormally high blood pH.

➤ **FIGURE 15.10** Human circulatory and respiratory processes operate only within a very narrow range of blood pH.

CHEMISTRY AROUND US • 15.2

BLOOD DONATION

The need for donated blood never ends. Every day, thousands of people need blood transfusions, especially trauma victims and surgical patients. Most of these individuals depend on blood that has been donated by volunteers. The requirements for donating blood can be met by most people, the most important one being general good health.

A blood donor must be at least 18 years old (16 with parental permission), and weigh at least 110 pounds. Donors may not have any bloodborne diseases, but may be taking medications such as aspirin, allergy medication, sleeping pills, ibuprofen, or the hepatitis B vaccine. People who are pregnant, anemic, or suffering from a cold or the flu are not allowed to donate. Anyone who has contracted malaria within 3 years prior to the time of donation, or who during that time has been in areas of the world where malaria is common, is also prohibited from donating.

Before donating blood, potential donors must fill out a questionnaire that requires a health history and the answers to some standard questions. Each donor is given a superficial medical examination in which pulse rate and temperature are measured, and a test for anemia is performed. If all the tests are satisfactory, about 1 pint of blood is removed from a donor in a process that takes about 45 minutes.

After donating, it is recommended that donors rest for a short time and consume some liquids. However, normal activities can usually be resumed about 10–15 minutes after the donation is completed. A 1-pint donation reduces the body's blood volume by about 10%, so strenuous activities should be avoided immediately after donating. The fluid volume lost during a donation is usually replaced by the body within 24 hours, but it can take several weeks for the body to replace the red blood cells. Because of this, donors must wait a minimum of 8 weeks before making another donation.

At some blood donation locations, it is possible to donate blood platelets instead of whole blood. Platelets are a component of blood that play an important role in the control of bleeding. In a process called *apheresis*, withdrawn blood is directed into a machine that removes the platelets and returns the blood to the donor. During this process, a donor has six times more platelets removed than are removed during a normal donation of a pint of whole blood.

The need for blood donors is great, it is relatively easy to qualify, and the act of donating blood may save a life. These are three good reasons to seriously consider becoming a volunteer donor.

http://www.brookscole.com/chemistry/seager_chemistry
Web site 15.5 provides information on a possible benefit of blood donation.

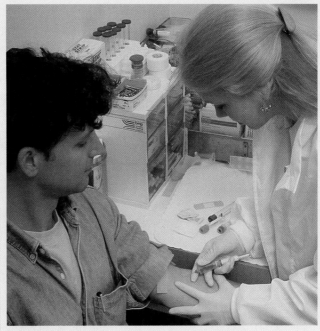

A volunteer blood donor.

bonate ions of the bicarbonate buffer system react with the excess H^+ and return the pH to the normal range:

$$HLac \rightleftharpoons H^+ + Lac^- \qquad (15.3)$$

$$H^+ + HCO_3^- \rightleftharpoons H_2CO_3 \qquad (15.4)$$

Reaction 15.4 is a net ionic reaction for the buffering process. When a total reaction is written, we see that the effect of the buffering is to replace lactic acid in the blood by carbonic acid, which is a weaker acid and so releases fewer H^+ ions to the blood than does lactic acid:

$$\underset{\substack{\text{stronger}\\\text{acid}}}{HLac} + HCO_3^- \longrightarrow \underset{\substack{\text{weaker}\\\text{acid}}}{H_2CO_3} + Lac^- \qquad (15.5)$$

This replacement of a stronger acid with a weaker acid is how biochemists often describe buffering reactions.

Although the bicarbonate system is not the strongest buffer system in the body (the most powerful and plentiful one is the proteins of plasma and cells), it is very important because the concentrations of both bicarbonate and carbonic acid are regulated by the kidneys and by the respiratory system. Without the respiratory and urinary processes, the capacity of buffers would eventually be exceeded.

➤ 15.8 Respiratory Control of Blood pH

The respiratory system helps control the acidity of blood by regulating the elimination of carbon dioxide and water molecules. These molecules, exhaled in every breath, come from carbonic acid as follows (see Figure 15.6):

$$H_2CO_3 \rightleftharpoons H_2O + CO_2$$

carbonic acid

(15.6)

The more CO_2 and H_2O that is exhaled, the more carbonic acid is removed from the blood, thus elevating the blood pH to a more alkaline level.

The respiratory mechanism for controlling blood pH begins in the brain with respiratory center neurons that are sensitive to blood CO_2 levels and pH. A significant increase in the CO_2 of arterial blood, or a decrease below about 7.38 of arterial blood pH, causes the breathing to increase both in rate and depth, resulting in **hyperventilation.** This increased ventilation eliminates more carbon dioxide, reduces carbonic acid and hydrogen-ion concentrations, and increases the blood pH back toward the normal level (see ■ Figure 15.11).

In the opposite situation, an increase in blood pH above normal causes **hypoventilation,** a reduced rate of respiration. Less CO_2 is exhaled, and the higher concentration of carbonic acid remaining in the blood lowers the pH back to normal.

Hyperventilation
Rapid, deep breathing.

Hypoventilation
Slow, shallow breathing.

➤ 15.9 Urinary Control of Blood pH

Because they can excrete varying amounts of acid and base, the kidneys, like the lungs, play a vital role in pH control. For example, with acidic blood, the excretion of hydrogen ions by the kidneys decreases the urine pH and simultaneously increases the blood pH back toward normal.

➤ **FIGURE 15.11** Respiratory control of blood pH.

> **FIGURE 15.12** Urinary control of blood pH in the kidney.

■ Figure 15.12 traces the reactions involved in the excretion of hydrogen ions by the kidneys:

1. Carbon dioxide diffuses from the blood capillaries into the kidney distal tubule cells.
2. Catalyzed by carbonic anhydrase, water and carbon dioxide react to give carbonic acid.
3. The carbonic acid ionizes to give hydrogen ions and bicarbonate ions. The hydrogen ions diffuse into the developing urine.
4. For every hydrogen ion entering the urine, a sodium ion passes into the tubule cells.
5. The sodium ions and bicarbonate ions enter the bloodstream capillaries.

The net result of these reactions is the conversion of CO_2 to HCO_3^- within the blood. Both the decrease in CO_2 and the increase in HCO_3^- tend to increase blood pH levels back to normal. The developing urine has picked up the hydrogen ions, which now react with buffering ions such as phosphate that are present in urine:

$$H^+ + HPO_4^{2-} \longrightarrow H_2PO_4^- \qquad (15.7)$$

The presence of the phosphate buffer system ($H_2PO_4^{2-}/HPO_4^{2-}$) usually keeps urine from going much below pH 6, but too great an excess of protons can exceed the buffer capacity of the system and result in quite acidic urine.

➤ 15.10 Acidosis and Alkalosis

When blood pH is normal and a state of acid–base balance exists, components of the bicarbonate buffer are present in plasma in a ratio of 20 parts of bicarbonate (HCO_3^-) to 1 part of carbonic acid (H_2CO_3).

When respiratory and urinary systems work together to maintain the blood at pH 7.4, they do so primarily by maintaining the HCO_3^-/H_2CO_3 ratio

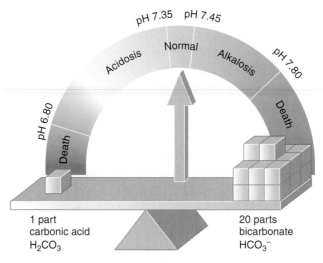

The normal ratio of 1 part H_2CO_3 to 20 parts HCO_3^- indicating that the body is in acid–base balance

➤ **FIGURE 15.13** Blood pH depends on the relative concentrations of carbonic acid and bicarbonate ion.

at 20:1. Any change in the ratio produces a change in pH. An increase in the ratio of bicarbonate to carbonic acid causes the pH to increase (alkalosis); and a decrease, which is much more common, causes the pH to decrease (acidosis) (see ■ Figure 15.13).

Two adjectives describe the general cause or origin of pH imbalances in body fluids: metabolic and respiratory. Respiratory acid–base imbalances (respiratory acidosis and respiratory alkalosis) result from abnormal breathing patterns. Metabolic acidosis and metabolic alkalosis are caused by factors other than abnormal breathing.

Respiratory Alkalosis

Respiratory alkalosis is caused by hyperventilating—breathing rapidly and deeply. Hysteria, anxiety, or prolonged crying may result in hyperventilation in which too much carbon dioxide is exhaled, which disturbs the blood's normal carbon dioxide/carbonic acid ratio. According to Le Châtelier's principle, Reaction 15.6 (written again below) shifts to the right as it attempts to restore the equilibrium:

$$\underset{\text{direction of shift to restore equilibrium}}{H_2CO_3 \rightleftharpoons H_2O + CO_2 \uparrow} \qquad \overset{loss\ of\ CO_2}{}$$

(15.8)

The equilibrium shift reduces the amount of carbonic acid present and rapidly increases the HCO_3^-/H_2CO_3 ratio to a value greater than 20:1. As a result, blood pH rises to 7.54 or higher within a few minutes. A corresponding loss of HCO_3^- also occurs, but it is not large enough to maintain the HCO_3^-/H_2CO_3 ratio at 20:1.

The treatment of respiratory alkalosis involves rebreathing one's own exhaled air by breathing into a paper bag (*not* a plastic bag), administering carbon dioxide, and treating the underlying causes of the hyperventilation.

Respiratory alkalosis
A condition of alkalosis caused by hyperventilation.

Respiratory Acidosis

Respiratory acidosis is sometimes the result of slow or shallow breathing (hypoventilation) caused by an overdose of narcotics or barbiturates. Anesthetists need to be particularly aware of respiratory acidosis, because most inhaled anesthetics depress respiration rates. Lung diseases, such as emphysema and pneumonia, or an object lodged in the windpipe, can also result in hypoventilation. In respiratory acidosis, too little carbon dioxide is exhaled, and its concentration within the blood increases. The carbon dioxide/carbonic acid equilibrium shifts to the left to restore the equilibrium:

$$\underset{\underset{\text{of equilibrium}}{\underset{\longleftarrow}{\text{direction of shift}}}}{H_2CO_3 \rightleftharpoons H_2O + CO_2} \overset{increased\ CO_2}{}$$

(15.9)

As a result of the shift, the concentration of carbonic acid in the blood increases, the HCO_3^-/H_2CO_3 ratio decreases to less than 20:1, and respiratory acidosis sets in. The treatment of respiratory acidosis involves identifying the underlying causes and possibly the intravenous administration of isotonic sodium bicarbonate solution or hemodialysis.

Metabolic Acidosis

Various body processes—collectively called metabolism—produce acidic substances that liberate H^+ in solution. The diffusion of these substances into the bloodstream causes a shift in the carbonic acid–bicarbonate equilibrium:

$$\underset{\underset{\text{restore equilibrium}}{\underset{\longleftarrow}{\text{direction of shift to}}}}{H_2CO_3 \rightleftharpoons H^+ + HCO_3^-} \overset{increased\ H^+}{}$$

(15.10)

The shift decreases the concentration of bicarbonate ions and increases the concentration of carbonic acid. As a result, the HCO_3^-/H_2CO_3 ratio decreases, as does the blood pH. **Metabolic acidosis** is often a serious problem in uncontrolled diabetes mellitus, and it also occurs temporarily after heavy exercise. Other causes include severe diarrhea and aspirin overdose.

Symptoms of metabolic acidosis, besides low blood pH and high levels of H_2CO_3, include the following: hyperventilation as the respiratory center tries to get excess CO_2 out of the lungs, increased urine formation as the kidneys try to remove excess acids, thirst (to replace urination water loss), drowsiness, headache, and disorientation. The treatment again depends on the cause; it may involve insulin therapy, intravenous bicarbonate, or hemodialysis. Note carefully that hyperventilation can be a symptom of metabolic acidosis and a cause of respiratory alkalosis, two different conditions requiring very different treatments.

Metabolic Alkalosis

In **metabolic alkalosis,** the body has lost acid in some form, perhaps from prolonged vomiting of the acidic stomach contents, or there has been an ingestion of alkaline substances. Excessive use of sodium bicarbonate (baking soda), for example, a common home remedy for upset stomach, increases the concentration of bicarbonate ions in the blood.

Respiratory acidosis
A condition of acidosis caused by hypoventilation.

Metabolic acidosis
A condition of acidosis resulting from causes other than hypoventilation.

Metabolic alkalosis
A condition of alkalosis resulting from causes other than hyperventilation.

| TABLE 15.4 | Some causes of blood acid–base imbalance |

Condition	Causes
Respiratory	
Acidosis: $CO_2 \uparrow$ pH $\downarrow$	Hypoventilation, blockage of diffusion within lungs, respiratory center depressants
Alkalosis: $CO_2 \downarrow$ pH $\uparrow$	Hyperventilation, excitement, trauma
Metabolic	
Acidosis: $H^+ \uparrow$ pH $\downarrow$	Kidney failure, prolonged diarrhea, ketone bodies from diabetes mellitus
Alkalosis: $H^+ \downarrow$ pH $\uparrow$	Kidney disease, prolonged vomiting, excessive intake of baking soda

A decrease in acid concentration within the body (from vomiting) or an increase in HCO_3^- levels (from excessive $NaHCO_3$) causes the HCO_3^-/H_2CO_3 ratio to rise above 20:1, and measurable alkalosis is observed. The respiratory center responds to the higher blood pH by slowing the breathing (hypoventilation). Other symptoms may include numbness, tingling, and headache. ■ Table 15.4 is a summary of causes of acidosis and alkalosis.

> **FOR FUTURE REFERENCE** DRUGS THAT MAY AFFECT THE HEART

Even though plasma, the liquid portion of the blood, comprises only about 7% of the total volume of fluids in the body, its proper circulation through the body is central to good health. Of course, good circulation depends, in part, on the proper functioning of the heart. It is well known that a medicine administered to treat one health problem might interact with other medicines or with systems in the body to create other health problems. The drugs listed below have been identified as having the potential to influence the efficient operation of the heart, or in some cases to damage the heart itself.

Drugs that may cause or contribute to abnormal heart rhythms (arrhythmias)

aminophylline	chlorpromazine	ketanserin	ranitidine
amiodarone	digoxin	lidocaine	sotalol
amitriptyline	diltiazem	maprotiline	terbutaline
antiarrhythmic drugs	disopyramide	methyldopa	theophylline
bepridil	diuretics	mexiletine	thiazide diuretics
beta-adrenergic-	doxepin	milrinone	thioridazine
blocking drugs	encainide	phenothiazines	trazodone
beta-adrenergic	fentolterol	prenylamine	tricyclic antidepressants
bronchodilators	flecainide	procainamide	verapamil
carbamazepine	isoproterenol	quinidine	

Drugs that may depress heart function (reduce pumping efficiency)

beta-adrenergic-	diltiazem	epinephrine	isoproterenol
blocking drugs	disopyramide	flecainide	nifedipine
cocaine	doxorubicin	fluorouracil	verapamil
daunorubicin			

Drugs that may reduce coronary artery blood flow (reduce oxygen supply to heart muscle)

amphetamines	cocaine	nifedipine	vasopressin
beta-adrenergic-	ergotamine	oral contraceptives	vinblasine
blocking drugs	fluorouracil	ritodrine	vincristine
(abrupt withdrawal)			

Drugs that may impair healing of heart muscle following a heart attack (myocardial infarction)

adrenocortical steroids
nonsteroidal anti-inflammatory drugs (NSAIDs)

Drugs that may cause heart valve damage

dexfenfluramine (Redux)	ergotamine	fen-phen (fenfluramine-	methysergide
minocycline		phenteramine)	

Drugs that may cause pericardial disease

actinomycin D	cyclophosphamide	methysergide	practolol
anthracyclines	cytarabine	minoxidil	procainamide
bleomycin	fluorouracil	phenylbutazone	sulfasalazine
cisplatin	hydralazine		

SOURCE: Adapted from Rybacki, J.J.; Long, J.W. *The Essential Guide to Prescription Drugs.* New York: Harper Perennial, 1999.

CONCEPT SUMMARY

A Comparison of Body Fluids. The fluids of the body are located inside tissue cells (intracellular fluid) or outside tissue cells (extracellular fluid). Volumetrically, most of the extracellular fluids are of two types—blood and interstitial fluid, the fluid that occupies spaces between cells and moves in the lymph vessels. Other extracellular body fluids that occur in smaller amounts are urine, digestive juices, and cerebrospinal fluid. Chemically, blood (plasma) and interstitial fluid (including lymph) are similar. Intracellular fluid is chemically different from the extracellular fluids.

Oxygen and Carbon Dioxide Transport. Nearly all oxygen that is transported from the lungs to tissue cells is transported by the blood in the form of oxyhemoglobin. Most carbon dioxide transported from the cells to the lungs is transported by the blood in the form of bicarbonate ions dissolved in the plasma. Smaller amounts also travel as carbaminohemoglobin and as CO_2 gas dissolved in plasma. The formation of the two hemoglobin compounds and the dissolving of various materials in the plasma are influenced by equilibria related to concentration differences of materials in the blood as it passes through capillaries in the lungs or near tissue cells.

Chemical Transport to the Cells. Substances transported by the blood to tissue cells move from the blood, through capillary walls, into the interstitial fluid, then through cell membranes, and finally into the cytoplasm of the cells. Waste products of the cells move in the opposite direction. The movement of fluid through capillary walls is governed by the blood pressure against the capillary walls and by osmotic pressure differences that arise from protein concentration differences between blood and interstitial fluid.

The Constituents of Urine. Normal urine is about 96% water and 4% dissolved organic and inorganic wastes. Because urine composition is an excellent indicator of a person's health, urinalysis is a routine procedure in physical checkups.

Fluid and Electrolyte Balance. Fluid balance in the body involves a balance between fluid intake and output, and a proper distribution of fluid between the three fluid areas. Fluid and electrolyte balance are interdependent. Water enters the body through food and drink and leaves in the form of urine, water vapor in exhaled air, perspiration, and feces. The amount of fluid in the body is maintained or restored primarily by variations in urine output, which is regulated chiefly by the hormones vasopressin and aldosterone.

Acid–Base Balance. Blood pH is normally in the range of 7.35–7.45. Acidosis and alkalosis exist when the blood pH becomes lower or higher, respectively, than these values.

Buffer Control of Blood pH. The blood contains three major buffer systems: the bicarbonate system, the phosphate system, and the plasma protein system. The most important is the bicarbonate system, which consists of a mixture of bicarbonate ions and carbonic acid.

Respiratory Control of Blood pH. The respiratory system influences blood pH by regulating the elimination of carbon dioxide and water. Because carbon dioxide and water are produced during the decomposition of carbonic acid, the greater the elimination of carbon dioxide and water from the body, the greater is the elimination of carbonic acid from the blood.

Urinary Control of Blood pH. The kidneys influence blood pH by varying the amount of hydrogen ions excreted in the urine. When the blood is too acidic, carbonic acid is converted to bicarbonate ions and hydrogen ions. The hydrogen ions are excreted, and the bicarbonate ions enter the blood.

Acidosis and Alkalosis. Acidosis and alkalosis can result from metabolic or respiratory causes. Respiratory acidosis and alkalosis are the result of abnormal breathing patterns. Alkalosis is caused by hyperventilation that accompanies hysteria, anxiety, or prolonged crying. Acidosis can result from hypoventilation brought on by such factors as an overdose of depressant narcotics or anesthesia. Metabolic acidosis can result from the metabolic production of acids in the body or the elimination of alkaline materials, as during severe diarrhea. Metabolic alkalosis results from the loss of acid in some form such as the prolonged vomiting of acidic stomach contents or the ingestion of alkaline materials.

LEARNING OBJECTIVES

When you have completed your study of this chapter, you should be able to:

1. Compare the chemical composition of plasma, interstitial fluid, and intracellular fluid.
2. Explain how oxygen and carbon dioxide are transported within the bloodstream.
3. Explain how materials move from the blood into the body cells and from the body cells into the blood.
4. List the normal and abnormal constituents of urine.
5. Discuss how proper fluid and electrolyte balance is maintained in the body.
6. Explain how acid–base balance is maintained in the body.
7. List the causes of acidosis and alkalosis.

KEY TERMS AND CONCEPTS

Alkalosis (15.6)
Carbaminohemoglobin (15.2)
Chloride shift (15.2)

Deoxyhemoglobin (hemoglobin) (15.2)
Extracellular fluid (15.1)
Hyperventilation (15.8)

Hypoventilation (15.8)
Interstitial fluid (15.1)
Intracellular fluid (15.1)

Metabolic acidosis (15.10)   Oxyhemoglobin (15.2)   Respiratory acidosis (15.10)
Metabolic alkalosis (15.10)   Plasma (15.1)   Respiratory alkalosis (15.10)

KEY REACTIONS

1. Formation of oxyhemoglobin (Section 15.2):

$$HHb + O_2 \rightleftarrows HbO_2^- + H^+$$

Reaction 15.1

2. Formation of carbaminohemoglobin (Section 15.2):

$$HHb + CO_2 \rightleftarrows HHbCO_2$$

Reaction 15.2

3. Decomposition of carbonic acid (Section 15.8):

$$H_2CO_3 \rightleftarrows H_2O + CO_2$$

Reaction 15.6

EXERCISES

A COMPARISON OF BODY FLUIDS (SECTION 15.1)

15.1 Where is most fluid in the body located? What is it called?

15.2 Which two of the following have nearly the same chemical composition: plasma, intracellular fluid, interstitial fluid?

15.3 In what way does the composition of the two similar fluids in Exercise 15.2 differ?

15.4 Which of the fluids of Exercise 15.2 contains the highest concentration of the following?
   a. protein
   b. K^+
   c. Na^+
   d. HPO_4^{2-}
   e. HCO_3^-
   f. Mg^{2+}
   g. Cl^-

15.5 What is the principal cation found in the following?
   a. blood plasma
   b. interstitial fluid
   c. intracellular fluid

15.6 What is the principal anion in the following?
   a. blood plasma
   b. interstitial fluid
   c. intracellular fluid

OXYGEN AND CARBON DIOXIDE TRANSPORT (SECTION 15.2)

15.7 Hemoglobin exists in several forms—oxyhemoglobin, deoxyhemoglobin, and carbaminohemoglobin. Match each of the following to the type of hemoglobin involved:
   a. Formed when CO_2 combines with hemoglobin
   b. The oxygen-transporting form of hemoglobin
   c. The form of hemoglobin designated as HHb

15.8 What term is given to the reversible flow of chloride ion across the red blood cell membrane during the oxygen and carbon dioxide transport process?

15.9 What percentages of the total O_2 and CO_2 transported by the blood are in the following forms?
   a. O_2 as dissolved gas in plasma
   b. O_2 as oxyhemoglobin
   c. CO_2 as carbaminohemoglobin
   d. CO_2 as dissolved gas in plasma
   e. CO_2 as bicarbonate ions

15.10 Discuss the significance of red blood cells in the O_2 and CO_2 transport reactions that occur at (a) the lungs and (b) the tissue cells.

15.11 Why are large capillary surface areas critical to O_2 and CO_2 transport processes?

15.12 Write the equilibrium equation representing the reaction between oxygen and hemoglobin. In which direction will the equilibrium shift when the following occurs?
   a. The pH is increased.
   b. The concentration (pressure) of O_2 is increased.
   c. The concentration of CO_2 is increased (see Figure 15.5).
   d. The blood is in a capillary near actively metabolizing tissue cells.

15.13 Use equilibrium discussions based on Figure 15.5 to explain two ways that the oxygenation of hemoglobin helps release CO_2 from red blood cells to be exhaled.

CHEMICAL TRANSPORT TO THE CELLS (SECTION 15.3)

15.14 Which factor, blood pressure or osmotic pressure, has the greater influence on the direction of fluid movement through capillary walls at (a) the venous end of a capillary and (b) the arterial end of a capillary?

15.15 Describe the changes in blood pressure and osmotic pressure of blood as the blood moves from the arterial end to the venous end of a capillary.

THE CONSTITUENTS OF URINE (SECTION 15.4)

15.16 What organic material is normally excreted in the largest amount in urine?

15.17 What cation and anion are normally excreted in the largest amounts in urine?

15.18 Under what circumstances would the presence of a normal constituent of urine be considered abnormal?

15.19 List the abnormal constituents of urine.

FLUID AND ELECTROLYTE BALANCE (SECTION 15.5)

15.20 List four routes by which water leaves the body.

15.21 Besides the water contained in food and drink, what other source provides water to the body?

15.22 What is the primary way fluid balance is maintained in the body?

15.23 Describe how vasopressin and aldosterone function to maintain fluid levels in the blood.

15.24 Trace the events of the regulation of fluid balance by the thirst mechanism.

ACID–BASE BALANCE (SECTION 15.6)

15.25 What is meant by the terms *acidosis* and *alkalosis*?

15.26 List three systems that cooperate to maintain blood pH in an appropriate narrow range.

BUFFER CONTROL OF BLOOD pH (SECTION 15.7)

15.27 List three major buffer systems of the blood.

15.28 Write equations to illustrate how the $H_2PO_4^-/HPO_4^{2-}$ buffer would react to added H^+ and added OH^-.

15.29 Hemoglobin (HHb) is a weak acid that forms the Hb^- anion. Write equations to show how the HHb/Hb^- buffer would react to added H^+ and added OH^-.

RESPIRATORY CONTROL OF BLOOD pH (SECTION 15.8)

15.30 Describe the respiratory mechanism for controlling blood pH.

15.31 Hyperventilation is a symptom of what type of acid–base imbalance in the body?

15.32 Hypoventilation is a symptom of what type of acid–base imbalance in the body?

URINARY CONTROL OF BLOOD pH (SECTION 15.9)

15.33 Write a short summary description of the role of the kidneys and urine in the control of blood pH.

15.34 What happens to CO_2 in the blood when the kidneys function to control blood pH?

15.35 What buffer system in the urine usually keeps the urine pH from dropping lower than 6?

15.36 What ionic shift maintains electron charge balance when the kidneys function to increase the pH of blood?

ACIDOSIS AND ALKALOSIS (SECTION 15.10)

15.37 What is the difference between the causes of respiratory acid–base imbalances and metabolic acid–base imbalances in the body?

15.38 What type of acid–base imbalance in the blood is caused by hypoventilation? By hyperventilation?

15.39 What type of acid–base imbalance might develop in an individual who has taken an overdose of a depressant narcotic?

15.40 Explain how prolonged vomiting can lead to alkalosis.

15.41 Explain how uncontrolled diarrhea can lead to acidosis.

CHEMISTRY AROUND US AND OVER THE COUNTER

15.42 What cellular materials does the drug EPO stimulate the body to produce?

15.43 Explain what is meant by the term *blood packing*.

15.44 What dangerous side effects may accompany the use of the drug EPO?

15.45 Name two substances present as key components in biological bandages.

15.46 How much blood is typically collected during blood donation?

15.47 What is the waiting period between blood donations?

▶ POINTS TO PONDER AND PURSUE

15.1 Explain why an hysterical individual might develop alkalosis.

15.2 Explain why a person suffering from metabolic acidosis might develop a thirst.

15.3 Ordinary water may be the most effective "sports drink." Explain why this might be true.

15.4 How does the concentration of hemoglobin of someone living in the Himalayas compare with that of someone living at a lower altitude?

15.5 According to Web site 15.2, where is EPO produced in the body?

15.6 Using Web site 15.3, list the three major types of wounds that don't repair very well.

15.7 Based on Web site 15.4, what percentage of body mass does intracellular water constitute?

15.8 According to Web site 15.5, what may be a significant benefit to blood donation?

http://www.brookscole.com/chemistry/seager_chemistry

Appendix A

The International System of Measurements

The International System of Units (SI units) was established in 1960 by the International Bureau of Weights and Measures. The system was established in an attempt to streamline the metric system, which included certain traditional units that had historical origins but that were not logically related to other metric units. The International System established fundamental units to represent seven basic physical quantities. These quantities and the fundamental units used to express them are given in Table A.1.

TABLE A.1 Fundamental SI units

Physical quantity	SI unit	Abbreviation
Length	meter	m
Mass	kilogram	kg
Temperature	kelvin	K
Amount of substance	mole	mol
Electrical current	ampere	A
Time	second	s
Luminous intensity	candela	cd

All other SI units are derived from the seven fundamental units. Prefixes are used to indicate multiples or fractions of the fundamental units (Table 1.2). In some cases, it has been found convenient to give derived units specific names. For example, the derived unit for force is kg m/s^2, which has been given the name newton (abbreviated N) in honor of Sir Isaac Newton (1642–1726). Some examples of derived units are given in Table A.2.

TABLE A.2 Examples of derived SI units

Physical quantity	Definition in fundamental units	Specific name	Abbreviation
Volume	m^3 (see NOTE below)	—	—
Force	kg m/s^2	newton	N
Energy	kg m^2/s^2 = N m	joule	J
Power	kg m^2/s^3 = J/s	watt	W
Pressure	kg/m s^2 = N/m^2	pascal	Pa
Electrical charge	A s	coulomb	C
Electrical potential	kg m^2/A s^3 = W/A	volt	V
Electrical resistance	kg m^2/A^2 s^3 = V/A	ohm	Ω
Frequency	1/s	hertz	Hz

NOTE: The liter (L), a popular volume unit of the metric system, has been redefined in terms of SI units as 1 dm^3 or 1/1000 m^3. It and its fractions (mL, μL, etc.) are still widely used to express volume.

Appendix B

Answers to Even-Numbered End-of-Chapter Exercises

CHAPTER 1

1.2 Early scientists noted that organized (living) systems were the source of these compounds; hence, the name "organic" was applied.

1.4 They all contain carbon.

1.6 Covalent

1.8 **a.** Organic; most organic compounds are flammable.
b. Inorganic; a high-melting solid is typical of ionic compounds.
c. Organic; many organic compounds are insoluble in water.
d. Organic; most organic compounds can undergo combustion.
e. Organic; a low-melting solid is characteristic of covalent materials.

1.10 The ability of carbon to bond to itself repeatedly; isomerism.

1.12 The s orbital of the hydrogen overlaps with the sp^3 orbital of carbon.

1.14 A p orbital and an sp^3 orbital both have a two-lobed shape. A p orbital has equal-sized lobes, whereas an sp^3 orbital has different-sized lobes.

1.16

a.

$$H-\overset{\overset{\displaystyle H}{|}}{\underset{\underset{\displaystyle H}{|}}{C}}-\overset{\overset{\displaystyle H}{|}}{C}=\overset{\overset{\displaystyle H}{|}}{C}-H$$

b.

$$H-\overset{\overset{\displaystyle H}{|}}{\underset{\underset{\displaystyle H}{|}}{C}}-\overset{\overset{\displaystyle O}{\|}}{C}-\overset{\overset{\displaystyle H}{|}}{\underset{\underset{\displaystyle H}{|}}{C}}-H$$

c.

$$H-\overset{\overset{\displaystyle H}{|}}{\underset{\underset{\displaystyle H}{|}}{C}}-\overset{\overset{\displaystyle O}{\|}}{C}-O-H$$

d.

$$H-\overset{\overset{\displaystyle H}{|}}{\underset{\underset{\displaystyle H}{|}}{C}}-\overset{\overset{\displaystyle H}{|}}{\underset{\underset{\displaystyle H}{|}}{N}}-\overset{\overset{\displaystyle H}{|}}{\underset{\underset{\displaystyle H}{|}}{C}}-H$$

1.18 b, d, e

1.20 Two structures (a and d) are correct. In structure b, the second carbon atom has only three bonds. In structure c, carbon atoms 2 and 3 have too many bonds. In structure e, the second carbon atom from the right has only three bonds.

1.22

a. $CH_3CH_2CH_2CH = CH - \overset{\overset{\displaystyle O}{\|}}{C} - CH_2CH_3$

b. $CH_3CH_2 - \overset{\overset{\displaystyle O}{\|}}{C} - NH_2$

1.24

a.

$$H-\overset{\overset{\displaystyle H}{|}}{\underset{\underset{\displaystyle H}{|}}{C}}-\overset{\overset{\displaystyle H}{|}}{\underset{\underset{\displaystyle H}{|}}{C}}-\overset{\overset{\displaystyle H}{|}}{\underset{\underset{\displaystyle H}{|}}{C}}-\overset{\overset{\displaystyle H}{|}}{N}-H$$

b.

$$H-\overset{\overset{\displaystyle H}{|}}{\underset{\underset{\displaystyle H}{|}}{C}}-\overset{\overset{\displaystyle H}{|}}{\underset{\underset{\displaystyle H}{|}}{C}}-\overset{\overset{\displaystyle O}{\|}}{C}-\overset{\overset{\displaystyle H}{|}}{\underset{\underset{\displaystyle H}{|}}{C}}-H$$

1.26 The molecular formula is C_9H_{20}. The condensed structural formula is $CH_3(CH_2)_7CH_3$.

1.28 **a.** 6
b. 5
c. 7

1.30 **a.** propyl
b. *t*-butyl
c. ethyl
d. *sec*-butyl

1.32 **a.** 2-methylbutane
b. 2,2-dimethylpropane
c. 2,3-dimethylpentane
d. 4,6,6-triethyl-2,7-dimethylnonane
e. 5-*sec*-butyl-2,4-dimethylnonane

1.34

a. $CH_3\overset{\overset{\displaystyle CH_3}{|}}{\underset{\underset{\displaystyle CH_3}{|}}{CH}}CHCH_3$

b. $CH_3CH_2\overset{\overset{\displaystyle CH_3CHCH_3}{|}}{CH}CH_2CH_2CH_2CH_3$

c. $CH_3CHCHCH_2CHCH_2CH_2CH_3$

with substituents: CH_3 above, CH_3CCH_3 (with CH_3 on top) above, and CH_3 below.

d. $CH_3CH_2CH_2CCH_2CH_2CH_3$

with CH_2CH_3 above and CH_3 below.

1.36

$CH_3CH_2CH_2CH_2CH_3$ $CH_3CHCH_2CH_3$ (with CH_3 above)
pentane   2-methylbutane

CH_3CCH_3 (with CH_3 above and CH_3 below)
2,2-dimethylpropane

1.38

$CH_3CH_2CH_2CH_2CH_2CH_3$ hexane

$CH_3CHCH_2CH_2CH_3$ (with CH_3 above)   2-methylpentane

$CH_3CH_2CHCH_2CH_3$ (with CH_3 above)   3-methylpentane

$CH_3CCH_2CH_3$ (with CH_3 above and CH_3 below)   2,2-dimethylbutane

$CH_3CHCHCH_3$ (with CH_3 above and CH_3 below)   2,3-dimethylbutane

1.40 a. CH_3CHCH_3 (with CH_3 below)

The longest carbon chain is four carbon atoms rather than three. The correct name is 2-methylbutane.

b. $CH_3CH_2CHCHCH_3$ (with CH_3 above and CH_3 below)

The chain should be numbered from the right to give 2,3-dimethylpentane.

c. $CH_3CHCH_2CHCH_3$ (with CH_2CH_3 above and CH_3 below)

The longest carbon chain is six carbon atoms and should be numbered from the right to give 2,4-dimethylhexane.

d. $CH_3CHCHCH_3$ (with Br above and CH_2CH_3 below)

The longest carbon chain is five carbon atoms rather than four. The correct name is 2-bromo-3-methylpentane.

1.42 a. cyclopropane
b. 1,2-dimethylcyclopentane
c. 1-ethyl-1-methylcyclohexane
d. 1-isopropyl-2-methylcyclobutane

1.44

a. cyclopentane with $CHCH_3$ substituent (CH_3 above)

b. cyclobutane with CH_3 and CH_3 substituents

c. cyclohexane with CH_2CHCH_3 (CH_3 above) and $CHCH_3$ (CH_3 below) substituents

1.46 b, c, and d

1.48 So that the carbon atoms with attached hydrogens can assume a tetrahedral shape.

1.50 c and d

c.

cyclobutane with two CH_3 groups   *trans*-1,2-dimethylcyclobutane

cyclobutane with two CH_3 groups   *cis*-1,2-dimethylcyclobutane

d. cyclopropane with CH_3 and $CHCH_3$ (CH_3)   *trans*-1-isopropyl-2-methylcyclopropane

H_3C cyclopropane with $CHCH_3$ (CH_3)   *cis*-1-isopropyl-2-methylcyclopropane

1.52 a. *trans*-1-ethyl-2-methylcyclopropane
b. *cis*-1-bromo-2-chlorocyclopentane
c. *trans*-1-methyl-2-propylcyclobutane
d. *trans*-1,3-dimethylcyclohexane

1.54 a. Liquid
b. No
c. Yes
d. Less dense

1.56 2-methylheptane

1.58 a. $C_3H_8 + 5O_2 \longrightarrow 3CO_2 + 4H_2O$
b. $2C_4H_{10} + 13O_2 \longrightarrow 8CO_2 + 10H_2O$
c. $2C_5H_{10} + 15O_2 \longrightarrow 10CO_2 + 10H_2O$

1.60 $2C_6H_{14} + 13O_2 \longrightarrow 12CO + 14H_2O$

1.62 The goal of combinatorial chemistry is to synthesize a large variety of organic compounds, and then rapidly test them for desirable properties.

1.64 Mineral oil and petroleum jelly

························ **CHAPTER 2** ························

2.2 a. 2-butene
b. 3-ethyl-2-pentene
c. 4,4-dimethyl-2-hexyne
d. 4-methylcyclopentene
e. 6-bromo-2-methyl-3-heptyne
f. 1-ethyl-2,3-dimethylcyclopropene
g. 6-methyl-1,4-heptadiene

2.4

a. $CH_3CH=CHCHCH_2CH_3$ (with CH_3 substituent)

b. $CH\equiv CCH_2CCH_3$ (with two CH_3 substituents)

c. $CH_2=CH-CH=CH_2$

d.

e.

2.6 $CH\equiv CCH_2CH_2CH_3$ 1-pentyne

$CH=CHCH=CHCH_3$ 1,3-pentadiene

cyclopentene

2.8 3,7,11-trimethyl-1,3,6,10-dodecatetraene

2.10

a. $CH_3CHCH_2CH=CHCH_3$ (with CH_3 substituent)

5-methyl-2-hexene
The chain is numbered from the right.

b. $CH_3CH_2CH=CHCH=CHCH_3$

2,4-heptadiene
The chain is numbered from the right.

c.

3-methylcyclobutene
The ring is numbered as shown.

2.12 The overlap of $2p$ orbitals forms a π (pi) bond containing two electrons.

2.14 Structural isomers have a different order of linkage of atoms. Geometric isomers have the same order of linkage of atoms but different three-dimensional arrangements of their atoms in space.

2.16

a. cis-4-octene

trans-4-octene

c. cis-1,4-dibromo-2-butene   *trans*-1,4-dibromo-2-butene

2.18 a.

b.

2.20 Both alkanes and alkenes are nonpolar compounds that are insoluble in water, less dense than water, and soluble in nonpolar solvents.

2.22 When H—X adds to an alkene, the hydrogen becomes attached to the carbon atom that is already bonded to more hydrogens.

2.24 a. $CH_3CH_2CH_2CH_3$

b. $Br-CH_2CHCH_2CH_3$ (with Br substituent)

c.

d.

2.26 a. Cl_2
   b. H_2 with Pt catalyst
   c. HCl
   d. H_2O with H_2SO_4 catalyst

2.28 The addition of Br_2 to these samples will cause the cyclohexane solution to become light orange with unreacted Br_2, whereas the 2-hexene will react with the Br_2 and remain colorless.

2.30 A monomer is a starting material used in the preparation of polymers, long-chain molecules made up of many repeating units. Addition polymers are long-chain molecules prepared from alkene monomers through numerous addition reactions. A copolymer is prepared from two monomer starting materials.

2.32 A carbon–carbon double bond

2.34

$$n CH_2 = CH \xrightarrow{\text{catalyst}} + CH_2 - CH +_n$$
$$\quad\quad\ |\quad\quad\quad\quad\quad\quad\quad\quad\quad |$$
$$\quad\quad Cl \quad\quad\quad\quad\quad\quad\quad\quad Cl$$

2.36 sp; two

2.38 Linear

2.40 Acetylene is used as a fuel for welding torches and in the synthesis of monomers for plastics and fibers.

2.42 Nonpolar, water-insoluble compounds that undergo addition reactions and combustion

2.44 Unhydridized p orbitals

2.46 Aromatic refers to compounds containing the benzene ring. Aliphatic substances do not contain a benzene ring.

2.48 In the cyclohexane ring, attached groups are located above and below the plane of the ring. In benzene, attached groups are located in the same plane as the ring.

2.50 a. isopropylbenzene
   b. 1,3-diethylbenzene

2.52 a. 2-phenyl-1-butene
   b. 3-phenylcyclopentene

2.54 a. *m*-bromophenol
   b. *p*-ethylaniline

2.56 a. 2,6-dibromobenzoic acid
   b. 3-chloro-5-ethyltoluene

2.58 a.

b.

c. $CH_3CHCHCH_3$ with CH_3 group and phenyl

2.60 Nonpolar and insoluble in water

2.62 Cyclohexene readily undergoes addition reactions. Benzene resists addition reactions and favors substitution reactions. Both undergo combustion.

2.64 a. phenol
   b. styrene
   c. aniline
   d. phenol

2.66 beta-carotene

2.68 Terpenes are synthesized from multiple units of isoprene, which contains five carbon atoms.

2.70 A carbocation is an ion of the form . It is formed when a H^+ ion bonds to a carbon of a carbon–carbon double bond.

2.72 nicotine

2.74 Polycyclic aromatic hydrocarbons contain "fused" benzene rings. They form in the pyrolysis of organic matter.

················ **CHAPTER 3** ················

3.2 a. 1-propanol
   b. 3-phenyl-1-propanol
   c. 2-chloro-3-methyl-1-butanol
   d. 1,4-butanediol
   e. 2-chloro-5-ethylcyclopentanol
   f. 1,2,4-cyclohexanetriol

3.4 a. methanol
   b. 2-propanol
   c. ethanol
   d. 1,2-ethanediol
   e. 1,2,3-propanetriol

3.6 a. $CH_3CCH_2CH_2CH_3$ with OH and CH_3 substituents

b. $CH_2CH_2CHCH_3$ with two OH groups

c.

3.8 a. 2-isopropylphenol
  b. 2,3-dimethylphenol

3.10 a.

  b.

3.12 a. Tertiary
  b. Primary
  c. Secondary

3.14

$CH_3CH_2CH_2CH_2-OH$ 1-butanol   primary

$CH_3CHCH_2CH_3$ 2-butanol   secondary

CH_3CHCH_2-OH 2-methyl-1-propanol   primary
  CH_3

CH_3CCH_3 2-methyl-2-propanol   tertiary
  CH_3

3.16 a. methanol, ethanol, 1-propanol
  b. butane, 1-propanol, ethylene glycol

3.18 a.

  b.

3.20 a.

  b. $CH_3CH=CCH_3$
          CH_2CH_3

3.22 a. $CH_3CH_2CH_2-O-CH_2CH_2CH_3$

  b.

  c.

3.24 a.

  b.

  c. CH_3CHCH_2-OH
          CH_3

3.26 a.

  b. $2CH_3CH_2CH_2-OH \xrightarrow{H_2SO_4}{140°C}$
          $CH_3CH_2CH_2-O-CH_2CH_2CH_3 + H_2O$

  c.

  d.

  e. $CH_3CH_2CH_2CH_2CH_2CH_2-OH + 2(O) \longrightarrow$

3.28

  a.

  b.

c. $CH_3CH_2CH_2CH_2 \overset{\overset{\displaystyle OH}{|}}{} \xrightarrow[180°C]{H_2SO_4} CH_3CH_2CH=CH_2 + H_2O$

$CH_3CH_2CH=CH_2 + H_2O \xrightarrow{H_2SO_4} CH_3CH_2\overset{\overset{\displaystyle OH}{|}}{C}HCH_3$

$CH_3CH_2\overset{\overset{\displaystyle OH}{|}}{C}HCH_3 + (O) \rightarrow CH_3CH_2\overset{\overset{\displaystyle O}{||}}{C}CH_3 + H_2O$

3.30 $CH_3\overset{\overset{\displaystyle O}{||}}{C}-\overset{\overset{\displaystyle O}{||}}{C}-OH$

3.32 Methanol is used as an automobile fuel and in the preparation of formaldehyde, a starting material for certain plastics.

3.34 a. glycerol
b. ethanol
c. ethylene glycol
d. isopropyl alcohol
e. menthol
f. ethanol

3.36 a. o-phenylphenol or 2-benzyl-4-chlorophenol
b. 4-hexylresorcinol or phenol
c. BHA or BHT

3.38 a. ethyl methyl ether
b. phenyl propyl ether
c. di-sec-butyl ether

3.40 a. 1-methoxypropane
b. 1-ethoxy-3-methylpentane
c. propoxycyclopropane
d. 1,3,5-trimethoxybenzene

3.42 a. $CH_3-O-\overset{\overset{\displaystyle CH_3}{|}}{C}HCH_3$

b. ⬡$-O-CH_2CH_2CH_3$

c. $CH_3\overset{\overset{\displaystyle OCH_3}{|}}{C}HCH_2CH_2CH_3$

d. cyclopentane with OCH_2CH_3 and OCH_2CH_3 substituents

e. phenyl $O-\overset{\overset{\displaystyle CH_3}{|}}{C}=CHCH_3$

3.44 Inertness

3.46 $CH_3CH_2CH_2-OH$, $CH_3CH_2-O-CH_3$, $CH_3CH_2CH_2CH_3$
This order is based on the decreasing ability to form hydrogen bonds with water.

3.48 $CH_3CH_2-O\overset{\displaystyle CH_2CH_3}{\underset{\cdots}{}}$ ... with O and H H

3.50 a. $CH_3CH_2CH_2-S-Hg-S-CH_2CH_2CH_3 + 2H^+$
b. $CH_3CH_2CH_2-S-S-CH_2CH_2CH_3 + H_2O$

c. cyclopentane with SH $+ CH_3-SH$

3.52 Alcohols are oxidized to aldehydes and ketones, whereas thiols are oxidized to disulfides.

3.54

$CH_3\overset{\overset{\displaystyle OH}{|}}{C}HCH_3 + H_2SO_4 \rightleftharpoons CH_3\overset{\overset{\displaystyle \overset{+}{O}H_2}{|}}{C}HCH_3 + HSO_4^-$

$CH_3\overset{\overset{\displaystyle \overset{+}{O}H_2}{|}}{C}HCH_3 \rightleftharpoons CH_3\overset{+}{C}HCH_3 + :\overset{\overset{\displaystyle H}{}}{\underset{\displaystyle H}{O}}$

$CH_2-\overset{+}{C}HCH_3 \rightleftharpoons CH_2=CHCH_3 + H_2SO_4$
with H and HSO_4^-

3.56 urushiol

3.58 Vitamin E serves as an antioxidant, protecting lipids and cell membranes from oxidation.

.... **CHAPTER 4** ....

4.2 a. aldehyde
b. neither
c. ketone
d. neither
e. ketone
f. neither

4.4 a. propanal
b. 3-bromopropanal
c. 3-phenylpropanal
d. 4-methyl-2-pentanone
e. 2-isopropylcyclopentanone

4.6

a. $CH_3CH_2CH_2-\overset{\overset{\displaystyle O}{||}}{C}-H$

b. $CH_3\overset{\overset{\displaystyle O}{||}}{\underset{\underset{\displaystyle CH_3}{|}}{C}H}-C-CH_2CH_3$

c. cyclohexanone with CH_2CH_3 and $CH_2CH_2CH_3$ substituents

d. $CH_3CHCH_2CH - \overset{\overset{\displaystyle O}{\|}}{C} - H$ (with Cl above first CH carbon and a benzene ring below)

4.8 a. $H - \overset{\overset{\displaystyle O}{\|}}{C} - CHCHCH_3$ (with CH_2CH_3 above and CH_3 below)

The longest chain contains five carbon atoms. The correct name is 2,3-dimethylpentanal.

b. The component is an aldehyde. Number from the carbonyl group.

$CH_3CHCH_2 - \overset{\overset{\displaystyle O}{\|}}{C} - H$ (with CH_3 above)

The correct name is 3-methylbutanal.

c.

(cyclopentanone ring with O at top, Br at two adjacent positions)

Number the ring so that the bromines have lower numbers. The correct name is 2,3-dibromocyclopentanone.

4.10 Acetone dissolves remaining traces of water, and together they are discarded. The remaining small amounts of acetone evaporate.

4.12 Propane is nonpolar. Ethanal is a polar molecule and exhibits greater interparticle forces.

4.14 a.

(cyclopentane ring with $\overset{\overset{\displaystyle O \cdots H}{\|}}{C} - CH_3$ and water molecule)

b.

(benzene ring with $\overset{\overset{\displaystyle O \cdots H}{\|}}{C} - H$ and water molecule)

4.16 The alcohol will have the higher melting point because of hydrogen bonding. Menthol is a solid, and menthone is a liquid.

4.18

a.

(cyclopropane ring with OH and CH_3) $+ (O) \longrightarrow$ (cyclopropane ring with O and CH_3) $+ H_2O$

b. $CH_3CHCHCH_3 + (O) \longrightarrow CH_3 - \overset{\overset{\displaystyle O}{\|}}{C} - CHCH_3 + H_2O$ (with OH above first and Br below on left; Br below on right)

c.

(benzene ring with $CH_2 - OH$ and CH_2CH_3) $+ (O) \longrightarrow$ (benzene ring with $\overset{\overset{\displaystyle O}{\|}}{C} - H$ and CH_2CH_3) $+ H_2O$

4.20 a. hemiketal
   b. hemiacetal
   c. neither
   d. hemiacetal

4.22 a. ketal
   b. acetal
   c. neither

4.24 a. ketal
   b. none of these
   c. hemiacetal

4.26 They are unstable, splitting into an alcohol and aldehyde.

4.28 The formation of a silver precipitate or mirror

4.30 b. (benzene ring with $\overset{\overset{\displaystyle O}{\|}}{C} - O^-NH_4^+$)

   d. $CH_3CH_2 - \overset{\overset{\displaystyle O}{\|}}{C} - O^-NH_4^+$

4.32 a. no
   b. no
   c. yes
   d. yes
   e. no

4.34 $\overset{\overset{\displaystyle OH}{|}}{CH_2} - \overset{\overset{\displaystyle OH}{|}}{CH} - \overset{\overset{\displaystyle OH}{|}}{CH} - \overset{\overset{\displaystyle OH}{|}}{CH} - \boxed{\overset{\overset{\displaystyle OH}{|}}{CH} - \overset{\overset{\displaystyle O}{\|}}{C} - H}$

4.36 $\overset{\overset{\displaystyle OH}{|}}{CH_2} - \overset{\overset{\displaystyle OH}{|}}{CH} - \overset{\overset{\displaystyle OH}{|}}{CH} - \overset{\overset{\displaystyle OH}{|}}{CH} - \boxed{\overset{\overset{\displaystyle O}{\|}}{C} - \overset{\overset{\displaystyle OH}{|}}{CH_2}}$ or

$\overset{\overset{\displaystyle OH}{|}}{CH_2} - \overset{\overset{\displaystyle OH}{|}}{CH} - \overset{\overset{\displaystyle OH}{|}}{CH} - \boxed{\overset{\overset{\displaystyle OH}{|}}{CH} - \overset{\overset{\displaystyle O}{\|}}{C}} - \overset{\overset{\displaystyle OH}{|}}{CH_2}$

4.38 a.

(two cyclohexane rings connected by $\overset{\overset{\displaystyle OH}{|}}{CH}$)

   b. $CH_3CH_2 - \overset{\overset{\displaystyle OCH_2CH_2CH_3}{|}}{\underset{\underset{\displaystyle CH_3}{|}}{C}} - OCH_3 \quad + H_2O$

c.

$$CH_3CH_2 - \overset{\overset{\displaystyle O-CHCH_3}{|}}{\underset{\underset{\displaystyle CH_3}{|}}{\underset{\displaystyle O-CHCH_3}{\overset{|}{C}}}} - H \quad + H_2O$$

$$\overset{\displaystyle CH_3}{|}$$

d.

$$\text{(cyclopentane)} - CH_2 - \overset{\overset{\displaystyle O}{||}}{C} - OH$$

e.

$$CH_3CH_2\overset{\overset{\displaystyle CH_3}{|}}{CH}CH_2 - OH$$

4.40 Alkenes give alkanes.
Alkynes give alkenes and alkanes.
Aldehydes give primary alcohols.
Ketones give secondary alcohols.

4.42 a.

$$\overset{\overset{\displaystyle OH}{|}}{CH_2} \qquad \overset{\overset{\displaystyle O}{||}}{C} - CH_3 + HO - CH_2CH_3$$

b.

$$\overset{\displaystyle CH_2 - OH}{\diagup} \qquad C = O + HO - CH_2CH_2CH_3 \atop H$$

4.44

a.

$$CH_3CH_2\overset{\overset{\displaystyle OH}{|}}{CH_2} + (O) \longrightarrow CH_3CH_2 - \overset{\overset{\displaystyle O}{||}}{C} - H$$

$$CH_3CH_2 - \overset{\overset{\displaystyle O}{||}}{C} - H + 2CH_3CH_2 - OH \overset{H^+}{\rightleftharpoons}$$

$$CH_3CH_2 - \overset{\overset{\displaystyle OCH_2CH_3}{|}}{\underset{\underset{\displaystyle OCH_2CH_3}{|}}{CH}} + H_2O$$

b.

$$CH_3\overset{\overset{\displaystyle CH_3}{|}}{CH} - \overset{\overset{\displaystyle O}{||}}{C} - CH_3 + H_2 \overset{Pt}{\longrightarrow} CH_3\overset{\overset{\displaystyle CH_3}{|}}{CH} - \overset{\overset{\displaystyle OH}{|}}{CH} - CH_3$$

$$CH_3\overset{\overset{\displaystyle CH_3}{|}}{CH} - \overset{\overset{\displaystyle OH}{|}}{CH} - CH_3 \overset{H_2SO_4}{\underset{180°}{\longrightarrow}} CH_3\overset{\overset{\displaystyle CH_3}{|}}{C} = CH - CH_3 + H_2O$$

4.46 a. menthone
   b. biacetyl
   c. cinnamaldehyde
   d. vanillin

4.48 dihydroxyacetone

4.50 Norethynodrel is effective because it mimics the natural hormone progesterone.

4.52 hot and spicy

5.2 Carboxylic acids with more than ten carbon atoms are called fatty acids. They were originally produced by the saponification of fats.

5.4 lactic acid

5.6 a. butanoic acid
   b. 3-methoxypropanoic acid
   c. 3-bromo-4-methylpentanoic acid
   d. 3-ethylbenzoic acid
   e. 4-bromo-2-methylbutanoic acid

5.8 a.

$$CH_3CH_2CH_2CH_2CH_2 - \overset{\overset{\displaystyle O}{||}}{C} - OH$$

b.

$$CH_3\overset{\overset{\displaystyle Br}{|}}{CH}CHCH_2 - \overset{\overset{\displaystyle O}{||}}{C} - OH \atop \underset{\displaystyle CH_3}{|}$$

c.

$$\overset{\overset{\displaystyle O}{||}}{C} - OH \atop CH_2CH_3$$

5.10 a. Acetic acid because it forms dimers
   b. Propanoic acid because it forms dimers
   c. Butyric acid because it has a higher molecular weight

5.12

$$CH_3CH_2 - \overset{\overset{\displaystyle O \cdots \cdot HO}{||}}{C} - OH \cdots \cdot O \atop \overset{\displaystyle - C - CH_2CH_3}{||}$$

5.14 The carboxyl group is hydrophilic; the carbon chain is hydrophobic.

5.16 b, d, a, c

5.18 Under cellular pH conditions, lactic acid ionizes to produce the anion lactate.

5.20 a.

$$H - \overset{\overset{\displaystyle O}{||}}{C} - O^-Na^+ + H_2O$$

b.

$$CH_3CH_2CH_2 - \overset{\overset{\displaystyle O}{||}}{C} - O^-K^+ + H_2O$$

5.22 a.

$$CH_3CH_2 - \overset{\overset{\displaystyle O}{||}}{C} - OH + KOH \longrightarrow$$

$$CH_3CH_2 - \overset{\overset{\displaystyle O}{||}}{C} - O^-K^+ + H_2O$$

b.

$$CH_3CH_2 - \overset{\overset{\displaystyle O}{||}}{C} - OH + H_2O \longrightarrow$$

$$CH_3CH_2 - \overset{\overset{\displaystyle O}{||}}{C} - O^- + H_3O^+$$

c. $CH_3CH_2-\overset{\overset{O}{\|}}{C}-OH + NaOH \longrightarrow$

$CH_3CH_2-\overset{\overset{O}{\|}}{C}-O^-Na^+ + H_2O$

5.24 a. sodium 3-bromobutanoate
b. calcium formate
c. potassium 3-phenoxypropanoate

5.26 a. $H-\overset{\overset{O}{\|}}{C}-O^-Na^+$

b. $(CH_3\overset{\overset{CH_3}{|}}{C}HCH_2-\overset{\overset{O}{\|}}{C}-O^-)_2Ca^{2+}$

c.

5.28 a. sodium stearate
b. acetic acid
c. sodium benzoate
d. zinc 10-undecylenate
e. sodium propanoate
f. citric acid/sodium citrate

5.30 b.

d.

f.

5.32 a. $CH_3-\overset{\overset{O}{\|}}{C}-O-CH_3 + CH_3-\overset{\overset{O}{\|}}{C}-OH$

b. $CH_3\overset{\overset{}{}}{\underset{\underset{CH_3}{|}}{C}}H-\overset{\overset{O}{\|}}{C}-O-CH_2-\bigcirc + H_2O$

c. $CH_3\overset{\overset{}{}}{\underset{\underset{CH_3}{|}}{C}}H-\overset{\overset{O}{\|}}{C}-O-\bigcirc + HCl$

5.34 a. $CH_3CH_2-\overset{\overset{O}{\|}}{C}-O-CH_3$

b. $CH_3CH_2-\overset{\overset{O}{\|}}{C}-O-\bigcirc$

c. $CH_3CH_2-\overset{\overset{O}{\|}}{C}-O-CH_2\overset{\overset{CH_3}{|}}{C}HCH_3$

5.36

5.38 a. ethyl formate
b. isopropyl caproate

5.40 a. ethyl butanoate
b. isopropyl 2-methylbenzoate

5.42 a. methyl propionate
b. methyl butyrate
c. methyl lactate

5.44 a. $CH_3\overset{\overset{O}{\|}}{C}-OCH_3$

b.

c. $CH_3\overset{\overset{CH_3}{|}}{C}HCHCH_2-\overset{\overset{O}{\|}}{C}-OCH_2CH_3$ with CH_3 below

5.46 $CH_3CH_2CH_2-\overset{\overset{O}{\|}}{C}-O-\bigcirc + H_2O \overset{H^+}{\longrightarrow}$

$CH_3CH_2CH_2-\overset{\overset{O}{\|}}{C}-OH + \bigcirc\text{-}OH$

butanoic acid   phenol

$$CH_3CH_2CH_2-\overset{\overset{\displaystyle O}{\|}}{C}-O-\bigcirc + NaOH \longrightarrow$$

$$CH_3CH_2CH_2-\overset{\overset{\displaystyle O}{\|}}{C}-O^-Na^+ + \overset{\displaystyle OH}{\bigcirc}$$

sodium butanoate   phenol

5.48 a. $CH_3CH_2\overset{\overset{\displaystyle CH_3}{|}}{CH}-\overset{\overset{\displaystyle O}{\|}}{C}-O^-Na^+ + HO-CH_2CH_3$

b. $CH_3(CH_2)_{10}-\overset{\overset{\displaystyle O}{\|}}{C}-OH + HO-CH_2CHCH_3$
$\bigcirc$

5.50 $+O-\overset{\overset{\displaystyle O}{\|}}{C}-\overset{\overset{\displaystyle O}{\|}}{C}-O-CH_2CH_2CH_2 \rightarrow_n$

5.52 $\overset{\overset{\displaystyle OH}{|}}{CH_2}-\overset{\overset{\displaystyle O}{\|}}{C}-CH_2-O-\overset{\overset{\displaystyle O}{\|}}{\underset{\underset{\displaystyle OH}{|}}{P}}-OH + H_2O$

5.54 $R-\overset{\overset{\displaystyle OH}{|}}{CH}-\overset{\overset{\displaystyle O}{\|}}{C}-OH$

5.56 atherosclerosis, esophageal and colorectal cancers

5.58 The reaction is carried out under basic conditions, where any carboxylic acid formed would be converted to the carboxylate salt.

········· **CHAPTER 6** ·········

6.2 a. Tertiary
b. Primary
c. Secondary

6.4 (1) $CH_3-CH_2-CH_2-NH_2$ primary
(2) $CH_3-\overset{\overset{\displaystyle }{}}{CH}-NH_2$ primary
$\overset{\underset{\displaystyle CH_3}{|}}{}$
(3) $CH_3-CH_2-NH-CH_3$ secondary
(4) $CH_3-\overset{\overset{\displaystyle }{}}{N}-CH_3$ tertiary
$\overset{\underset{\displaystyle CH_3}{|}}{}$

6.6 a. isopropylamine
b. ethylphenylamine
c. ethylmethylpropylamine

6.8 a. 4-aminocyclohexene
b. 1,3-diaminobutane
c. 4-amino-2-butanone

6.10 a. *N*-isopropylaniline
b. *N*-ethyl-*N*-methylaniline

6.12
a. $CH_2=\overset{\overset{\displaystyle NH_2}{|}}{C}CH_2CH_2CH_3$
$\overset{\underset{\displaystyle CH_3}{|}}{}$

b. $\overset{\displaystyle NH_2}{\bigcirc}$
$CH_2CH_2CH_3$

c. $\bigcirc\overset{\overset{\displaystyle N-CH_3}{|}}{\underset{\underset{\displaystyle CH_3}{}}{}}$

6.14 Amines form hydrogen bonds with water.

6.16 Tertiary amines do not form hydrogen bonds among themselves.

6.18 a. $CH_3CH_2-\overset{\overset{\displaystyle H}{|}}{N}-CH_3$
$\overset{\displaystyle H}{\underset{\displaystyle H}{O}}$

b. $\overset{\overset{\displaystyle H}{|}}{N}\cdots\overset{\displaystyle H}{N}$

6.20 $CH_3CH_2-NH-CH_2CH_3 + H_2O \rightleftharpoons$
$CH_3CH_2-\overset{\overset{\displaystyle H}{|}}{\underset{\underset{\displaystyle H}{|}}{N^+}}-CH_2CH_3 + OH^-$

6.22 a. $CH_3CH_2CH_2CH_2-NH_3^+Cl^-$

b. $\overset{\displaystyle NH_3^+ + OH^-}{\bigcirc}$

c. $\bigcirc NH_3^+ CH_3\overset{\overset{\displaystyle O}{\|}}{C}-O^-$

d. $CH_3\overset{\overset{\displaystyle NH_2}{|}}{CH}CH_3 + NaCl + H_2O$

e. $CH_3\overset{\overset{\displaystyle }{}}{CH}-\overset{\overset{\displaystyle O}{\|}}{C}-NH-CH_3 + HCl$
$\overset{\underset{\displaystyle CH_3}{|}}{}$

f. $\bigcirc\overset{\overset{\displaystyle O}{\|}}{C}-NH_2 + \bigcirc\overset{\overset{\displaystyle O}{\|}}{C}-OH$

6.24 The amine salts are much more soluble in water (or the aqueous fluids in the body).

6.26 (1) $CH_3 - \overset{\overset{\textstyle O}{\|}}{C} - Cl + CH_3 - NH - CH_3 \longrightarrow$

$$CH_3 - \overset{\overset{\textstyle O}{\|}}{C} - \underset{\underset{\textstyle CH_3}{|}}{N} - CH_3 + HCl$$

(2) $CH_3 - \overset{\overset{\textstyle O}{\|}}{C} - O - \overset{\overset{\textstyle O}{\|}}{C} - CH_3 + CH_3 - NH - CH_3 \longrightarrow$

$$CH_3 - \overset{\overset{\textstyle O}{\|}}{C} - \underset{\underset{\textstyle CH_3}{|}}{N} - CH_3 + HO - \overset{\overset{\textstyle O}{\|}}{C} - CH_3$$

6.28 a. NaOH
    b. NH_3
    c. HCl

6.30 Synapse

6.32 Norepinephrine and serotonin

6.34 Norepinephrine and dopamine

6.36 Epinephrine is used clinically in local anesthetics to constrict blood vessels in the area of injection.

6.38 Plants

6.40 a. caffeine
    b. atropine
    c. nicotine
    d. codeine
    e. quinine
    f. morphine

6.42 a. 2-ethylbutanamide
    b. *N*-phenylbutanamide
    c. *N*-isopropylbenzamide
    d. *N,N*-dimethylpropanamide

6.44 a.

b. $CH_3\overset{\overset{\textstyle O}{\|}}{C} - NHCH_3$

c. $CH_3CHCH_2 - \overset{\overset{\textstyle O}{\|}}{C} - NHCH_3$

6.46 a.

b. $CH_3\underset{\underset{\textstyle CH_3}{|}}{CH} - \overset{\overset{\textstyle O}{\|}}{C} - \underset{\underset{\textstyle H}{|}}{N} - H \cdots O = \underset{\underset{\textstyle CHCH_3}{|}}{C} - NH_2$
with $CHCH_3$ bearing CH_3

6.48 a. $CH_3CH_2 - \overset{\overset{\textstyle O}{\|}}{C} - O^-Na^+ + CH_3 - NH_2$

b. $CH_3CH_2 - \overset{\overset{\textstyle O}{\|}}{C} - OH + CH_3 - NH_3{}^+Cl^-$

6.50

$- \overset{\overset{\textstyle O}{\|}}{C} - O^-Na^+ + CH_3CH_2 - NH - CH_2CH_3$

6.52 Histamine

6.54 Decongestants, antihistamines, and allergy preventives

6.56 Acetaminophen does not irritate the intestinal tract and is not correlated with Reye's syndrome. However, acetaminophen has no anti-inflammatory activity.

6.58 L-dopa

.... **CHAPTER 7** ....

7.2 a. A structural material in plants
    b. An energy source in our diet
    c. A stored form of energy in animals
    d. A stored form of energy in plants

7.4 A carbohydrate is a polyhydroxy aldehyde or ketone, or a substance that yields such compounds upon hydrolysis.

7.6 They don't have four different groups attached; C number 1 has only three groups; C number 3 has four groups, but two are hydrogens.

7.8 a. No chiral carbon atoms
    b. $CH_3CH_2 - \overset{*}{\underset{\underset{\textstyle OH}{|}}{CH}} - \overset{\overset{\textstyle O}{\|}}{C} - CH_3$

    c.

7.10 The chiral carbon is at the intersection of the lines. The H and OH are above the plane of the paper, and the CHO and CH_2CH_3 are below the plane.

7.12 a. The D form is given; the L enantiomer is

$$\begin{array}{c} CH_2OH \\ | \\ C=O \\ | \\ HO-\!\!\!-\!\!\!-H \\ | \\ HO-\!\!\!-\!\!\!-H \\ | \\ CH_2OH \end{array}$$

b. The L form is given; the D enantiomer is

$$
\begin{array}{c}
\text{CHO} \\
\text{HO}\!-\!\!-\!\!\text{H} \\
\text{HO}\!-\!\!-\!\!\text{H} \\
\text{H}\!-\!\!-\!\!\text{OH} \\
\text{CH}_2\text{OH}
\end{array}
$$

7.14 a. Two chiral carbon atoms; four stereoisomers
  b. Three chiral carbon atoms; eight stereoisomers

7.16 8 aldopentoses; 4 are in the D form, and 4 are in the L form.

7.18 Optically active molecules rotate the plane of polarized light.

7.20 a. aldohexose
  b. ketopentose

7.22 Some carbohydrates have a sweet taste; these are sugars.

17.24 a. The α form is given in the text; the β form is

  b. The β form is given in the text; the α form is

7.26

α-talose   β-talose

7.28 Nonreducing sugar

7.30

a.

$$
\begin{array}{c}
\text{COOH} \\
\text{HO}\!-\!\!-\!\!\text{H} \\
\text{HO}\!-\!\!-\!\!\text{H} \quad + \text{Cu}_2\text{O} \\
\text{CH}_2\text{OH}
\end{array}
$$

b.

c.

7.32 D-glucose is the same compound as blood sugar.

7.34 The hexoses glucose and galactose differ in the positions of H and OH on carbon number 4.

7.36 Fructose can be used as a low-calorie sweetener because the caloric content per gram is about the same as for sucrose. However, since fructose is so much sweeter, less is needed to provide the same degree of sweetness.

7.38 a. Sucrose
  b. Lactose
  c. Lactose
  d. Maltose
  e. Sucrose
  f. Sucrose

7.40 a. sucrose + water $\xrightarrow{\text{acid}}$ glucose + fructose
  b. maltose + water $\xrightarrow{\text{acid}}$ glucose + glucose
  c. lactose + water $\xrightarrow{\text{acid}}$ galactose + glucose
  d. cellobiose + water $\xrightarrow{\text{acid}}$ glucose + glucose

7.42 Perform a Benedict's test. Sucrose is not a reducing sugar, but honey contains reducing sugars.

7.44 The right ring in lactose has a hemiacetal group that can open to form the aldehyde and alcohol; hence, it is a reducing sugar. Sucrose has no hemiacetal bonds, so it can't reduce Cu^{2+}.

7.46 All are reducing sugars because in each structure, a hemiacetal group is located in the right-hand ring.

7.48 a. galactose, glucose, fructose
  b. No. There is no hemiacetal or hemiketal group present.
  c. α(1 → 6) and α1 → β2

7.50 a. Amylose
  b. Cellulose
  c. Amylopectin
  d. Cellulose
  e. Glycogen

7.52 saccharin, acesulfame-K, aspartame

7.54 Heavily sugared foods would supply the needed calories but might lack other necessary nutrients.

7.56 Fiber may cause people to feel full and therefore eat less.

·················· **CHAPTER 8** ··················

8.2 Energy storage, structural components, some hormones

8.4 a. Nonsaponifiable
  b. Saponifiable
  c. Saponifiable
  d. Saponifiable
  e. Saponifiable
  f. Nonsaponifiable

8.6 A micelle is a spherical cluster with the nonpolar portions of fatty acids on the interior and the polar portions on the outside. Dispersion forces hold the nonpolar portions together.

8.8 Linoleic acid and linolenic acid are essential fatty acids because they are needed by the body but are not synthesized within the body.

8.10 Fats and oils are both triglycerides (esters). Fats are solid at room temperature; oils are liquid at room temperature. Fats tend to have more saturated fatty acids, whereas oils tend to have more unsaturated fatty acids.

8.12 Butter, lard, eggs, olive oil, canola

8.14 Triglyceride B

8.16 a. Saturated fatty acids and glycerol
b. Unsaturated fatty acids and glycerol
c. Saturated fatty acid salts and glycerol
d. Unsaturated fatty acid salts and glycerol

8.18 Hydrogenation

8.20

a.

$$CH_2-O-\overset{\overset{O}{\|}}{C}-(CH_2)_{16}CH_3$$

$$CH-O-\overset{\overset{O}{\|}}{C}-(CH_2)_{16}CH_3 + 3H_2O \xrightarrow{H^+}$$

$$CH_2-O-\overset{\overset{O}{\|}}{C}-(CH_2)_{16}CH_3$$

$$\begin{array}{l} CH_2-OH \\ CH-OH + 3HO-\overset{\overset{O}{\|}}{C}-(CH_2)_{16}CH_3 \\ CH_2-OH \qquad\qquad \text{stearic acid} \\ \text{glycerol} \end{array}$$

b. Same reaction as in part a.

c.

$$CH_2-O-\overset{\overset{O}{\|}}{C}-(CH_2)_6(CH_2-CH=CH)_2(CH_2)_4CH_3$$

$$CH-O-\overset{\overset{O}{\|}}{C}-(CH_2)_6(CH_2-CH=CH)_3CH_2CH_3 + 5H_2 \xrightarrow{Ni}$$

$$CH_2-O-\overset{\overset{O}{\|}}{C}-(CH_2)_6(CH_2-CH=CH)_2(CH_2)_4CH_3$$

$$CH_2-O-\overset{\overset{O}{\|}}{C}-(CH_2)_{16}-CH_3$$

$$CH-O-\overset{\overset{O}{\|}}{C}-(CH_2)_{16}CH_3$$

$$CH_2-O-\overset{\overset{O}{\|}}{C}-(CH_2)_6(CH_2-CH=CH)_2(CH_2)_4CH_3$$

d.

$$CH_2-O-\overset{\overset{O}{\|}}{C}-(CH_2)_{16}CH_3$$

$$CH-O-\overset{\overset{O}{\|}}{C}-(CH_2)_{16}CH_3 + 3NaOH \longrightarrow$$

$$CH_2-O-\overset{\overset{O}{\|}}{C}-(CH_2)_{16}CH_3$$

$$\begin{array}{l} CH_2-OH \\ CH-OH + 3Na^+\,{}^-O-\overset{\overset{O}{\|}}{C}-(CH_2)_{16}CH_3 \\ CH_2-OH \qquad\qquad\quad \text{soap} \\ \text{glycerol} \end{array}$$

8.22 $$CH_3(CH_2)_{12}-\overset{\overset{O}{\|}}{C}-O-(CH_2)_{15}CH_3$$

8.24
g — fatty acid
l
y
c — fatty acid
e
r
o — phosphoric acid
l
    alcohol

8.26 Lecithins serve as components in cell membranes and aid in lipid transport.

8.28 They differ in the identity of the amino alcohol portion. Lecithins contain choline, and cephalins contain ethanolamine or serine.

8.30
s
p
h
i
n — fatty acid
g
o
s
i — phosphoric acid
n
e   choline

8.32 Glycolipids, brain tissue

8.34 Phosphoglycerides, sphingomyelin, and cholesterol

8.36 The model is composed of a flexible bilayer structure with protein molecules floating in the bilayer. The lipid molecules are free to move in a lateral direction.

8.38 Lowering high blood cholesterol levels reduces the risk of heart disease and heart attack.

8.40 Bile salts emulsify lipids, providing a much greater surface area for the hydrolysis reactions of digestion.

8.42 Cholesterol

8.44 The two groups are glucocorticoids and mineralocorticoids. Cortisol, a glucocorticoid, acts to increase the glucose and glycogen concetrations in the body. Aldosterone, a mineralocorticoid, regulates the retention of Na^+ and Cl^- during urine formation.

8.46 Testosterone; estradiol, estrone, and progesterone

8.48 Estrogens are involved in egg development; progesterone causes changes in the wall of the uterus to prepare it to accept a fertilized egg and maintain the resulting pregnancy.

8.50 Arachidonic acid

8.52 Induction of labor, treatment of asthma, treatment of peptic ulcers

8.54 Testosterone and estrogen

8.56 Progestin helps protect against endometrial cancer, which is a risk when estrogen is taken alone.

8.58 It can cause gastrointestinal side effects such as diarrhea and can rob the body of the fat-soluble vitamins.

8.60 Cigarette smoking, obesity, and an inactive lifestyle

········· **CHAPTER 9** ·········

9.2 $CH_3CH_2CH_2 \overset{*}{-} \overset{O}{\underset{\parallel}{C}} - OH$

9.4 a. $(H_3\overset{+}{N}) - \underset{\underset{(OH)}{\overset{|}{(CH-CH_3)}}}{\overset{(H)}{\overset{|}{C}}} - (COO^-)$

b. $(H_3\overset{+}{N}) - \underset{\underset{(COO^-)}{\overset{|}{(CH_2)}}}{\overset{(H)}{\overset{|}{C}}} - (COO^-)$

c. $(H_3\overset{+}{N}) - \underset{\underset{(OH)}{\overset{|}{(CH_2)}}}{\overset{(H)}{\overset{|}{C}}} - (COO^-)$

d. $(H_3\overset{+}{N}) - \underset{\overset{|}{(CH_2)}}{\overset{(H)}{\overset{|}{C}}} - (COO^-)$ (benzyl ring)

9.6 a.

L-cysteine

D-cysteine

b.

L-glutamate

D-glutamate

9.8 They are crystalline solids with relatively high melting points and high water solubilities.

9.10

a. $H_2N-CH-COO^-$ with CH_2, CH_2, $C=O$, NH_2

b. $HN-CH-COO^-$ with CH_2 CH_2 / CH_2

9.12 a.

$H_3\overset{+}{N}-CH-COO^- + OH^- \longrightarrow H_2N-CH-COO^- + H_2O$ (with CH_2, OH side chains)

b.

$H_3\overset{+}{N}-CH-COO^- + H_3O^+ \longrightarrow H_3\overset{+}{N}-CH-COOH + H_2O$ (with CH_2, OH side chains)

9.14

$H_3\overset{+}{N}-CH-\overset{O}{\underset{\parallel}{C}}-NH-CH-\overset{O}{\underset{\parallel}{C}}-NH-CH-\overset{O}{\underset{\parallel}{C}}-O^-$ with side chains $CHCH_3 / CH_3$, CH_2 / COO^-, CH_2 / SH

Val-Asp-Cys

9.16 Six

9.18 Each peptide contains a disulfide bridge in which two cysteines have been linked.

9.20 Normally proteins do not pass through membranes, and the appearance of an abnormal protein in blood or urine indicates cellular damage, which has released those substances.

9.22 When a protein is at a pH corresponding to the isoelectric point, the net charge on the protein is 0. The protein is not as soluble with a 0 charge as when it possesses a net negative or positive charge.

9.24 Catalysis, structural, storage, protection, regulatory, nerve impulse transmission, motion, transport

9.26 a. Fibrous; collagen is a structural protein.
  b. Globular; lactate dehydrogenase is an enzyme and needs to be water soluble.

9.28 Simple proteins contain only amino acid residues, whereas conjugated proteins contain additional components.

9.30 The sequence of the amino acid residues in the protein

9.32

$$-NH-\underset{\underset{R}{|}}{CH}-\overset{\overset{O}{\|}}{C}-NH-\underset{\underset{R'}{|}}{CH}-\overset{\overset{O}{\|}}{C}-NH-\underset{\underset{R''}{|}}{CH}-\overset{\overset{O}{\|}}{C}-NH-\underset{\underset{R'''}{|}}{CH}-\overset{\overset{O}{\|}}{C}-$$

9.34 The gathering of two or more helicial proteins into a structure analogous to the strands in a cable

9.36 In secondary structures, the hydrogen bonding is between the atoms in the backbone; in tertiary structures, the hydrogen bonding is between atoms in the side chains.

9.38 a. Hydrogen bonding
  b. Salt bridges
  c. Hydrophobic interactions
  d. Hydrophobic interactions

9.40 The nonpolar R groups in phenylalanine, alanine, and methionine will point inward; the polar R groups in glutamate and lysine will point toward the water.

9.42 Ionic attractions, disulfide bridges, hydrogen bonds, and hydrophobic forces between subunits

9.44 A subunit is a polypeptide chain that is part of a larger protein.

9.46 The products of hydrolysis are the separate, individual amino acid molecules. The product of denaturation has the same primary protein backbone with different secondary, tertiary, and quaternary structure.

9.48 The primary structure of the protein is the same.

9.50 Cooking egg white is an irreversible denaturation. Whipping air into them is a reversible denaturation.

9.52 The heavy-metal ions denature the protein in the egg white by interacting with the —SH groups and the carboxylate (—COO⁻) groups. Vomiting should then be induced to remove the denatured proteins so they are not digested, releasing the metal ions.

9.54 Bromocriptine, cancer chemotherapy drugs, ergotamine, lithium, methotrexate, illegal drugs

9.56 Memory loss

.... **CHAPTER 10** ....

10.2 Enzymes are highly specific in the reactions they catalyze, and their activity as catalysts can be regulated.

10.4 Enzymes are specific for a general type of reaction and in some cases are specific for a single reactant.

10.6 International Enzyme Commission system

10.8

Enzyme	Substrate
a. urease	urea
b. fumarase	fumarate
c. fructase	fructose
d. arginase	arginine
e. lactase	lactose

10.10 a. Succinate dehydrogenase: catalyzes the dehydrogenation (oxidation) of succinate
  b. L-amino acid reductase: catalyzes the reduction of L-amino acids
  c. cytochrome oxidase: catalyzes the oxidation of cytochrome

10.12 An enzyme is a catalytically active structure made up of a protein portion called an apoenzyme and a nonprotein portion called a cofactor.

10.14 Mg^{2+}, Zn^{2+}, Fe^{2+}, Ca^{2+}

10.16 The binding of a substrate to an enzyme at the active site forms a combined substance called the enzyme-substrate complex.

10.18 According to the lock-and-key theory, the enzyme surface accommodates only substrates having a specific size and shape. According to the induced-fit theory, the enzyme is somewhat flexible and adapts its shape to accommodate or fit an incoming substrate molecule.

10.20 Any observation that allows the rate of product formation or reactant depletion to be measured will allow enzyme activity to be determined.

10.22 One international unit (IU) is the amount of enzyme that will convert 1 micromol (μ mol) of substrate to product in 1 minute. The international unit is useful in medical diagnosis because it represents a standard quantity of enzyme (or enzyme activity level) against which the level present in a patient can be compared.

10.24 a. The enzyme activity (rate of reaction) increases with increased substrate concentration until all the enzyme is utilized, at which point no further increase in rate occurs.
  b. The activity (rate) increases with an increase in enzyme concentration.
  c. There is an optimum pH for the reaction, so at a pH above or below the optimum value the rate will decrease.
  d. The activity (rate) increases up to a maximum as the temperature increases and then decreases with further temperature increase.

10.26 Prepare a series of reactions with constant enzyme amount and increasing substrate amounts. Plot reaction rate versus substrate concentration. A constant rate at higher substrate concentrations is V_{max}.

10.28 pepsin, 1.5; sucrase, 6.2; arginase, 9.0

10.30 An irreversible inhibitor forms a strong covalent bond with an enzyme. A reversible inhibitor does not attach permanently but binds reversibly to an enzyme.

10.32 a. Cyanide poisoning is treated with $Na_2S_2O_3$ (sodium thiosulfate), which oxidizes CN^- to SCN^-. The thiocyanate does not bind to the iron in the enzyme.

b. Heavy-metal poisoning is treated by administering chelating agents such as EDTA. The chelating agents bond to the metal ions to form water-soluble complexes that are excreted from the body in the urine.

10.34 1. Zymogens or proenzymes
2. Allosteric regulation
3. Genetic control over the synthesis of enzymes

10.36 Proenzyme

10.38 Modulators may increase activity (activators) or decrease the activity (inhibitors).

10.40 Genetic control of enzyme activity means that a cell synthesizes an enzyme when conditions temporarily require it. Such a controlled synthesis is called enzyme induction.

10.42 a. CK assays are used in diagnosing heart attack.
b. ALP assays are useful in diagnosing liver or bone disease.
c. Amylase assays are useful in diagnosing diseases of the pancreas.

10.44 An LDH assay is a good initial diagnostic test because the isoenzyme forms of LDH are so widespread in the human system. Thus, the locations of diseases can be pinpointed.

10.46 *Inborn errors of metabolism* is the term used for hereditary diseases characterized by the absence of enzymes.

10.48 Extremozymes are enzymes isolated from microorganisms that thrive in extreme environments.

10.50 The amount of ointment or cream applied should be equal in surface area to the tip of a finger.

10.52 Ethanol is an antidote for ethylene glycol poisoning because it can compete for the enzyme that catalyzes the first step in the oxidation of ethylene glycol to oxalic acid.

···················· ■ **CHAPTER 11** ■ ····················

11.2 Deoxyribose in DNA, ribose in RNA

11.4 a. Purine
b. Pyrimidine
c. Pyrimidine
d. Pyrimidine
e. Purine

11.6

11.8 By convention, in the DNA segment AGTCAT, the 5′ end is the A and the 3′ end is the T.

11.10 The chemical analysis of DNA from a variety of sources consistently gave equal numbers of adenine and thymine bases and also equal numbers of cytosine and guanine bases. Those data suggested a pairing between adenine and thymine and a pairing between cytosine and guanine.

11.12 a. 16
b. 15

11.14 A chromosome is a tightly packed bundle of DNA and protein that is involved in cell division. A human cell normally contains 46 chromosomes.

11.16 Helicase catalyzes the separation and unwinding of the DNA double helix.

11.18 The daughter strand forms from its 5′ end toward the 3′ end.

11.20
a. C T T A A G
   G A A T T C
b. Old strand 5′ C T T A A G 3′ ⟩ daughter DNA molecule
   New strand 3′ G A A T T C 5′
   New strand 5′ C T T A A G 3′ ⟩ daughter DNA molecule
   Old strand 3′ G A A T T C 5′

11.22 Nucleus, cytoplasm, mitochondria

11.24 The ratio of guanine to cytosine in RNA does not have to be 1:1 because parts of the single strand do not fold back to form helices. The 1:1 ratio is necessary only in those parts that do form helices. Thus, the overall ratio can be different from 1:1.

11.26 tRNA

11.28 Transcription is the transferring of genetic DNA information to messenger RNA. Translation is the expression of messenger RNA information in the form of an amino acid sequence in a protein.

11.30 5′ C A A U A U G G C 3′

11.32 In the DNA of eukaryotic cells, segments called introns carry no amino acid coding, whereas exons are coded segments. Both introns and exons are copied (transcribed) and produce hnRNA. A series of enzyme-catalyzed reactions cut out the introns and splice the remaining segments to produce mRNA.

11.34 a. histidine
b. serine
c. serine

11.36 a. False; should be three bases.
b. True
c. False; 61 of 64 codons represent amino acids.
d. False; each living species is thought to use the same genetic code.
e. True
f. False; stop signals for protein synthesis are represented by three codons: UAA, UAG, and UGA.

11.38 Initiation, elongation, termination

11.40 The A site (aminoacyl site) is where an incoming tRNA carrying the next amino acid will bond. The P site (peptidyl site) is where the tRNA carrying the peptide chain is located.

11.42 A change in the DNA of a cell wherein a wrong base sequence is produced on the DNA.

11.44 The protein synthesized would contain a proline rather than an alanine residue at that site in the protein.

11.46 Recombinant DNA is DNA of one organism that contains segments of DNA from another organism.

11.48 Human insulin will become widely available for treating diabetes. Human growth hormone will be used for treating pituitary disorders. Recombivax Hb, a vaccine for hepatitis B, will be marketed.

11.50 Reverse transcriptase inhibitors and protease inhibitors

11.52 It is a retrovirus and attacks the helper T cells, which provide immunity to diseases.

11.54 Coupling PCR techniques, which allow the replication of a small DNA sample, with DNA fingerprinting would enable scientists to identify very small samples of human DNA.

11.56 Mapping the complete human genome would identify the gene responsible for synthesizing a specific protein (or hormone, etc.).

11.58 Cells of the mammary glands of an adult sheep

···············   **CHAPTER 12**   ···············

12.2 Cellulose

12.4

Carbohydrates	Energy source
	Provide materials for cell and tissue synthesis
Proteins	Provide amino acids for cell repair and maintenance
	Provide ingredients for the synthesis of enzymes and hormones
Lipids	Energy source; solvent for fat-soluble vitamins
	Source of essential fatty acids

12.6 a. Carbohydrates, lipids, proteins
  b. Carbohydrates, lipids, proteins
  c. Carbohydrates, proteins
  d. Carbohydrates, lipids, proteins
  e. Proteins, lipids
  f. Lipids, proteins, carbohydrates

12.8 fat, 65 g; saturated fat, 20 g; total carbohydrate, 300 g; fiber, 25 g

12.10 Water-soluble vitamins can be excreted in water-based fluids much more readily than fat-soluble vitamins. Thus, large doses of fat-soluble vitamins accumulate in the body.

12.12 a. vitamin C
  b. vitamin B_1
  c. vitamin B_{12}
  d. niacin

12.14 Minerals are classified as major or trace on the basis of the amount present in the body.

12.16 The general functions of trace minerals in the body are to serve as components of vitamins (Co), enzymes (Zn, Se), hormones (I), or specialized proteins (Fe, Cu).

12.18 $6CO_2 + 6H_2O \rightarrow C_6H_{12}O_6 + 6O_2$

12.20 Energy from the sun powers the process of photosynthesis in plants. As the carbohydrates of plants are converted to CO_2 and H_2O, the released energy is trapped in molecules of ATP.

12.22 A metabolic pathway is a sequence of reactions that produces one product or accomplishes one process.

12.24 a. Stage III
  b. Stage I
  c. Stage III
  d. Stage II

12.26 The production of ATP

12.28 ATP molecules store the energy released during the oxidation of food and deliver the energy of the parts of the cell where energy is needed.

12.30 P_i is the symbol for the phosphate ion. PP_i represents the pyrophosphate ion.

12.32 The hydrolysis of phosphoenolpyruvate liberates 14.8 kcal/mol, while the hydrolysis of glycerol 3-phosphate liberates only 2.2 kcal/mol.

12.34 The triphosphate portion

12.36 Mitochondria are the sites for most ATP synthesis in the cells.

12.38 Mitochondria are cigar-shaped organelles with both an outer and an inner membrane. The inner membrane has many folds and contains the enzymes for the electron transport chain and oxidative phosphorylation. The enzymes for the citric acid cycle are attached to or near to the surface of the inner membrane.

12.40 a. FAD
  b. Coenzyme A
  c. NAD^+

12.42 $^-OOC - CH_2CH_2 - COO^-$ is oxidized. FAD is reduced.

12.44

(reduced form)

12.46 Vitamin B_{12}, vitamin D, calcium, iron, and zinc

12.48 People with anorexia have a fear of becoming fat, a preoccupation with food, a poor body image, low self-esteem and other psychological problems, and amenorrhea (in women).

···············   **CHAPTER 13**   ···············

13.2 a. Glucose
  b. Glucose and galactose
  c. Glucose and fructose

13.4 Blood sugar level is the concentration of glucose in the blood. The normal fasting level is the blood glucose concentration after a fast of 8–12 hours.

13.6 a. A blood sugar level below the normal range.
  b. A blood sugar level above the normal fasting level.
  c. A blood sugar level at which glucose is not reabsorbed by the kidneys.
  d. An elevated blood sugar level resulting in the excretion of glucose in the urine.

13.8 If the blood sugar is too low, severe hypoglycemia can cause convulsions and shock.

13.10 Glucose gives pyruvate.

13.12 Steps 1 and 3 require ATP, Steps 7 and 10 produce ATP.

13.14

$$
\begin{array}{ccc}
\text{1 CHO} & & \text{1 CH}_3 \\
\text{2 H} — \text{OH} & & \text{2 C} = \text{O} \\
\text{3 HO} — \text{H} & \longrightarrow & \text{3 COO}^- \\
\text{4 H} — \text{OH} & & \text{4 COO}^- \\
\text{5 H} — \text{OH} & & \text{5 C} = \text{O} \\
\text{6 CH}_2\text{OH} & & \text{6 CH}_3
\end{array}
$$

13.16 Glucose 6-phosphate is a feedback inhibitor and slows down the first reaction of glycolysis.

13.18 Aerobic denotes the presence of oxygen, whereas anaerobic means without oxygen.

13.20 Lactate formation from pyruvate regenerates a supply of NAD^+, which is necessary for glycolysis to continue.

13.22 The two ATPs produced are sufficient to maintain life.

13.24 The citric acid cycle is the principal process for generating NADH and $FADH_2$, which are necessary for ATP synthesis. The citric acid cycle is also important as a source of intermediates in the biosynthesis of amino acids.

13.26 a. acetyl CoA
   b. CO_2
   c. NADH and $FADH_2$

13.28 a. three
   b. one
   c. one
   d. two

13.30

$$
\begin{array}{ccc}
\text{COO}^- & & \text{H} \quad \text{COO}^- \\
\text{CH}_2 & + \text{FAD} \longrightarrow & \text{C} \\
\text{CH}_2 & & \parallel \quad + \text{FADH}_2 \\
\text{COO}^- & & \text{C} \\
& & {}^-\text{OOC} \quad \text{H}
\end{array}
$$

succinate   fumarate

13.32 a. Step 1, citrate synthetase
   Step 3, isocitrate dehydrogenase
   Step 4, α-ketoglutarate dehydrogenase
   b. When supplies of NADH and ATP are abundant, the cell has sufficient energy for its needs, and so the operation of the citric acid cycle is inhibited.
   c. As the cell uses ATP and makes ADP, the enzymes controlling the cycle are activated to make more ATP.

13.34 NADH and $FADH_2$

13.36 Cytochromes pass electrons along the electron transport chain to oxygen, which combines with H^+ ions to form water.

13.38 NADH and $FADH_2$ are oxidized to NAD^+ and FAD, respectively. ADP is phosphorylated to ATP.

13.40 1.5 molecules of ATP are formed during a sequence of reactions.

13.42 60 ATPs

13.44 94 mol of ATP

13.46 $F_1 - $ATPase

13.48 a. 2   **b.** 2   **c.** 28

13.50 The liver and muscles

13.52 UTP

13.54 Liver cells have the essential enzyme glucose 6-phosphatase, which is lacking in muscle tissue.

13.56 The liver

13.58 Lactate, glycerol (derived from the hydrolysis of fats), and certain amino acids

13.60 a. Insulin promotes the utilization of blood glucose, causing the level to decrease. Glucagon activates the breakdown of glycogen, thereby raising blood glucose levels.
   b. Insulin promotes glycogen formation, while glucagon stimulates the breakdown.

13.62 Lactase deficiency causes lactose intolerance, which may be characterized by a bloated feeling, cramps, and diarrhea when lactose-containing foods are eaten.

13.64 Strenuous exercise results in a slight decrease in blood pH, which stimulates the body to breathe heavily to bring in more oxygen.

13.66 Carbohydrate loading is a manipulation of the diet to achieve a maximum storage of energy in glycogen supplies.

13.68 In Type I diabetes, which usually appears in children before the age of 10, practically no insulin is produced. Type II diabetes results from a gradual decrease in insulin production.

13.70 Hyperglycemia, glucosuria, and a lowered blood pH may develop; if left unchecked, diabetes can eventually lead to blindness, coma, and death.

CHAPTER 14

14.2 Glycerol, fatty acids, and monoglycerides

14.4 Chylomicrons are lipoprotein aggregates found in the lymph and the bloodstream.

14.6 Chylomicrons, very low-density lipoproteins, low-density lipoproteins, high-density lipoproteins

14.8 Epinephrine initiates fat mobilization by stimulating the hydrolysis of triglycerides to fatty acids and glycerol.

14.10 Resting muscle and liver cells

14.12 Glycerol can be converted to pyruvate and contribute to cellular energy production or be converted to glucose.

14.14 A fatty acid is first converted to a fatty acyl CoA, a high-energy compound. This occurs in the cytoplasm of the cell.

14.16 FAD and NAD^+ are coenzymes that pick up hydrogen atoms to form $FADH_2$ and NADH as fatty acids are oxidized.

14.18

$$CH_3(CH_2)_{14} - \overset{\overset{\displaystyle O}{\|}}{C} - OH + HS - CoA \underset{ATP \quad AMP + 2P_i}{\longrightarrow}$$

$$CH_3(CH_2)_{14} - \overset{\overset{\displaystyle O}{\|}}{C} - SCoA + H_2O$$

14.20 The word *cycle* implies that the product is identical to the starting material, which is not the case. The fatty acyl product of a trip through the sequence has two fewer carbon atoms than the starting material.

14.22 a.

$$CH_3CH_2CH_2CH = CH - \overset{\overset{\displaystyle O}{\|}}{C} - S - CoA + FADH_2$$

b.

$$CH_3CH_2CH_2CH = CH - \overset{\overset{\displaystyle O}{\|}}{C} - S - CoA + H_2O \longrightarrow$$

$$CH_3CH_2CH_2\overset{\overset{\displaystyle OH}{|}}{C}HCH_2 - \overset{\overset{\displaystyle O}{\|}}{C} - S - CoA$$

c.

$$CH_3CH_2CH_2\overset{\overset{\displaystyle OH}{|}}{C}HCH_2 - \overset{\overset{\displaystyle O}{\|}}{C} - S - CoA + NAD^+ \longrightarrow$$

$$CH_3CH_2CH_2\overset{\overset{\displaystyle O}{\|}}{C}CH_2 - \overset{\overset{\displaystyle O}{\|}}{C} - S - CoA + NADH + H^+$$

d.

$$CH_3CH_2CH_2\overset{\overset{\displaystyle O}{\|}}{C}CH_2 - \overset{\overset{\displaystyle O}{\|}}{C} - S - CoA + CoA - SH \longrightarrow$$

$$CH_3CH_2CH_2 - \overset{\overset{\displaystyle O}{\|}}{C} - S - CoA + CH_3 - \overset{\overset{\displaystyle O}{\|}}{C} - S - CoA$$

14.24 10 ATP

14.26 64 ATP molecules

14.28 Acetoacetate, β-hydroxybutyrate, and acetone

14.30 a. The liver
b. The brain, heart, and skeletal muscles

14.32 Two of the ketone bodies are carboxylic acids, which lower blood pH.

14.34 Acetyl CoA

14.36 The required enzymes are missing.

14.38 Amino acid pool

14.40 Protein turnover

14.42 Purines and pyrimidines, heme, choline, ethanolamine

14.44 a. Transamination
b. Deamination
c. Transamination
d. Transamination and deamination

14.46

$$CH_3 - \overset{\overset{\displaystyle O}{\|}}{C} - COO^- + H_3\overset{+}{N} - CH - COO^- \longrightarrow$$

with side chain $CH_2 - CH-CH_3$ with CH_3

pyruvate   leucine

$$CH_3 - \overset{\overset{\displaystyle NH_3^+}{|}}{C}H - COO^- + \overset{\overset{\displaystyle O}{\|}}{C} - COO^-$$

with side chain $CH_2 - CH - CH_3$ with CH_3

14.48

$$H_3\overset{+}{N} - CH_2 - \overset{\overset{\displaystyle O}{\|}}{C} - O^- + NAD^+ \longrightarrow$$

$$H - \overset{\overset{\displaystyle O}{\|}}{C} - \overset{\overset{\displaystyle O}{\|}}{C} - O^- + NADH + NH_4^+$$

14.50 The liver

14.52 Carbamoyl phosphate

14.54 $NH_4^+ + CO_2 + 3ATP + 2H_2O + COO^- \longrightarrow$

$$\overset{\overset{\displaystyle CH_2}{|}}{\underset{\overset{\displaystyle CH - NH_3^+}{|}}{}}$$
$$COO^-$$
aspartate

$$H_2N - \overset{\overset{\displaystyle O}{\|}}{C} - NH_2 + {}^-OOC - CH = CH - COO^-$$
urea   fumarate

$$+ 2H^+ + 2ADP + 2P_i + AMP + PP_i$$

14.56 Yes; isoleucine, tryptophan, phenylalanine, and tyrosine

14.58 a. Glucogenic
b. Ketogenic
c. Ketogenic

14.60 Amino acids that can be synthesized by the body are called nonessential. Those amino acids that must be obtained from the diet because the body cannot produce them are called essential amino acids.

14.62 The glycolysis pathway and the citric acid cycle

14.64 Cholesterol is involved in the formation of cell membranes, the insulation of nerves, the synthesis of several hormones, and the digestion of food.

14.66 The caloric count of vegetables is much lower than that of meats.

14.68 Food protein in contrast to protein from supplements.

14.70 The diet of an infant with PKU is controlled to eliminate phenylalanine and is supplemented with tyrosine.

15.2 Plasma and interstitial fluid

15.4 **a.** Intracellular fluid
 b. Intracellular fluid
 c. Plasma and interstitial fluid
 d. Intracellular fluid
 e. Plasma and interstitial fluid
 f. Intracellular fluid
 g. Plasma and interstitial fluid

15.6 **a.** Cl^-
 b. Cl^-
 c. HPO_4^{2-}

15.8 Chloride shift

15.10 Red blood cells are the site for all the O_2 and CO_2 exchange reactions both within the lungs and at the tissue cells.

15.12 $HHb + O_2 \rightleftarrows HbO_2^- + H^+$
 a. Right
 b. Right
 c. Left
 d. Left

15.14 **a.** Osmotic pressure
 b. Blood pressure

15.16 Urea

15.18 If the concentration was abnormally high

15.20 Kidneys (urine), lungs (water vapor), skin (perspiration), and intestines (feces)

15.22 Variations in urine output

15.24 As fluid level decreases, saliva output decreases, giving the "thirst" feeling. Drinking water raises fluid level.

15.26 Blood buffers, urinary system, and respiratory system

15.28 $HPO_4^{2-} + H^+ \rightarrow H_2PO_4^-$
  $H_2PO_4^- + OH^- \rightarrow H_2O + HPO_4^{2-}$

15.30 Exhaled H_2O and CO_2 are derived from H_2CO_3. The more CO_2 and H_2O that are exhaled, the more carbonic acid is removed from the blood, thus elevating the blood pH.

15.32 Respiratory alkalosis

15.34 CO_2 diffuses into the distal tubule cells to form carbonic acid, which then ionizes to H^+ and HCO_3^-. Thus, CO_2 concentration is lowered in the blood.

15.36 As H^+ enters the urine, Na^+ is reabsorbed into the blood from the urine.

15.38 Respiratory acidosis is caused by hypoventilation. Respiratory alkalosis is caused by hyperventilation.

15.40 Vomiting results in a loss of acid from the body, which can lead to a rise in blood pH (alkalosis).

15.42 Red blood cells

15.44 There is an increased chance for blood clots to form, which can lead to strokes or heart attacks.

15.46 One pint

Appendix C

Solutions to Learning Checks

1.1 Look for the presence of carbon atoms.

    a. Inorganic   **d.** Inorganic
    b. Organic   **e.** Organic
    c. Organic   **f.** Inorganic

1.2 Refer to Table 1.1 for help.
    a. Organic
    b. Organic
    c. Inorganic

1.3 Compound (a) is a structural isomer because it has the same molecular formula but a different structural formula.

1.4
    a.

$$CH_3-\overset{\overset{\textstyle OH}{|}}{CH}-CH_2-CH_2-CH_3$$

    b.

$$CH_3-\overset{\overset{\textstyle }{}}{\underset{\underset{\textstyle CH_3}{|}}{CH}}-CH_2-\overset{\overset{\textstyle O}{\|}}{C}-OH$$

1.5 **a.** The number of hydrogen atoms is twice the carbon atoms plus two: C_8H_{18}.

    b. $CH_3-CH_2-CH_2-CH_2-CH_2-CH_2-CH_2-CH_3$

1.6 **a.** Same molecule: In both molecules, the five carbons are bonded in a continuous chain.

    b. Same molecule: In both molecules, there is a continuous chain of four carbons with a branch at position 2.

$$\underset{1}{CH_3}-\underset{2}{\overset{\overset{\textstyle CH_3}{|}}{CH}}-\underset{3}{CH_2}-\underset{4}{CH_3} \qquad \underset{1}{CH_3}-\underset{2}{\overset{\overset{\textstyle \underset{3}{CH_2}-\underset{4}{CH_3}}{|}}{CH}}-CH_3$$

    c. Structural isomers: Both molecules have a continuous chain of five carbons, but the branch is located at different positions.

1.7 **a.**

$$\underset{2}{CH_2}\overset{\overset{\textstyle \underset{1}{CH_3}}{|}}{}-\underset{3}{CH_2}-\underset{4}{CH_3} \qquad \text{a butane}$$

    b.

$$\overset{\underset{1}{CH_3}}{\underset{\underset{2}{CH_2}}{|}}-\underset{3}{\overset{}{CH}}\overset{\underset{5}{CH_3}}{\underset{\underset{4}{CH_2}}{|}} \qquad \text{a pentane}$$
$$\underset{CH_3}{|}$$

    c.

$$\underset{1}{CH_3}-\underset{2}{\overset{\overset{\textstyle }{}}{CH}}-CH_3 \qquad \text{or}$$
$$\underset{3}{CH_3}-\underset{4}{CH}-CH_3$$

$$CH_3-\overset{}{CH}-CH_3$$
$$\underset{\underset{1}{CH_3}}{|}-\underset{2}{\overset{}{CH}}-CH_3 \qquad \text{a butane}$$

1.8 **a.**

$$\overset{3}{CH_3}-\overset{4}{\overset{\overset{\textstyle }{}}{CH}}-\overset{5}{CH_2}-\overset{6}{CH_2}-CH_3$$
$$\underset{\underset{2}{CH_2}}{|}$$
$$\underset{\underset{1}{CH_3}}{|}$$

A CH_3 group is located at position 3. If the chain had been numbered beginning at the right, the CH_3 group would have been at position 4.

    b. Numbering from the left, groups are located at positions 5, 5, 7. From the right, the groups are at positions 2, 4, 4. The first difference occurs with the first number, so numbering from the right (2, 4, 4) is correct.

1.9 Proceeding from the left, the groups are a methyl, propyl, isopropyl, *sec*-butyl, and ethyl.

1.10 **a.** The chain is numbered from the right to give 2-methylhexane.

    b.

$$\underset{8}{CH_3}-\underset{7}{CH_2}-\underset{6}{CH_2}-\underset{5}{\overset{\overset{\textstyle \underset{3}{CH_2}-CH_3}{|}}{CH}}-\underset{4}{CH_2}-\underset{}{\overset{\overset{\textstyle }{}}{CH}}-CH_3$$
$$\overset{}{\underset{\underset{2}{CH_2}-\underset{1}{CH_3}}{|}}$$

The chain is numbered from the right to give 5-ethyl-3-methyloctane.

    c. The chain is numbered from the right to give 4-isopropyl-2,3-dimethylheptane.

1.11 **a.**

$$CH_3-\overset{\overset{\textstyle CH_3}{|}}{\underset{\underset{\textstyle CH_3}{|}}{C}}-CH_2-\overset{\overset{\textstyle CH_3}{|}}{CH}-CH_3$$

    b.

$$CH_3-CH_2-\overset{\overset{\textstyle \overset{\overset{\textstyle CH_3}{|}}{CH}-CH_3}{|}}{CH}-CH_2-CH_2-CH_3$$

c.
$$CH_3 - CH - CH - CH - CH_2 - CH_2 - CH_3$$
with $CH_2 - CH_3$ above the second carbon (CH), CH_3 below the third carbon (CH), and CH_3 below the fourth carbon (CH).

1.12 **a.** 1,4-dimethylcyclohexane

b. The correct name is ethylcyclopropane. When only one group is attached to a ring, the position is not designated.

c. The name 1-ethyl-2-methylcyclopentane is correct, whereas 2-ethyl-1-methylcyclopentane is incorrect because the ring numbering begins with the carbon attached to the first group alphabetically.

1.13 **a.** (1) *Trans* because the two Br's are on opposite sides of the ring.
(2) *Cis*; both Cl's are on the same side.
(3) *Cis*; the two groups are on the same side.

b. In showing geometric isomers of ring compounds, it helps to draw the ring in perspective:

························ **CHAPTER 2** ························

2.1 In each of these alkenes, the double-bonded carbons occur at positions 1 and 2.

    a. 3-bromo-1-propene
    b. 2-ethyl-1-pentene
    c. 3,4-dimethylcyclohexene

2.2 **a.** The chain is correctly numbered from the right to give 2-methyl-1,3-butadiene.

b. The chain is correctly numbered from the left to give 2-methyl-1,3,6-octatriene.

    c. 7-bromo-1,3,5-cycloheptatriene

2.3 **a.** This structure does not exhibit geometric isomerism because there are two H's attached to the carbon at position 1.

b.

cis and trans structures:

cis: $CH_3-CH(CH_3)$ and H on one carbon, H and CH_3 on other carbon ($C=C$)

trans: $CH_3-CH(CH_3)$ and H, CH_3 and H on the double bond ($C=C$)

    c. This structure does not exhibit geometric isomerism because there are two methyl groups attached to the left double-bonded carbon:

2.4
a.
$$CH_3 - C - CH_2$$
with Br, Br on top carbon and CH_3 below.

b. cyclopentane with Cl and Cl substituents

2.5 **a.**
$$CH_3 - CH - CH_2 - CH_3$$
with CH_3 below.

b. cyclopentane with CH_3 and H (= methylcyclopentane)

2.6 **a.** The major product will be the one where H attaches to the CH carbon:

$$CH_3 - CH_2 - \underset{Br}{\overset{CH_3}{C}} - CH_2 - CH_3$$

b. Position 1 on the ring has an attached hydrogen, whereas position 2 does not have any attached hydrogens:

cyclohexene with CH_3 and CH_2CH_3 groups, positions 2 and 1 with H

Thus, H attaches at position 1 to give

cyclohexane with CH_3, CH_2CH_3, Br, H = cyclohexane with CH_3, CH_2CH_3, Br

2.7 **a.** Markovnikov's rule predicts that H will attach at position 1 to give:

$$CH_3CH_2CH_2\underset{H}{\overset{OH}{C}} - CH_2 = CH_3CH_2CH_2CHCH_3$$
with OH

b. cyclopentane with H, OH = cyclopentane with OH

2.8 The double bond becomes a single bond:

$$-CH_2 - \underset{1}{CH} - CH_2 - \underset{2}{CH} - CH_2 - \underset{3}{CH} - CH_2 - \underset{4}{CH} -$$
each CH with CH_3 above

2.9 Each chain is correctly numbered from the right.
    a. 2-pentyne   **b.** 5-methyl-2-hexyne

2.10 **a.** Numbers or the term *meta* may be used:
    1,3-diethylbenzene   or   *m*-diethylbenzene
    b. The compound must be named as a derivative of cyclopentane. The correct name is 1-chloro-3-phenylcyclopentane.
    c. Numbers must be used when there are three groups: 1,2,3-tribromobenzene.
    d. If the compound is named as a derivative of benzoic acid, then the methyl group is at position 2. The name is 2-methylbenzoic acid.

············ **CHAPTER 3** ······················

3.1 a.

$$CH_3CHCH_2-OH \quad \text{(with } CH_3 \text{ branch)}$$

2-methyl-1-propanol

b.

$$CH_3CHCHCH_2CHCH_3 \quad \text{(with } OH, CH_2CH_3, CH_3 \text{ branches)}$$

2,5-dimethyl-4-heptanol

c.

(cyclopentane ring with CH_3 and OH)

3.2 a. 1,4-butanediol **b.** 1,2-cyclohexanediol

3.3 Because the compound is named as a phenol, the OH group is at position 1. The name is 2-ethylphenol or *o*-ethylphenol.

3.4 In each case, count the number of carbon atoms attached to the hydroxy-bearing carbon.

  a. Primary **b.** Secondary **c.** Secondary

3.5 b, c, a

3.6 Methanol molecules form hydrogen bonds, whereas molecules of propane do not.

3.7 a. There are two ways of removing water from this structure, but they both produce the same product:

$$CH_3CH=CHCH_2CH_3$$

  b. Dehydration will occur in the direction that gives the highest number of carbon groups attached to the double-bonded carbons:

3.8 a. Ether formation is favored by H_2SO_4 at 140°C.
  b. Alkene formation predominates when H_2SO_4 is used at 180°C.

3.9

first product   second product

3.10

3.11 The alcohols in (a) and (c) react because they are secondary and primary, respectively. Alcohol (b) is tertiary and does not react.

3.12 Working backward, you could prepare an ether product from an alcohol. Thus, the starting alkene must first be converted to an alcohol.

$$CH_2=CH_2 + H_2O \xrightarrow{H_2SO_4} CH_3CH_2-OH$$

$$2CH_3CH_2-OH \xrightarrow[140°C]{H_2SO_4} CH_3CH_2-O-CH_2CH_3 + H_2O$$

3.13 Use Table 3.2 to answer these questions:
  a. Solvent
  b. Alcoholic beverages
  c. Rubbing alcohol
  d. Antifreeze
  e. Food moisturizer
  f. Cough drops

3.14 a. Mouthwash antiseptic
  b. Throat lozenges

3.15 a. ethyl phenyl ether
  b. diisopropyl ether

3.16 a. 1,2-dimethoxycyclopentane
  b. 3-isopropoxypentane

3.17 a. Ether molecules do not form hydrogen bonds, whereas alcohol molecules do form hydrogen bonds.
  b. They are quite unreactive.

3.18 Oxidation of a thiol produces a disulfide:

$$CH_3CH-S-S-CHCH_3 + H_2O \quad \text{(with } CH_3 \text{ branches)}$$

3.19 a.

$$2CH_3CH-SH \quad \text{(with } CH_3 \text{ branch)}$$

  b.

3.20

$$CH_3CH-S-Pb-S-CHCH_3 + 2H^+ \quad \text{(with } CH_3 \text{ branches)}$$

············ **CHAPTER 4** ······················

4.1 a. The aldehyde contains five carbon atoms, and the name is pentanal.
  b. The aldehyde carbon is position 1. The name is 3,4-dimethylpentanal.

4.2 a. The chain is numbered from the right. The name is 4,6-dibromo-2-heptanone.
  b. The carbonyl carbon atom is position 1. The name is 2,5-dimethylcyclopentanone.

4.3 a.

$$CH_3CHCH_2-C-OH \quad \text{(with } CH_3 \text{ and } O \text{)}$$

  b. This is a ketone, and it does not undergo oxidation.

4.4 a.

$$CH_3CH_2CHCH_2CH_3 \quad \text{(with } OH \text{)}$$

b.

4.5 **a.**

hemiacetal   acetal

b.

hemiketal   ketal

4.6 **a.**

$$CH_3CH_2CH_2 - \overset{\overset{O}{\|}}{C} - H + 2CH_3CH_2 - OH$$

b.

$+ 2CH_3 - OH$

4.7 **a.** This compound is not a hemiacetal or hemiketal because there is no OH group. The presence of the ring hydrogen makes it an acetal.

Hydrolysis breaks these bonds

b. An OH group is missing, so this compound is an acetal or ketal. Because the carbon with two attached oxygens does not have a hydrogen present, the compound is a ketal.

Hydrolysis breaks these bonds

············· **CHAPTER 5** ·····················

5.1 **a.** The longest chain beginning with the carboxyl group has four carbon atoms. The name is 2-methylbutanoic acid.

     b. The carboxyl group is at position 1. The name is 3,4-dibromobenzoic acid.

5.2

$$CH_3 - \overset{\overset{O}{\|}}{C} - \overset{\overset{O}{\|}}{C} - O^-$$

5.3

$$CH_3CH_2CH_2 - \overset{\overset{O}{\|}}{C} - O^-Na^+ + H_2O$$

5.4 **a.** This is the lithium salt of 2-methylbenzoic acid. The name is lithium 2-methylbenzoate.

     b. This is the sodium salt of butanoic acid. The name is sodium butanoate.

5.5 Each of these reactions requires H^+ as a catalyst. The products are as follows:

a.

$$CH_3CH_2 - \overset{\overset{O}{\|}}{C} - OCH_3 + H_2O$$

b.

$+ H_2O$

c.

$+ H_2O$

5.6 **a.**

$+ CH_3 - OH \longrightarrow$

b.

$$CH_3CH_2 - \overset{\overset{O}{\|}}{C} - Cl + CH_3CH - OH \longrightarrow$$
$$\underset{|}{CH_3}$$

5.7 **a.** The common name for the acid used to prepare this ester is formic acid. The alkyl group attached to oxygen is isopropyl. Thus, the common name is isopropyl formate. The IUPAC name for formic acid is methanoic acid. Thus, the IUPAC name for the ester is isopropyl methanoate.

     b. The common and IUPAC names are identical when the carboxylic acid is benzoic acid. The name of the ester is methyl benzoate.

5.8

$+ HO - CH_2CH_2CH_3$

5.9 **a.** $CH_3CH_2 - \overset{\overset{O}{\|}}{C} - O^-Na^+ + HO - \underset{|}{CH}CH_3$
                                                 CH_3

b.

$+ HO - CH_2CH_3$

········· **CHAPTER 6** ·········

6.1 In each case, count the number of carbon atoms directly attached to the nitrogen.

    a. Secondary   **b.** Secondary   **c.** Tertiary

6.2 Identify the groups attached to the nitrogen.
    a. propylamine
    b. methylphenylamine
    c. trimethylamine

6.3   **a.** The butane chain is numbered from the right. The name is 1,3-diaminobutane.
    b. If the base word reflects the fact that this compound is an alcohol, then the OH group is attached at position 1. The name is 2-aminocyclopentanol.

6.4   **a.** *N*-ethylaniline
    b. *N*-ethyl-*N*-methylaniline
    c. 3-ethyl-2-methylaniline

6.5   **a.**

    b.

6.6   **a.**

$$CH_3 - \overset{\overset{H}{\underset{|}{|}}}{\underset{\underset{CH_3}{|}}{N}} - CH_2CH_3 + OH^-$$

    b.

$$\bigcirc - NH_3^+ + OH^-$$

6.7   **a.**

    b.

6.8   **a.**

$$CH_3 - \overset{O}{\overset{||}{C}} - NH - CH_2CH_2CH_3 + CH_3 - \overset{O}{\overset{||}{C}} - OH$$

    b.

$$CH_3CH_2 - \overset{O}{\overset{||}{C}} - NH_2 + HCl$$

6.9   **a.** This amide is derived from butanoic acid. The name is butanamide.
    b. hexanamide

6.10   **a.** This amide is a derivative of the two-carbon acid acetic acid (common name) or ethanoic acid (IUPAC name). The amide is named *N,N*-diethylacetamide (common name) or *N,N*-diethylethanamide (IUPAC name).
    b. This amide is a derivative of butyric acid (common name) or butanoic acid (IUPAC name). The amide is named *N*-isobutylbutyramide (common name) or *N*-isobutylbutanamide (IUPAC name).
    c. The acid portion of this amide came from benzoic acid. The amide is named *N*-ethyl-*N*-phenylbenzamide.
    d. The acid portion of this amide came from 2-methylbenzoic acid. The amide is named 2,*N*-dimethylbenzamide.

6.11   **a.**

    b.

6.12   **a.** Under acidic conditions, amide hydrolysis produces a carboxylic acid and the salt of an amine.

$$\bigcirc - \overset{O}{\overset{||}{C}} - OH \quad + CH_3CH_2 - NH_3^+Cl^-$$

    b. Under basic conditions, amide hydrolysis produces an amine and the salt of a carboxylic acid.

$$\bigcirc - \overset{O}{\overset{||}{C}} - O^-Na^+ \quad + CH_3CH_2 - NH_2$$

········· **CHAPTER 7** ·········

7.1 In each case, look for the presence of four different groups attached to the colored atom.
    a. The carbon is not chiral because of the two attached CH_2OH groups.
    b. Yes
    c. The carbon is not chiral because it is attached to two H's.
    d. Yes

7.2   Each of the atoms marked with an asterisk has four different attached groups and is chiral.

a.

$$CH_2OH$$
$$|$$
$$C{=}O$$
$$|$$
$$HO{-}\overset{*}{C}{-}H$$
$$|$$
$$H{-}\overset{*}{C}{-}OH$$
$$|$$
$$CH_2OH$$

b.   CHO
$$|$$
$$\overset{*}{C}H{-}OH$$
$$|$$
$$\overset{*}{C}H{-}OH$$
$$|$$
$$\overset{*}{C}H{-}OH$$
$$|$$
$$CH_2OH$$

c.

7.3   The chiral carbon atoms (three of them) are marked with an asterisk. Thus, there are 2^3, or 8, possible stereoisomers:

$$\overset{OH}{|}\ \ \overset{OH}{|}\ \ \overset{OH}{|}\ \ \overset{OH}{|}$$
$$CH_2{-}\underset{*}{CH}{-}\underset{*}{CH}{-}\underset{*}{CH}{-}CHO$$

7.4   The chiral carbon atom is the one with the attached nitrogen. The chiral carbon atom is placed at the intersection of the two lines, and the COOH is placed at the top:

$$COOH \qquad\qquad COOH$$

H ———— NH₂   H₂N ———— H

$$CHCH_3 \qquad\qquad CHCH_3$$
$$|\qquad\qquad\qquad |$$
$$CH_3 \qquad\qquad CH_3$$
D form   L form

7.5   In each case, look at the direction of the OH group on the bottom chiral carbon.

a. L   **b.** D   **c.** L

7.6   **a.** aldopentose
b. ketotetrose
c. aldotetrose

7.7

HOCH₂, O

α

OH

OH   OH

7.8   **a.**

CH₂OH   — acetal group

HO   O

— glycosidic linkage

OH

OCH₂CH₃

OH

---

b.

ketal group

glycosidic linkage

.... **CHAPTER 8** ....

8.1   One of several possibilities is the following structure:

$$\overset{O}{\overset{\|}{}}$$
$$CH_2{-}O{-}C{-}(CH_2)_{16}CH_3 \qquad \text{(stearic acid component)}$$
$$|$$
$$\overset{O}{\overset{\|}{}}$$
$$CH{-}O{-}C{-}(CH_2)_7CH{=}CH(CH_2)_7CH_3 \qquad \text{(oleic acid component)}$$
$$|$$
$$\overset{O}{\overset{\|}{}}$$
$$CH_2{-}O{-}C{-}(CH_2)_{14}CH_3 \qquad \text{(palmitic acid component)}$$

8.2

$$CH_2{-}OH \qquad HOOC{-}(CH_2)_7CH{=}CH(CH_2)_7CH_3$$
$$|$$
$$CH{-}OH \ + \ HOOC{-}(CH_2)_{14}CH_3$$
$$|$$
$$CH_2{-}OH \qquad HOOC{-}(CH_2)_{16}CH_3$$

8.3

$$CH_2{-}OH \qquad Na^+{}^-OOC{-}(CH_2)_{16}CH_3$$
$$|$$
$$CH{-}OH \ + \ 2Na^+{}^-OOC{-}(CH_2)_7CH{=}CH(CH_2)_7CH_3$$
$$|$$
$$CH_2{-}OH$$

8.4   During complete hydrogenation, all CH=CH groups become CH₂—CH₂ groups:

$$\overset{O}{\overset{\|}{}}$$
$$CH_2{-}O{-}C{-}(CH_2)_7CH_2CH_2(CH_2)_7CH_3$$
$$|$$
$$\overset{O}{\overset{\|}{}}$$
$$CH{-}O{-}C{-}(CH_2)_7CH_2CH_2CH_2CH_2CH_2CH_2CH_2CH_2CH_3$$
$$|$$
$$\overset{O}{\overset{\|}{}}$$
$$CH_2{-}O{-}C{-}(CH_2)_7CH_2CH_2CH_2CH_2CH_2(CH_2)_4CH_3$$

Combining all the CH₂ groups gives

$$\overset{O}{\overset{\|}{}}$$
$$CH_2{-}O{-}C{-}(CH_2)_{16}CH_3$$
$$|$$
$$\overset{O}{\overset{\|}{}}$$
$$CH{-}O{-}C{-}(CH_2)_{16}CH_3$$
$$|$$
$$\overset{O}{\overset{\|}{}}$$
$$CH_2{-}O{-}C{-}(CH_2)_{16}CH_3$$

8.5

$$CH_2-O-\overset{\overset{\displaystyle O}{\|}}{C}-(CH_2)_{14}CH_3$$

$$CH-O-\overset{\overset{\displaystyle O}{\|}}{C}-(CH_2)_7CH=CH(CH_2)_7CH_3$$

$$CH_2-O-\overset{\overset{\displaystyle O}{\|}}{\underset{\underset{\displaystyle O^-}{|}}{P}}-OCH_2CH_2-NH_3{}^+$$

.... **CHAPTER 9** ....

9.1 **a.**

$$H_3\overset{+}{N}-\underset{\underset{\displaystyle CH_2-OH}{|}}{CH}-\overset{\overset{\displaystyle O}{\|}}{C}-OH$$

b.

$$H_2N-\underset{\underset{\displaystyle CH_2-OH}{|}}{CH}-\overset{\overset{\displaystyle O}{\|}}{C}-O^-$$

9.2

$$H_3\overset{+}{N}-\underset{\underset{\displaystyle \overset{|}{\underset{\displaystyle \bigcirc}{}}}{\underset{\displaystyle CH_2}{|}}}{CH}-\overset{\overset{\displaystyle O}{\|}}{C}-NH-\underset{\underset{\displaystyle SH}{\underset{\displaystyle CH_2}{|}}}{CH}-\overset{\overset{\displaystyle O}{\|}}{C}-NH-$$

N-terminal
residue

$$-\underset{\underset{\displaystyle OH}{\underset{\displaystyle CH_2}{|}}}{CH}-\overset{\overset{\displaystyle O}{\|}}{C}-NH-\underset{\underset{\displaystyle CH_2CH_3}{\underset{\displaystyle CHCH_3}{|}}}{CH}-\overset{\overset{\displaystyle O}{\|}}{C}-O^-$$

C-terminal
residue

9.3 **a.** Hydrophobic attractions between nonpolar groups
   b. Hydrogen bonding
   c. Salt bridges between the ionic groups

.... **CHAPTER 10** ....

10.1 **a.** maltose
   b. peptides
   c. glucose 6-phosphate

10.2 Turnover number is the number of substrate molecules acted on per minute by one molecule of enzyme. One enzyme international unit is the amount of enzyme that catalyzes 1 μmol of substrate to react per minute.

10.3 **a.** An increase in enzyme concentration increases the rate of an enzyme-catalyzed reaction.
   b. An increase in substrate concentration increases the rate until the enzyme is saturated. Above the saturation point, rate of reaction is constant.
   c. An increase in temperature increases the rate of the enzyme-catalyzed reaction until the optimum temperature is reached. Above the optimum temperature, the rate decreases.
   d. An increase in pH increases the reaction rate until the optimum pH is reached. Above the optimum pH, the rate decreases.

10.4 **a.** The structure of a competitive inhibitor resembles that of the substrate. The structure of a noncompetitive inhibitor bears no resemblance to that of the substrate.
   b. A competitive inhibitor binds at the active site. A noncompetitive inhibitor binds at some other region of the enzyme.
   c. Increasing substrate concentration reverses the effect of a competitive inhibitor but has no effect on a noncompetitive inhibitor.

10.5 **a.** Several enzymes are synthesized in inactive forms called zymogens. Under the proper reaction conditions, zymogens are activated to carry out their catalytic role.
   b. The catalytic ability of allosteric enzymes may be either increased or decreased by the binding of modulators.
   c. Certain enzymes are synthesized in greater amounts when needed and in smaller amounts when not needed.

.... **CHAPTER 11** ....

11.1

phosphodiester
linkages

11.2 5′ T–T–A–C–G   3′   original strand
          3′ A–A–T–G–C   5′   complementary strand

The complementary strand written in the 5′ to 3′ direction is C–G–T–A–A.

11.3 5′ A–T–T–A–G–C–C–G   3′   DNA template
          3′ U–A–A–U–C–G–G–C   5′   mRNA

The mRNA written in the 5′to 3′ direction is C–G–G–C–U–A–A–U.

11.4 f–Met–His–His–Val–Leu–Cys

···················· **CHAPTER 12** ····················

12.1

$$\begin{array}{ccc} \text{OH} & & \text{O} \\ | & & \| \\ \text{CH} - \text{COO}^- + \text{NAD}^+ \longrightarrow & \text{C} - \text{COO}^- + \text{NADH} + \text{H}^+ \\ | & & | \\ \text{CH}_2 - \text{COO}^- & & \text{CH}_2 - \text{COO}^- \end{array}$$

12.2 a. $^-\text{OOC}-\text{CH}=\text{CH}-\text{COO}^- + \text{FADH}_2$

   b.

$$\begin{array}{c} \text{O} \\ \| \\ \text{CH}_3 - \text{C} - \text{COO}^- + \text{NADH} + \text{H}^+ \end{array}$$

···················· **CHAPTER 13** ····················

13.1 a. Hexokinase, the first enzyme of the glycolysis pathway, is controlled by concentrations of glucose 6-phosphate through feedback inhibition.

   b. Phosphofructokinase is an allosteric enzyme involved in the production of fructose 1,6-bisphosphate. The enzyme is inhibited by high concentrations of ATP and citrate, and it is activated by high concentrations of ADP and AMP.

   c. Pyruvate kinase controls the last step of glycolysis. It is inhibited by higher concentrations of ATP.

13.2 NAD^+ produced in the various fates of pyruvate can be used in the glycolysis pathway, enabling further glycolysis to take place.

13.3 a. Three: Each FADH_2 produces 1.5 molecules of ATP.
   b. Five: Each NADH produces 2.5 molecules of ATP.

c. 20: Each acetyl CoA produces 10 molecules of ATP.

13.4 Glucose 6-phosphatase is present in the liver but not in the muscles.

···················· **CHAPTER 14** ····················

14.1 a. Each acetyl CoA contains two carbon atoms from capric acid. Thus, five acetyl CoA molecules are formed.

   b. Ten carbon atoms require four trips through the fatty acid spiral. Thus, four molecules of FADH_2 are formed.

   c. Four trips through the spiral also produce four NADH molecules and four H^+ ions.

14.2 A 16-carbon fatty acid would yield eight molecules of acetyl CoA. Seven trips through the spiral are necessary and would produce 7 FADH_2 and 7 NADH + 7 H^+.

8 acetyl CoA × 10 ATP/acetyl CoA	=	80 ATP
7 FADH_2 × 1.5 ATP/FADH_2	=	10.5 ATP
7 NADH × 2.5 ATP/NADH	=	17.5 ATP
		108 ATP

Two high-energy bonds in ATP are consumed in the activation step.

$$\frac{-2}{106 \text{ ATP}}$$

14.3

$$\text{HO} - \bigcirc - \text{CH}_2 - \overset{\overset{\displaystyle \text{O}}{\|}}{\text{C}} - \text{COO}^-$$

$$+ \ ^-\text{OOC} - \text{CH}_2 - \text{CH}_2 - \overset{\overset{\displaystyle \text{NH}_3^+}{|}}{\text{CH}} - \text{COO}^-$$

14.4 a. The urea cycle converts toxic ammonia into less toxic urea for excretion from the body.
   b. carbamoyl phosphate

···················· **CHAPTER 15** ····················

15.1 a. Lungs
   b. Tissue cells
   c. Lungs
   d. Tissue cells

Glossary

Absolute specificity   The characteristic of an enzyme that it acts on one and only one substrate.

Acetal   A compound that contains the $-\overset{\displaystyle OR}{\underset{\displaystyle OR}{\overset{|}{\underset{|}{C}}}}-H$ arrangement of atoms.

Acetone breath   A condition in which acetone can be detected in the breath.

Acidosis   A low blood pH.

Activator   A substance that binds to an allosteric enzyme and increases its activity.

Active site   The location on an enzyme where a substrate is bound and catalysis occurs.

Addition polymer   A polymer formed by the linking together of many alkene molecules through addition reactions.

Addition reaction   A reaction in which a compound adds to a multiple bond.

Adipose tissue   A kind of connective tissue where triglycerides are stored.

Aerobic   In the presence of oxygen.

Alcohol   A compound in which an —OH group is connected to an aliphatic carbon atom.

Alcoholic fermentation   The conversion of glucose to ethanol.

Aldehyde   A compound that contains the $-\overset{\displaystyle O}{\overset{||}{C}}-H$ group; the general formula is $R-\overset{\displaystyle O}{\overset{||}{C}}-H$.

Aliphatic compound   Any organic compound that is not aromatic.

Alkaloids   A class of nitrogen-containing organic compounds obtained from plants.

Alkalosis   An abnormally high blood pH.

Alkane   A hydrocarbon that contains only single bonds.

Alkene   A hydrocarbon that contains one or more double bonds.

Alkoxy group   The —O—R functional group.

Alkyl group   A group differing by one hydrogen from an alkane.

Alkyl halide   See **Haloalkane**.

Alkyne   A hydrocarbon that contains one or more triple bonds.

Allosteric enzyme   An enzyme with quaternary structure whose activity is changed by the binding of modulators.

Alpha-amino acid   An organic compound containing both an amino group and a carboxylate group, with the amino group attached to the carbon next to the carboxylate group.

α-helix   The helical structure in proteins that is maintained by hydrogen bonds.

Amide   An organic compound having the functional group $-\overset{\displaystyle O}{\overset{||}{C}}-\overset{|}{N}-$.

Amide linkage   The carbonyl carbon–nitrogen single bond of the amide group.

Amine   An organic compound derived by replacing one or more of the hydrogen atoms of ammonia with alkyl or aromatic groups, as in RNH_2, R_2NH, and R_3N.

Amino acid pool   The total supply of amino acids in the body.

Amino acid residue   An amino acid that is a part of a peptide, polypeptide, or protein chain.

Amphetamines   A class of drugs structurally similar to epinephrine, used to stimulate the central nervous system.

Anabolism   All reactions involved in the synthesis of biomolecules.

Anaerobic   In the absence of oxygen.

Anomers   Stereoisomers that differ in the three-dimensional arrangement of groups at the carbon of an acetal, ketal, hemiacetal, or hemiketal group.

Anomeric carbon   An acetal, ketal, hemiacetal, or hemiketal carbon atom giving rise to two stereoisomers.

Antibiotic   A substance produced by one microorganism that kills or inhibits the growth of other microorganisms.

Antibody   A substance that helps protect the body from invasion by foreign materials known as antigens.

Anticodon   A three-base sequence in tRNA that is complementary to one of the codons in mRNA.

Antigen   A foreign substance that can stimulate the production of antibodies.

Antihistamine   A drug that relieves the symptoms of allergies and colds by countering the action of histamine.

Antioxidant   A substance that prevents another substance from being oxidized.

Apoenzyme   A catalytically inactive protein formed by removal of the cofactor from an active enzyme.

Aromatic compound   Any organic compound that contains the characteristic benzene ring or similar feature.

Benedict's reagent   A mild oxidizing solution containing Cu^{2+} ions used to test for the presence of aldehydes.

β-oxidation process   A pathway in which fatty acids are broken down into molecules of acetyl CoA.

β-pleated sheet   A secondary protein structure in which protein chains are aligned side by side in a sheetlike array held together by hydrogen bonds.

Biochemistry   A study of the compounds and processes associated with living organisms.

Biomolecule   A general term referring to an organic compound essential to life.

Blood sugar level   The amount of glucose present in blood, normally expressed as milligrams per 100 mL of blood.

Branched alkane   An alkane in which at least one carbon atom is not part of a continuous chain.

Carbaminohemoglobin   Hemoglobin combined with carbon dioxide.

Carbocation   An ion of the form $-\overset{+}{\underset{|}{C}}-$.

Carbohydrate   A polyhydroxy aldehyde, ketone, or substance that yields such compounds on hydrolysis.

Carbonyl group   The $-\overset{\overset{\text{O}}{\|}}{C}-$ group.

Carboxyl group   The $-\overset{\overset{\text{O}}{\|}}{C}-OH$ group.

Carboxylate ion   The $R-\overset{\overset{\text{O}}{\|}}{C}-O^-$ ion that results from the dissociation of a carboxylic acid.

Carboxylic acid   An organic compound that contains the $-\overset{\overset{\text{O}}{\|}}{C}-OH$ functional group.

Carboxylic acid anhydride   An organic compound that contains the $-\overset{\overset{\text{O}}{\|}}{C}-O-\overset{\overset{\text{O}}{\|}}{C}-$ functional group.

Carboxylic acid chloride   An organic compound that contains the $-\overset{\overset{\text{O}}{\|}}{C}-Cl$ functional group.

Carboxylic ester   A compound with the $-\overset{\overset{\text{O}}{\|}}{C}-OR$ functional group.

Catabolism   All reactions involved in the breakdown of biomolecules.

Cellular respiration   The entire process involved in the use of oxygen by cells.

Central dogma of molecular biology   The well-established process by which genetic information stored in DNA molecules is expressed in the structure of synthesized proteins.

Cephalin   A phosphoglyceride containing ethanolamine or serine.

Chemiosmotic hypothesis   The theory that a proton flow across the inner mitochondrial membrane during operation of the electron transport chain provides energy for ATP synthesis.

Chiral   A descriptive term for compounds or objects that cannot be superimposed on their mirror image.

Chiral carbon   A carbon atom with four different groups attached.

Chloride shift   The movement of chloride ions from one fluid to another to maintain charge neutrality.

Chromosome   A tightly packaged bundle of DNA and protein that is involved in cell division.

Chylomicron   A lipoprotein aggregate found in the lymph and the bloodstream.

Cis-   On the same side (as applied to geometric isomers).

Citric acid cycle   A series of reactions in which acetyl CoA is oxidized to carbon dioxide, and reduced forms of coenzymes FAD and NAD^+ are produced.

Codon   A sequence of three nucleotide bases that represents a code word on mRNA molecules.

Coenzyme   An organic molecule required by an enzyme for catalytic activity.

Cofactor   A nonprotein molecule or ion required by an enzyme for catalytic activity.

Common catabolic pathway   The reactions of the citric acid cycle plus those of the electron transport chain and oxidative phosphorylation.

Competitive inhibitor   An inhibitor that competes with substrate for binding at the active site of an enzyme.

Complementary DNA strands   Two strands of DNA in a double-helical form such that adenine and guanine of one strand are matched and hydrogen-bonded to thymine and cytosine, respectively, of the second strand.

Complete protein   Food that contains all essential amino acids in the proportions needed by the body.

Complex carbohydrates   The polysaccharides amylose and amylopectin, collectively called starch.

Complex lipid   An ester-containing lipid with more than two types of components: an alcohol, fatty acids, plus other components.

Condensation polymerization   The process by which monomers combine with the simultaneous elimination of a small molecule.

Condensed structural formula   A structural molecular formula showing the general arrangement of atoms but without showing all the covalent bonds.

Conformations   The different arrangements of atoms in space achieved by rotation about single bonds.

Conjugated protein   A protein made up of amino acid residues and other organic or inorganic components.

Copolymer   An addition polymer formed by the reaction of two different monomers.

Cori cycle   The process in which glucose is converted to lactate in muscle tissue, lactate is reconverted to glucose in the liver, and glucose is returned to the muscle.

C-terminal residue   An amino acid on the end of a chain that has an unreacted or free carboxylate group.

Cycloalkane   An alkane in which carbon atoms form a ring.

Cytochrome   An iron-containing enzyme located in the electron transport chain.

Daily Reference Values (DRV)   A set of standards for nutrients and food components (such as fat and fiber) that have important relationships with health; used on food labels as part of the Daily Values.

Daily Values (DV)   Reference values developed by the FDA specifically for use on food labels. The Daily Values represent two sets of standards: Reference Daily Intakes (RDI) and Daily Reference Values (DRV).

Dehydration reaction   A reaction in which water is chemically removed from a compound.

Denaturation   The process by which a protein loses its characteristic native structure and function.

Deoxyhemoglobin (hemoglobin)   Nonoxygenated hemoglobin.

Deoxyribonucleic acid (DNA)   A nucleic acid found primarily in the nuclei of cells.

Dextrorotatory   Describes substances that rotate plane-polarized light to the right.

Diabetes mellitus   A family of diseases in which the uptake of blood glucose by tissues is impaired.

Dimer   Two identical molecules bonded together.

Dipeptide   A compound formed when two amino acids are bonded by an amide linkage.

Disaccharide   A carbohydrate formed by the combination of two monosaccharides.

Disulfide   A compound containing an —S—S— group.

Disulfide bridge   A bond produced by the oxidation of —SH groups on two cysteine residues. The bond loops or holds two peptide chains together.

Electron transport chain   A series of reactions in which protons and electrons from the oxidation of foods are used to reduce molecular oxygen to water.

Elimination reaction   A reaction in which two or more covalent bonds are broken and a new multiple bond is formed.

Enantiomers   Stereoisomers that are mirror images.

Enzyme   A protein molecule that catalyzes chemical reactions.

Enzyme activity   The rate at which an enzyme catalyzes a reaction.

Enzyme induction   The synthesis of an enzyme in response to a cellular need.

Enzyme inhibitor   A substance that decreases the activity of an enzyme.

Enzyme international unit   A quantity of enzyme that catalyzes the conversion of 1 μmol of substrate per minute under specified conditions.

Essential amino acid   An amino acid that cannot be synthesized within the body at a rate adequate to meet metabolic needs.

Essential fatty acid   A fatty acid that cannot be synthesized by the body.

Ester   A compound in which the —OH of an acid is replaced by an —OR.

Ester linkage   The carbonyl carbon–oxygen single bond of the ester group.

Esterification   The process of forming an ester.

Ether   A compound that contains a $-\overset{|}{\underset{|}{C}}-O-\overset{|}{\underset{|}{C}}-$ functional group.

Eukaryotic cell   A cell containing membrane-enclosed organelles, particularly a nucleus.

Exon   A segment of a eukaryotic DNA molecule that is coded for amino acids.

Expanded structural formula   A structural molecular formula showing all the covalent bonds.

Extracellular fluid   Body fluid located outside cells.

Extremozyme   A nickname for certain enzymes isolated from microorganisms that thrive in extreme environments.

Fat   A triglyceride that is a solid at room temperature.

Fat mobilization   The hydrolysis of stored triglycerides followed by the entry of fatty acids and glycerol into the bloodstream.

Fatty acid   A long-chain carboxylic acid found in fats.

Feedback inhibition   A process in which the end product of a sequence of enzyme-catalyzed reactions inhibits an earlier step in the process.

Fermentation   A reaction of sugars, starch, or cellulose to produce ethanol and carbon dioxide.

Fiber   Indigestible plant material composed primarily of cellulose.

Fibrous protein   A protein made up of long rod-shaped or stringlike molecules that intertwine to form fibers.

Fischer projection   A method of depicting three-dimensional shapes for chiral molecules.

Fluid-mosaic model   A model of membrane structure in which proteins are embedded in a flexible lipid bilayer.

Functional group   A unique reactive combination of atoms that differentiates molecules of organic compounds of one class from those in another.

Furanose ring   A five-membered sugar ring system containing an oxygen atom.

Gene   An individual section of a chromosomal DNA molecule that is the fundamental unit of heredity.

Genome   A summation of all the genetic material (chromosomes) of a cell; a person's genetic blueprint.

Geometric isomers   Molecules that differ in the three-dimensional arrangements of their atoms in space and not in the order of linkage of atoms.

Globular protein   A spherical-shaped protein that usually forms stable suspensions in water or dissolves in water.

Glucogenic amino acid   An amino acid whose carbon skeleton can be converted metabolically to intermediates used in the synthesis of glucose.

Gluconeogenesis   The synthesis of glucose from noncarbohydrate molecules.

Glucosuria   A condition in which elevated blood sugar levels result in the excretion of glucose in the urine.

Glycogenesis   The synthesis of glycogen from glucose.

Glycogenolysis   The breakdown of glycogen to glucose.

Glycolipid   A complex lipid containing sphingosine, a fatty acid, and a carbohydrate.

Glycolysis   A series of reactions by which glucose is oxidized to pyruvate.

Glycoside   Another name for a carbohydrate containing an acetal or ketal group.

Glycosidic linkage   The carbon–oxygen–carbon linkage that joins the components of a glycoside to the ring.

Haloalkane   A derivative of an alkane in which one or more hydrogens are replaced by halogens; also called an alkyl halide.

Haworth structure   A method of depicting three-dimensional carbohydrate structures.

Hemiacetal   A compound that contains the functional group

$$-\overset{\displaystyle OH}{\underset{\displaystyle}{C}}-H$$

OR

Hemiketal   A compound that contains the group

$$-\overset{\displaystyle}{C}-\overset{\displaystyle OH}{C}-\overset{\displaystyle}{C}-$$

OR

Heterocyclic ring   A ring in which one or more atoms is an atom other than carbon.

Heterogeneous nuclear RNA (hnRNA)   RNA produced when both introns and exons of eukaryotic cellular DNA are transcribed.

High-energy compound   A substance that on hydrolysis liberates a great amount of free energy.

Homologous series   Compounds of the same functional class that differ by a $-CH_2-$ group.

Hormone   A chemical messenger secreted by specific glands and carried by the blood to a target tissue where it triggers a particular response.

Hybrid orbital   An orbital produced from the combination of two or more nonequivalent orbitals of an atom.

Hydration   The addition of water to a multiple bond.

Hydrocarbon   An organic compound that contains only carbon and hydrogen.

Hydrogenation   A reaction in which the addition of hydrogen takes place.

Hydrolysis   Bond breakage by reaction with water.

Hydrolysis reaction   Any reaction with water; for salts, a reaction of the acidic cation and/or basic anion of the salt with water.

Hydroxy group   The $-OH$ functional group.

Hyperglycemia   A higher-than-normal blood sugar level.

Hyperventilation   Rapid, deep breathing.

Hypoglycemia   A lower-than-normal blood sugar level.

Hypoventilation   Slow, shallow breathing.

Inborn error of metabolism   A disease in which a genetic change causes a deficiency of a particular protein, often an enzyme.

Induced-fit theory   The theory of enzyme action proposing that the conformation of an enzyme changes to accommodate an incoming substrate.

Inhibitor   A substance that decreases reaction rates.

Inorganic chemistry   The study of the elements and all noncarbon compounds.

Insulin shock   A severe hypoglycemia produced by the injection of too much insulin.

Interstitial fluid   The fluid surrounding individual tissue cells.

Intracellular fluid   Body fluid located inside cells.

Intron   A segment of a eukaryotic DNA molecule that carries no codes for amino acids.

Invert sugar   A mixture of equal amounts of glucose and fructose.

Isoelectric point   The characteristic solution pH at which an amino acid has a net charge of zero.

Isoenzyme A slightly different form of the same enzyme produced by different tissues.

Isomerism A property in which two or more compounds have the same molecular formula but different arrangements of atoms.

Ketal A compound that contains the $-\overset{\displaystyle |}{\underset{\displaystyle |}{C}}-\overset{\displaystyle OR}{\underset{\displaystyle OR}{\overset{\displaystyle |}{\underset{\displaystyle |}{C}}}}-\overset{\displaystyle |}{\underset{\displaystyle |}{C}}-$ group.

Ketoacidosis A low blood pH due to elevated levels of ketone bodies.

Ketogenic amino acid An amino acid whose carbon skeleton can be converted metabolically to acetyl CoA or acetoacetyl CoA.

Ketone A compound that contains the $-\overset{\displaystyle |}{\underset{\displaystyle |}{C}}-\overset{\displaystyle O}{\overset{\displaystyle \|}{C}}-\overset{\displaystyle |}{\underset{\displaystyle |}{C}}-$ group; the general formula is $R-\overset{\displaystyle O}{\overset{\displaystyle \|}{C}}-R'$.

Ketone bodies Three compounds—acetoacetate, β-hydroxybutyrate, and acetone—formed from acetyl CoA.

Ketonemia An elevated level of ketone bodies in the blood.

Ketonuria The presence of ketone bodies in the urine.

Ketosis A condition in which ketonemia, ketonuria, and acetone breath exist together.

Lactate fermentation The production of lactate from glucose.

Lactose intolerance The inability to digest milk and other products containing lactose.

Lecithin A phosphoglyceride containing choline.

Levorotatory Describes substances that rotate plane-polarized light to the left.

Lipid A biological compound that is soluble only in nonpolar solvents.

Lipid bilayer A structure found in membranes, consisting of two sheets of lipid molecules arranged so that the hydrophobic portions are facing each other.

Lock-and-key theory The theory of enzyme specificity proposing that a substrate has a shape fitting that of the enzyme's active site, as a key fits a lock.

Macronutrient A substance needed by the body in relatively large amounts.

Major mineral A mineral found in the body in quantities greater than 5 g.

Markovnikov's rule In the addition of H—X to an alkene, the hydrogen becomes attached to the carbon atom that is already bonded to more hydrogens.

Messenger RNA (mRNA) RNA that carries genetic information from the DNA in the cell nucleus to the site of protein synthesis in the cytoplasm.

Metabolic acidosis A condition of acidosis resulting from causes other than hypoventilation.

Metabolic alkalosis A condition of alkalosis resulting from causes other than hyperventilation.

Metabolic pathway A sequence of reactions used to produce one product or accomplish one process.

Metabolism The summation of all reactions occurring in an organism.

Micelle A spherical cluster of molecules in which the polar portions of the molecules are on the surface and the nonpolar portions are located in the interior.

Micronutrient A substance needed by the body only in small amounts.

Mineral A metal or nonmetal used in the body in the form of ions or compounds.

Mitochondrion A cellular organelle where reactions of the common catabolic pathway occur.

Modulator A substance that binds to an enzyme at a location other than the active site and alters the catalytic activity.

Monomer The starting material that becomes the repeating unit of polymers.

Monosaccharide A simple carbohydrate most commonly consisting of three to six carbon atoms.

Mutagen A chemical that induces mutations by reacting with DNA.

Mutation Any change resulting in an incorrect base sequence on DNA.

Native state The natural three-dimensional conformation of a functional protein.

Neurotransmitter A substance that acts as a chemical bridge in nerve impulse transmission between nerve cells.

Noncompetitive inhibitor An inhibitor that binds to the enzyme at a location other than the active site.

Nonessential amino acid An amino acid that can be synthesized within the body in adequate amounts.

Normal alkane Any alkane in which all the carbon atoms are aligned in a continuous chain.

N-terminal residue An amino acid on the end of a chain that has an unreacted or free amino group.

Nucleic acid A biomolecule involved in the transfer of genetic information from existing cells to new cells.

Nucleic acid backbone The sugar–phosphate chain that is common to all nucleic acids.

Nucleotide The repeating structural unit or monomer of polymeric nucleic acids.

Nutrition An applied science that studies food, water, and other nutrients and the ways living organisms use them.

Oil A triglyceride that is a liquid at room temperature.

Okazaki fragment A DNA fragment produced during replication as a result of strand growth in a direction away from the replication fork.

Optically active molecule   A molecule that rotates the plane of polarized light.

Optimum pH   The pH at which enzyme activity is highest.

Optimum temperature   The temperature at which enzyme activity is highest.

Organelle   A membrane-enclosed structure in the interior of a cell.

Organic chemistry   The study of carbon-containing compounds.

Organic compound   A compound that contains the element carbon.

Oxidative deamination   An oxidation process resulting in the removal of an amino group.

Oxidative phosphorylation   A process coupled with the electron transport chain whereby ADP is converted to ATP.

Oxyhemoglobin   Hemoglobin combined with oxygen.

Peptide   Amino acid polymers of short chain length.

Peptide linkage (peptide bond)   The amide linkage between amino acids that results when the amino group of one acid reacts with the carboxylate group of another.

Phenol   A compound in which an —OH group is connected to a benzene ring. The parent compound is also called phenol.

Phenyl group   A benzene ring with one hydrogen absent, C_6H_5—.

Phosphoglyceride   A complex lipid containing glycerol, fatty acids, phosphoric acid, and an alcohol component.

Phospholipid   A phosphorus-containing lipid.

Phosphoric anhydride   A compound that contains the

$$-O-\overset{\overset{\displaystyle O}{\|}}{\underset{\underset{\displaystyle O^-}{|}}{P}}-O-\overset{\overset{\displaystyle O}{\|}}{\underset{\underset{\displaystyle O^-}{|}}{P}}-O- \text{ group.}$$

Plasma   The liquid portion of whole blood.

Plasmid   Circular, double-stranded DNA found in the cytoplasm of bacterial cells.

Polycyclic aromatic compound   A derivative of benzene in which carbon atoms are shared between two or more benzene rings.

Polyfunctional compound   A compound with two or more functional groups.

Polymer   A very large molecule made up of repeating units.

Polymerization   A reaction that produces a polymer.

Polypeptide   An amino acid polymer of intermediate chain length, containing up to 50 amino acid residues.

Polyribosome   A complex of mRNA and several ribosomes; also called a polysome.

Polysaccharide   A carbohydrate formed by the combination of many monosaccharide units.

Polysome   See **Polyribosome**.

Polyunsaturated   A term usually applied to molecules with several double bonds.

Primary alcohol   An alcohol in which the carbon bearing the —OH group is attached to one other carbon atom.

Primary amine   An amine having one alkyl or aromatic group bonded to nitrogen, as in R—NH_2.

Primary protein structure   The linear sequence of amino acid residues in a protein chain.

Prion   An infectious protein particle capable of causing disease.

Prokaryotic cell   A simple unicellular organism that contains no nucleus and no membrane-enclosed organelles.

Prostaglandin   A substance derived from unsaturated fatty acids, with hormonelike effects on a number of body tissues.

Prosthetic group   The nonamino acid part of a conjugated protein.

Protein   An amino acid polymer made up of more than 50 amino acids.

Protein turnover   The continuing process in which body proteins are hydrolyzed and resynthesized.

Pyranose ring   A six-membered sugar ring system containing an oxygen atom.

Quaternary ammonium salt   An ionic compound containing a positively charged ion in which four alkyl or aromatic groups are bonded to nitrogen, as in R_4N^+.

Quaternary protein structure   The arrangement of subunits that forms a larger protein.

Recombinant DNA   DNA of an organism that contains genetic material from another organism.

Reducing sugar   A sugar that can be oxidized by Cu^{2+} in solution.

Reference Daily Intakes (RDI)   A set of standards for protein, vitamins, and minerals used on food labels as part of the Daily Values.

Relative specificity   The characteristic of an enzyme that it acts on several structurally related substances.

Renal threshold   The blood glucose level at which glucose begins to be excreted in the urine.

Replication   The process by which an exact copy of a DNA molecule is produced.

Replication fork   A point where the double helix of a DNA molecule unwinds during replication.

Respiratory acidosis   A condition of acidosis caused by hypoventilation.

Respiratory alkalosis   A condition of alkalosis caused by hyperventilation.

Restriction enzyme   A protective enzyme found in some bacteria that catalyzes the cleaving of all but a few specific types of DNA.

Retrovirus   A virus that stores its genetic program in RNA instead of DNA.

Ribonucleic acid (RNA)   A nucleic acid found mainly in the cytoplasm of cells.

Ribosomal RNA (rRNA)   RNA that constitutes about 65% of the material in ribosomes, the sites of protein synthesis.

Ribosome   A subcellular particle that serves as the site of protein synthesis in all organisms.

Saponification   The basic cleavage of an ester linkage.

Satiety   A state of satisfaction or fullness.

Saturated hydrocarbon   Another name for an alkane.

Secondary alcohol   An alcohol in which the carbon bearing the —OH group is attached to two other carbon atoms.

Secondary amine   An amine having two alkyl or aromatic groups bonded to nitrogen, as in R_2NH.

Secondary protein structure   The arrangement of protein chains into patterns as a result of hydrogen bonds between amide groups of amino acid residues in the chain. The common secondary structures are the α-helix and the β-pleated sheet.

Semiconservative replication   A replication process that produces DNA molecules containing one strand from the parent and a new strand that is complementary to the strand from the parent.

Simple carbohydrates   Monosaccharides and disaccharides commonly called sugars.

Simple lipid   An ester-containing lipid with just two types of components: an alcohol and one or more fatty acids.

Simple protein   A protein made up entirely of amino acid residues.

Soap   A salt of a fatty acid often used as a cleaning agent.

Sphingolipid   A complex lipid containing sphingosine, a fatty acid, phosphoric acid, and choline.

Stereochemical specificity   An enzyme's ability to distinguish between stereoisomers.

Stereoisomers   Compounds with the same structural formula but different spatial arrangements of atoms.

Steroid   A compound containing four rings fused in a particular pattern.

Structural isomers   Compounds that have the same molecular formula but in which the atoms bond in different patterns.

Substrate   The substance that undergoes a chemical change catalyzed by an enzyme.

Subunit   A polypeptide chain having primary, secondary, and tertiary structural features that are a part of a larger protein.

Sulfhydryl group   The —SH functional group.

Terpene   A biological compound consisting of five-carbon isoprene units.

Tertiary alcohol   An alcohol in which the carbon bearing the —OH group is attached to three other carbon atoms.

Tertiary amine   An amine having three alkyl or aromatic groups bonded to nitrogen, as in R_3N.

Tertiary protein structure   A specific three-dimensional shape of a protein resulting from interactions between R groups of the amino acid residues in the protein.

Thioester bond   The carbon-sulfur bond of the

$$
\begin{array}{c} \text{O} \\ \parallel \\ \text{—C—S—} \end{array}
$$

functional group.

Thiol   A compound containing an —SH group.

Tollens' reagent   A mild oxidizing solution containing silver ions used to test for aldehydes.

Trace mineral   A mineral found in the body in quantities less than 5 g.

Trans-   On opposite sides (as applied to geometric isomers).

Transaminase   An enzyme that catalyzes the transfer of an amino group.

Transamination   The enzyme-catalyzed transfer of an amino group to a ketoacid.

Transcription   The transfer of genetic information from a DNA molecule to a molecule of messenger RNA.

Transfer RNA (tRNA)   RNA that delivers individual amino acid molecules to the site of protein synthesis.

Translation   The conversion of the code carried by messenger RNA into an amino acid sequence of a protein.

Triglyceride (triacylglycerol)   A triester of glycerol in which all three alcohol groups are esterified.

Turnover number   The number of molecules of substrate acted on by one molecule of enzyme per minute.

Unsaturated hydrocarbon   A hydrocarbon containing a C=C or C≡C group.

Urea cycle   A metabolic pathway in which ammonia is converted to urea.

Vector   A carrier of foreign DNA into a cell.

Virus   An infectious particle composed of only proteins and DNA or RNA.

Vitamin   An organic nutrient that the body cannot produce in the small amounts needed for good health.

Vitamin international unit   A measure of vitamin activity, determined by biological methods.

Wax   An ester of a long-chain fatty acid and a long-chain alcohol.

Zwitterion   A dipolar ion that carries both a positive and a negative charge as a result of an internal acid–base reaction in an amino acid molecule.

Zymogen or proenzyme   The inactive precursor of an enzyme.

Index

Photo and Illustration Credits. **xi** © Michael C. Slabaugh; **xii** Courtesy of DuPont deNemours Company; **xiii** © Freddy Storneil/Tony Stone Worldwide; **xiv** © Chris Baker/Tony Stone Worldwide; **xv** © Christian Liewig/Tony Stone Worldwide; **xvi** © Amythest/Custom Medical Stock; **xviii** Clark Taylor; **xix** © Bruce Ayres/ Tony Stone Images, Inc.; **1** Clark Taylor; **2** NASA; **4** Clark Taylor; **11** © Mark D. Phillips/Photo Researchers; **13** © Maren Slabaugh; **26** © Simon Fraser/ Science Photo Library/Photo Researchers; **27** © Michael C. Slabaugh; **28** West; **29 top** Tim Davis/Photo Researchers; **29 bottom** © Michael C. Slabaugh; **39** © Jeffrey Greenberg/ Photo Researchers Inc.; **42** © Michael C. Slabaugh; **44** © Maren Slabaugh; **49** © Joel Gordon, courtesy of West Publishing Company/CHEMISTRY by Radel & Navidi, 2d ed; **53** © Michael C. Slabaugh; **55 top** © Robert Daemmrich/Tony Stone Images; **55 left** © Frank Cezus/ Tony Stone Worldwide; **55 right** Clark Taylor; **58** Stock Montage; **61** © Maren Slabaugh; **63** © A. Dix/Photo Researchers Inc.; **73** © Alan Levinson/Tony Stone Worldwide; **74** Tom and Pat Leeson/Photo Researchers; **83** Spencer L. Seager; **86** © David Madison/Tony Stone Worldwide; **87 top** © Maren Slabaugh; **87 bottom** Clark Taylor; **88** Clark Taylor; **89** © Michael C. Slabaugh; **91 top** © David Woods/The Stock Market; **91 bottom** Clark Taylor; **93** © Michael C. Slabaugh; **94** West; **95** © Dr. E. R. Degginger; **105** Clark Taylor; **108** © Maren Slabaugh; **112** Spencer L. Seager; **113** © Ray Ellis/Photo Researchers; **114** © Michael C. Slabaugh; **115** Spencer L. Seager; **122** © B. Kramer/Custom Medical Stock Photo. All rights reserved; **123** © Michael C. Slabaugh; **124** © David Leifer/OffShoot Stock; **133** © David Grossman/Phototake; **134 top** © Michael C. Slabaugh; **134 bottom left** © Breck Reeves; **134 bottom right** West; **138** © Maren Slabaugh; **141 top** © K. Cavanagh/Photo Researchers; **141 bottom** © Michael C. Slabaugh; **142** © Gerard Smith/Photo Researchers Inc.; **144** Courtesy of DuPont deNemours Company; **147** Clark Taylor; **152** © Van Bucher/Photo Researchers; **161** © Bruce Ayres/ Tony Stone Images, Inc.; **168** West; **171** West; **172** © Joel Gordon, courtesy of West Publishing Company/CHEMISTRY by Radel & Navidi, 2d ed; **173** © Jim Steinberg/Photo Researchers; **191** © 1994/Custom Medical Stock Photo. All rights reserved; **192** © Freddy Storneil/Tony Stone Worldwide; **205** West; **209** Clark Taylor; **211** Clark Taylor; **212** © Neal and Mary Mishler/Tony Stone Worldwide; **213 left** Spencer L. Seager; **213 right** © Tony Stone Worldwide; **215** © Dr. E. R. Degginger; **216** © Michael C. Slabaugh; **217** © Felicia Martinez/PhotoEdit; **225** MacNeal Hospital/© Tony Stone Worldwide; **226 top** © C. James Webb/Phototake; **226 bottom** © Tony Stone Worldwide; **231** © Michael C. Slabaugh; **234** West; **236** © Bonnie Kamin; **237** © Maren Slabaugh; **240** Elizabeth Morales-Denny. Adapted from *Understanding Human Anatomy and Physiology* by Ann Stalheim-Smith and Greg K. Fitch (Copyright © West Publishing, 1993). Used with permission; **242** © Blair Setz/Science Source/Photo Researchers; **243** © Dept. of Clinical Radiology, Salisbury District Hospital/Science Photo Library/Photo Researchers; **244** © Maren Slabaugh; **246** © Michael C. Slabaugh; **253** © David M. Grossman/Phototake; **267** © Michael C. Slabaugh; **271** © Jackie Lewin, Royal Free Hospital/Science Photo Library/Photo Researchers; **275** © Dr. E. R. Degginger; **283** Clark Taylor; **284** © Michael C. Slabaugh; **285** © Dr. E. R. Degginger; **292** © Dr. E. R. Degginger; **293** © Peter Ryan, Scripps Institute/ Photo Researchers Inc.; **297** © Michael C. Slabaugh; **309** © Yoav Levy/Phototake; **310** © Chris Baker/Tony Stone Worldwide; **314 left** A. Barrington Brown/Science Source/Photo Researchers; **314 right** Photo courtesy of Dr. Ed Barrett and Carolina Biological Supply Company; **316** © Maren Slabaugh; **317** © L. Lisco and D. W. Fawcett/Visuals Unlimited; **326** NIBSC/Science Photo Library/Photo Researchers; **334** © Richard Nowitz/ Phototake; **335** © Mauritius, GMBH/Phototake; **343** © Michael Newman/PhotoEdit; **344** © Michael C. Slabaugh; **347** Clark Taylor; **360 left** © T. Kanariki and D. W. Fawcett/Visuals Unlimited; **360 right** Elizabeth Morales-Denny. Adapted from Chiras. Daniel D., *Human Biology: Health, Homeostasis, and the Environment* (Copyright © West Publishing, 1991), p. 72; **361** © Michael C. Slabaugh; **371** Clark Taylor; **374** West; **377** © Sherri Woods/Photo Researchers Inc.; **378** UPI/Bettmann; **380** © Christian Liewig/Tony Stone Worldwide; **388** © Robert Brenner/PhotoEdit; **389** © SIU Biomedical Comm./Custom Medical Stock Photo. All rights reserved; **390** © Dr. E. R. Degginger; **397** Clark Taylor; **400** Sandra McMahon. Adapted from Hamilton, Eva May Nunnelly, Eleanor Noss Whitney, and Frances Sienkiewicz Sizer, *Nutrition: Concepts and Controversies*, 5th ed. (Copyright © West Publishing, 1991). p. 134; **401** © Michael C. Slabaugh; **414** © Custom Medical Stock Photo. All rights reserved; **416** © Maren Slabaugh; **425** © Simon Fraser/Photo Researchers Inc.; **427** Professors P. M. Motta and S. Correr/Science Photo Library/Photo Researchers; **433** © Jon Meyer/Custom Medical Stock Photo. All rights reserved; **435** © Amythest/Custom Medical Stock.

Table of atomic weights and numbers

Name	Symbol	Atomic Number	Atomic Weight	Name	Symbol	Atomic Number	Atomic Weight
Actinium	Ac	89	(227)	Mendelevium	Md	101	(260)
Aluminum	Al	13	26.98	Mercury	Hg	80	200.6
Americium	Am	95	(243)	Molybdenum	Mo	42	95.94
Antimony	Sb	51	121.8	Neodymium	Nd	60	144.2
Argon	Ar	18	39.95	Neon	Ne	10	20.18
Arsenic	As	33	74.92	Neptunium	Np	93	(237)
Astatine	At	85	(210)	Nickel	Ni	28	58.69
Barium	Ba	56	137.3	Niobium	Nb	41	92.91
Berkelium	Bk	97	(247)	Nitrogen	N	7	14.01
Beryllium	Be	4	9.012	Nobelium	No	102	(259)
Bismuth	Bi	83	209.0	Osmium	Os	76	190.2
Boron	B	5	10.81	Oxygen	O	8	16.00
Bohrium	Bh	107	(264)	Palladium	Pd	46	106.4
Bromine	Br	35	79.90	Phosphorus	P	15	30.97
Cadmium	Cd	48	112.4	Platinum	Pt	78	195.1
Calcium	Ca	20	40.08	Plutonium	Pu	94	(242)
Californium	Cf	98	(251)	Polonium	Po	84	(209)
Carbon	C	6	12.01	Potassium	K	19	39.10
Cerium	Ce	58	140.1	Praseodymium	Pr	59	140.9
Cesium	Cs	55	132.9	Promethium	Pm	61	(145)
Chlorine	Cl	17	35.45	Protactinium	Pa	91	(231)
Chromium	Cr	24	52.00	Radium	Ra	88	(226)
Cobalt	Co	27	58.93	Radon	Rn	86	(222)
Copper	Cu	29	63.55	Rhenium	Re	75	186.2
Curium	Cm	96	(248)	Rhodium	Rh	45	102.9
Dubnium	Db	105	(262)	Rubidium	Rb	37	85.47
Dysprosium	Dy	66	162.5	Ruthenium	Ru	44	101.1
Einsteinium	Es	99	(252)	Rutherfordium	Rf	104	(261)
Erbium	Er	68	167.3	Samarium	Sm	62	150.4
Europium	Eu	63	152.0	Scandium	Sc	21	44.96
Fermium	Fm	100	(257)	Seaborgium	Sg	106	(266)
Fluorine	F	9	19.00	Selenium	Se	34	78.96
Francium	Fr	87	(223)	Silicon	Si	14	28.09
Gadolinium	Gd	64	157.3	Silver	Ag	47	107.9
Gallium	Ga	31	69.72	Sodium	Na	11	22.99
Germanium	Ge	32	72.59	Strontium	Sr	38	87.62
Gold	Au	79	197.0	Sulfur	S	16	32.07
Hafnium	Hf	72	178.5	Tantalum	Ta	73	180.9
Hassium	Hs	108	(269)	Technetium	Tc	43	(98)
Helium	He	2	4.003	Tellurium	Te	52	127.6
Holmium	Ho	67	164.9	Terbium	Tb	65	158.9
Hydrogen	H	1	1.008	Thallium	Tl	81	204.4
Indium	In	49	114.8	Thorium	Th	90	(232)
Iodine	I	53	126.9	Thulium	Tm	69	168.9
Iridium	Ir	77	192.2	Tin	Sn	50	118.7
Iron	Fe	26	55.85	Titanium	Ti	22	47.87
Krypton	Kr	36	83.80	Tungsten	W	74	183.9
Lanthanum	La	57	138.9	Uranium	U	92	(238)
Lawrencium	Lr	103	(262)	Vanadium	V	23	50.94
Lead	Pb	82	207.2	Xenon	Xe	54	131.3
Lithium	Li	3	6.941	Ytterbium	Yb	70	173.0
Lutetium	Lu	71	175.0	Yttrium	Y	39	88.91
Magnesium	Mg	12	24.31	Zinc	Zn	30	65.38
Manganese	Mn	25	54.94	Zirconium	Zr	40	91.22
Meitnerium	Mt	109	(268)				

A value in parentheses is the mass number of the isotope of longest half-life for radioactive elements.

Features within the Text